Multiprocessor Systems-on-Chips

The Morgan Kaufmann Series in Systems on Silicon

Series Editors: Peter Ashenden, Ashenden Designs Pty. Ltd. and Adelaide University, and Wayne Wolf, Princeton University

The rapid growth of silicon technology and the demands of applications are increasingly forcing electronics designers to take a systems-oriented approach to design. This has led to new challenges in design methodology, design automation, manufacture and test. The main challenges are to enhance designer productivity and to achieve correctness on the first pass. *The Morgan Kaufmann Series in Systems on Silicon* presents high quality, peer-reviewed books authored by leading experts in the field who are uniquely qualified to address these issues.

The Designer's Guide to VHDL, Second Edition
Peter J. Ashenden

The System Designer's Guide to VHDL-AMS
Peter J. Ashenden, Gregory D. Peterson, and Darrell A. Teegarden

Readings in Hardware/Software Co-Design
Edited by Giovanni De Micheli, Rolf Ernst, and Wayne Wolf

Modeling Embedded Systems and SoCs
Axel Jantsch

ASIC and FPGA Verification: A Guide to Component Modeling
Richard Munden

Multiprocessor Systems-on-Chips
Edited by Ahmed Amine Jerraya and Wayne Wolf

Forthcoming Titles

Functional Verification
Bruce Wile, John Goss, and Wolfgang Roesner

Rosetta User's Guide: Model-Based Systems Design
Perry Alexander, Peter J. Ashenden, and David L. Barton

Rosetta Developer's Guide: Semantics for Systems Design
Perry Alexander, Peter J. Ashenden, and David L. Barton

Multiprocessor Systems-on-Chips

Edited by

Ahmed Amine Jerraya

TIMA Laboratory

Wayne Wolf

Princeton University

AMSTERDAM • BOSTON • HEIDELBERG • LONDON
NEW YORK • OXFORD • PARIS • SAN DIEGO
SAN FRANCISCO • SINGAPORE • SYDNEY • TOKYO

Morgan Kaufmann Publishers is an imprint of Elsevier

Senior Editor Denise E. M. Penrose
Publishing Services Manager Andre Cuello
Project Manager Brandy Palacios
Editorial Assistant Valerie Witte
Cover Design Chen Design Associates, San Francisco
Composition Best-Set Typesetter Ltd.
Technical Illustration Graphic World Publishing Services
Copyeditor Graphic World Publishing Services
Proofreader Graphic World Publishing Services
Indexer Graphic World Publishing Services
Interior printer Maple Press
Cover printer Phoenix Color

Morgan Kaufmann Publishers is an Imprint of Elsevier
500 Sansome Street, Suite 400, San Francisco, CA 94111

This book is printed on acid-free paper.

Library of Congress Cataloging-in-Publication Data
Application Submitted

ISBN: 0-12385-251-X

For all information on all Morgan Kaufmann publications visit our website at *www.mkp.com*

Printed in the United States of America

Dedication

For Houda, Lydia and Nelly—Ahmed Amine Jerraya

For Nancy and Alec—Wayne Wolf

About the Editors

Ahmed Amine Jerraya (*ahmed.jerraya@imag.fr*) is Research Director with CNRS and is currently managing research dealing with Multiprocessor System-on-Chips at TIMA Laboratory, France. He received a degree in engineering from the University of Tunis in 1980 and the D.E.A., "Docteur Ingénieur", and the "Docteur d'Etat" degrees from the University of Grenoble in 1981, 1983, and 1989 respectively, all in computer sciences. From April 1990 to March 1991, he was a Member of the Scientific Staff at Nortel in Canada, working on linking system design tools and hardware design environments. He served as General Chair for the Conference DATE in 2001 and published more than 200 papers in International Conferences and Journals. He received the Best Paper Award at the 1994 ED&TC for his work on Hardware/Software Co-simulation.

Wayne Wolf (*wolf@princeton.edu*) is Professor of Electrical Engineering at Princeton University. Before joining Princeton, he was with AT&T Bell Laboratories, Murray Hill, New Jersey. He received the B.S., M.S., and Ph.D. degrees in Electrical Engineering from Stanford University in 1980, 1981, and 1984, respectively. His research interests include embedded computing, VLSI systems, and multimedia information systems. He is the author of *Computers as Components* and *Modern VLSI Design* (for which he won the ASEE/CSE and HP Frederick E. Terman Award). Wolf has been elected to Phi Beta Kappa and Tau Beta Pi. He is a Fellow of the IEEE and ACM and a member of the SPIE and ASEE.

Contents

1   The What, Why, and How of MPSoCs   1

Ahmed Amine Jerraya and Wayne Wolf

4   Architecture of Embedded Microprocessors   81

Eric Rotenberg and Aravindh Anantaraman

7   Design of Communication Architectures for High-Performance and Energy-Efficient Systems-on-Chips 187

Sujit Dey, Kanishka Lahiri, and Anand Raghunathan

8   Design Space Exploration of On-Chip Networks: A Case Study   223

Bishnupriya Bhattacharya, Luciano Lavagno, and Laura Vanzago

PART II   SOFTWARE   249

11   Cost-Efficient Mapping of Dynamic Concurrent Tasks in Embedded Real-Time Multimedia Systems   313

Peng Yang, Paul Marchal, Chun Wong, Stefaan Himpe, Francky Catthoor, Patrick David, Johan Vounckx, and Rudy Lauwereins

12   ILP-Based Resource-Aware Compilation   337

Jens Palsberg and Mayur Naik

Contents

PART III METHODOLOGY AND APPLICATIONS 355

13 Component-Based Design for Multiprocessor Systems-on-Chip 357

Wander O. Cesário and Ahmed A. Jerraya

14   MPSoCs for Video   395

*Santanu Dutta, Jens Rennert, Teihan Lv, Jiang Xu, Shengqi Yang,
and Wayne Wolf*

Preface

This book had its origins in the Multiprocessor System-on-Chip (MPSoC) Workshop, which has been held every summer since 2001. We started the workshop to bring together a broad range of people who need to be involved in SoC design. MPSoCs are much more complex than ASICs and so require traditionally separate disciplines to converge. Computer architecture, real-time operating systems, embedded software, computer-aided design, and circuit design are all fields that must contribute to successful MPSoC designs.

An outstanding cross-section of experts from these disciplines attended the workshop and gave tutorial lectures that introduced their fields to the broader audience. We started to talk about how to make use of all the information that was presented at the workshop and make it available to a broader audience. After some debate we decided to organize a book of contributed chapters.

The contents of the book evolved somewhat as we developed it. Some of the workshop speakers didn't have enough room in their schedules for the considerable effort it takes to write a chapter. As a result, we ended up asking several other groups of people to write chapters. This allowed us to bring more perspectives into the discussion of MPSoC design.

Thanks to a great deal of effort, the book is now a reality. We believe that it will be useful to both professionals who want to know more about multiprocessor SoCs and students who are studying the subject. We hope that this collection of material will be informative and long-lasting. We'd like to thank the IEEE Circuits and Systems Society and the European Design and Automation Association for their sponsorship of the MPSoC Workshop over the years. We would like to thank all the participants—speakers and attendees—of the MPSoC workshops. We would like to thank all of the authors for their work in creating these chapters. We would like to thank the book's reviewers: Nikil Dutt, University of California, Irvine; Axel Jantsch, Royal Institute of Technology (Stockholm, Sweden); Dan Phillips, Rochester Institute of Technology; Miodrag Potkonjak, UCLA; Gerald E. Sobelman, University of Minnesota; and James M. Ziobro, Rochester Institute of Technology. And we would like to thank the staff at Morgan Kaufman/Elsevier for their tireless efforts to make this book a reality.

Ahmed Amine Jerraya, Grenoble, France
Wayne Wolf, Princeton, New Jersey

1

CHAPTER

The What, Why, and How of MPSoCs

Ahmed Amine Jerraya and Wayne Wolf

1.1 INTRODUCTION

Multiprocessor systems-on-chips (MPSoCs) are the latest incarnation of very large-scale integration (VLSI) technology. A single integrated circuit can contain over 100 million transistors, and the International Technology Roadmap for Semiconductors predicts that chips with a billion transistors are within reach. Harnessing all this raw computing power requires designers to move beyond logic design into computer architecture. The demands placed on these chips by applications require designers to face problems not confronted by traditional computer architecture: real-time deadlines, very low-power operation, and so on. These opportunities and challenges make MPSoC design an important field of research.

The other chapters in this book examine various aspects of MPSoC design in detail. This chapter surveys the field and tries to put the problems of MPSoC design into perspective. We will start with a brief introduction to MPSoC applications. We will then look at the hardware and software architectures of MPSoCs. We will conclude with a survey of the remaining chapters in the book.

1.2 WHAT ARE MPSoCS?

We first need to define *system-on-chip* (SoC). An SoC is an integrated circuit that implements most or all of the functions of a complete electronic system. The most fundamental characteristic of an SoC is complexity. A memory chip may have many transistors, but its regular structure makes it a component and not a system.

Exactly what components are assembled on the SoC varies with the application. Many SoCs contain analog and mixed-signal circuitry for input/output (I/O). Although some high-performance I/O applications require a separate analog interface chip that serves as a companion to a digital SoC, most of an SoC is digital because that is the only way to build such complex functions reliably. The system may contain memory, instruction-set processors (central processing units [CPUs]), specialized logic, busses, and other digital functions. The architecture of the system is generally tailored to the application rather than being a general-purpose chip: we will discuss the motivations for custom, heterogeneous architectures in the next section.

Systems-on-chips can be found in many product categories ranging from consumer devices to industrial systems:

+ Cell phones use several programmable processors to handle the signal-processing and protocol tasks required by telephony. These architectures must be designed to operate at the very low-power levels provided by batteries.

+ Telecommunications and networking use specialized systems-on-chips, such as network processors, to handle the huge data rates presented by modern transmission equipment.

+ Digital televisions and set-top boxes use sophisticated multiprocessors to perform real-time video and audio decoding and user interface functions.

+ Television production equipment uses systems-on-chips to encode video. Encoding high-definition video in real time requires extremely high computation rates.

+ Video games use several complex parallel processing machines to render gaming action in real time.

These applications do not use general-purpose computer architectures either because a general-purpose machine is not cost-effective or because it would simply not provide the necessary performance. Consumer devices must sell for extremely low prices. Today, digital video disc (DVD) players sell for US $50, which leaves very little money in the budget for the complex video decoder and control system that playing DVDs requires. At the high end, general-purpose machines simply can't keep up with the data rates for high-end video and networking; they also have a hard time providing reliable real-time performance.

So what is an MPSoC? It is simply a system-on-chip that contains multiple instruction-set processors (CPUs). In practice, most SoCs are MPSoCs because it is too difficult to design a complex system-on-chip without making use of

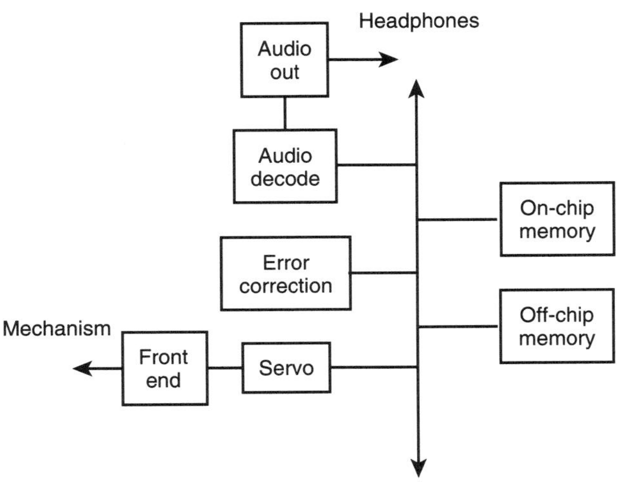

1-1   Architecture of a CD/MP3 player.

FIGURE

multiple CPUs. We will discuss the rationale for multiprocessors in more detail in the next section.

Figure 1-1 shows a block diagram for a typical compact disc/MPEG layer-3 (CD/MP3) player, a chip that controls a CD drive and decodes MP3 audio files. The architecture of a DVD player is more complex but has many similar characteristics, particularly in the early stages of processing. This block diagram abstracts the interconnection between the different *processing elements* (PEs). Although interconnect is a significant implementation concern, we want first to focus on the diversity of the PEs used in an SoC. At one end of the processing chain is the mechanism that controls the CD drive. A small number of analog inputs from the laser pickup must be decoded both to be sure that the laser is on track and to read the data from the disc. A small number of analog outputs controls the lens and sled to keep the laser on the data track, which is arranged as a spiral around the disc. Early signal conditioning and simple signal processing is done in analog circuitry because that is the only cost-effective means of meeting the data rates. However, most of the control circuitry for the drive is performed digitally. The CD player is a triumph of signal processing over mechanics—a very cheap and low-quality mechanism is controlled by sophisticated algorithms to very fine tolerances. Several control loops with 16 or more taps are typically performed by a digital signal processor (DSP) in order to control the CD drive mechanism. Once

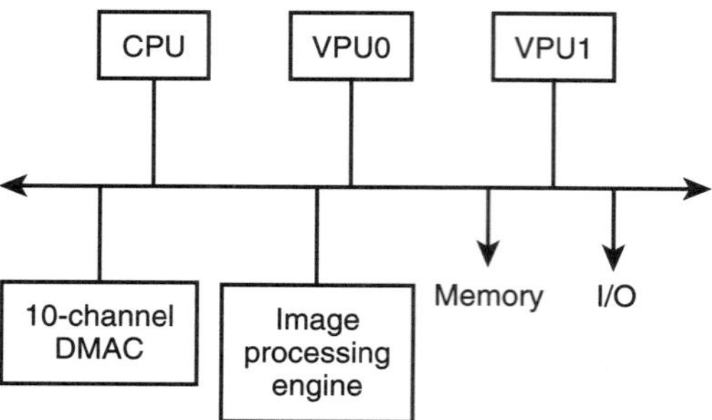

1-2

Architecture of the Sony Playstation 2 Emotion Engine.

FIGURE

the raw bits have been read from the disc, error correction must be performed. A modified Reed-Solomon algorithm is used; this task is typically performed by a special-purpose unit because of the performance requirements. After error correction, the MP3 data bits must be decoded into audio data; typically other user functions such as equalization are performed at the same time. MP3 decoding can be performed relatively cheaply, so a relatively unsophisticated CPU is all that is required for this final phase. An analog amplifier sends the audio to headphones.

Figure 1-2 shows the architecture of the Emotion Engine chip from the Sony PlayStation 2 [1]. The Emotion Engine is one of several complex chips in the PlayStation 2. It includes a general-purpose CPU that executes the millions of instructions per second (MIPS) instruction set and two vector processing units, VPU0 and VPU1. The two vector processing units have different internal architectures. The chip contains 5.8 million transistors, runs at 300 MHz, and delivers 5.5 Gflops.

Why do we care about performance? Because most of the applications for which SoCs are used have precise performance requirements. In traditional interactive computing, we care about speed but not about deadlines. Control systems, protocols, and most real-world systems care not just about average performance but also that tasks are done by a given deadline. The vast majority of SoCs are employed in applications that have at least some real-time deadlines. Hardware designers are used to meeting clock performance goals, but most deadlines span many clock cycles.

Why do we care about energy? In battery-operated devices, we want to extend the life of the battery as long as possible. In non-battery-operated devices, we still care because energy consumption is related to cost. If a device utilizes too much power, it runs too hot. Beyond a certain operating temperature, the chip must be put in a ceramic package. Ceramic packages are much more expensive than plastic packages.

The fact that an MPSoC is a multiprocessor means that software design is an inherent part of the overall chip design. This is a big change for chip designers, who are used to coming up with hardware solutions to chip design problems. In an MPSoC, either hardware or software can be used to solve a problem; which is best generally depends on performance, power, and design time. Designing software for an MPSoC is also a big change for software designers. Software that will be shipped as part of a chip must be extremely reliable. That software must also be designed to meet many design constraints typically reserved for hardware, such as hard timing constraints and energy consumption. This melding of hardware and software design disciplines is one of the things that makes MPSoC design interesting and challenging.

The fact that most MPSoCs are heterogeneous multiprocessors makes them harder to program than traditional symmetric multiprocessors. Regular architectures are much easier to program. Scientific multiprocessors have also gravitated toward a shared-memory model for programmers. Although these regular, simple architectures are simple for programmers, they are often more expensive and less energy efficient than heterogeneous architectures. The combination of high reliability, real-time performance, small memory footprint, and low-energy software on a heterogeneous multiprocessor makes for a considerable challenge in MPSoC software design.

Many MPSoCs need to run software that was not developed by the chip designers. Because standards guarantee large markets, multichip systems are often reduced to SoCs only when standards emerge for the application. However, users of the chip must add their own features to the system to differentiate their products from competitors who use the same chip. This requires running software that is developed by the customer, not the chip designer. Early VLSI systems with embedded processors generally used very crude software environments that would have been impossible for outside software designers to use. Modern MPSoCs have better *development environments*, but creating a different software development kit for each SoC is in itself a challenge.

1.3   WHY MPSoCS?

The typical MPSoC is a *heterogeneous multiprocessor*: there may be several different types of PEs, the memory system may be heterogeneously distributed around the machine, and the interconnection network between the PEs and the memory

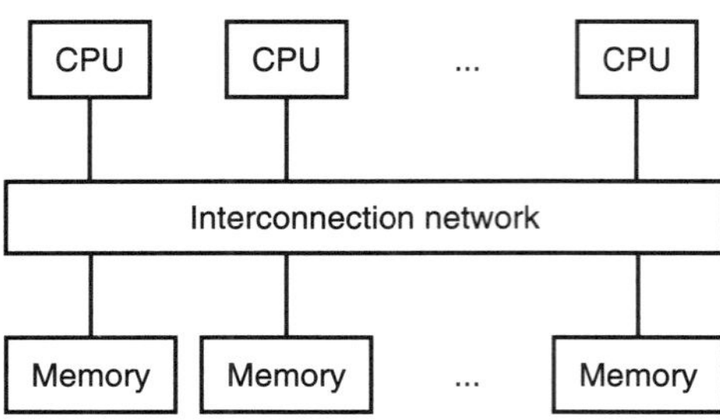

1-3

A generic shared-memory multiprocessor.

FIGURE

may also be heterogeneous. MPSoCs often require large amounts of memory. The device may have embedded memory on-chip as well as relying on off-chip commodity memory.

We introduced two examples of SoCs in the last section and they implement, in fact, heterogeneous multiprocessors. In contrast, most scientific multiprocessors today are much more regular than the typical MPSoC. Figure 1-3 shows the traditional view of a *shared-memory multiprocessor* [2]—a pool of processors and a pool of memory are connected by an interconnection network. Each is generally regularly structured, and the programmer is given a regular programming model. A shared-memory model is often preferred because it makes life simpler for the programmer. The Raw architecture [3] is a recent example of a regular architecture designed for high-performance computation.

Why not use a single *platform* for all applications? Why not build SoCs like field programmable gate arrays (FPGAs), in which a single architecture is built in a variety of sizes? And why use a multiprocessor rather than a uniprocessor, which has an even simpler programming model?

Some relatively simple systems are, in fact, uniprocessors. The personal digital assistant (PDA) is a prime example. The architecture of the typical PDA looks something like a PC, with a CPU, peripherals, and memory attached to a bus. A PDA runs many applications that are small versions of desktop applications, so the resemblance of the PDA platform to the PC platform is important for software development.

However, uniprocessors may not provide enough performance for some applications. The simple database applications such as address books that run on PDAs can easily be handled by modern uniprocessors. But when we move to real-time video or communications, multiprocessors are generally needed to keep up with

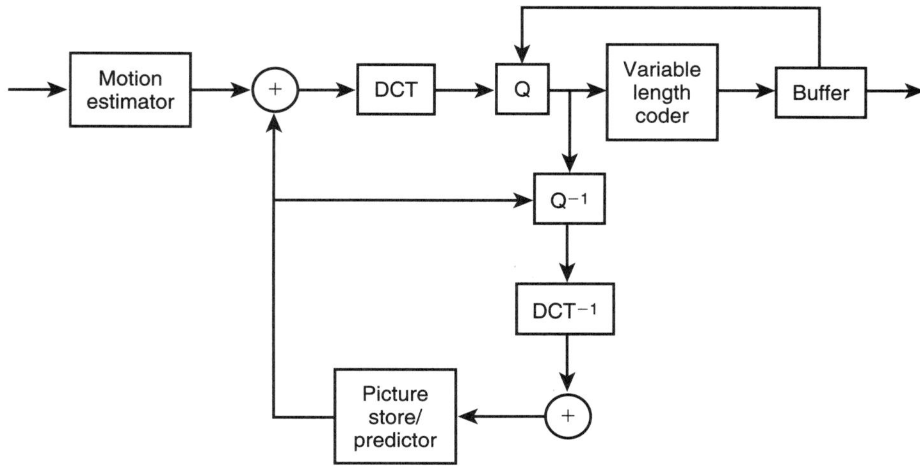

1-4

FIGURE

Block diagram of MPEG-2 encoding.

the incoming data rates. Multiprocessors provide the computational concurrency required to handle concurrent real-world events in real time.

Embedded computing applications typically require real concurrency, not just the apparent concurrency of a multitasking operating system running on a uniprocessor. *Task-level parallelism* is very important in embedded computing. Most of the systems that rely on SoCs perform complex tasks that are made up of multiple phases. For example, Figure 1-4 shows the block diagram for MPEG-2 encoding [4]. Video encoding requires several operations to run concurrently: motion estimation, discrete cosine transform (DCT), and Huffman coding, among others. Video frames typically enter the system at 30 frames/sec. Given the large amount of computation to be done on each frame, these steps must be performed in parallel to meet the deadlines. This type of parallelism is relatively easy to leverage since the system specification naturally decomposes the problem into tasks. Of course, the decomposition that is best for specification may not be the best way to decompose the computation for implementation on the SoC. It is the job of software or hardware design tools to massage the decomposition based on implementation costs. But having the original parallelism explicitly specified makes it much easier to repartition the functionality during design.

But why not use a symmetric multiprocessor to provide the required performance? If we could use the same architecture for many different applications, we could manufacture the chips in even larger volumes, allowing lower prices.

Programmers could also more easily develop software since they would be familiar with the platforms and they would have a richer tool set. And a symmetric processor would make it easier to map an application onto the architecture.

However, we cannot directly apply the scientific computing model to SoCs. SoCs must obey several constraints that do not apply to scientific computation:

+ They must perform real-time computations.

+ They must be area-efficient.

+ They must be energy-efficient.

+ They must provide the proper I/O connections.

All these constraints push SoC designers toward heterogeneous multiprocessors. We can consider these constraints in more detail.

Real-time computing is much more than high-performance computing. Many SoC applications require very high performance—consider high-definition video encoding, for example—but they also require that the results be available at a predictable rate. Rate variations can often be solved by adding buffer memory, but memory incurs both area and energy consumption costs. Making sure that the processor can produce results at predictable times generally requires careful design of all the aspects of the hardware: instruction set, memory system, and system bus. It also requires careful design of the software, both to take advantage of features of the hardware and to avoid common problems like excessive reliance on buffering.

Real-time performance also relies on predictable behavior of the hardware. Many mechanisms used in general-purpose computing to provide performance in an easy programming model make the system's performance less predictable. Snooping caching, for example, dynamically manages cache coherency but at the cost of less predictable delays since the time required for a memory access depends on the state of several caches. One way to provide predictable performance and high performance is to use a mechanism that is specialized to the needs of the application: specialized memory systems or application-specific instructions, for example. And since different tasks in an application often have different characteristics, different parts of the architecture often need different hardware structures.

Heterogeneous multiprocessors are more area-efficient than symmetric multiprocessors. Many scientific computing problems distribute homogeneous data across multiple processors; for example, they may decompose a matrix in parallel using several CPUs. However, the task-level parallelism that embedded computing applications display is inherently heterogeneous. In the MPEG block diagram, as with other applications, each block does something different and has different computational requirements.

Although application heterogeneity does not inherently require using a different type of processor for each task, doing so can have significant advantages. A special-purpose PE may be much faster and smaller than a programmable processor; for example, several very small and fast motion estimation machines have been developed for MPEG. Even if a programmable processor is used for a task, specialized CPUs can often improve performance while saving area. For example, matching the CPU datapath width to the native data sizes of the application can save a considerable amount of area. Choosing a cache size and organization to match the application characteristics can greatly improve performance.

Memory specialization is an important technique for designing efficient architectures. A general-purpose memory system can try to handle special cases on the fly using information gathered during execution, but they do so at a considerable cost in hardware. If the system architect can predict some aspect of the memory behavior of the application, it is often possible to reflect those characteristics in the architecture. Cache configuration is an ideal example—a considerably smaller cache can often be used when the application has regular memory access patterns.

Most SoC designs are power-sensitive, whether due to environmental considerations (heat dissipation) or to system requirements (battery power). As with area, specialization saves power. Stripping away features that are unnecessary for the application reduces energy consumption; this is particularly true for leakage power consumption. Scientific multiprocessors are standard equipment that are used in many different ways; each installation of a supercomputer may perform a different task. In contrast, SoCs are mass-market devices due to the economics of VLSI manufacturing. The design tweaks that save power for a particular architecture feature can therefore be replicated many times during manufacturing, amortizing the cost of designing those power-saving features.

SoCs also require specialized I/O. The point of an SoC is to provide a complete system. One would hope that input and output devices could be implemented in a generic fashion given enough transistors; to some extent, this has been done for FPGA I/O pads. But given the variety of physical interfaces that exist, it can be difficult to create customizable I/O devices effectively.

One might think that increasing transistor counts might argue for a trend away from heterogeneous architectures and toward regularly structured machines. But applications continue to soak up as much computational power as can be supplied by Moore's law. Data rates continue to go up in most applications, for example, data communication, video, audio. Furthermore, new devices increasingly combine these applications. A single device may perform wireless communication, video compression, and speech recognition. SoC designers will start to favor regular architectures only when the performance pressure from application eases and the performance available from integrated circuits catches up. It does not appear that customers' appetites will shrink any time soon.

1.4   CHALLENGES

Before delving into MPSoC design challenges in more detail, let us take a few moments to summarize a few major challenges in the field.

Software development is a major challenge for MPSoC designers. The software that runs on the multiprocessor must be high performance, real time, and low power. Although much progress has been made on these problems, much remains to be done. Furthermore, each MPSoC requires its own software development environment: compiler, debugger, simulator, and other tools.

Task-level behavior also provides a major and related challenge for SoC software. As mentioned above, task-level parallelism is both easy to identify in SoC applications and important to exploit. *Real-time operating systems* (RTOSs) provide scheduling mechanisms for tasks, but they also abstract the process. The detailed behavior of a task—how it accesses memory, its flow of control—can influence its execution time and therefore the system schedule that is managed by the RTOS. We need a better understanding of how to abstract tasks properly to capture the essential characteristics of their low-level behavior for system-level analysis.

Networks-on-chips have emerged over the past few years as an architectural approach to the design of single-chip multiprocessors. A network-on-chip uses packet networks to interconnect the processors in the SoC. Although a great deal is known about networks, traditional network design assumes relatively little about the characteristics of the traffic on the network. SoC applications can often be well characterized; that information should be useful in specializing the network design to improve cost/performance/power.

FPGAs have emerged as a viable alternative to application-specific integrated circuits (ASICs) in many markets. FPGA fabrics are also starting to be integrated into SoCs. The FPGA logic can be used for custom logic that could not be designed before manufacturing. This approach is a good complement to software-based customization. We need to understand better where to put FPGA fabrics into SoC architectures so they can be most effectively used. We also need tools to help designers understand performance and allocation using FPGAs as processing elements in the multiprocessor.

As SoCs become more sophisticated, and particularly as they connect to the Internet, security becomes increasingly important. Security breaches can cause malfunctions ranging from annoying to life-threatening. Hardware and software architectures must be designed to be secure. Design methodologies must also be organized to ensure that security considerations are taken into account and that security-threatening bugs are not allowed to propagate through the design.

Finally, MPSoCs will increasingly be connected into networks of chips. Sensor networks are an example of networks of chips; automotive and avionics systems have long used networks to connect physically separated chips. Clearly, networking must be integrated into these chips. But, more important, MPSoCs that are organized into networks of chips do not have total control over the system. When designing a single chip, the design team has total control over what goes onto the chip. When those chips are assembled into networks, the design team has less control over the organization of the network as a whole. Not only may the SoC be used in unintended ways, but the network of the configuration may change over time as nodes are added and deleted. Design methodologies for these types of chips must be adapted to take into account the varying environments in which the chips will have to operate.

1.5   DESIGN METHODOLOGIES

In the preceding sections, we explained the "What" and "Why" of MPSoC architectures and presented the main challenges designers must face for the "How" to design an MPSoC. It is clear that many kinds of tools and specialized design methodologies are needed. Although advanced tools and methodologies exist today to solve partial problems, much remains to be done when considering heterogeneous MPSoC architectures as a whole. On this new design scenario, fast design time, higher level abstractions, predictability of results, and meeting design metrics are the main goals.

Fast design time is very import in light of typical applications for MPSoC architectures—game/network processors, high-definition video encoding, multimedia hubs, and base-band telecom circuits, for example—that have particularly tight time-to-market and time window constraints.

System-level modeling is the enabling technology for MPSoC design. Register-transfer level (RTL) models are too time consuming to design and verify when considering multiple processor cores and associated peripherals—a higher abstraction level is needed on the hardware side. When designers use RTL abstractions, they can produce, on average, the equivalent of 4 to 10 gates per line of RTL code.

Thus, hypothetically, in order to design a 100 million-gate MPSoC circuit using only RTL code—even if we assume that 90% of this code can be reused—more than 1 million lines of code would need to be written to describe the remaining 10 million gates for the custom part. Of course, this tremendous design effort is unrealistic for most MPSoC target markets.

MPSoCs may use hundreds of thousands of lines of dedicated software and complex software development environments; programmers cannot use mostly low-level programming languages anymore—higher level abstractions are needed on the software side too. Design components for MPSoC are heterogeneous: they come from different design domains, have different interfaces, and are described using different languages at different refinement levels and have different granularities. A key issue for every MPSoC design methodology is the definition of a good system-level model that is capable of representing all those heterogeneous components along with local and global design constraints and metrics.

High-level abstractions make global MPSoC design methodologies possible by hiding precise circuit behavior—notably accurate timing information—from system-level designers and tools. However, MPSoCs are mostly targeted for real-time applications in which accurate performance information must be available at design time in order to respect deadlines. Thus, when considering the huge design space allowed by MPSoC architectures, high-level design metrics and performance estimation are essential parts in MPSoC design methodologies. This notwithstanding, high-level estimation metrics and evaluation remains an active research subject since a system's design metrics are not easy to compose from design metrics of its components.

MPSoC design is a complex process involving different steps at different abstraction levels. Design steps can be grouped into two major tasks: design space exploration (hardware/software partitioning, selection of architectural platform and components) and architecture design (design of components, hardware/software interface design). The overall design process must consider strict requirements, regarding time-to-market, system performance, power consumption, and production cost. The reuse of predesigned components from several vendors—for hardware and software parts—is necessary for reducing design time, but their integration into a system also presents a variety of challenges.

A complete design flow for MPSoCs includes refinement processes that require multiple competences and tools because of the complexity and diversity of the current applications. Current (and previous) work tries to reduce the gap between different design steps [5] and to master the integration of heterogeneous components [6], including hardware and software parts. Existing approaches deal only with a specific part of the MPSoC design flow. A full system level flow is quite complex, and to our knowledge very few existing tools cover both system design space exploration and system architecture design.

1.6   HARDWARE ARCHITECTURES

We can identify several problems in MPSoC architecture starting from the bottom and working to the highest architectural levels:

+ Which CPU do you use? What instruction set and cache should be used based on the application characteristics?

+ What set of processors do you use? How many processors do you need?

+ What interconnect and topology should be used? How much bandwidth is required? What *quality-of-service* (QoS) characteristics are required of the network?

+ How should the memory system be organized? Where should memory be placed and how much memory should be provided for different tasks?

The following discussion describes several academic and industrial research projects that propose high-performance MPSoC architectures for high-performance applications.

Philips Nexperia™ DVP [7] is a flexible architecture for digital video applications. It contains two software-processing units, a very long instruction word (VLIW) media processor (32-bit or 64-bit @ 100 to 300+ MHz) and a MIPS core (32-bit or 64-bit @ 50 to 300+ MHz). It also contains a library of dedicated hardware processing units (image coprocessors, DSPs, universal asynchronous receiver transmitter [UART], 1394, universal serial bus [USB], and others). Finally, it integrates several system busses for specialized data transfer, a peripheral interface (PI) bus and a digital video platform (DVP) memory bus for intensive shared data transfer. This fixed memory architecture, besides the limited number and types of the incorporated CPU cores, reduces the application field of the proposed platform. However, the diversity of the hardware components included in the Device Block library makes possible the design of application-specific computation.

The Texas Instruments (TI) OMAP platform is an example of an architecture model proposed for the implementation of wireless applications. The two processing units consist of an ARM9 core (@ 150 MHz) and a C55x DSP core (@ 200 MHz). Both of them have a 16-Kb I-cache, an 8-Kb D-cache, and a two-way set associative global cache. There is a dedicated memory and traffic controller to handle data transfers. However, the proposed architecture for data transfer is still simple and the concurrency still limited (only two processing cores).

Virtex-II Pro™ from Xilinx is a recent FPGA architecture that embeds 0, 1, 2, or 4 PowerPC cores. Each PowerPC core occupies only about 2% of the total die area. The rest of the die area can be used to implement system busses, interfaces, and hardware IPs. The IP library and development tools provided by Xilinx especially support the IBM CoreConnect busses.

Given the examples of commercially available MPSoC chips, we see that most of them:

+ limit the number and types of integrated processor cores

+ provide a fixed or not well-defined memory architecture

+ limit the choice of interconnect networks and available IPs

+ do not support the design from a high abstraction level.

1.7   SOFTWARE

The software for MPSoC needs to be considered from the following three viewpoints: the programmer's viewpoint, the software architecture and design reuse viewpoint, and the optimization viewpoint.

1.7.1   Programmer's Viewpoint

From the viewpoint of the application programmer, an MPSoC provides a parallel architecture that consists of processors connected with each other via a communication network. Thus, to exploit the parallel architecture, parallel programming is required. Conventionally, there are two types of parallel programming model: shared-memory programming and *message-passing* programming. OpenMP and *message-passing interface* (MPI) are examples, respectively.

When using conventional parallel programming models for SoC, we face conventional issues in parallel programming, e.g., shared memory versus message passing. We also need to identify the difference between conventional parallel programming and MPSoC programming. We need to exploit the characteristics specific to MPSoC to use the parallel programming models in a more efficient way.

Differences appear in two aspects of MPSoC software design: application and architecture. Conventional parallel programming models need to support any type of program. Thus, they have a huge number of programming features, e.g., the user-defined complex data types and multicasting in MPI. However, MPSoC is application-specific. That is, an MPSoC supports only one or a set of fixed applications. Thus, for parallel programming, the designer does not need full-featured parallel programming models. Instead, we need to be able to find a suitable or application-specific subset of a parallel programming model for MPSoC programming.

MPSoC architectures have two main characteristics that differ from conventional multiprocessor architectures. One is heterogeneity and the other is massive parallelism. MPSoC can have different types of processors and any arbitrary topology of interconnection of processors. An MPSoC can also have a massive number of (fine-grained) processors, or processing elements. Thus, considering heterogeneous multiprocessor architecture with massive parallelism, parallel programming for MPSoC is more complicated than conventional parallel programming. To cope with this difficulty, it is necessary to obtain efficient models and methods of MPSoC programming.

1.7.2   Software Architecture and Design Reuse Viewpoint

We define software architecture to be the component that enables application software to run on the MPSoC architecture. The software architecture includes the *middleware* (for communication), the operating system (OS), and the *hardware abstraction layer* (HAL). HAL is the software component that is directly dependent on the underlying processor and peripherals. Examples of HAL include context switching, bus drivers, configuration code for the memory management unit (MMU), and interrupt service routines (ISRs).

From the viewpoint of application software, the architecture provides a virtual machine on which the application software runs. The basic role of software architecture is to enable (1) communication between computation units in the application software, i.e., software tasks; (2) task scheduling to provide task-level parallelism described in parallel programming models; and (3) external event processing, e.g., processing interrupts.

The *application programming interfaces* (APIs) of the middleware, OS, and HAL provide an abstraction of the underlying hardware architecture to upper layers of software. The HAL API gives an abstraction of underlying processor and processor local architecture to upper layers of software including application software, middleware, and OS. The middleware and OS API gives an abstraction of underlying multiprocessor architecture to application software.

The hardware abstraction represented by the API can play the role of a contract between software and hardware designers. That is, from the viewpoint of the software designer, the hardware design guarantees at least the functionality represented by the API. Thus, the software designer can design application software based on the API, independently of the hardware design. Thus, in terms of design flow, software and hardware design can be performed concurrently. Such a concurrent design reduces the design cycle in conventional design flow, in which software and hardware design is sequential.

In terms of software design reuse, the software architecture may enable several levels of software design reuse. First, the middleware API or OS API enables reuse of application software. For instance, in the case that application software is written in an OS API, e.g., POSIX API, the same application software can be reused in other MPSoC designs in which the same OS API is provided. The same applies to the HAL API. If the same HAL API is used in OS and middleware code, OS and middleware can be reused over different designs.

Key challenges in software architecture for SoC are as follows. Determining which abstraction of MPSoC architecture is most suitable at each of the design steps. Determining how to obtain application-specific optimization of software architecture.

Since MPSoC has cost and performance constraints, the software architecture needs to be designed to minimize its overhead. For instance, instead of using a full-featured OS, the OS needs only to support the functions required by the application software and MPSoC architecture.

1.7.3   Optimization Viewpoint

MPSoC is mostly used in cost-sensitive real-time systems, e.g., cellular phone, high-definition digital television (HDTV), game stations. Thus, the designer has stringent cost requirements (chip area, energy consumption) as well as real-time performance requirements. To satisfy those requirements, MPSoC software needs to be optimized in terms of code size, execution time, and energy consumption.

Two of the major factors that affect the cost and performance of software are processor architecture and memory hierarchy. The processor architecture affects the cost and performance in terms of parallelism and application-specific processor architecture. In terms of parallelism, the processor can offer *instruction-level parallelism* (ILP) dynamically (superscalar) and statically (VLIW) and thread-level parallelism statically (clustered VLIW) and dynamically (simultaneous multi-threading [SMT]). A great deal of compiler research has exploited the above-mentioned parallelism in the processor architecture. Recently, the dynamic reconfiguration (explicitly parallel instruction computing [EPIC]) processor [8] has provided another degree of parallelism in the temporal domain.

Application-specific processor architectures can provide orders of magnitude better performance than general-purpose processor architectures. The DSP and application-specific instruction set processor (ASIP) are examples of application-specific processors. Recently, configurable processor architectures, e.g., Tensilica Xtensa, are receiving more and more attention. Compared with the DSP and ASIP, configurable processors have a basic set of general processor instructions and add application-specific instructions to accelerate the core functionality of the intended application. From the viewpoint of the designer who optimizes the software performance and cost, the design of an application-specific processor and corresponding compiler is a challenging design task, although there are a few computer-assisted design (CAD) tools to assist in the effort. More automated solutions, such as an automatic instruction selection method, are still needed.

Another factor of MPSoC software performance and cost is memory architecture. In the case of uniprocessor architecture, cache parameters such as size, associativity, replacement policies, and others affect the software performance. In the case of MPSoC architecture, there are two common types of memory architecture: shared memory and distributed memory. Shared-memory architecture usually requires local caches and cache coherency protocols to prevent the performance bottlenecks. Distributed memory can be classified into several levels according to the degree of distribution. For instance, a massively parallel MPSoC architecture, which consists of arrays of small processors (e.g., arithmetic and logic units [ALUs]) that contain small memory elements, has a finer grained distribution of memory than an MPSoC architecture, which consists of general-purpose processors, each of which has a distributed memory element.

Most of the above-mentioned issues of processor architecture and memory hierarchy have been studied in the domains of processor architecture, compiler, and multiprocessor architecture. In the domain of MPSoC, researchers consider the same problem in a different context with more design freedom in hardware architecture and with a new focus on energy consumption.

For instance, in the case of memory hierarchy design, conventional design methods assume that a set of regular structures are given. Application software code is then transformed to exploit the given memory hierarchy. However, in the case of MPSoC, the designer can change the memory hierarchy in a way specific to the given application. Thus, further optimization is possible with such hardware design freedom.

Another focus is low-power software design. Conventional software design on multiprocessor architecture focuses on performance improvement, i.e., fast or scalable performance. In the software design for MPSoC, the designer needs to reconsider the existing design problems with energy consumption in mind. For instance, when a symmetric multiprocessor architecture is used for MPSoC design, processor affinity can be defined to include the energy consumption as well as the system runtime.

1.8 THE REST OF THE BOOK

This book is divided into three major sections: (1) hardware, (2) software, and (3) applications and methodologies. Each section starts with its own introduction that describes the chapters in more detail.

The hardware section looks at architectures and the constraints imposed on those architectures. Many MPSoCs must operate at low power and energy levels; several chapters discuss low-power design. Adequate communication structures are also important given the relatively high data rates and real-time requirements in many embedded applications. The multiprocessors on SoCs may be more specialized than in traditional architectures, but SoC design must clearly rest on the foundations of computer architecture. Several chapters discuss processor and multiprocessor architectures.

The software section considers several levels of abstraction. Scheduling the tasks on the multiprocessor is critical to efficient use of the processors. RTOSs and more custom software synthesis methods are important mechanisms for managing the tasks in the system. Compilers are equally important, and several chapters discuss compilation for size, time constraints, memory system efficiency, and other objectives.

The applications and methodology section looks at sample applications such as video computing. It also considers design methodologies and the abstractions that must be built to manage the design process.

Hardware

PART I

*T*he levels of integration provided by modern VLSI technology have fundamentally altered the hardware design process. When designing a 50-million or 100-million chip, design teams can no longer afford to design a gate-at-a-time or a wire-at-a-time. Chip architects and designers must learn to think in much larger blocks of intellectual property (IP). Some of those blocks will be synthesized from higher-level hardware description language (HDL) sources. Some blocks will be manually designed in order to provide maximum performance and minimum power consumption for critical modules. In either case, most of these IP blocks will be designed for reuse on multiple chips.

The design of multiprocessors also requires designers to change their ways of thinking. Traditional hardware design focuses on a single unit in which communication occurs at very low levels of abstraction. Multiprocessors are composed of loosely-coupled processing elements (PEs) that communicate via well-defined interfaces. Those interfaces are reflected in hardware structures that may themselves be complex modules. Those interfaces must, of course, be designed with the system software in mind.

MPSOC designers must choose their components and design architectures to meet tough requirements. Energy/power consumption is often the toughest constraint in modern designs. High performance is always important, since it makes little sense to incur the expense of system-on-chip design without a compelling performance requirement. And cost is still an important factor that can make or break the feasibility of an MPSOC project.

Chapter 2, by Mary Jane Irwin, Luca Benini, Vijaykrishnan Narayanan, and Mahmut Kandemir, concentrates on energy and power consumption. It both looks at the sources of energy consumption in nanometer-scale VLSI and potential approaches to mitigating energy and power consumption problems. The traditional source of energy consumption in CMOS is dynamic dissipation but leakage is becoming increasingly important. Controlling leakage in particular encourages architectural solutions.

Chapter 3, by Luca Benini and Giovanni De Micheli, describes network-on-chip-based design. Networks-on-chips are becoming increasingly popular because they help solve problems at both the circuit level and the architecture level. An on-chip network can be built from carefully designed components that encapsulate high-performance circuits. Packet-based communication also helps structure the application's communication. Because different applications require different structures and characteristics, Benini and De Micheli discuss methodologies for designing networks-on-chips.

Chapter 4, by Eric Rotenberg and Aravindh Anandtaraman, surveys processor architectures. Many of the architectural techniques from general-purpose computing have found their way into embedded processors. But because of the strict characteristics of embedded applications—hard real-time, low-energy operation—system architects must carefully consider how the mechanisms inside the CPUs they use as building blocks will affect their embedded software and overall application characteristics.

Chapter 5, by Chris Rowen, looks at CPUs from a different angle. Configurable processors are IP blocks that have been generated for a particular application. By customizing the instruction set of a processor, architects can create a CPU that much more closely matches the characteristics of the application. Since the CPU modifications are described as instruction set extensions, they can be made easily and quickly. However, the system designer is still responsible for characterizing the application to determine what instructions are most cost-effective.

Chapter 6, by Rolf Ernst, considers directly the problem of performance modeling and analysis. The system architect must consider not only the performance of individual processes but also how the processes interact through the system schedule. Careful system modeling looks at both the detailed operation of the processes on the processing elements as well as the overall system behavior.

Chapter 7, by Kanishka Lahiri, Sujit Dey, and Anand Raghunathan, looks at communication design for MPSOCs. They describe a methodology that breaks on-chip communication design into several tasks: topology design, on-chip protocol design, and communication mapping. All these steps take both performance and energy into account.

Chapter 8, by Bishnupriya Bhattacharya, Luciano Lavagno, and Laura Vanzango, uses a case study to consider the design of on-chip networks. They work through a design of the physical layer of Hiperlan/2. They use this detailed example to illustrate their methodology for on-chip network design.

Techniques for Designing Energy-Aware MPSoCs

Mary Jane Irwin, Luca Benini, N. Vijaykrishnan,
and Mahmut Kandemir

2.1 INTRODUCTION

Power and energy consumption have become significant constraints in modern day microprocessor systems. Whereas energy-aware design is obviously crucial for battery-operated mobile and embedded systems, it is also important for desktop and server systems due to packaging and cooling requirements. In such systems, power consumption has grown from a few watts per chip to over 100 watts. As desktop/server (and even embedded) systems evolve from the uniprocessor space to the *multiprocessor system-on-chips* (MPSoCs) space, energy-aware design will take on new dimensions.

Techniques for energy and power consumption reduction have been successfully applied at all levels of the design space in uniprocessor systems: circuit, logic gate, functional unit, processor, system software, and application software levels. The primary focus has been on reducing active (dynamic) power. As technology continues to scale up accompanied by reductions in the supply and threshold voltages, the percentage of the power budget due to standby (leakage) energy has driven the development of additional techniques for reducing standby energy as well [9,10].

Figure 2-1 illustrates the components of power consumption in a simple CMOS circuit. In well-designed circuits, $I_{\text{s-c}}$ is a fixed percentage (less than 10%) of I_{on}. Thus, *active power consumption* is usually estimated as:

$$P_{\text{Act}} = C_{\text{avg}} V_{\text{dd}}^2 (Act) f_{\text{clock}}$$

where C_{avg} is average capacitive load of a transistor, V_{dd} is the power supply, f_{clock} is the operating frequency, and *Act* is the activity factor that accounts for the

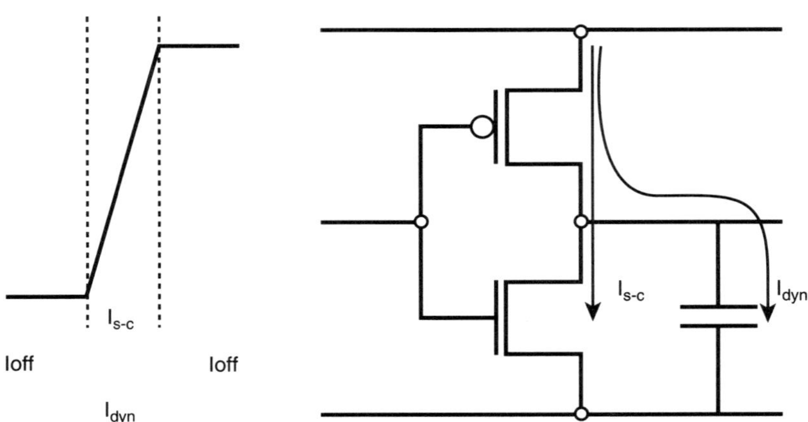

2-1

FIGURE

Components of power consumption. I_{s-c}, short-circuit current; Ioff, leakage current; I_{dyn}, dynamic switching power.

number of devices that are actually switching (drawing current from the power supply) at a given time. This definition of active power consumption illustrates the significant benefit of supply voltage scaling—a quadratic reduction in active power consumption. However, it is important for high performance that $V_{DD} > 3V_T$ so that there is sufficient current drive. Thus, as supply voltages scale, threshold voltages must also scale, causing leakage power to increase. As an example, the leakage current increases from $20\,pA/\mu m$ when using a Taiwan Semiconductor Manufacturing Corporation (TSMC) CL018G process with a threshold voltage of $0.42\,V$ to $13,000\,pA/\mu m$ when using a TSMC CL013HS process with a threshold voltage of $0.25\,V$. *Standby power consumption* due to subthreshold leakage current is estimated as:

$$P_{leak} = V_{dd} I_{off} K_L$$

where I_{off} is the current that flows between the supply rails in the absence of switching and K_L is a factor that accounts for the distribution/sizing of P and N devices, the stacking effect, the idleness of the device, and the design style used. In a CMOS-based style, at most half of the transistors are actually leaking whereas the remaining are ON (in the resistive region). As oxide thicknesses decrease, gate tunneling leakage will also become a significant source of standby power. This component of standby power is not included in the above equation and is not a target of the techniques presented in this chapter.

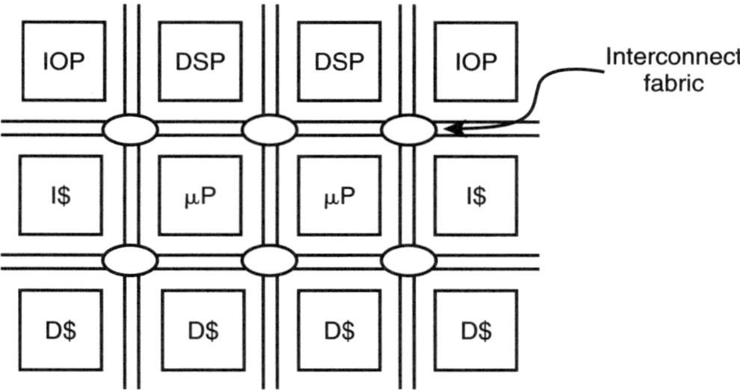

2-2 FIGURE Model MPSoC. I$, instruction cache; D$, data cache.

This chapter surveys a number of energy-aware design techniques for controlling both active and standby power consumption that are applicable to the MPSoC design space and points to emerging areas that will need special attention in the future. Our model MPSoC system is depicted in Figure 2-2. The chapter is organized as follows. Section 2.2 covers techniques that can be used to design energy-aware MPSoC processor cores. Section 2.3 focuses on the on-chip memory hierarchy, Section 2.4 on the on-chip communication system, and Section 2.5 on energy-aware software techniques that can be used in MPSoC designs. We close with conclusions.

2.2 ENERGY-AWARE PROCESSOR DESIGN

Many techniques for controlling both active and standby power consumption have been developed for processor cores [11]. Almost all of these will transition to the MPSoC design space, with some of them taking on additional importance and expanded design dimensions. Figure 2-3 shows the processor power design space for both active and standby power [12]. The second column lists techniques that are applied at design time and thus are part of the circuit fabric. The last two columns list techniques that are applied at run time, the middle column for those cases when the component is idle, and the last column when the component is in active use. (Run time techniques can additionally be partitioned into those that reduce leakage current while retaining state and those that are state-destroying.) The underlying circuit fabric must provide the "knobs" for controlling the run

	Constant Throughput/Latency		Variable Throughput/Latency
	Design Time	**Sleep Mode**	**Run Time**
Active (Dynamic) [C V_2 f]	Logic design Trans sizing Multiple V_{dd}'s	Clock gating	DFS DVS DTM
Standby (Leakage) [V I_{off}]	Stack effect Multiple V_T's Multiple t_{ox}'s	Sleep trans Multi-V_{dd} Variable V_T Input control	Variable V_T

2-3

FIGURE

Processor power design space. DFS, dynamic frequency scaling; DVS, dynamic voltage scaling; DTM, dynamic thermal management.

time mechanisms—by either the hardware (e.g., in the case of clock gating) or the system software (e.g., in the case of dynamic voltage scaling [DVS]). In the case of software control, the cost of transitioning from one state to another (in terms of both energy and time) and the relative energy savings need to be provided to the software for decision making.

2.2.1   Reducing Active Energy

As has already been pointed out, the best knob for controlling power is setting the supply voltage appropriately to meet the computational load requirements. Although lowering the supply voltage has a quadratic impact on active energy, it decreases systems performance since it increases gate delay, as shown in Figure 2-4 [4]. Multiple supply voltages can be used very effectively in MPSoCs since they contain multiple processors of different types with different performance requirements (e.g., as in Fig. 2-2: an MPSoC will contain high-speed digital signal processors (DSPs), moderate-speed processor cores, and low-speed I/O processors [IOPs]). Choosing the appropriate supply voltage at design time for the entire component will minimize (or even eliminate) the overhead of level converters that are needed whenever a module at a lower supply drives a module at a higher supply [13].

The most popular of the techniques for reducing active power consumption at run time is DVS combined with *dynamic frequency scaling* (DFS). Most embed-

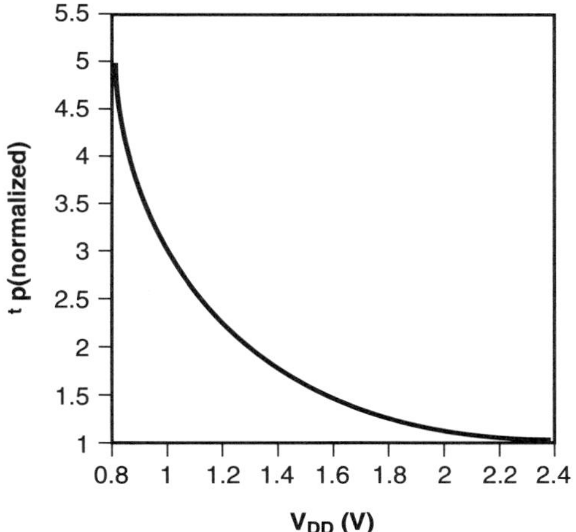

2-4   Delay as a function of *VDD*.

FIGURE

ded and mobile processors contain this feature (triggered by thermal on-chip sensors when thermal limits are being approached or by the run-time system when the CPU load changes). DFS + DVS requires a power supply control loop containing a buck converter to adjust the supply voltage and a programmable Phase locked loop (PLL) to adjust the clock [10]. As long as the supply voltage is increased before increasing the clock rate or decreased after decreasing the clock rate, the system only need stall when the PLL is relocking on the new clock rate (estimated to be around 20 msec). Future MPSoCs in which all (or selected) processing cores are DFS + DVS capable would require each to have its own converter and PLL (or ring oscillator) as well as system software capable of determining the optimum setting for each core, taking into account the computational load. The overall system will have to be designed such that cores are tolerant of periodic dropouts of other cores that are stalling when transitioning clock rates. An additional complication is that the voltage converter and PLL, being analog components, are susceptible to substrate noise induced by digital switching. Thus, uniprocessor systems typically distance the digital and analog components as much as possible to reduce this effect. Whether this technique will carry over to MPSoCs, in which the ability to separate the analog and digital components physically becomes much more complex, remains an open question.

2.2.2   Reducing Standby Energy

Several techniques have recently evolved for controlling subthreshold current, as shown in Figure 2-3. Since increasing the threshold voltage, V_T, decreases subthreshold leakage current (exponentially), adjusting V_T is one such technique. As shown in Figure 2-5, a 90-mV reduction in V_T increases leakage by an order of magnitude. Unfortunately, increasing V_T also negatively impacts gate delay. As with multiple supply voltages, multiple threshold voltages can be employed at design time or run time. At run time, multiple levels of V_T can be provided using *adaptive body-biasing* whereby a negative bias on V_{SB} increases V_T, as shown in Figure 2-5 [11]. Simultaneous DVS, DFS, and variable V_T has been shown to be an effective way to trade off supply voltage and body biasing to reduce total power—both active and standby—under variable processor loads [14]. Once again, whether similar techniques will carry over to MPSoCs remains an open question.

Another technique that will apply to reducing standby power in MPSoC's is the use of *sleep transistors* like those shown in Figure 2-6. Standby power can be greatly reduced by gating the supply rails for idle components. In normal mode (non-idle), the sleep transistors must present as small a resistance as possible (via sizing) so as not to negatively affect performance. In sleep mode

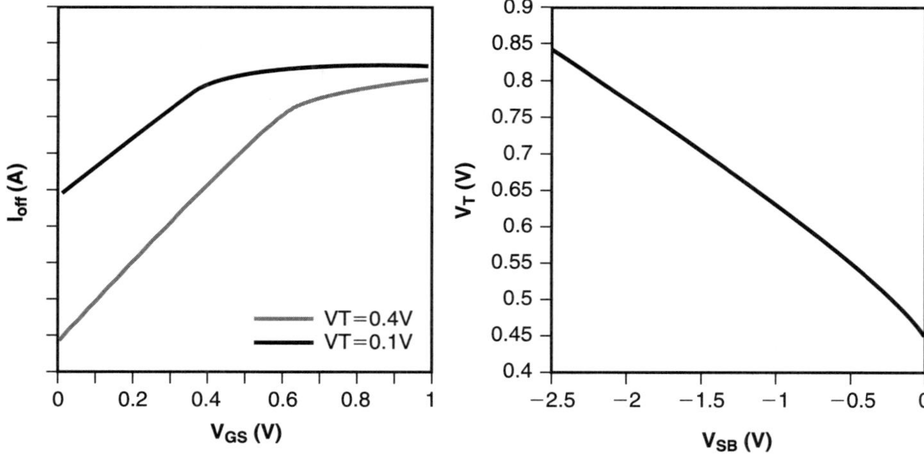

2-5   VT effects.

FIGURE

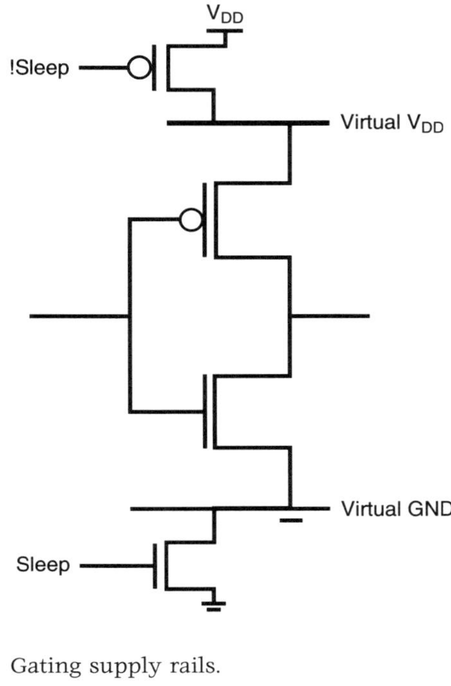

2-6

Gating supply rails.

FIGURE

(idle), the transistor stack effect [10] reduces leakage by orders of magnitude. Alternatively, standby power can be completely eliminated by switching off the supply to idle components. An MPSoC employing such a technique will require system software that can determine the optimal scheduling of tasks on cores and can direct idle cores to switch off their supplies while taking into account the cost (in terms of both energy and time) of transitioning from the on-to-off and off-to-on states.

2.3 ENERGY-AWARE MEMORY SYSTEM DESIGN

In an MPSoC system, memories constitute a significant portion of the overall chip resources, as there are various memory structures ranging from private memories for individual processors to large shared cache structures. Energy is expended in these memories due to data accesses (reads/writes), coherence activity required to maintain consistency between shared data, and leakage energy expended in just storing the data.

2.3.1   Reducing Active Energy

Many techniques have been proposed in the past to reduce cache energy consumption. Among these are partitioning large caches into smaller structures to reduce the dynamic energy [15,16] and the use of a memory hierarchy that attempts to capture most accesses in the smallest size memory. By accessing the tag and data array in series, Alpha 21164's L2 cache [17] can access the selected cache bank for energy efficiency. In Inoue et al. [18] and Powell et al. [19], *cache way-prediction* is used to reduce energy consumption of set-associative caches. *Selective way caches* [20] varies the number of ways for different application requirements. In Kin et al. [21], a small *filter cache* is placed prior to the L1 cache to reduce energy consumption. *Dynamic zero compression* [22] employs single-bit access for zero-valued byte in the cache to reduce energy consumption. Many of these techniques applied in the context of single processors are also applicable for the design of caches associated with the individual processors of the MPSoC system. In addition to the hardware design techniques, software optimizations that reduce the number of memory accesses through code and data transformations [23] can be very useful in reducing energy consumption.

2.3.2   Reducing Standby Energy

However, most of the above techniques do little to alleviate the leakage energy problem as the memory cells in all partitions and all levels of the hierarchy continue to consume leakage power as long as the power supply is maintained to them, irrespective of whether they are used or not. Various circuit technologies have been designed specifically to reduce leakage power when the component is not in use. Some of these techniques focus on reducing leakage during idle cycles of the component by turning off the supply voltage. One such scheme, *gated-*V_{dd}*,* was integrated into the architecture of caches [24] to shut down portions of the cache dynamically. This technique was applied at a cache block granularity in Kaxiras et al. [25] and used in conjunction with software to remove dead objects in Chen et al. [26]. However, all these techniques assume that the state (contents) of the supply-gated cache memory is lost. Although totally eliminating the supply voltage results in the state of the cache memory being lost, it is possible to apply a state-preserving leakage optimization technique if a small supply voltage is maintained to the memory cell. Many alternate implementations have recently been proposed at the circuit level to achieve such a state-preserving leakage control mechanism [27–29]. As an abstraction of these techniques, the choice between the state-preserving and state-destroying techniques depends on the relative overhead of the additional leakage required to maintain the state as opposed to the cost of restoring the lost state from other levels of the memory hierarchy.

An important requirement to reduce leakage energy using either a state-preserving or a state-destroying leakage control mechanism is the ability to identify unused resources (or data contained in them). In Yang et al. [24], the cache size is reduced (or increased) dynamically to optimize the utility of the cache. In Kaxiras et al. [25], the cache block is supply-gated if it has not been accessed for a period of time. In Zhou et al. [30], hardware tracks the hypothetical miss rate and the real miss rate by keeping the tag line active when deactivating a cache line. Then the turn-off interval can be dynamically adjusted based on such information. In Flautner et al. [31] and Kim et al. [32], dynamic supply voltage scaling is used to reduce the leakage in the unused portions of the memory. In contrast to the other schemes, their *drowsy cache* scheme also preserves data when a cache line is in low leakage mode. The usefulness and practicality of such state-preserving voltage scaling schemes for embedded power-optimized memories is demonstrated in Qin and Rabaey [33]. The focus in Heo et al. [34] is on reducing bitline leakage power using leakage-biased bitlines. The technique turns off precharging transistors of unused subbanks to reduce bitline leakage, and actual bitline precharging is delayed until the subbank is accessed. All these techniques for cache power optimization can also be applied to reduce the leakage energy in the MPSoC.

2.3.3   Influence of Cache Architecture on Energy Consumption

In addition to applying these techniques for active and standby energy reduction, the choice of cache architecture plays an important role in the amount of utilized memory storage space and, consequently, the leakage energy. There are two popular alternatives for building a cache architecture for on-chip multiprocessors. The first one is to adopt a single multi-ported cache architecture that is shared by the multiple processors. There are two major advantages of this alternative: (1) constructive interference can reduce overall miss rates, and (2) inter-processor communication is easy to implement. However, a large multi-ported cache can consume significant energy. Additionally, this is not a scalable option. The second alternative is to allow each processor to have its own private cache [35–37]. The

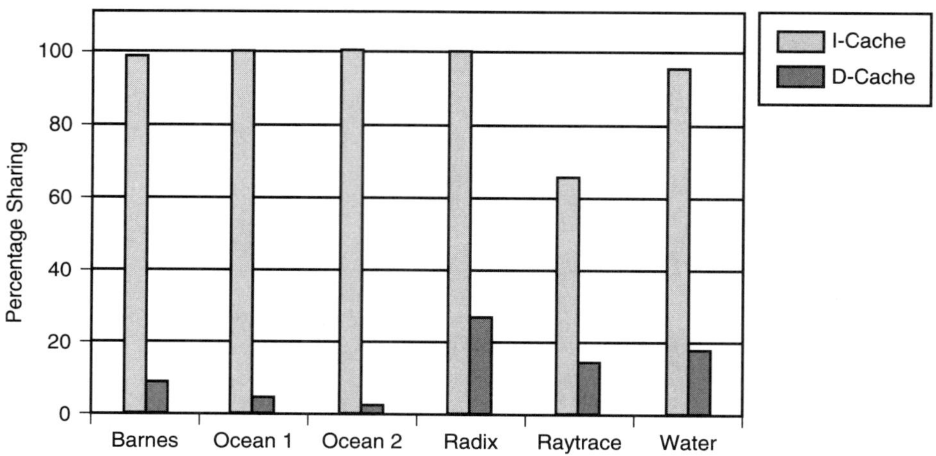

2-7

FIGURE   Percentage sharing of instructions and data in an 8-processor MPSoC.

main benefits of the private cache option are low power per access, low access latency, and good scalability. Its main drawback is duplication of data and instructions in different caches. In addition, a complex cache coherence protocol is needed to maintain consistency.

In this section, we explore a cache configuration, referred to as the *crossbar-connected cache* (CCC) that tries to combine the advantages of the two options discussed above without their drawbacks. Specifically, the shared cache is divided into multiple banks using an $N \times M$ crossbar to connect the N processors and the M cache banks. In this way, the duplication problem is eliminated (as logically there is a single cache), a consistency mechanism is no longer needed, and the solution is scalable. These advantages are useful in reducing the energy consumption.

Figure 2-7 illustrates the percentage of sharing for instructions and data when SPLASH-2 benchmarks [38] are executed using eight processors each with private L1 caches. Percentage sharing is the percentage of data or instructions shared by at least two processors when considering the entire execution of the application. The first observation is that instruction sharing is extremely high. This is a direct result of the data parallelism exploited in these benchmarks, that is, different processors typically work on (mostly) different sets of data using the same set of instructions. Instruction sharing is above 90% for all applications except *raytrace*. These results clearly imply a large amount of instruction duplication across

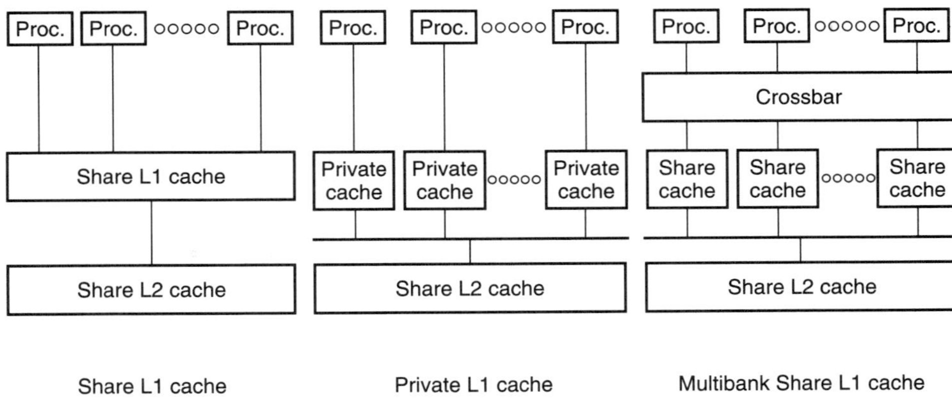

2-8

Different cache organizations: multiported shared cache, private caches, and CCC.

FIGURE

private caches. Consequently, a strategy that eliminates this duplication can bring large benefits (as there will be more available space), i.e., the effective cache capacity is increased by eliminating instruction duplication. Eliminating duplications also eliminated the extra leakage energy expended in powering duplicated cache lines. When one looks at the data sharing in Figure 2-7, the picture changes. The percentage sharing for data is low, averaging around 11%. However, employing CCC can still be desirable from the cache coherence viewpoint (since complex coherence protocols associated with private caches can be a major energy consumer). Based on these observations, using CCC might be highly beneficial in practice.

Figure 2-8 illustrates the three different cache organizations evaluated here: a multi-ported shared cache, private caches, and CCC. A common characteristic of these configurations is that they all have a shared L2 cache. For performance reasons, a CCC architecture would need a crossbar interconnect between the processors and L1 cache banks. Compared with a bus-based architecture, the crossbar-based architecture provides simultaneous accesses to multiple cache banks (which, in turn, helps to improve performance). However, due to the presence of the crossbar, the pipeline of the CCC processors needs to be restructured. For example, if the pipeline stages of the private cache (and multi-ported cache) architecture are instruction fetch (IF), instruction decode (ID), execution (EX), and write-back (WB), the pipeline stages in the CCC system are instruction crossbar transfer (IX), IF, ID, data crossbar transfer (DX), EX, and WB. In the IX stage, the processor transfers its request to an instruction cache bank through the

Tag#	Line#	Offset

Address format for private cache (and multi-ported shared cache)

Tag#	Line#	Bank#	Offset

Address format for multibank share cache

2-9   Address formats used by the different cache organizations.

FIGURE

crossbar. If more than one request targets the same bank, only one request can continue with the next pipeline stage (IF), and all other processors experience a pipeline stall. In the DX stage, if the instruction is load or store, another transfer request through the crossbar is issued.

Figure 2-9 shows the address formats used by the private cache system (and also the shared multiported cache system) and the CCC architecture. In the latter, a new field, called the bank number, is added. This field is used to identify the bank to be accessed. Typically, continuous cache blocks are distributed across different banks in order to reduce bank conflicts. Although CCC is expected to be preferable over the other two architectures, it can cause processor stalls when concurrent accesses to the same bank occur. To alleviate this, there are two possible optimizations. The first optimization is to use more cache banks than the number of processors. For instance, for a target MPSoC with four processors, utilize 4, 8, 16, or 32 banks. The rationale behind this is to reduce conflicts for a given bank. The second optimization deals with the references to the same block. When two such references occur, instead of stalling one of the processors, just read the block once and forward it to both the processors. This optimization should be more successful when there is a high degree of inter-processor block sharing.

To compare the cache energy consumption of the three architectures depicted in Figure 2-8, we ran a set of simulations assuming 70-nm technology (in which standby energy is expected to be comparable to active energy). We used the cache leakage control strategy proposed in Flautner et al. [31] in both the L1 and L2 caches whereby a cache line that has not been used for some time is placed into a leakage control mode. The experimental results obtained through cycle-accurate simulations and the SPLASH-2 benchmark suite indicate that the energy benefits of CCC range from 9% to 26% with respect to the private cache option. These savings come at the expense of a small degradation in performance. These results show that the choice of cache architecture is very critical in designing energy-aware MPSoCs.

2.3.4   Reducing Snoop Energy

In bus-based symmetric multi-processors (SMPs), all bus-side cache controllers snoop the bus (*bus snooping*) to ensure data coherence. Specifically, snoops occur when writes from the bus-side of the cache are issued to already cached blocks or on cache misses that require a cache fill. Thus, another class of memory system energy optimizations is that associated with snoops.

Since the cache snoop activity primarily involves tag comparisons and mostly incur tag misses, tag and data array accesses are separated, unlike normal caches in which both are accessed simultaneously to improve performance. Thus, energy optimizations include the use of dedicated tag arrays for snoops and the serialization of tag and data array accesses. For example, servers based on both Intel Xeon 2 and Alpha 21164 fetch tag and data serially.

An optimization that goes beyond reducing the data array energy of these tag misses is presented in Moshovos et al. [39]. In this work, a small, cache-like structure, called JETTY, is introduced between the bus and the backside of each processor's cache to filter the vast majority of snoops that would not find a locally cached copy. Energy is reduced, as accesses are mostly confined to JETTY, and thus accesses to the energy consuming cache tag arrays are decreased. Their work demonstrates that JETTY filters 74% (average) of all snoop-induced tag accesses that would miss, resulting in an average energy reduction of 29% in the cache.

Another approach to reducing the snoop energy is proposed in Saldhana and Lipasti [40]. In this work, the authors suggest a serial snooping of the different caches as opposed to a parallel snoop of all the caches attached to the bus. Whenever a request is generated on the bus, the snooping begins with the node closest to the requester and then propagates to the next successive node in the path only if the previous nodes have not satisfied the request. The tradeoff in this approach is the increased access latency due to the sequential snooping in order to reduce energy.

In Ekman et al. [41], an evaluation of JETTY and the serial snooping schemes is performed in the context of on-chip multiprocessors. The authors concluded that serial snooping is not very effective, as all the caches need to be searched when none of the caches have the requested data.

This performance penalty also translates to an energy penalty. Furthermore, they also found the JETTY approach not as attractive in the case of MPSoCs because the private cache associated with each processor in a MPSoC is comparable in size to the JETTY structure.

Another important impact of the snooping activity is the traffic generated on the interconnects. The memory accesses and organization have a significant impact on the interconnect activity and power consumption. The next section explores on-chip communication energy issues.

2.4 ENERGY-AWARE ON-CHIP COMMUNICATION SYSTEM DESIGN

Computation and storage energy directly benefit from device scaling (smaller gates, smaller memory cells), but unfortunately the energy for global communication does not scale down. Projections assuming aggressive physical and circuit level optimizations for global wires [42] show that global communication on chip will require increasingly higher energy, the majority of it being active energy. Hence, communication-related energy is a significant concern in future MPSoC systems that will create many new challenges that have not been addressed in traditional high-performance on-chip interconnect design.

In this section we will explore various communication energy minimization approaches, moving from simple architectures (shared busses and point-to-point links), to complex advanced interconnects (networks on-chip [NoCs]) that are applicable to MPSoCs. Also, refer to Chapter 3 in this volume for additional material related to this section. The simplest on-chip communication channel is a bundle of wires, often called a *bus*. To be more precise, however, a bus can be significantly more complex than a bundle of wires. On-chip busses are highly shared communication infrastructures, with complex logic blocks for arbitration and interfacing. In an effort to maintain consistency with the surveyed literature, in the following we will use the term "bus" just as a synonym for a set of bundled wires. It is, however, important to remember that implementing encoding schemes in real-life designs requires maintaining compliance with complex SoC bus protocols. This is not an easy task, as outlined in Osborne et al. [43].

2.4.1 Bus Encoding for Low Power

Bus data communication has traditionally adopted straightforward encodings and modulation. The transfer function of the encoder and modulator of binary non-return-to-zero input streams was implicitly assumed to be unitary. In other words, no modulation and encoding were applied to the binary data before sending it on the bus. Low-energy communication techniques remove this implicit assumption, by introducing the concept of *signal coding for low power*.

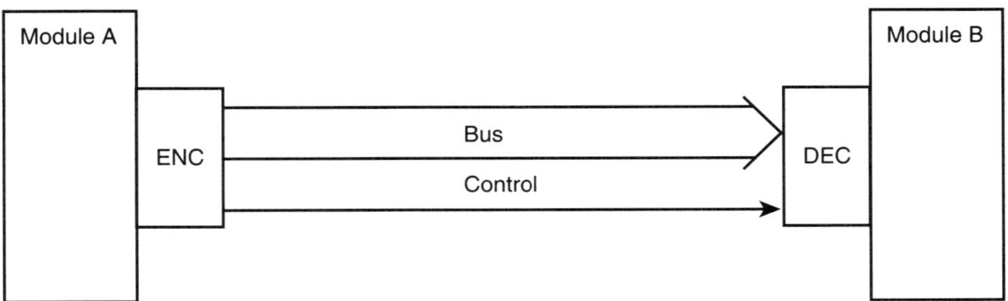

2-10

FIGURE

Bus encoding: one-directional communication.

The basic idea behind these approaches is to encode data sent through the communication channel to minimize its average switching activity, which is proportional to dynamic power consumption in CMOS technology. Ramprasad et al. [44] studied the data encoding for minimum switching activity problem and obtained upper and lower bounds on transition activity reduction for any encoding algorithm. This important theoretical result can be summarized as follows: *the savings obtainable by encoding depend on the entropy rate[1] of the data source and on the amount of redundancy in the code. The higher the entropy rate, the lower the energy savings that can be obtained by encoding with a given code redundancy.*

Even though the work by Ramprasad et al. [44] provides a theoretical framework for analyzing encoding algorithms, it does not provide general techniques for obtaining effective encoders and decoders. The complexity and energy cost of encoding and decoding circuits must be taken into account when evaluating any bus-encoding scheme.

Several authors have proposed low-transition activity encoding and decoding schemes. To illustrate the characteristics of these schemes in a simple setting, we consider a point-to-point, one-directional bus connecting two modules (e.g., a processor and its memory), as shown in Figure 2-10. Data from the source module are encoded, transmitted on the bus, and decoded at the destination. A practical instance of this configuration is the address bus for the processor/memory system. If the bus has very large parasitic capacitance, the energy dissipated in driving the

---

[1] In intuitive terms, the entropy of a data stream is related to the probability distribution of its symbols: the more evenly distributed, the higher the entropy. The entropy rate is the limit for $n \to \infty$ of the entropy of streams obtained by concatenating n symbols of the original data stream.

lines during signal transitions dominates the total energy cost of communication (including the cost of encoding and decoding). However, as discussed in the following, this assumption has to be carefully validated whenever a new encoding scheme is proposed.

Encoding for Random White Noise

A few encoding schemes have been studied starting from the assumption that the data sent on the bus can be modeled as random white noise (RWN), i.e., having a maximum entropy rate. Under this assumption, it is possible to focus solely on how to exploit redundancy to decrease switching activity, because all irredundant codes have the same switching activity (this result is proved in the paper by Ramprasad et al. [44]). Data are formatted in words of equal width, and a single word is transmitted every clock cycle.

In its simplest form, redundant encoding requires adding *redundant* wires to the bus. This can be seen as extending the word's width by one or more redundant bits. These bits inform the receiver about how the data were encoded before the transmission (see Fig. 2-10.) Low-energy encodings exploit the correlation between the word currently being sent and the previously transmitted one. The rationale is that energy consumption is related to the number of switching lines, i.e., to the Hamming distance between the words. Thus, transmitting identical words will consume no power, but alternating a word and its complement would produce the largest power dissipation, because all bus lines would be switching.

A conceptually simple and powerful scheme was proposed by Stan and Burleson [45] and called *bus invert* (BI) encoding. To reduce the switching, the transmitter computes the Hamming distance between the word to be sent and the previously transmitted one. If the distance is larger than half the word width, the word to be transmitted is inverted, i.e., complemented. An additional wire carries the *bus invert* information, which is used at the receiver end to restore the data.

This encoding scheme has some interesting properties. First, the worst-case number of transitions of an n-bit bus is $n/2$ at each time frame. Second, if we assume that data are uniformly randomly distributed, it is possible to show that the average number of transitions with this code is lower than that of any other encoding scheme with just one redundant line [45].

An unfortunate property of the 1-bit redundant *bus-invert code* is that the average number of transitions per line increases as the bus gets wider and asymptotically converges to 0.5, which is also the average switching per line of an unencoded bus. Moreover, the average number of transitions per line is already close to 0.5 for 32-bit busses. Thus, this encoding provides small energy saving for busses of typical width.

A solution to this problem is to partition the bus into fields and to use bus inversion in each field independently. If a word is partitioned in m fields, then m control lines are needed. Although this scheme can be much more energy efficient compared with 1-bit bus invert, m-bit bus invert is no longer the best among m-redundant codes. Nevertheless, it is conceptually simpler than other encoding schemes based on redundancy, and thus its implementation overhead (in terms of power) is small.

Extensions to the bus invert encoding approach include the use of limited-weight codes and transition signaling. A k-limited-weight code is a code having at most k 1's per word. This can be achieved by adding appropriate redundant lines [46]. Such codes are useful in conjunction with transition signaling, i.e., with schemes in which 1's are transmitted as a 0-1 (or 1-0) transition and 0's by the lack of a transition. Thus, a k-limited-weight code would guarantee at most k transitions per time frame (if we neglect the transitions on the redundant lines).

As a general note on bus invert and all redundant encoding schemes, although a redundant code can be implemented using additional bus lines, there are other options. In particular, it is possible to communicate with redundant codes without increasing the number of bus lines if the bus transmission rate is increased by $(m + n)/n$ with respect to the rate of the unencoded input stream, where n is the data word width and m is the number of redundant bits. This approach is often called *time redundancy* as opposed to *space redundancy* (i.e., adding extra lines). Obviously, many hybrid schemes can be envisioned.

Encoding for Correlated Data

Even though the RWM data model is useful for developing redundant codes with good worst-case behavior, in many practical cases data words have significant correlations in time and space. From an information theoretical viewpoint, the data source is not maximum entropy. This fact can be profitably exploited by advanced encoding schemes that outperform codes developed under the RWN model [46], even without adding redundant bus lines.

A typical example of a highly correlated data stream is the address stream in between a processor and its private memory. Addresses show a high degree of sequentiality. This is typical for instruction addresses (within basic blocks) and for data addresses (when data are organized in arrays). Therefore, in the limiting case of addressing a stream of data with consecutive addresses, Gray coding would be beneficial [47], because the Hamming distance between any pair of consecutive words is one, and thus the transitions on the address bus are minimized.

By using Gray encoding, instruction addresses need to be converted to Gray addresses before being sent on the bus. The conversion is necessary because offset addition and arithmetic address manipulation is best done with standard binary encoding [48]. Moreover, address increments depend on the word width n. Since most processors are byte-addressable, consecutive words require an increment by $n/8$, e.g., by 4 [8] for 32-bit (64-bit) processors. Thus the actual encoding of interest is a partitioned code, whose most significant field is Gray encoded and whose least significant field has $(\log n/8)$ bits.

Musoll et al. [49] proposed a different partitioned code for addresses, which exploits the locality of reference, namely, most software programs favor *working zones* of their address space. The proposed approach partitions the address into an offset within a working zone and an identifier of the current working zone. In addition, a bit is used to denote a hit or a miss of the working zone. When there is a miss, the full address is transmitted through the bus. In the case of a hit, the bus is used to transmit the offset (using 1-hot encoding, and transition signaling), and additional lines are used to send the identifier of the working zone (using binary encoding). As an improvement over working zone, an irredundant code with similar properties, called *sector-based encoding* has been proposed by Aghaghiri et al. [50].

The $T\,0$ code [51] uses one redundant line to denote when an address is consecutive to the previously transmitted one. In this case, the transmitter does not need to transmit the address and freezes the information on the bus, thus avoiding any switching. The receiver updates the previous address. When the address to be sent is not consecutive, then it is transmitted *tout court*, and the redundant line is de-asserted to inform the receiver to accept the address as is. When one is transmitting a sequence of consecutive addresses, this encoding requires no transition on the bus, compared with the single transition (per transmitted word) of the Gray code. In this context, an irredundant code with zero transition activity for address streams was also demonstrated, called INC-XOR [44], and later improved on by Aghaghiri et al. [52].

The highly sequential nature of addresses is just one simple example of spatio-temporal correlation on address busses. For instance, DRAM address busses use a time multiplexed addressing protocol, whose transition activity can be reduced by a tailored encoding scheme, as outlined in Cheng and Pedram [53]. Furthermore, several encoding schemes have been proposed for dealing with more general correlations [44,54–56] than those found in address busses. Unfortunately, many of these approaches assume that data stream statistics can be collected and analyzed at design time, making them far less useful for MPSoCs.

Practical Guidelines

The literature on low-power encoding has flourished in the last decade, and choosing the best encoding scheme is today a challenging task. A few practical guidelines can be of help. First, in estimating the power savings obtained on the bus, the capacitive load of bus lines should be estimated with great care. In fact, in sub-micron technologies, coupling capacitance between adjacent wires dominates with respect to substrate capacitances; hence one must account for the impact of multiple transitions on adjacent lines. In this area, several coupling-aware low-power encoding schemes have recently been proposed [57,58].

Second, bus load capacitance estimates should be used as a selection criteria for encoding schemes. For typical on-chip bus capacitances, many encoding schemes are not practical, because encoders and decoders are too big and power-hungry. A good rule of thumb is to convert the expected switched capacitance reduction on the target bus $SW_{red} = C_{line}W_{bus}\alpha_{red}$ (i.e., the product between bus capacitance, number of bus lines, and bus switching activity reduction) into an equivalent number of inverter loads switching $Nb = SW_{red}/C_{in,INV}$ (i.e., how many inverter input switches are needed to switch the same amount of capacitance that we save with encoding). If the complexity of encoder and decoder, expressed in terms of inverter-equivalents, is similar to N_{eq}, then it is very likely that most of the transition activity savings are swamped by encoder and decoder power consumption. Also, it is very important to account for the speed and area penalty of the pure error-detecting circuit (codec). In general, most of the encoding schemes outlined above are well suited for off-chip busses with very large capacitances (e.g., 10 pF or more). Only the simplest schemes are suitable for typical on-chip busses (e.g., 1 pF or less).

2.4.2 Low Swing Signaling

An effective approach to high-speed energy-efficient communication is low swing signaling [59]. Even though it requires the design of receivers with good adaptation to line impedance and high sensitivity (often achieved by means of simplified sense-amplifiers), power savings on the order of 10× have been estimated with reduced interconnect swings of a few hundreds of mV in a 0.18-μm process [60]. The use of low swing signaling poses a critical challenge for design: communication reliability has to be provided in spite of the decreased noise margins and the strong technology limitations, under limited power budgets.

With present technologies, most chips are designed under the assumption that electrical waveforms can always carry correct information on chip. As technology scales, communication is likely to become inherently unreliable because of the increased sensitivity of interconnects to on-chip noise sources, such as crosstalk and power-supply noise. As a consequence, solutions for combined energy minimization and communication reliability control have to be developed for

NoCs. Redundant bus encoding provides a degree of freedom for spanning the energy-reliability tradeoff [61,62]. The key point of this approach is to model on-chip interconnects as noisy channels and to exploit the error detection capability coding schemes, which would provide a link transfer reliability in excess with respect to the constraint, to decrease the voltage swing, resulting in an overall energy saving (compared with the unencoded link) in spite of the overhead associated with the code implementation. The energy efficiency of a code is tightly related to its error recovery technique, namely, error correction or retransmission of corrupted data. This issue resembles the tradeoff investigation between forward error correction (FEC) and automatic repeat request (ARQ), well known to network engineers, but for on-chip communication networks this study is still in its early stage.

Bertozzi et al. [61] have explored the energy-reliability tradeoff for on-chip low-swing busses. Starting from a standard on-chip bus (AMBA), various reliability-enhancement encoding techniques are explored. Hamming codes are explored in both their error correction and error detection embodiments. Other linear codes, namely, Cyclic Redundancy Check (CRC) codes, are considered as well. For error-detecting codes, an AMBA-compliant retransmission mechanism is also proposed. Experimental results in Bertozzi et al. [61] point out that the detection capability of a code plays a major role in determining its energy efficiency, because the higher the error detection capability the lower we can drive the voltage swing, knowing that the increased error rate caused by the lowered signal-to-noise ratio would not lead to system failures as long as the errors are detected. As far as error recovery is concerned, error correction is beneficial in terms of recovery delay but has two main drawbacks: it limits the detection capability of a code, and it makes use of high-complexity decoders. On the contrary, when the higher recovery delay (associated with the retransmission time) of retransmission mechanisms can be tolerated, they provide a higher energy efficiency, thanks to the lower swings and simpler codecs they can use while preserving communication reliability.

In many practical cases, communication noise depends on operating conditions and environmental factors. It is then possible to devise adaptive schemes that dynamically adjust signal levels (i.e., the voltage swing on bus lines) depending on noise levels. In the approach proposed by Worm et al. [62], the frequency of detected errors by an error-detecting code is monitored at run time. When the frequency rises too much, the voltage swing is increased, whereas it is decreased if error frequency becomes very small. This closed-loop control of voltage provides much increased robustness and flexibility, even though the complexity of the encoding and decoding circuitry is increased, and variable voltage swing circuits are also required.

2.4.3   Energy Considerations in Advanced Interconnects

Network architecture heavily influences communication energy. As hinted in the previous section, shared-medium networks (busses) are currently the most common choice, but it is intuitively clear that busses are not energy-efficient as network size scales up [63], as it must for MPSoCs. In bus-based communication, data are always broadcast from one transmitter to all possible receivers, whereas in most cases messages are destined to reach only one receiver, or a small group. Bus contention, with the related arbitration overhead, further contributes to the energy overhead.

Studies on energy-efficient on-chip communication indicate that hierarchical and heterogeneous architectures are much more energy-efficient than busses [64,65]. Zhang et al. [65] developed a *hierarchical generalized mesh* whereby network nodes with a high communication bandwidth requirement are clustered and connected through a programmable generalized mesh consisting of several short communication channels joined by programmable switches. Clusters are then connected through a generalized mesh of global long communication channels. Clearly such architecture is heterogeneous because the energy cost of intra-cluster communication is much smaller than that of inter-cluster communication. Although the work of Zhang et al. [65] demonstrates that power can be saved by optimizing network architecture, many network design issues are still open, and we need tools and algorithms to explore the design space and to tailor network architecture to specific applications or classes of applications.

Network architecture is only one facet of network layer design, the other major facet being network control. A critical issue in this area is the choice of a switching scheme for indirect network architectures. From the energy viewpoint, the tradeoff is between the cost of setting up a circuit-switched connection once and for all, and the overhead of switching packets throughout the entire communication time on a packet-based connection. In the former case the network control overhead is "lumped" and incurred once, whereas in the latter case, it is distributed over many small contributions, one for each packet. When communication flow between network nodes is extremely persistent and stationary, circuit-switched schemes are likely to be preferable, whereas packet switched schemes should be more energy-efficient for irregular and non-stationary communication patterns. Needless to say, circuit switching and packet switching are just two extremes of a spectrum, with many hybrid solutions in between.

Packetization and Energy Efficiency

The current trend in on-chip network design is toward packet-switched architectures [65–66]. This choice is mostly driven by the need for utilizing limited global wiring resources as efficiently as possible. (It is a well-known fact that circuit-switched networks have a low utilization with respect to packet-switched networks.) Ye et al. [67] have studied the impact of packet size on energy and performance for a regular on-chip network (a mesh) connecting homogeneous processing elements, in a cache-coherent MPSoC. Packetized data flow on the MPSoC network is affected by: (1) the number of packets in the network, (2) the energy consumed by each packet on one hop, and (3) the number of hops each packet travels. Different packetization schemes will have different impact on these factors and, consequently, affect the network power consumption. Several findings are reported in Ye et al. [67] and are summarized below.

Larger packet size will increase the energy consumed per packet, because there are more bits in the payload. Furthermore, larger packets will occupy the intermediate node switches for a longer time and cause other packets to be re-routed to non-shortest datapaths. This leads to more contention that will increase the total number of hops needed for packets traveling from source to destination. As a result, as packet size increases, energy consumption on the interconnect network will increase.

Although an increase in packet size will increase the energy dissipated by the network, it will decrease the energy consumption in cache and memory. Because larger packet sizes will decrease the cache miss rate, both cache energy consumption and memory energy consumption will be reduced. The total energy dissipated by the MPSoC comes from non-cache instructions (instructions that do not involve cache access) executed by each processor, the caches, and the shared memories, as well as the interconnect network. In order to assess the impact of packet size on the total system energy consumption, all MPSoC energy contributors should be considered together.

Ye et al. [67] showed that the overall MPSoC energy initially decreases as packet size increases. However, when the packets are too large, the total MPSoC energy starts increasing again. This is because the interconnect network energy outgrows the decrease of energy on cache, memory, and non-cache instructions. As a result, there is an optimal packet size, which, for the architecture analyzed by Ye et al., is around 128 bytes. It is important to stress, however, that different processor architectures and system interconnects may have different optimal packet sizes.

Energy and Reliability in Packet-Switched NoCs

As MPSoC communication architectures evolve from shared busses to packet-based micro-networks, packetization issues have to be considered when assessing the energy efficiency of error recovery techniques. In general, packets are often broken into message flow control units or *flits*. In the presence of channel width constraints, multiple physical channel cycles may be used to transfer a single flit. A *phit* is the unit of information that can be transferred across a physical channel in a single cycle. Flits represent logical units of information, as opposed to phits, which correspond to physical quantities, i.e., the number of bits that can be transferred in a single cycle. In many implementations, a flit is set to be equal to a phit.

Communication reliability can be guaranteed at different levels of granularity. We might refer control bits (i.e., a checksum) to an entire packet, thus minimizing control bits overhead, although this would prevent us from stopping the propagation of corrupted flits, as routing decisions would be taken in advance with respect to data integrity checking. In fact, control bits would be transmitted as the last flit of the packet. In this scenario, the cost for redundancy would be paid in the time domain. The alternative solution is to provide reliability at the flit level, thus refining control granularity but paying for redundancy in the space domain (additional wiring resources for check bits). The considerations that follow will refer to this latter scenario, wherein two different solutions are viable:

+ The error recovery strategy can be *distributed* over the network. Each communication switch is equipped with error detecting/correcting circuitry, so that error propagation can be immediately stopped. This is the only way to avoid routing errors: should the header get corrupted, its correct bit configuration could be immediately restored, preventing the packet from being forwarded across the wrong path to the wrong destination. Unfortunately, retransmission-oriented schemes need power-hungry buffering resources at each switch, so their advantage in terms of higher detection capability has to be paid for with circuit complexity and power dissipation.

+ Alternatively, an *end-to-end* approach to error recovery is feasible: only end-nodes are able to perform error detection/correction. In this case, retransmission may not be convenient at all, especially when source and destination are far apart from each other, and retransmitting corrupted packets would stimulate a large number of transitions, beyond giving rise to large delays. For this scenario, error correction is the most efficient solution, even though the proper course of action has to be taken to handle incorrectly routed packets (retransmission time-outs at the source node, deadlock avoidance, and so on).

Another consideration regards the way retransmissions are carried out in an NoC. Traditional shared bus architectures can be modified to perform retransmissions in a "stop and wait" fashion: the master drives the data bus and waits for the slave to carry out sampling on one of the following clock edges. If the slave detects corrupted data, a feedback has to be given to the master, scheduling a retransmission. To this purpose, an additional feedback line can be used, or built-in mechanisms of the bus protocol can be exploited.

In packet-switched networks, data packets transmitted by the master can be seen as a continuous flow, so the retransmission mechanism must be either "go-back-N" or "selective repeat." In both cases, each packet has to be acknowledged (ACK), and the difference lies in the receiver (switch or network interface) complexity. In a "go-back-N" scheme, the receiver sends a not ACK (NACK) to the sender relative to a certain incorrectly received packet. The sender reacts by retransmitting the corrupted packet as well as all other following packets in the data flow. This alleviates the receiver from the burden to store packets received out of order and to reconstruct the original sequence. On the other hand, when this capability is available at the receiver side (at the cost of further complexity), retransmissions can be carried out by selectively requiring the corrupted packet without the need to retransmit also successive packets. The tradeoff here is between switch and network interface complexity and number of transitions on the link lines.

2.5   ENERGY-AWARE SOFTWARE

Although employing energy-efficient circuit/architecture techniques is a must for energy-aware MPSoCs, there are also important roles for the software to play. Perhaps one of the most important of these is to parallelize a given application across on-chip processors. It is to be noted that, in this context, it may not be acceptable to use a conventional application parallelization strategy from the high-end computing domain since such strategies focus only on performance and do not account for the energy behavior of the optimized code. More specifically, such high-end computing-oriented techniques can be very extravagant in their use of processors (or machine resources in general), i.e., they tend to use a large number of parallel processors in executing applications even though most of these processors bring only marginal performance benefits.

However, a compiler designed for energy-aware MPSoCs may not have this luxury. In other words, it should strive to strike an acceptable balance between increased parallelism (reduced execution cycles) and increased energy consumption as the number of processors is increased. Therefore, our belief is that it is very important to determine the ideal number of processors to use in executing a given MPSoC application. In addition, it is conceivable to assume that different parts of the same application can demand different numbers of processors to generate the best behavior at run-time. Consequently, it may not be sufficient to fix the number of processors to use at the same value throughout the execution. During execution of a piece of code, unused processors can be placed in a low power state for saving energy [68].

Most published work on parallelism for high-end machines [69] is static, that is, the number of processors that execute the code is fixed for the entire execution. For example, if the number of processors that execute a code is fixed at eight, all parts of the code (for example, all loop nests) are executed using the same eight processors. However, this may not be a very good strategy from the viewpoint of energy consumption. Instead, one can consume much less energy by using the minimum number of processors for each loop nest (and shutting down the unused processors). In adaptive parallelization [70], the number of processors is tailored to the specific needs of each code section (e.g., a nested loop in array-intensive applications). In other words, the number of processors that are active at a given period of time changes dynamically as the program executes. For instance, an adaptive parallelization strategy can use four, six, two, and eight processors to execute the first four nests in a given code. There are two important issues that need to be addressed for energy-aware parallelization (assuming that inter-processor communication occurs through shared memory):

+ *Determining the ideal number of processors to use.* There are at least two ways to determine the number of processors needed per nest. The first option is to adopt a fully dynamic strategy whereby the number of processors (for each nest) is decided on in the course of execution (at run-time). Kandemir et al. [71] present such a dynamic strategy. Although this approach is expected to generate better results once the number of processors has been decided (as it can take run-time code behavior and dynamic resource constraints into account), it may also incur some performance overhead during the process of determining the number of processors. This overhead can, in some cases, offset the potential benefits of adaptive parallelism. In the second option, the number of processors for each nest is decided statically at compile-time. This approach has a compile-time overhead, but it does not lead to any run-time penalty. It should be emphasized, however, that although this approach determines the number of processors statically at compile-time, the activation/deactivation of processors and their caches occurs dynamically at run-time. Examples of this strategy can be found in Kadayif et al. [68,72].

+ *Selection criterion for the number of processors.* An optimizing compiler can target different objective functions such as minimizing the execution time of the compiled code, reducing executable size, improving power or energy behavior of the generated code, and so on. In addition, it is also possible to adopt complex objective functions and compilation criteria such as minimizing the sum of the cache and main memory energy consumptions while keeping the total execution cycles bounded by M. For example, the framework described in Kandemir et al. [71] accepts as objective function and constraints any linear combination of energy consumptions and execution cycles of individual hardware components.

2.6 CONCLUSIONS

This chapter has surveyed a number of energy-aware design techniques for controlling both active and standby power consumption that are applicable to MPSoCs. In particular, it covered techniques used to design energy-aware MPSoC processor cores, the on-chip memory hierarchy, the on-chip communication system, and MPSoC energy-aware software techniques. The vast majority of the techniques are derived from existing uniprocessor energy-aware systems that take on new dimensions in the MPSoC space. As energy-aware MPSoC design is a relatively new area, many open questions remain.

Other issues to be considered but not discussed in this chapter are the possible promise of globally asynchronous, locally synchronous (GALS) clocking for MPSoCs, as well as the inextricable link between energy-aware design techniques and their impact on system reliability. As shown in Figure 2-11, design tradeoffs

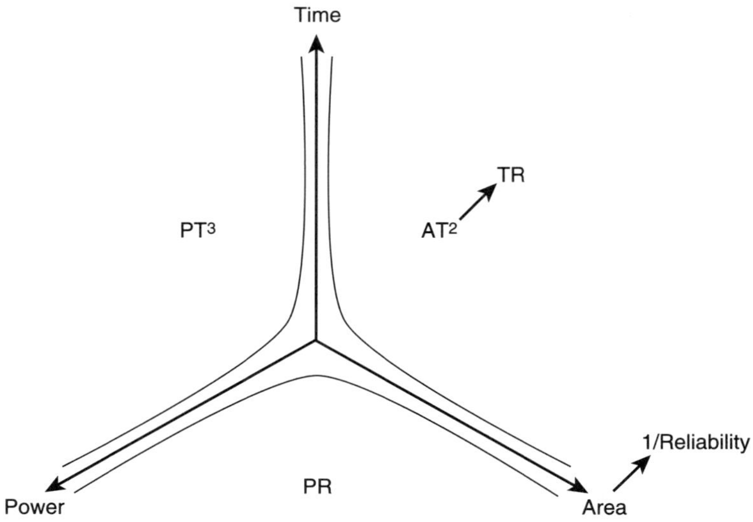

2-11   Evolving design tradeoffs. T, time; R, reliability; A, area; P, power; PT, product of power and time.

FIGURE

have changed over the years from a focus on area to a focus on performance and area to a focus on performance and power. In the future, the metrics of performance, power, and reliability and their tradeoffs will be of increasing importance.

ACKNOWLEDGMENTS

The authors have been supported in part by NSF Career Award Grants CCR-0093082 and CCR-0093085, NSF grant ITR-0082064, and a MARCO 98-DF-600 grant from GSRC. The authors would also like to acknowledge the contributions of various graduate students from their labs who worked on several projects whose results are abstracted here.

Networks on Chips: A New Paradigm for Component-Based MPSoC Design

Luca Benini and Giovanni De Micheli

3.1 INTRODUCTION

We consider *multiprocessor systems on chips* (MPSoCs) that will be designed and produced in *deep submicron technologies* (DSMs) with minimal features in the (100 to 25) nm range. Such systems will be designed using pre-existing components, such as processors, controllers, and memory arrays. The major design challenge will be to provide a functionally correct, reliable operation of the interacting components. The physical interconnections on chip will be a limiting factor for performance and possibly for energy consumption.

In this chapter, we address a methodology for the design of the interconnection among components that satisfies the needs for modular and robust design, under the constraints provided by the physical interconnect.

3.1.1 Technology Trends

In the projections of future silicon technologies, the operating frequency and transistor density will continue to grow, making energy dissipation and heat extraction a major concern. At the same time, supply voltages will continue to decrease, with adverse impact on signal integrity. The voltage reduction, even though beneficial, will not suffice to mitigate the energy consumption problem, in which a major contribution will be due to leakage. Thus, SoCs will incorporate *dynamic power management* (DPM) techniques in various forms to satisfy energy consumption bounds [81].

Global wires, connecting different functional units, are likely to have propagation delays largely exceeding the clock period [82]. Whereas signal pipelining on interconnections will become common practice, correct design will require knowing the signal delay with reasonable accuracy. Indeed, a negative side effect of technology downsizing will be the spreading of physical parameters (e.g., variance of wire delay per unit length) and its relative importance compared with the timing reference signals (e.g., clock period).

Synchronization of future chips with a single clock source and negligible skew will be extremely hard or impossible. The most likely synchronization paradigm for future chips is *globally asynchronous, locally synchronous* (GALS), with many different clocks. Global wires will span multiple clock domains, and synchronization failures in communicating between different clock domains will be rare but unavoidable events [83].

3.1.2   Nondeterminism in SoC Abstraction Models

As SoCs complexity scales, it will be more difficult, if not impossible, to capture their functionality with fully deterministic models of operation. For example, a transaction request may not be satisfied within a fixed worst-case time window, unless this window is so large as to make the constraint useless. On the other hand, average response time will be a more realistic objective to pursue. Similarly, high-level abstraction of components will lead to nondeterministic models. For example, the time for context restoration in a processor waking up from sleep mode cannot always be assessed deterministically at design time, because it often depends on the amount of state information that needs to be restored to resume operation.

SoC designs will rely on tens, if not hundreds, of information processing components that can dialogue among each other. Global control of the information traffic will be unlikely to succeed, because of the need to keep track of the states of each component. Thus components will initiate autonomously data transfer according to their needs. The global communication pattern will be fully distributed, with little or no global coordination. Overall, SoC design will be based on both deterministic and stochastic models. There will be a defnite trend toward quantifying design objectives (e.g., performance, power) by probabilistic metrics (e.g., average values, variance). This trend will lead to a major change in design methodologies. The problem of large-scale design under stochastic models is not new to engineers. Networks are examples of systems that are well understood and successfully designed with stochastic techniques and models [84,85].

3.1.3   A New Design Approach to SoCs

To tackle the above-mentioned issues we can borrow models, techniques, and tools from the network design field and apply them to SoCs design. In other words, we view a SoC as a *micronetwork* of components. The network is the abstraction of the communication among the components and has to satisfy *quality of service* (QoS) requirements (e.g., *reliability*, performance, and energy bounds) under the limitation of intrinsic unreliable signal transmission and nonnegligible communication delays on wires. We postulate that SoC interconnect design can be done using the *micronetwork stack* paradigm, which is an adaptation of the *protocol stack* [85] (Fig. 3-1). Thus the electrical, logic, and functional properties of the interconnection scheme can be abstracted.

SoCs differ from wide-area networks because of local proximity and because they are much more predictable at design time. Local, high-performance networks (such as those developed for large-scale multiprocessors) have closer characteristics. Nevertheless, SoCs place computational and storage elements very close to each other. The use of the same substrate for computing, storage, and communication, the lack of bottlenecks (such as input/output [I/O] pads), and the freedom in designing wiring channels yield different design constraints and opportunities.

Overall, the most distinctive characteristics of micronetworks are energy constraints and design-time specialization. Whereas computation and storage energy

3-1   Micronetwork stack.

FIGURE

beneft greatly from device scaling (smaller gates, smaller memory cells), the energy for global communication does not scale down. On the contrary, projections based on current delay optimization techniques for global wires [86] show that global communication on chip will require increasingly higher energy consumption. Hence, communication energy minimization will be a growing concern in future technologies. Furthermore, network traffic control and monitoring can help in better managing the power consumed by networked computational resources. For instance, clock speed and voltage of end nodes can be varied according to available network bandwidth.

Design-time specialization is another facet of the SoC network design problem that raises many new challenges. Macroscopic networks emphasize general-purpose communication and modularity. Communication network design has traditionally been decoupled from specific end applications and strongly influenced by standardization and compatibility constraints with legacy network infrastructures. In SoC networks, these constraints are less restrictive. The communication network fabric is designed on silicon from scratch. Standardization is needed only for specifying an abstract network interface for the end nodes, but the network architecture itself can be tailored to the application, or class of applications, targeted by the SoC design. Hence, we envision a vertical design flow in which every layer of the micronetwork stack is specialized and optimized for the target application domain. Such an *application-specific on-chip network synthesis* paradigm represents, in our view, an open and exciting research field. Needless to say, specialization does not imply complete loss of flexibility. From a design standpoint, network reconfigurability will be key in providing plug-and-play use of components, since they will interact with the others through (reconfigurable) protocols.

3.2   SIGNAL TRANSMISSION ON CHIP

We consider now on-chip communication and its abstraction as a micronetwork. We analyze the physical layer of the micronetwork stack in this section. Wires are the physical realization of communication channels in SoCs. (For our purposes, we view busses as ensembles of wires.) Even though components may be hierarchically designed, we shall focus on components and their interconnection with *global wires*. We neglect the internals of components and their wiring by *local wires*, because the properties of such wires can be made to scale with the technology, and it is likely that present design styles and methods will still apply [86].

3.2.1   Global Wiring and Signaling

Global wires are the physical implementation of the communication channels. The technologies under consideration will exceed 10 wiring levels. Most likely, global wires will be routed on the top metal layers provided by the technology. The pitch (i.e., width plus separation) of such wires is reverse scaled, i.e., wiring pitch increases (with respect to device pitch) with higher wiring levels. Wire widths also grow with wiring levels. Wires at top levels can be much wider and thicker than low-level wires [87]. There are several good reasons for routing global communication on top wires. Increased pitch reduces cross-coupling and offers more opportunities for wire shielding, thereby improving noise immunity. Increased width reduces wire resistance (even considering the skin effect), while at the same time increased spacing around the wire prevents capacitance growth. At the same time, inductive effects will grow relative to resistance and capacitance [88]. As a result, future global wires are likely to be modeled as lossy transmission lines, as opposed to today's lumped or distributed RC models.

Physical layer signaling techniques for lossy transmission lines have been studied for a long time by high-speed board designers and microwave engineers [83,89]. Non-negligible line resistance causes signal attenuation and dispersion in frequency of fast signals, thereby increasing intersymbol interference, and ultimately limiting communication bandwidth. The impact of attenuation and frequency dispersion can be reduced by splitting wires in several sections with signal regeneration buffers in between. Buffering significantly improves bandwidth, and it is also beneficial for latency, especially when the RC component is dominant [82]. Line inductance requires careful impedance matching at the transmitting or receiving end of the line, in order to avoid signal reflections that largely increase intersymbol interference and noise [83]. Traditional rail-to-rail voltage signaling with capacitive termination, as used today for on-chip communication, is definitely not well suited for high-speed, low-energy communication on future global interconnect [83].

In light of these trends, we believe that physical-layer signaling techniques for on-chip global communication will evolve in the near future. High-speed communication is achieved by reducing the voltage swing, which also has a beneficial effect on power dissipation. A reduced swing, current-mode transmitter requires careful receiver design, with good adaptation to line impedance, and high-sensitivity sensing, possibly with the help of sense amplifiers.

Example 1. *Low swing drivers and receivers are already common in current integrated circuits. The bitlines of fast static random access memory (SRAM) use a differential, precharged, low-swing signaling scheme to improve speed and energy efficiency. The transmitter is made of the precharge transistors (one for each biline), and* *the selected cell. The receiver is the sense amplifier at the end of bitline pairs. During SRAM readout, the addressed cell is connected to the bitline, and it draws a small, almost constant discharge current. (The cell's transistors are in saturation.) A small differential deviation (e.g., 100 mV) from the precharged value on the bitlines is sufficient for the sense amplifier to detect the information stored in the addressed cell and to restore it to full-swing logic values. A well-designed SRAM read circuit based on this signaling scheme is more than 10 times faster and approximatly two orders of magnitude more energy efficient than single-ended, rail-to-rail signaling [83].*

Example 2. *Dynamic voltage scaling can be applied to interconnect wires, with the objectives of lowering the voltage swing as much as possible, while preserving correctness of the transmitted information. Worm et al. [90] devised a feedback scheme by which voltage swing on wires (busses) for a communication link is determined on-line by monitoring the error rate of the information. In particular, this scheme supports packetized information transmission, with retransmission of the corrupted packets. As retransmission rate increases over (decreases below) a threshold, the voltage swing is increased (decreased).*

This scheme is useful in setting the voltage swing according to the operating conditions of the SoC (environmental noise, temperature), as well as for self-calibrating the SoC with respect to technological parameters, which may have a wide spread in future technologies.

In general, transmitter and receiver design for on-chip communication should follow the well-known channel equalization paradigm employed in digital communication, as shown in Figure 3-2. In this approach, the channel is abstracted by a transfer function $F_C(f)$ in the frequency domain. The transmitter and the receiver should *equalize* the transfer function by jointly implementing its inverse: $F_T(f) \cdot F_R(f) = F_C^{-1}(f)$. Note that, in general, effective channel equalization may require nonlinear transformations of the signal by transmitter and receiver. Signal *modulation* is probably the simplest nonlinear transformation that could be applied to optimize channel efficiency. Currently, no modulation is employed for standard on-chip communication, but several modulation schemes are being investigated, e.g., ref. 91. Modulation can greatly enhance the effective communication

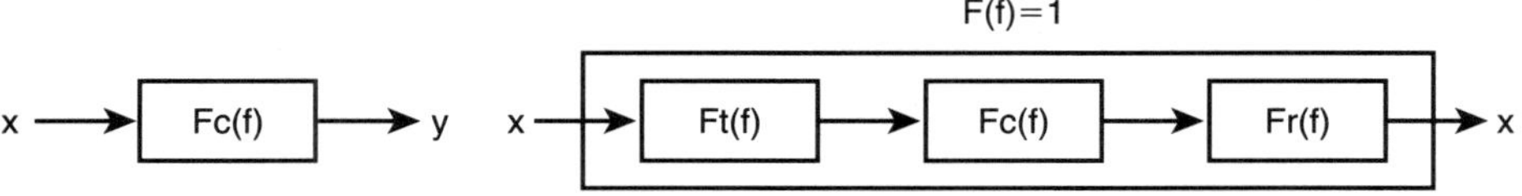

3-2

FIGURE   Channel transfer function and equalization.

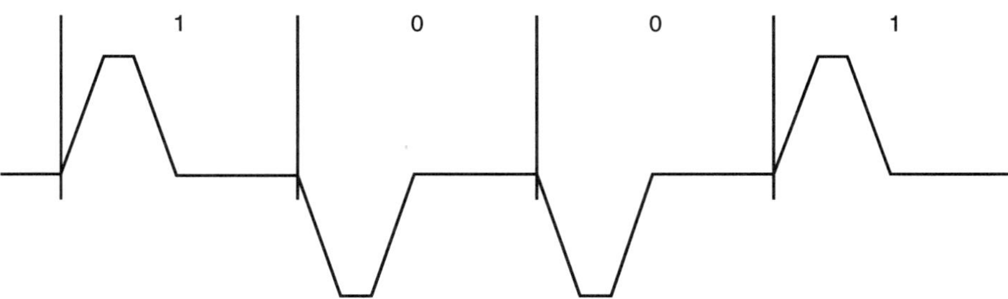

3-3   Current pulse signaling: a simple AM scheme.

FIGURE

bandwidth: (1) by shifting the information content of a signal out of frequency regions where the channel does not transmit reliably, and (2) by enabling multiple channels to be carried over the same wire.

Example 3. *Current-pulse signaling [83], as shown in Figure 3-3, is a simple example of amplitude modulation in which the carrier is a square current waveform with duty cycle much smaller than 50%. The amplitude of the waveform is modulated by multiplying it by −1 (a negative current pulse) when transmitting a zero and by +1 (a positive current pulse) when transmitting a one. A more complex example of on-chip modulation has been presented by Yoshimura et al. [91], who introduced a new multimaster bus architecture in which several transmitters share the same bus, by modulating their output data using a* code-division multiple access (CDMA) scheme. *The frequency spreading and despreading is achieved by multiplying the information by a pseudorandom sequence generated by* linear feedback shift registers (LFSRs) *at both the transmitting and receiving ends. Thus, superimposed signals can be discriminated by just knowing the characteristics of the LFSRs. In this approach, multiple simultaneous bus driving is achieved by using charge pumps in the bus drivers.*

Nonlinear transformations can also help in increasing communication energy efficiency. In most current on-chip signaling schemes, power is consumed during signal transitions. Thus, signal encoding (a nonlinear transformation) for reduced transition activity can help optimize communication energy. Numerous encoding techniques have been proposed for this purpose [92].

3.2.2   Signal Integrity

With present technologies, most chips are designed under the assumption that electrical waveforms can always carry correct information on chip. We believe that guaranteeing error-free information transfer (at the physical level) on global on-chip wires will become harder, if not impossible, in future technologies for several reasons:

+ Signal swings are going to be reduced, with a corresponding reduction in voltage noise margins.

+ *Crosstalk* is bound to increase, and the complexity of avoiding crosstalk by identifying all potential on-chip noise sources will make it unlikely to succeed fully.

+ *Electromagnetic interference* (EMI) by external sources will become more of a threat because of the smaller voltage swings and smaller dynamic storage capacitances.

+ The probability of occasional *synchronization failures* and/or metastability will rise. These erroneous conditions are possible during system operation, because of transmission speed changes, local clock frequency changes, timing noise (jitter), and so on.

+ Alpha particles (from the package) and collisions of thermal neutrons (from cosmic ray showers) can create spurious pulses, which can affect signals on chip and/or discharge dynamic storage capacitances. This problem, known as *soft errors*, used to be relevant only for large DRAMs. It is now conjectured to be a potential hazard for large SoCs as well [93].

Example 4. *A soft error occurs when a particle induces the creation of electron-hole pairs, corresponding to an electric charge that is larger than a critical charge. The critical charge is a characteristic of the cell node (e.g., memory or register node), technology, and operating voltage. Typical values range from 10 to 100 pC in current technologies. The shrinking of layout geometries and the corresponding reduction of node capacitances, as well as the reduction in supply voltages, contribute to a reduction of the critical charge. Recent experiments [94] have shown that the soft error rate can grow by two orders of magnitudes in the current decade.*

Soft errors may be permanent or transient. Transient soft errors can be seen as voltage pulses that propagate through a circuit when the pulse width is larger than the transition times of logic gates. The arrival of a pulse at a latch boundary when a clock transition occurs may cause the wrong information to be stored. Nicolaidis et al. [95] developed and commercialized error correcting techniques based on timing redundancy. Latches are replaced by latch pairs that have identical inputs and whose outputs feed comparators. One latch has a delay line on its input. Thus, the comparator detects spurious pulses that arrive at the latch boundary. Standard error recovery techniques can then be used when a soft error is detected.

From the aforementioned considerations, we conclude that it will not be possible to abstract the physical layer of on-chip networks as a fully reliable, fixed-delay channel. At the micronetwork stack layers above the physical one, we can view the effects of synchronization losses, crosstalk, noise, ionizing radiation, and so on as a source of local transient malfunctions. We can abstract malfunctions as *upsets*. This is a simple yet powerful model that parallels the *stuck-at-fault* model in the testing field. In particular, an upset is the abstraction of a malfunction that causes the wrong binary information to be read or stored. Upset modeling consists of deriving analytical models of crosstalk and radiation-induced disturbances and then deriving the upset probability. Upset probability can vary over different physical channels and over time. It is important to understand that upset probability is negligible (but not null) in current technologies, but it will not be so in future technologies.

A number of new design challenges are emerging at the micronetwork physical layer, and a paradigm shift is needed to address them effectively. Indeed, current design styles consider wiring-related effects as undesirable parasitics and try to reduce or cancel them by specific and detailed physical design techniques. It is important to realize that a well-balanced design should not overdesign wires so that their behavior approaches an ideal one, because the corresponding cost in performance, energy efficiency, and modularity may be too high. Physical layer design should find a compromise between competing quality metrics and provide a clean and complete abstraction of channel characteristics to micronetwork layers above.

3.3 MICRONETWORK ARCHITECTURE AND CONTROL

Network design entails the specification of *network architectures* and *control protocols*. The architecture specifies the topology and physical organization of the interconnection network, whereas the protocols specify how to use network resources during system operation. Even though architecture and protocol design are tightly intertwined, we shall first focus on different network architectures and then address protocol design issues.

Network design is influenced by many design metrics. Whereas both micronetwork and general network design must meet performance requirements, on-chip network implementations will be differentiated by the need to satisfy tight energy bounds. Design choices are strongly impacted by physical channel characteristics as well as by expected workloads. In many cases it is possible to extract some information on the usual communication patterns and workload. This information can be used to optimize some design parameters. For example, for SoC executing embedded software programs, the characteristics of the embedded instruction stream can be modeled and used for the memory processor interconnection design [96]. When no expected workload information is available, network design should emphasize robustness.

3.3.1   Interconnection Network Architectures

On-chip networks are closely related to interconnection networks for high-performance parallel computers with multiple processors, where processors are individual chips. Similarly to multiprocessor interconnection networks, nodes are physically close, and link reliability is high. Furthermore, multiprocessor interconnections have traditionally been designed under stringent bandwidth and latency constraints to support effective parallelization. Similar constraints will drive the design of micronetworks. For these reasons we survey different on-chip network architectures using the same taxonomy proposed by Duato et al. [97] for multiprocessor interconnections. We show that current on-chip interconnections can be classified in a similar way and that their evolution follows a path similar to that of multiprocessor interconnection architectures.

Shared-Medium Networks

Most current SoC interconnection architectures fall within the *shared-medium* class [98–102]. These are the simplest interconnect structures, in which the transmission medium is shared by all communication devices. Only one device can drive the network at a time. Every device connected to the network has a network interface, with requester, driver, and receiver circuits. The network is usually passive, and it does not generate control or data messages. A critical issue in the design of shared medium networks is the *arbitration strategy* that assigns the mastership of the medium and resolves access conflicts. A distinctive property of these networks is the support for broadcast of information, which is highly advantageous when communication is highly asymmetric, i.e., the flow of information originates from few transmitters to many receivers.

Within current technologies, the most common embodiment of the on-chip, shared medium structure is the *backplane bus*. This is a very convenient, low-overhead interconnection for a small number of active processors (i.e., bus masters) and a large number of passive modules (i.e., bus slaves) that only respond to requests from bus masters. The information units on a bus belong to three classes, namely, *data*, *address*, and *control*. Data, address, and control information can either be time-multiplexed on the bus, or it can travel over dedicated busses/wires, spanning the tradeoff between performance and hardware cost (area). Most on-chip busses value performance and use a large number of dedicated wires.

A critical design choice for on-chip busses is synchronous versus asynchronous operation. In synchronous operation, all bus interfaces are synchronized with a common clock, whereas in asynchronous busses all devices operate with their own clock and use a handshaking protocol to synchronize with each other. The tradeoffs involved in the choice of synchronization approach are complex and depend on a number of auxiliary constraints, such as testability, ease of debugging and simulation, and the presence of legacy components. Currently, all commercial on-chip busses are synchronous, but the bus clock is slower than the clock of fast masters. Hence, simplicity and ease of testing/debugging is prioritized over sheer performance.

Bus arbitration mechanisms are required when several processors attempt to use the bus simultaneously. Arbitration in current on-chip busses is performed in a centralized fashion by a bus arbiter module. A processor wishing to communicate must first gain bus mastership from the arbiter. This process implies a control transaction and communication performance loss; hence arbitration should be as fast as possible and as rare as possible. Together with arbitration, the response time of slow bus slaves may cause serious performance losses, because the bus remains idle while the master waits for the slave to respond. To minimize the waste of bandwidth, split-transaction protocols have been devised for high-performance busses. In these protocols, bus mastership is released just after a request has completed, and the slave has to gain access to the bus to respond, possibly several bus cycles later. Thus the bus can support multiple outstanding transactions. Needless to say, bus masters and bus interfaces for split-transaction busses are much more complex than those of simple atomic-transaction busses. Even though most current on-chip busses support only atomic transactions, some split transaction organizations have begun to appear [103].

Example 5. *The AMBA 2.0 bus standard has been designed for the ARM processor family [98]. It is fairly simple and widely used today. Within AMBA, the high-performance bus protocol (AHB) connects high-performance modules, such as ARM cores and on-chip RAM. The bus supports separate address and data transfers. A bus transaction is initiated by a bus master, which requests access from a central arbiter. The arbiter decides priorities when there are conflicting requests. The arbiter is implementation-specific, but it must adhere to the ASB protocol, namely, the master issues a request to the arbiter. When the bus is available, the arbiter issues a grant to the master. Arbitration address and data transfer are pipelined in AM*BA AHB* to increase the bus effective bandwidth, and burst transactions are allowed to amortize the performance penalty of arbitration. However, multiple outstanding transactions are supported only to a very limited degree: the bus protocol allows a split transaction,* *whereby a burst can be temporarily suspended (by a slave request). New bus transactions can be initiated during a split burst.*

The Lucent Daytona chip [103] is a multiprocessor on a chip that contains four 64-bit processing elements. These elements generate transactions of different sizes. Thus, a special 128-bit split-transaction bus was chosen. The bus was designed to minimize the average latency when simultaneous requests are made. Large transfers are partitioned into smaller packets enabling bus bandwidth to be better utilized. Each transaction is associated with an ID. Thus, multiple outstanding transactions can be serviced and prioritized under program control. This feature is critical for system performance. In fact, the recently announced most advanced version of the AMBA bus protocol (AMBA AXI) fully supports multiple outstanding transactions.

Shared-medium network organizations are well understood and widely used, but their scalability is seriously limited. The bus-based organization is still convenient for current SoCs that integrate less than 5 processors and rarely more than 10 bus masters. Multiprocessor architects realized long ago that bus-based systems are not scalable, as the bus invariably becomes the bottleneck when more processors are added. Another critical limitation of shared-medium networks is their energy ineffciency. In these architectures, every data transfer is broadcast. Broadcasts must reach each possible receiver at a large energy cost. Future integrated systems will contain tens to hundreds of units that generate information to be transferred. For such systems, a bus-based network would become the critical performance and power bottleneck.

Direct and Indirect Networks

The *direct* or *point-to-point* network is a network architecture that overcomes the scalability problems of shared-medium networks. In this architecture, each node is directly connected with a subset of other nodes in the network, called *neighboring* nodes. Nodes are on-chip computational units, but they contain a network interface block, often called a *router*, which handles communication-related tasks. Each router is directly connected with the routers of the neighboring nodes. Unlike shared-medium architectures, as the number of nodes in the system increases, the total communication bandwidth also increases. Direct interconnect networks are therefore very popular for building large-scale systems.

Indirect or *switch-based* networks are an alternative to direct networks for scalable interconnection design. In these networks, a connection between nodes has to go through a set of *switches*. The network adapter associated with each node connects to a port of a switch. Switches do not perform information processing. Their only purpose is to provide a programmable connection between their ports, or, in other words, to set up a communication path that can be changed over time

[97]. Interestingly enough, the distinction between direct and indirect networks is blurring, as routers (in direct networks) and switches (in indirect networks) become more complex and absorb each other's functionality.

Most current *field-programmable gate arrays* (FPGAs) consist of a homogeneous fabric of programmable elements connected by a switch-based network. FPGAs can be seen as the archetype of future programmable SoCs: they contain a large number of interconnected computing elements. Current FPGA communication networks differ from future SoC micronetworks in *granularity* and *homogeneity*.

The concept of dynamic reconfigurability of FPGAs also applies to the design of micronetworks. SoCs benefit from programmability in the field (e.g., to match environmental constraints) and to run time reconfiguration (e.g., to adapt to a varying workload). Reconfigurable micronetworks exploit programmable routers and/or switches. Their embodiment may leverage multiplexers whose control signals are set by configuration bits in local storage, as in the case of FPGAs.

Processing elements in FPGAs implement simple bit-level functional blocks. Thus communication channels in FPGAs are functionally equivalent to wires that connect logic gates. Since future SoCs will house complex processing elements, interconnect will carry much coarser quanta of information. The different granularity of computational elements and communication requirements has far-reaching consequences on the complexity of the network interface circuitry associated with each communication channel. Interface circuitry and network control policies must be kept extremely simple for FPGAs, whereas they can be much more complex when supporting coarser grain information transfers. The increased complexity will also introduce larger degrees of freedom for optimizing communication.

Example 6. *The Xilinx SpartanII FPGA chips are rectangular arrays of configurable logic blocks (CLBs). Each block can be programmed to perform a specific logic function. CLBs are connected via a hierarchy of routing channels. Thus each chip has an indirect network over a homogeneous fabric.*

The RAW architecture [104] is a fully programmable SoC, consisting of an array of identical computational tiles with local storage. Full programmability means that the compiler can program both the function of each tile and the interconnections among them. The name RAW stems from the fact that the "raw" hardware is fully exposed to the compiler. To accomplish programmable communication, each tile has a router. The compiler programs the routers on all tiles to issue a sequence of commands that determine exactly which set of wires connect at every cycle. Moreover, the compiler pipelines the long wires to support high clock frequency. Thus, RAW can be viewed as a direct network.

The Xilinx VirtexII are FPGAs with various configurable elements to support reconfigurable digital signal processor (DSP) design. The internal configurable rectangular array contains CLBs, RAMs, multipliers, and clock managers. Programmable inter-connection is achieved by routing switches. Each programmable element is connected to a switch matrix, allowing multiple connections to the general routing matrix. All programmable elements, including the routing resources, are controlled by values stored in static memory cells. Thus, VirtexII can be also seen as an indirect network over a heterogeneous fabric.

From an energy viewpoint, direct networks have the potential for being more energy efficient than shared medium networks, because energy per transfer on a point-to-point communication channel is smaller than that on a large shared-medium architecture. Clearly, communication between two end nodes may go through several point-to-point links, and the comparison should take this effect into account [105]. When local communication between neighboring nodes dominates, the energy advantage of point-to-point networks is hardly disputable, and some recent experimental work has quantified such an advantage [106]. It is also interesting to observe that point-to-point networks are generally much more demanding in terms of area than shared-medium networks. In this case the positive correlation between area and power dissipation breaks down.

Hybrid Networks

Direct and indirect networks yield the scalability required to support communication in future SoCs, and they can provide, with the support of effective control techniques (discussed in the next subsection), the performance levels required to avoid communication bottlenecks. Current FPGAs show convincing evidence that multistage, scalable networks can be implemented on-chip, and current high-performance multiprocessors demonstrate that these advanced network topologies can sustain extremely high performance levels. However, the communication infrastructure of both FPGAs and multiprocessors is strongly oriented toward homogeneity in information transfer.

Homogeneous interconnection architectures have important advantages: they facilitate modular design, and they are easily scaled up by replication. Nevertheless, homogeneity can be an obstacle to flexibility and fine tuning of architectures to application characteristics. Although homogeneous network architectures may be the best choice for general-purpose computing, systems developed for a particular application (or a class of applications) can benefit from a more heterogeneous communication infrastructure that provides high bandwidth in a localized fashion only where it is needed to eliminate bottlenecks.

The advantages of introducing a controlled amount of nonuniformity in communication network design are widely recognized. Many heterogeneous, or *hybrid* interconnection architectures, have been proposed and implemented. Notable examples are *multiple-backplane* and *hierarchical* (or *bridged*) busses. These organizations allow clustering of tightly coupled computational units with high communication bandwidth and provide lower bandwidth (and/or higher latency) intercluster communication links. Compared with a homogeneous, high-bandwidth architecture, they can provide comparable performance using a fraction of the communication resources, and at a fraction of the energy. Thus, energy efficiency is a strong driver toward hybrid architectures.

Example 7. *The AMBA 2.0 standard [98] specifies three protocols: the high-performance bus (AHB), the advanced system bus (ASB), and the advanced peripheral bus (APB). The latter is designed for lower performance transactions related to peripherals. An AMBA-based chip can instantiate multiple bus regions that cluster components with different communication bandwidth requirements. Connections between two clusters operating with different protocols (and/or different clock speed) are supported through bus bridges. Bridges perform protocol matching and provide the buffering and synchronization required to communicate across different clock domains. Note that bridges are also beneficial from an energy viewpoint, because the high-transition rate portion of the bus is decoupled from the low-speed, low-transition rate, peripheral section.*

3.3.2   Micronetwork Control

Effective utilization of the physical realization of micronetwork architectures depends on protocols, i.e., on network control algorithms, which are often distributed. We shall describe control algorithms following the micronetwork stack of Figure 3-1 from the bottom up. Network protocols can be implemented either in software or in hardware. In our analysis, this distinction is immaterial, because we focus only on the various control functions and not on how to implement them. In synthesis, network control is responsible for dynamically managing network resources during system operation, striving to provide the required quality of service.

Datalink Layer

As seen in Section 3.2 and in Figure 3-1, on-chip global wires are the physical support for communication. The datalink layer abstracts the physical layer as an unreliable digital link, where the probability of bit upsets is non-null. Such upset probabilities are increasing as technology scales down. Furthermore, we have seen that reliability can be traded off for energy. The main purpose of datalink protocols is to increase the reliability of the link up to a minimum required level, under the assumption that the physical layer by itself is not sufficiently reliable. Another important function of the datalink layer is to regulate the access to a shared-medium network, in which contention for a communication channel is possible. Thus we can see this layer as the superposition of two sublayers. The lower sublayer (i.e., closest to the physical channel) is called *media access control* (MAC), whereas the higher sublayer is the *datalink control* (DLC). The MAC regulates access to the medium, whereas the DLC increases channel reliability, e.g., by providing error detection and/or correction capabilities.

Error detecting and *correcting codes* (ECCs) can be used in different ways. When only error detection is used, error recovery involves the retransmission of the faulty bit or word. When using error correction, some (or all) errors can be corrected at the receiving end. Error detection and/or correction require an encoder/decoder pair at the channel's end, whose complexity depends on the encoding being used. Obviously, error detection is less hardware-intensive than error detection and correction. In both cases, a small delay has to be accounted for in the encoder and decoder. Data retransmission has a price in terms of latency. Moreover, both error detection and correction require additional (redundant) signal lines. When ECC techniques are used in conjunction with packetized information, the redundant lines can be avoided by adding the redundant information at the tail of the packet, thus trading off space for delay.

There are several error detecting and correcting codes. For example, Berger codes can detect all unidirectional codes. Other codes, like Hamming codes, address the issues of bidirectional errors, such as those found in noisy channels. Error detection may reveal single or multiple errors in a data fragment (e.g., word). Similarly, error correction can restore the correct information caused by single or multiple errors. When errors go undetected, their effect can be catastrophic. When errors are detected and cannot be corrected (e.g., multiple errors with a single-error correction scheme), data retransmission has to be used. In the general case, one often has to admit that some errors may go undetected, and the designer's goal is to make this event very unlikely. In practice, one can choose a *mean time to failure* (MTTF) as a design specification. For example, one can chose an encoding scheme that has an MTTF of 10 years, much above the SoC projected lifespan.

Latency and energy consumption for data transmission using ECC are very important design parameters. Although the former has been studied for long time, the latter has been the object of research only recently, especially in view of the fact that energy consumption in micronetworks has to be curtailed. Overall, energy depends on the switching activity in the communication link (including the redundant lines), on the encoder and decoder, on the retransmission scheme, and of course on the noisiness of the channel.

Example 8. *Bertozzi et al. [107] applied error correcting and detecting codes to an AMBA 2.0 AHB bus and compared the energy consumption in five cases: (1) original* *unencoded data, (2) single-error correction (SEC), (3) single-error correction and double-error detection (SECDEC), (4) multiple-error detection (ED), and (5) parity. Hamming codes were used. Note that in case 3, a detected double error requires retransmission. In case 4, using (n, k) linear codes, $2^n - 2^k$ error patterns of length n can be detected. In all cases, some errors may go undetected and be catastrophic. Using the property of the codes, it is possible to map the MTTF requirement into bit upset probabilities, thus comparing the effectiveness of the encoding scheme in a given noisy channel (characterized by the upset probability) in meeting the MTTF target.*

Bertozzi et al. [107] investigated the energy efficiency of various encoding schemes. We summarize here one interesting case, in which three assumptions apply. First, wires are long enough so that the corresponding energy dissipation dominates encoding/decoding energy. Second, voltage swing can be lowered until the MTTF target is met. Third, upset probabilities are computed using a white gaussian noise model [108].

Figure 3-4 shows the average energy per useful bit as a function of the MTTF (which is the inverse of the residual word error probability). In particular, for reliable SoCs (i.e., for MTTF = 1 year), multiple-error detection with retransmission is shown to be more efficient than error-correcting schemes. We refer the reader to Bertozzi et al. [107] for results under different assumptions.

An additional source of errors, not considered in Section 3.2, is contention in shared-medium networks. Contention resolution is fundamentally a nondeterministic process, because it requires synchronization of a distributed system, and

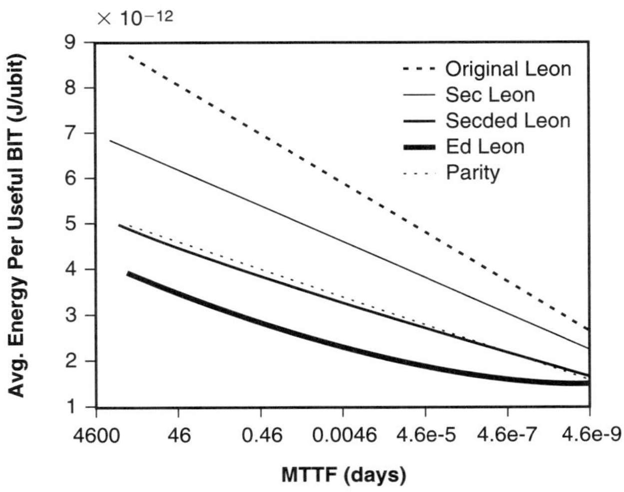

3-4

FIGURE

Energy efficiency for various encoding schemes on an AMBA bus connecting a memory controller to an I-cache of a LEON processor.

for this reason it can be seen as an additional noise source. In general, nondeterminism can be virtually eliminated at the price of some performance penalty. For instance, centralized bus arbitration in a synchronous bus eliminates contention-induced errors, at the price of a substantial performance penalty caused by the slow bus clock and by bus request/release cycles.

Future high-performance shared-medium on-chip micronetworks may evolve in the same direction as high-speed local area networks, in which contention for a shared communication channel can cause errors, because two or more transmitters are allowed to send data concurrently on a shared medium. In this case, provisions must be made for dealing with contention-induced errors.

An effective way to deal with errors in communication is to packetize data. If data are sent on an unreliable channel in packets, error containment and recovery is easier, because the effect of errors is contained by packet boundaries, and error recovery can be carried out on a packet-by-packet basis. At the *datalink* layer, error correction can be achieved by using standard error-correcting codes that add redundancy to the transferred information. Error correction can be complemented by several packet-based error detection and recovery protocols, such as *alternating-bit*, *go-back-N*, and *selective repeat*, which have been developed for macroscopic networks [84,85]. Several parameters in these protocols (e.g., packet size, number of outstanding packets, and so on) can be adjusted depending on the goal to achieve maximum performance at a specified residual error probability and/or within given energy consumption bounds. The choice of protocol and optimization of its parameters for micronetworks is an open problem. We stress that contention for communication resources is unavoidable in all scalable network architectures. Contention-free communication is achievable either at the price of a large performance penalty (through global synchronization and centralized arbitration), or at the price of a large and poorly scalable contention-free interconnection architecture (e.g., a complete crossbar). Hence, almost every practical on-chip network will have to account for the effects of contention caused by shared channels. Packetization of information flow is needed to deal with contention effectively: it limits the effects of contention-related errors, but it is also instrumental in implementing fair sharing of communication resources. For these reasons, we believe that future on-chip micronetworks will be packet-based, and designers will need guidelines and tools to decide packetization granularity and to trade off packet data and control content.

Example 9. *The scalable programmable integrated network (SPIN) is an on-chip micronetwork [109,110] that defines packets as sequences of 36-bit words. The packet header fits in the first word. A byte in the header identifies the destination (hence, the network can be scaled up to 256 terminal nodes), and other bits are used for packet tagging and routing information. The packet payload can be of variable size. Every* *packet is terminated by a trailer, which does not contain data, but a checksum for error detection. Packetization overhead in SPIN is two words. The payload should be significantly larger than two words to amortize the overhead. A SPIN on-chip network consists of two macrocell circuits, a router and two wrappers respecting the virtual component interface (VCI) standard. A prototype implementation is described in Adriahantenana and Greiner [109].*

Network Layer

The datalink layer implements a reliable (packet-based) link between processing elements connected to a common link. The network layer implements the end-to-end delivery control in advanced network architectures with many communication channels. In most current on-chip networks, all processing elements are connected to the same channel, namely, the on-chip bus. Under these conditions, the network layer is empty. However, when processing elements are connected through a collection of links, we must decide how to set up connections between successive links and how to route information from its source to the final destination, through a series of nodes in between. These key tasks, called *switching* and *routing*, are specific to the network layer, and they have been extensively studied both in the context of multiprocessor interconnects [97] and in the context of general communication networks [84,85].

Switching algorithms can be grouped into four classes: *circuit switching, packet switching, cut-through switching*, and *wormhole switching*. With circuit switching, a path from the source to the destination is reserved prior to the transmission of data, and the network links on the paths are released only after the data transfer has been completed. Circuit switching is advantageous when traffic is characterized by infrequent and long messages, because communication latency and throughput on a fixed path are generally very predictable. With circuit switching, network resources are kept busy for the duration of the communication, and the time for setting up a path can produce a sizable initial latency penalty. Hence, circuit switching is not widespread in packet networks in which atomic messages are data packets of relatively small size: communication path setup and reset would cause unacceptable overhead and degrade channel utilization.

Packet switching addresses the limitations of circuit switching for packetized traffic. The data stream at the source is divided into packets, and the time interval between successive packets may change. A packet is completely buffered at each intermediate switch (or router) before it is forwarded to the next. This approach is also called *store-and-forward*, and it introduces nondeterministic delays due to message queueing at each intermediate node. Packet switching leads to significantly higher utilization of communication resources with respect to circuit switching, at the price of an increase in nondeterminism. Furthermore, many packets belonging to a message can be in the network simultaneously, and out-of-order delivery to the destination is possible.

Packet switching is based on the assumption that an entire packet has to be completely received before it can be routed to an output channel. This assumption implies that the switch, or the router, must provide significant storage resources. Furthermore, each intermediate node adds latency to the message, because it needs to receive a packet fully before forwarding it toward the next node. Cut-through switching strategies have been developed to reduce buffering overhead and latency. In cut-through switching schemes, packets are routed to the output channel, and they can start leaving the switch before they are completely received in the input buffer. Improved performance and memory usage are counterbalanced by increased message blocking probability under heavy traffic, because a single packet may occupy input and output ports in several switches. As a result, nondeterminism in message delivery increases.

Wormhole switching was originally designed for parallel computer clusters [97] because it achieves small network delay and requires little buffer usage. In wormhole switching, each packet is further segmented into *flits* (flow control units). The header flit reserves the routing channel of each switch, the body flits will then follow the reserved channel, and the tail flit will later release the channel reservation. One major advantage of wormhole routing is that it does not require the complete packet to be stored in the switch while waiting for the header flit to route to the next stages. Wormhole switching not only reduces the store-and-forward delay at each switch, but it also requires much less buffer spaces. One packet may occupy several intermediate switches at the same time. Because of its low latency and low storage requirement advantages, wormhole is an ideal switching technique for on-chip multiprocessor interconnect networks. A number of recently proposed network-on-chip prototype implementations are indeed based on wormhole packet switching [109,111–113].

Different switching approaches trade off better average delivery time and channel utilization for increased variance and decreased predictability. The impact of switching on energy effciency has not been explored in detail. Depending on the application domain, nondeterminism can be more or less tolerable. In many cases, hybrid or adaptive switching approaches can achieve an optimal balance between average performance measures and predictability and energy.

Switching is tightly coupled to routing. Routing algorithms establish the path followed by a message through the network to its final destination. The classification, evaluation, and comparison of on-chip routing schemes [97] involves the analysis of several tradeoffs, such as predictability versus average performance, router complexity and speed versus achievable channel utilization, and robustness versus aggressiveness. A coarse distinction can be made between *deterministic* and *adaptive* routing algorithms. Deterministic approaches always supply the same path between a given source-destination pair, and they are the best choice for uniform or regular traffic patterns. In contrast, adaptive approaches use information on network traffic and channel conditions to avoid congested regions of the network. They are preferable in the presence of irregular traffic or in networks with unreliable nodes and links. Among other routing schemes, probabilistic broadcast algorithms [114] have been proposed for *networks on chip* (NoCs).

Optimal routing for on-chip micronetworks is another open and important research theme. We predict that future approaches will emphasize speed and decentralization of routing decisions. Robustness and fault tolerance will also be highly desirable. These factors, and the observation that traffic patterns for special-purpose SoCs tend to be irregular, seem to favor adaptive routing. However, when traffic predictability is high and nondeterminism is undesirable, deterministic routing may be the best choice.

Example 10. *The SPIN micronetwork adopts wormhole routing [109]. Routing decisions are set by the network architecture, the fat-tree. Packets from a node (a tree leaf) are routed toward the tree root until they reach a switch that is a common ancestor with the destination node. At that point, the packet is routed toward the destination following the unique path between ancestor and destination node.*

Reconfigurable networks, such as today's FPGAs, or the heterogeneous interconnects of many reconfigurable computing platforms, adopt a deterministic and static routing approach. Routing (i.e., switch programming) is performed at reconfiguration time, and routing decisions are taken on a much coarser time scale than in packet-switched networks. From a networking viewpoint, switch programming can be viewed as setting up a number of circuit-switched connections that last a very long time.

Routing of signals in the RAW [104] machine is under software control. The compiler routes the signals along the optimal pathways by precisely scheduling the signals to meet the demand of the applications. Thus routing is adaptive, and connections in RAW are referred to as "soft wires."

The routing scheme in the Aethereal micronetwork has two modes of operation: it supports both best effort services and guaranteed throughput. Thus links can be configured according to the type of application that is running and its requirements between two end nodes. Best effort packets use wormhole source routing, whereas the guaranteed throughput mode uses time-division multiplexed circuit switching [113].

Some other routing schemes have found direct applications in NoCs. *Hot potato* or *deflection* routing is based on the idea of delivering a packet to an output channel of a switch at each cycle. It requires a switch with an equal number of input and output channels. Therefore, input packets can always find at least one output exit, and no *deadlock* occurs. However, *livelock* is a potential problem in hot potato routing. Proper deflection rules need to be defined to avoid livelock problems.

Example 11. *A contention-aware hot-potato routing scheme was proposed by Nilsson et al. [47]. It is based on a two-dimensional mesh NoC and on the switch architecture described by Kumar et al. [116]. Each switch node also serves as network interface to a node processor (also called a resource). Therefore, it has five inputs and five outputs. Each input has a buffer that can contain one packet. One input and one output are used for connecting the node processor. An internal first-in first-out (FIFO) is used to store the packets when output channels are all occupied. The routing decision at every node is based on the "stress values," which indicate the traffic loads of the neighbors. The stress value can be calculated based on the number of packets coming into the neighboring nodes at a unit time, or based on the running average of the number of packets coming to the neighbors over a period of time. The stress values are propagated between neighboring nodes. This scheme is effective in avoiding "hot spots" in the network. The routing decision steers the packets to less congested nodes.*

Ye et al. [117] perfected this scheme by using a wormhole-based contention-look-ahead routing algorithm that can "foresee" the contention and delays in the coming stages using a direct connection from the neighboring nodes. It is also based on a mesh network topology. The major difference from Nilsson et al. [115] is that information is handled in flits, and thus large and/or variable size packets can be handled with limited input buffers. Therefore, this scheme combines the advantages of wormhole switching and hot-potato routing.

The choice of routing algorithms has an impact on the overall system energy efficiency [114,117], i.e., there is a tradeoff between energy efficiency and performance. Consider, for instance, packet flooding schemes, sometimes adopted for adaptively routing information though the fastest route between two nodes. This approach can be beneficial for performance, but it is highly energy-inefficient.

Transport Layer

On top of the network layer, the transport layer decomposes messages into packets at the source. It also resequences and reassembles them at the destination. Packetization granularity is a critical design decision, because the behavior of most network control algorithms is quite sensitive to packet size. In most macroscopic networks, packets are standardized to facilitate internetworking, extensibility, and compatibility of networking hardware produced by different manufacturers.

Packet standardization constraints can be relaxed in SoC micronetworks, which can be customized at design time.

Example 12. *The size of the packets has a direct impact on both performance and energy consumption. Thus, an interesting problem is the search for optimal packet size. To some extent, optimal packetization depends on the network architecture and on the system function (including application software). Ye et al. [118] performed a set of experiments on mesh networks of computing nodes, each node including processing and local cache memory. The experiments consisted of simulating the execution of benchmark programs using the RSIM simulator. The qualitative meaning of these experiments can be summarized as follows.*

Larger packets correlate to lower cache miss rates but to higher miss penalties, due to the increase of packetization, memory access, and contention delays. The cache miss rate decreases and then stabilizes with the increase of packet size, whereas the miss penalty increases overlinearly. Therefore, there exists an optimal packet size to achieve best performance.

The energy spent on the interconnect network increases as the packet size increases, because the packet energy per hop increases as well as the number of hops. On the other hand, storage energy decreases as packets get larger, because of several factors including lower cache miss rates. Overall, the energy consumption has a minimum value for some packet length, such a minimum depending on the relative importance of storage and interconnect energy consumption. Note that the optimal packet length for performance may differ from the optimal packet length from an energy standpoint (Fig. 3-5).

Other functions of the transport layer are to control the flow of data into the network, to allocate network resources, and to negotiate quality of service. Data flow can be controlled through *admission control*, commonly used in

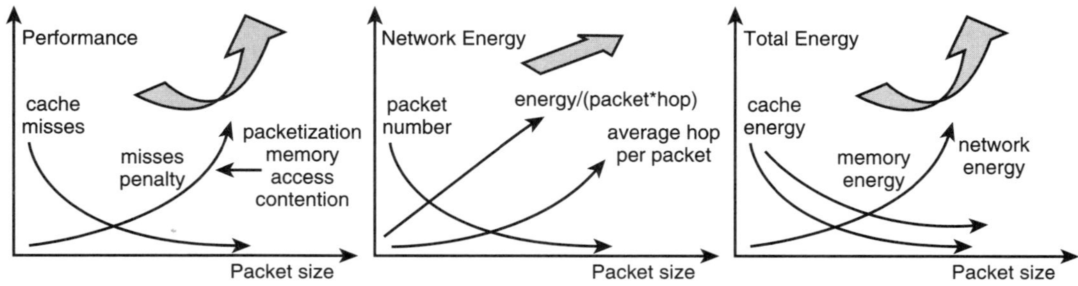

3-5   Latency and energy trends as packet size varies.

FIGURE

circuit-switched networks, that negate access to the network when there is no available path from source to destination. In packet-switched networks, access can be more finely tuned through *congestion control* that slows down forwarding of data packets to prevent congestion downstream. Alternatively, *traffic shaping* regulates the rate of entrance of data packets directly at the source. In packet-switched networks with predictable packet flow (for instance, in *virtual-circuit* networks in which all packets from a source to a destination follow the same route), it is possible to negotiate guarantees on network performance (throughput, latency) upon connection setup.

In general, flow control and negotiation can be based on either deterministic or statistical procedures. Deterministic approaches ensure that traffic meets specifications and provide hard bounds on delays or message losses. The main disadvantage of deterministic techniques is that they are based on worst cases, and they generally lead to significant underutilization of network resources. Statistical techniques are more efficient in terms of utilization, but they cannot provide worst-case guarantees. Similarly, from an energy viewpoint, we expect deterministic schemes to be more expensive than statistical schemes.

Example 13. *The SONICS Silicon Backplane Micronetwork [101,119] supports a time-division multiple access (TDMA) protocol. When a node wants to communicate, it needs to issue a request to the arbiter during a time slot and it may be granted access in the following time slot, if arbitration is favorable. Hence, arbitration introduces a nondeterministic waiting time in transmission. To reduce nondeterminism, the micronetwork protocol provides a form of slot reservation: nodes can reserve a fraction of the available time slots, and therefore they allocate bus bandwidth in a deterministic fashion. Reserved slots are assigned on a periodic pattern (a timing wheel); hence a master is guaranteed to have access to all its reserved slots within the revolution period of the timing wheel.*

The main shortcoming of the deterministic scheme is that the revolution period of the timing wheel is a large multiple of the bus clock period; hence long latencies can be experienced for bus access if the bus master requests are misaligned in time with respect to its reserved slots. A randomized TDMA scheme, called Lotterybus [120] has been proposed to achieve lower average bus access latency, at the price of the loss of worst-case latency guarantees.

Summarizing the overview of network architectures and control issues, we can conclude that the theoretical framework developed for large-scale (local, wide area) and on-chip interconnection networks provides a convenient environment for reasoning about on-chip micronetworks as well. It is important to note that the design space of micronetworks is currently hardly explored, and a significant amount of work is needed to predict the tradeoff curves in this space. We also believe that there is significant room for innovation: on-chip micronetwork architectures and protocols can be specialized for specific system configurations and classes of applications. Furthermore, as pointed out several times above, the impact of network design and control decisions on communication energy is an important research theme that will become critical as communication energy scales up in SoC architectures.

3.4   SOFTWARE LAYERS

The hardware infrastructure described in the previous sections provides a basic communication service to the network's end nodes. Most of the end nodes in future on-chip networks will be programmable, ranging from general-purpose microprocessors, to application-specific processors, to reconfigurable logic. Other nodes, such as I/O blocks and memories, will serve as slaves for the processor nodes, but they will also support some degree of reconfigurability. At the application level, programmers need a programming model and software services to exploit effectively the computational power and the flexibility of these highly parallel heterogeneous architectures. This section focuses on the system software and programming environment providing the hardware abstraction layer at the interface between the NoC architecture and the application programmer. First, we will discuss the programming models, then we will overview middleware architecture, and finally we will describe the development support infrastructure.

3.4.1   Programming Model

Programming models provide an abstract view of the hardware to application programmers. Abstraction is key for two main reasons: (1) it hides hardware complexity and (2) it enhances code longevity and portability. Complex application programming on a large scale would be impossible without good programming models. On the other hand, the abstract view of the hardware enforced by programming model implies an efficiency loss in exploiting advanced hardware features. Hence, striking the best balance between abstraction and direct hardware control is an extremely critical issue. In traditional parallel computing, the two most common and successful programming models are *shared memory* and message passing [121,122]. In the shared memory model, communication is implicitly performed when parallel tasks access shared locations in a shared address space. In the message passing models, tasks have separate address spaces, and intertask communication is carried out via explicit messages (*send and receive* primitives).

Writing parallel code, or parallelizing existing sequential programs in the shared memory model, is generally easier than writing message passing code. However, shared memory requires significant hardware support to achieve high performance (e.g., snoopy caches), and message passing code, albeit harder to write, can achieve high performance even in architectures that implement very simple communication channels. From the scalability viewpoint, the shared memory models become less viable when communication latency is high or highly nonuniform (depending, for instance, on the source and destination); thus, message passing is the best solution for large-scale and highly distributed parallel computers [121,122].

In our view, message passing is the programming model of choice for NoC application software. Our position is motivated by several reasons. First, software for application-specific systems on-chip is developed starting from specification languages that emphasize explicit communication between parallel executing tasks. For instance, complex signal processing applications are often developed using dataflow models, in which data flows from one processing kernel to the other [123]. Second, embedded code development is traditionally focused on achieving high performance on limited hardware resources, and message passing can achieve higher performance at the price of increased development effort. Third, by making communication explicit, message passing pinpoints the main sources of unpredictability in program execution, namely, communication latency and throughput. Hence, message passing programs are generally more predictable [124], a highly desirable characteristic for embedded applications. Finally, since the main motivation for moving to NoC architectures is to provide better architectural scalability even in the face of increasing communication latencies, it is natural to adopt a programming style that fully supports large-scale parallelism and makes it easier to focus on communication latency starting from the top abstraction layer in the design hierarchy.

The message passing model is supported by many formal frameworks (e.g., *communicating sequential processes* [CSP]), languages (e.g., Occam), and application programming interfaces (e.g., the *message passing interface* [MPI]) [121,124]. Even though the theoretical importance of formal models and languages cannot be overemphasized, in the design practice we believe that traditional languages (such as C, C++, or Java) will maintain dominance thanks to the adoption of standardized and portable communication application programming interfaces (APIs). In the domain of multiprocessor embedded signal processing systems, MPI has been proposed as a message passing support library [125]. A natural evolution of traditional MPI for embedded applications is real-time MPI [126], an enhanced MPI standard that emphasizes run time predictability. MPI appears to be an interesting candidate for NoC application level programming; however, significant challenges are still open. First, MPI provides a number of messaging primitives, and many of them require quite complex system and software support; many MPI implementations are built on top of lower level hardware-assisted messaging primitives, and hence their performance may not be sufficient for the NoC environment. In our view, only a restricted subset of MPI functions are strictly required for NoC software development. In some cases, when performance constraints are extremely tight, lower level messaging interfaces, such as *active messages* can be used [122].

3.4.2   Middleware Architecture

Programming models and their implementations (languages and APIs) are the front-end interface between application programmers and the target hardware platform. However, they are only the top layer of the complex software infrastructure that lies between hardware and applications. The main purpose of middleware is to provide safe and balanced access to hardware resources. First, tasks should be allocated and scheduled to the available NoC end nodes (processors). Similar allocation and scheduling decisions should be taken for transferring data from external memory to on-chip embedded memory blocks (and vice versa). Multiprocessor scheduling is a well-developed discipline [127], but memory allocation and memory transfer scheduling for embedded memories is still under very active investigation [128]. An ancillary function in task management is task creation and destruction when task sets are dynamic.

Synchronization is another key middleware function. In single-processor systems, task synchronization is simplified because tasks are serialized on the processor, and therefore no true task parallelism is manifest in the system. In multiprocessor systems, there is obviously true task parallelism. Synchronization becomes much harder and requires some form of hardware support (e.g., atomic transactions). Fortunately, this problem has been extensively studied for traditional multiprocessors and parallel computers, and a number of synchronization primitives, such as semaphores, critical sections, and monitors have been proposed and thoroughly understood [121]. For NoC, synchronization primitives should be as light weight as possible both from the performance and from the code size perspective.

Middleware is also in charge of managing (shared) peripherals, such as on-chip and off-chip memories, I/O controllers, slave coprocessors, and so on. Guaranteeing mutual exclusive access to peripherals is another facet of the synchronization function, but resource management also encompasses other tasks, such as power management and reconfiguration management. We can expect that most peripherals in future NoCs will be reconfigurable and/or power manageable. Furthermore, even the communication network can be expected to be reconfigurable at run time (for example, through dynamic update of routing tables). Management is relatively straightforward in a single-processor environment, but it becomes a challenging task when control is distributed. In many parallel computing platforms, peripheral management is performed by a centralized controller. Even though this choice simplifies the implementation of management policies, it creates a bottleneck in the system for scalability and robustness. For this reason, distributed management policies are desirable, and they should be actively investigated in the future, possibly leveraging results from large-scale distributed environment (e.g., wireless networks).

Finally, the NoC middleware includes low-level hardware control functions that abstract as much as possible platform-dependent details and build a hardware abstraction layer, with a small number of standard access functions. This should be the only part that must be modified when porting the middleware to a new NoC architecture. The middleware architecture and its layers are shown in Figure 3-6. Note that the number of interface functions between layers grows as we get closer to application code.

In principle, most modern *operating systems* (OSs), such as Linux or Unix, support all the above-mentioned functions. Unfortunately, these OSs are too complex and time- and memory-consuming to be acceptable in an embedded software environment. As a consequence, we should employ light-weight, customizable middleware services instead of a full-blown OS. Several embedded OSs have been designed in a modular fashion, and they can be tailored to the needs of applications. Most commercial embedded OSs are single processor even though some natively multiprocessor OSs are available both commercially and in the open source domain [129,130]. However, the performance of their communication primitives has not been validated yet for NoC platforms. In this area, an interesting research topic is the customization of OS features and automatic or semiautomatic retargeting of middleware to different platforms [131].

In our view, there are two main challenges in the development of the new generation of NoC middleware. First, finding the sweet spot balancing hardware and software support for communication services. Second, providing a flexible mix of guaranteed and best effort services. These two challenges are tightly interrelated but not completely overlapping. Latency and throughput requirements for NoCs are extremely tight (remember that an advanced NoC must successfully compete against high-bandwidth on-chip busses); thus, high-performance hardware-assisted schemes, minimizing slow software-based network transaction management, seem to be preferable. On the other hand, most general hardware-assisted communication schemes require dedicated communication processors,

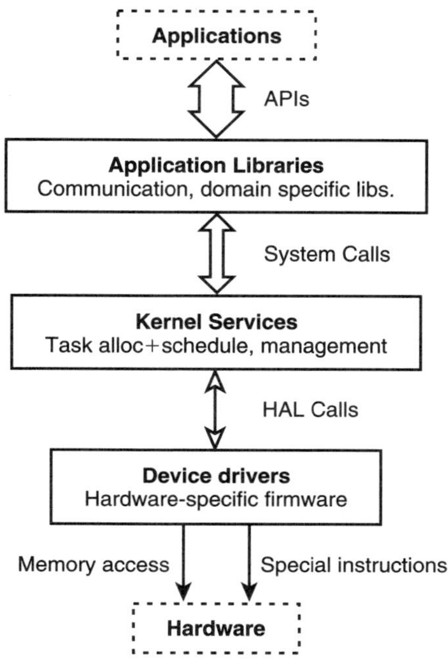

3-6   Middleware architecture.

FIGURE

which implies a nonnegligible cost overhead [122]. Alternatively, user-level communication access schemes can eliminate part of the overhead associated with middle-wareassisted network transactions, at the price of significant losses in software robustness and error containment capability. One possible compromise solution is to design communication-oriented, light-weight kernels that provide limited services but high performance on critical communication functions. This approach has been taken in the Virtuoso real-time kernel [132], which represents the state of the art in NoC-oriented operating systems. Virtuoso has been specifically designed for performance-constrained multiprocessor signal-processing systems, and it supports a fully distributed application development style, whereby tasks live in separate address spaces and communicate via simple message passing (CSP-like) through explicitly instantiated channels. As opposed to traditional operating systems, communication is handled with the same priority and reduced overhead used for handling interrupts.

Even though a communication-centric approach in middleware design can be effective in reducing the overhead in supporting communication services, additional care should be taken in ensuring guaranteed quality services, in which "good-on-average" performance (i.e., *best effort* services) is not enough. Typical examples of applications requiring performance guarantees are streaming audio and video processing, as well as many signal processing tasks. Guaranteed QoS support requires careful consideration across all levels of NoC design hierarchy, including application and middleware. Assuming that the NoC hardware backbone can support some form of QoS (e.g., upper bounds on message delivery times), the middleware framework is required to provide similar guarantees. One interesting way to tackle this challenge is to support quality of service negotiation followed by prioritized messaging starting from the application level [132]. In this paradigm a hard-real-time task initially negotiates a QoS with the NoS interface, and then it is assigned a level of priority for both CPU and network access. Lower priority packets waiting for network interface access can be preempted by high-priority packets (coming from a high-priority task), thereby preventing priority inversion on the NoC interface.

3.4.3   Software Development Tools

The tools employed by programmers to develop and tune application software and middleware are a critical part of the NoC design infrastructure. Even though a detailed survey of this area is beyond the scope of the chapter, we will briefly overview the main building blocks of the software development infrastructure, as depicted in Figure 3-7. Specification and development tools have the main purpose of facilitating code development and increasing the programmer's productivity. In this area, we can group interactive, graphical programming environments, code versioning systems, code documentation systems, and general computer-aided software engineering (CASE) frameworks. Existing CASE systems can be adapted to support NoC programming. An opportunity for innovation in this area is the integration of software development and target architecture models to guide the mapping of software to available computation and communication resources [133].

High-level software optimization tools take as input the source code generated by CASE systems and produce optimized source code. In this area we can group numerous tools for embedded software optimization like parallelizing compilers [134] and interactive code transformation methodologies, such as, for instance, the *data transfer and storage exploration* (DTSE) proposed by IMEC [135]. In an NoC environment these tools should focus on removing or reducing communication bottlenecks, either by eliminating redundant data transfers, or by hiding latency and maximizing available bandwidth. Most of the

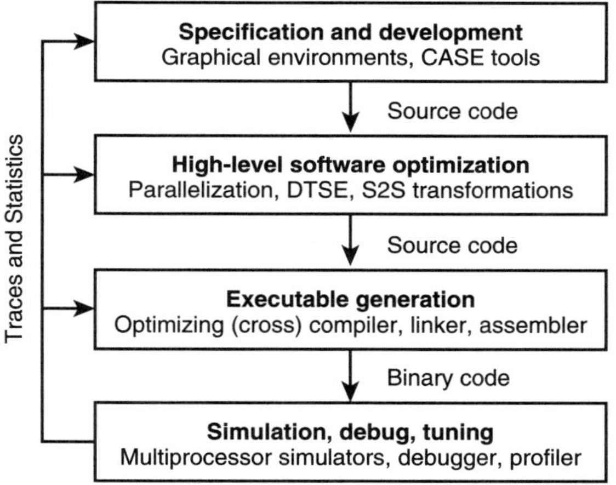

3-7   Software development infrastructure.

FIGURE

transformations performed at this stage should be weakly dependent on specific hardware targets.

Target-specific optimizations are confined to the third software development stage, namely, executable generation. The back end of the optimizing compiler should also link application object files with middleware libraries [136]. At this stage, intertask communication and synchronization points cannot be removed or transformed, and optimization opportunities are confined within the task code. Instruction-level parallelism can be discovered and exploited at this stage. High- and low-level optimizations during code generation could be seen as two separate stages in a complex software optimizer with architecture-neutral front end and architecture-specific back end. Concerning NoC specific optimizations, high-level communication optimization is performed in the front end, whereas fine tuning of communication primitives and function calls to architectural features can be performed in the back end.

The last class of tools includes system simulation, profiling, and code analysis/debugging. These tools are critical for functional debugging, as well as performance tuning. Several challenges must be addressed at this level. First, system simulation for NoCs is extremely challenging from the computational viewpoint. Hence, several simulation engines should be available, spanning the accuracy versus speed tradeoff. Modern hardware-software specification languages and simulators, such as SystemC [137], can support hardware descriptions at multiple abstraction levels. NoC simulation platforms have recently been presented [138], and this area of research and development is currently very active. Profilers play a strategic role in functional debugging, code tuning, and feedback-directed code optimization, in which profiling information is fed back as inputs to the optimization tools [139–145].

3.5 CONCLUSIONS

This chapter considers the challenges of designing SoCs with tens or hundreds of processing elements in (100 to 25) nm silicon technologies. Challenges include dealing with design complexity and providing reliable, high-performance operation with small energy consumption.

We have claimed that modular, component-based design of both hardware and software is needed to design complex SoCs. Starting from the observation that interconnect technology will be the limiting factor for achieving the operational goals, we developed a communication-centric view of design. We postulated that efficient communication on SoCs can be achieved by reconfigurable micronetworks, whose layered design can exploit methods and tools used for general networks. At the same time, micronetworks can be specialized and optimized for the specific SoC being designed.

We examined the different layers in micronetwork design, and we outlined the corresponding research problems. Despite the numerous challenges, we remain optimistic that adequate solutions can be found for such problems. At the same time, we believe that a layered micronetwork design methodology is likely to be the only path for mastering the complexity of SoC designs in the years to come.

ACKNOWLEDGMENTS

This chapter was written with support from the MARCO/ARPA GSRC.

4

Architecture of Embedded Microprocessors

Eric Rotenberg and Aravindh Anantaraman

4.1 INTRODUCTION

System-on-chip (SoC) designs are powered by one or more general-purpose microprocessor units (MPUs), digital signal processors (DSPs), and fixed-function-coprocessors. Therefore, an understanding of processor architecture provides a context for choosing the right processor(s) for a SoC, in terms of nominal and worst-case performance, cost, power, and other constraints. This chapter provides a tutorial on the architecture of MPUs used in SoCs and embedded systems in general.

We also include commentary on various themes regarding the evolution of general-purpose embedded processors. For example, many contemporary architectural features of general-purpose embedded processors originated in high-performance desktop processors. One reason is that, like desktop processors, embedded processors are characterized by general instruction sets and performance is naturally enhanced through generic instruction-level pipelining techniques. Another theme is the distinction between processors for open embedded systems (e.g., personal digital assistants [PDAs]) versus deeply embedded, closed systems (e.g., automotive). Dual tracks are explicit in ARM's family of 32-bit MPUs, which account for 75% of the 32-bit embedded MPU market [146]. In this chapter, we explore underlying reasons for the parallel evolution of embedded and desktop processors and underlying reasons for dual tracks targeting *open* versus *closed embedded systems*. Regarding the latter aspect, closed embedded systems constrain microarchitectural evolution due to the need for timing predictability [147]; we conclude the chapter by describing recent research aimed at bridging the dual tracks [148].

General-purpose DSPs form another important class of embedded processors that deliver high MIPS/watt on data streaming applications. Data-level parallelism and regular control flow in streaming applications has led engineers to base many DSPs on the very long instruction word (VLIW) execution model [149], which encodes regular parallelism in an efficient and predictable way. Efficiency is desirable from the standpoint of power, and predictability is desirable from the standpoint of real-time constraints. The architecture of general-purpose DSPs is beyond the scope of this chapter.

4.2   EMBEDDED VERSUS HIGH-PERFORMANCE PROCESSORS: A COMMON FOUNDATION

Embedded processors are general-purpose in a different sense than the high-performance processors used in personal computers. A personal computer is expected to run arbitrary software (Fig. 4-1): productivity tools (e-mail, word processors, spreadsheets, presentations, and so on), computer-aided design (CAD), games, multimedia, and the operating system (OS) itself. In contrast, a *closed* embedded system runs a fixed set of tasks, or task-set. Nonetheless, there are many embedded systems, each with its own, unique task-set (Fig. 4-2). Thus, a single processor powering all of these embedded systems must support arbitrary software, even though any one system has a fixed task-set.

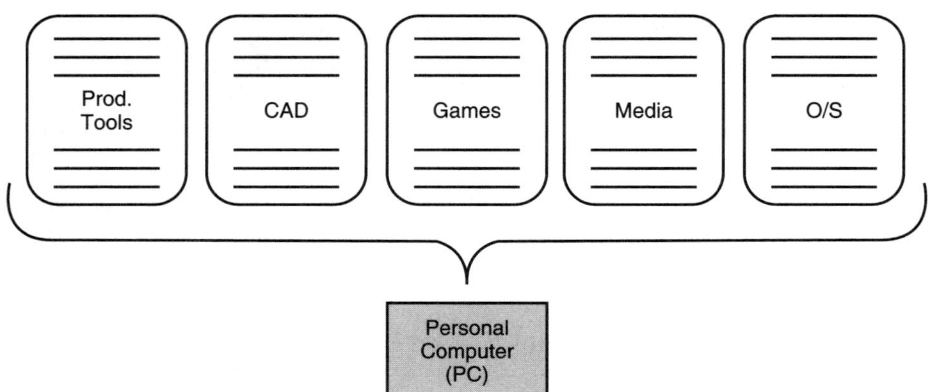

4-1   A personal computer requires a general-purpose processor to support arbitrary software.

FIGURE

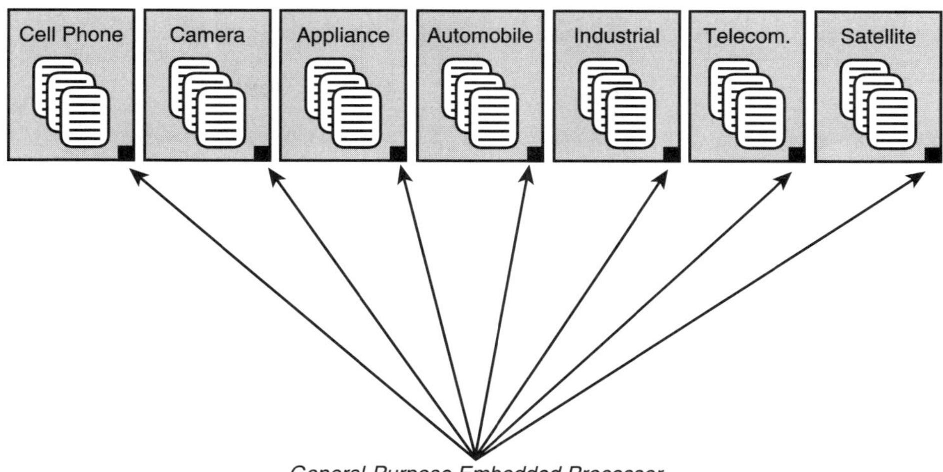

4-2

FIGURE

A particular embedded system has a fixed task-set, but the existence of many diverse embedded systems requires a general-purpose processor to drive them all.

The defining trait of general-purpose processors is a high degree of programmability via a general *instruction-set architecture* (ISA) that efficiently supports high-level languages such as C/C++. The ISA is a contract between hardware and software that defines (1) the logical (visible) state provided by the hardware, including registers and memory; and (2) instructions (arithmetic and logical operations, loads, stores, branches) and their functional behavior in terms of modifying logical state.

Although there are many different ISAs in existence and they have been widely researched, ISAs have not fundamentally changed since the birth of electronic computers in the 1940s and 1950s. Specifically, they still follow the sequential execution model commonly referred to as the *Von Neumann architecture* [150]. In the Von Neumann architecture, the instruction pointed to by a single *program counter* (PC) is the next instruction to be fetched/executed. Executing an instruction involves fetching one or more operands from the logical state (registers, memory, or both), performing an operation on the operands, and updating the logical state with the result. After executing an instruction, the PC is updated (most instructions just increment the PC, but branch instructions may redirect it) to point to the next instruction to be fetched/executed, and the *instruction fetch/execute cycle* repeats.

The microarchitecture (i.e., hardware organization) of general-purpose microprocessors reflects the generic nature of the instruction fetch/execute cycle. The

| IF | ID | EX | MEM | WB |

4-3   Simple five-stage pipeline.

FIGURE

microarchitecture is not tailored to any particular algorithm or set of algorithms. This is in contrast to domain-specific processors such as network processors [151–153] and graphics processors [154–156]. Their targeted applications enable them to exploit parallelism specifically at the *algorithmic level* (e.g., packets in network processors or triangles in graphics processors). As a result, diverse and exotic microarchitectures exist in these domains.

In contrast, a general-purpose microarchitecture is confined to exploiting parallelism generically and, hence, at the *instruction level*. What this means is that instruction execution is divided into multiple stages, forming an instruction pipeline. A simple pipeline that was typical of some of the first single-chip microprocessors circa 1980 (e.g., the Berkeley *reduced instruction set computing* [RISC] I [157,158] and Stanford MIPS [159]) is the five-stage pipeline shown in Figure 4-3. In the first stage, the instruction is fetched from memory (instruction fetch [IF]). It is then decoded and its operands are read from the register file (instruction decode [ID]). Next, it is executed using the operands (execute [EX]) or, in the case of loads and stores, a memory address is computed. In the next stage (memory [MEM]), loads and stores access memory using the computed address, either retrieving a value from memory (loads) or storing a value to memory (stores). In the final stage, the register file is updated with the result of execution (writeback [WB]).

Pipelining increases performance by simultaneously processing multiple instructions in different stages, as shown in Figure 4-4. While instruction 1 is completing (WB), three other instructions are in various stages of processing (instruction 2 in MEM, 3 in EX, 4 in ID), and instruction 5 is just entering the pipeline (IF). It takes five cycles to fill the pipeline. However, after filling the pipeline, a new instruction completes every cycle thereafter. This assumes that the pipeline can be kept flowing smoothly. In Section 4.3, we discuss conditions that stall the pipeline and techniques to minimize these stalls. These techniques are characteristic of high-performance microprocessors, and the simplest techniques are beginning to appear even in embedded processors.

The microarchitectures of contemporary high-performance processors are incredibly complex, with very deep pipelines and simultaneous execution of more than 100 in-flight instructions [160,161]. It is hard to imagine that there is

Cycle

Instruction	1	2	3	4	5	6	7	8	9
1	IF	ID	EX	MEM	WB				
2		IF	ID	EX	MEM	WB			
3			IF	ID	EX	MEM	WB		
4				IF	ID	EX	MEM	WB	
5					IF	ID	EX	MEM	WB

4-4

Instruction-level parallelism through pipelining.

FIGURE

anything in common between embedded processors and their high-performance counterparts, yet, in fact, they share a common foundation due to their general-purpose usage, general ISAs based on the Von Neumann architecture, and resulting confinement to generic instruction-level pipelining.

The difference between embedded and high-performance processors lies in their stages of evolution. Contemporary embedded processors lag some 10 years behind their high-performance counterparts, in terms of complexity. Whereas high-performance processor designs push and exceed the limits of technology, minimal embedded processor designs fully exploit the power and cost scaling advantages of new generations of CMOS technology. The question is, will embedded processors continue to follow in the footsteps of their high-performance predecessors? If current trends continue, the answer seems to be affirmative, but with a crucial caveat. Special constraints of embedded systems, most notably providing real-time guarantees, may cause dual evolutionary paths: one unrestricted in terms of borrowing high-performance techniques and the other more restricted. As we will show later, this evolutionary split is already evident in ARM products.

4.3 PIPELINING TECHNIQUES

Pipelines improve instruction throughput as long as the flow of instructions is not disrupted. In this section, we review the various types of pipeline hazards that may cause instructions to stall, as well as techniques that are used to minimize these stalls. This overview provides a context for discussing existing embedded processors (e.g., ARM, MIPS, PowerPC) in Section 4.4, as well as future embedded processors.

4.3.1   Bypasses

A *data hazard* exists when an instruction depends on data produced by a previous instruction, and the previous instruction has not yet completed (result has not been written to the register file). As shown in Figure 4-5, instruction 2 depends on a value produced by instruction 1, and so it must stall for three cycles in the ID stage. It is not until cycle 6 that the value is available in the register file and can be read by instruction 2 in the ID stage. All instructions after instruction 2 also stall, whether data-dependent or not.

The true data dependence between instructions 1 and 2 can be resolved without any stall cycles, if the value produced at the end of the EX stage by instruction 1 is *bypassed* directly to instruction 2, circumventing the register file. This is achieved via a bypass bus from the output of the arithmetic and logic unit (ALU) in the EX stage back to the pipeline latch between the ID and EX stages (i.e., a loop from the output of the ALU to its input), called the EX−EX bypass. As shown in Figure 4-6, the EX−EX bypass eliminates the previous stall cycles.

	cyc. 1	2	3	4	5	6	7	8	9	10
Instr. 1	IF	ID	EX	MEM	WB					
Instr. 2		IF	ID	(stall)	(stall)	(stall)	EX	MEM	WB	
Instr. 3			IF	(stall)	(stall)	(stall)	ID	EX	MEM	WB

4-5

FIGURE

Stall cycles caused by a true data dependence between instructions 1 and 2.

	cyc. 1	2	3	4	5	6	7	8	9	10
Instr. 1	IF	ID	EX	MEM	WB					
Instr. 2		IF	ID	EX	MEM	WB				
Instr. 3			IF	ID	EX	MEM	WB			

4-6

FIGURE

Bypassing eliminates stall cycles due to true data dependences. (EX-EX and MEM-EX data bypasses are shown with arrows.)

Instruction	1	2	3	4	5	6	7	8	9
1 *(load)*	IF	ID	EX	MEM	WB				
2 *(load-use)*		IF	ID	(stall)	EX	MEM	WB		
3			IF	(stall)	ID	EX	MEM	WB	

Cycle

4-7

FIGURE

The MEM-EX bypass (indicated with an arrow) reduces the load-use stall to one cycle.

	cyc. 1	2	3	4	5	6	7	8	9
Instr. 1 *(load)*	IF	ID	EX	MEM	WB				
Instr. 3		IF	ID	EX	MEM	WB			
Instr. 2 *(load-use)*			IF	ID	EX	MEM	WB		

4-8

FIGURE

Static scheduling reorders instructions to eliminate load-use stall altogether.

If instruction 3 also depends on instruction 1, then another bypass from the MEM stage to the EX stage (MEM−EX bypass) prevents instruction 3 from stalling.

The MEM−EX bypass also enables load instructions to bypass their results quickly to later dependent instructions. Not all stall cycles can be eliminated in this case, but the MEM−EX bypass reduces the length of the stall to only one cycle, as shown in Figure 4-7. To eliminate *load-use stall* cycles completely, the compiler can try to separate a load instruction from its use, by scheduling independent instructions between the two, as shown in Figure 4-8. This technique is known as *static scheduling*. Alternatively, hardware can reorder instructions on-the-fly such that the stalled load-dependent instruction does not stall later independent instructions. This technique is called *dynamic scheduling* or *out-of-order execution* and will be covered in Section 4.3.4.

4.3.2   Branch Prediction

A *control hazard* exists when the PC cannot be determined in the next cycle, stalling instruction fetching in the IF stage. Control hazards are caused by branch

Cycle

Instruction	1	2	3	4	5	6	7	8	9	10
branch	IF	ID	EX	MEM	WB					
		IF	(stall)	(stall)	ID	EX	MEM	WB		

4-9

FIGURE

Stall cycles due to a control hazard. The IF stage stalls until the branch executes (indicated with an arrow).

instructions, since the outcome of a branch (taken or not-taken) is not known until the end of the EX stage. As shown in Figure 4-9, a control hazard introduces two stall cycles in the simple pipeline. An optimization is to not re-fetch the next sequential instruction in cycle 4, saving one stall cycle if the branch is not-taken.

Branch prediction is a technique whereby the next PC is predicted in the IF stage in order to minimize instruction fetch disruptions. *Static branch prediction* assigns a fixed prediction to each branch instruction. A static prediction can be based on simple heuristics [162,163], or the compiler can use profiling (past executions of the program) to convey accurate predictions to the hardware via an extra bit in the branch opcode [164,165]. The simplest heuristic is always to predict not-taken (for all branches), i.e., continue sequential fetching until a taken branch is executed. An improved heuristic is to predict forward branches as not-taken and backward branches as taken. The basis for this heuristic is that backward branches typically demarcate loops, in which case the backward branch is more often taken than not-taken.

Dynamic branch prediction is typically much more accurate than static branch prediction because it uses recent past outcomes of a branch to predict its future outcomes [163]. A branch history table maintains a counter for each branch. A 1-bit counter predicts that the next outcome of the branch will be the same as the last outcome. A 2-bit counter significantly improves accuracy by adding inertia, i.e., two wrong predictions are needed before changing the prediction. A table of 2-bit counters indexed by the PC is called the bimodal predictor, as it can accurately predict branches biased in either the taken or not-taken directions. A wave of more advanced branch predictors was developed in the 1990s based on the observation that branches are often correlated with other branches and/or produce arbitrary repeating patterns [166–168]. These advanced branch predictors are crucial components in contemporary high-performance microprocessors. (Very deep pipelines expose a high opportunity cost for mispredicting branches: tens of cycles of fetching, decoding, and executing useless instructions down the wrong path. This is discussed further in Section 4.3.5.)

	cyc. 1	2	3	4	5	6	7	8	9	10
branch	IF	ID	EX	MEM	WB					
		IF	ID	EX	MEM	WB				

4-10

FIGURE

A correct branch prediction eliminates stall cycles due to the control hazard. The solid arrow indicates a correct branch prediction generated in the IF stage during cycle 1 and used in the IF stage during cycle 2.

	cyc. 1	2	3	4	5	6	7	8
branch	IF	ID	EX	MEM	WB			
predicted target		IF	ID	*EX*	*MEM*	*WB*		
			IF	*ID*	*EX*	*MEM*	*WB*	
actual target				IF	ID	EX	MEM	WB

4-11

FIGURE

An incorrect branch prediction (dashed arrow) re-exposes the performance penalty of a control hazard (solid arrow). The incorrect branch prediction generated in the IF stage during cycle 1 causes incorrect instructions to be fetched during cycles 2 and 3. The misprediction is discovered during cycle 3 in the EX stage, which signals the IF stage to redirect instruction fetching during cycle 4.

In addition to predicting the direction of a branch (taken or not-taken), the target PC must also be generated ahead of time, in the IF stage, if the branch predicted is taken. This is achieved by storing the pre-decoded targets of recently seen branches in a PC-indexed hardware table called the *branch target buffer* (BTB) [169].

With branch prediction (either static or dynamic), stall cycles due to a control hazard are eliminated if the prediction is correct, as shown in Figure 4-10. However, if the prediction is incorrect, then instructions following the branch must be squashed and instruction fetching redirected along the correct path, as shown in Figure 4-11. In this case, the performance penalty of the control hazard is re-exposed (two stall cycles). Moreover, there may be additional complexity associated with squashing wrong-path instructions that actually increases the penalty. However, cycles saved by correct predictions typically far outweigh misprediction overheads. In power-constrained embedded systems, performance gains of branch prediction must be weighed against the waste of power caused by squashing wrong-path instructions.

4.3.3   Caches

In the simple five-stage pipeline, memory is accessed in the IF stage (to fetch the next instruction) and in the MEM stage (to load or store data). The *memory wall* refers to the growing problem that processors are much faster than main memory [170]. Caches attempt to bridge the processor—memory performance gap. A *cache* is a small high-speed memory that keeps recently referenced instructions and data close to the processor, based on the expectation that they will be referenced again soon. Caches exploit *temporal* and *spatial locality*. Temporal locality is the notion that recently referenced items are likely to be accessed again soon; hence it is beneficial to keep these items close to the processor for quick access. Spatial locality is the notion that items neighboring a recently referenced item are also likely to be accessed soon, so it is beneficial to prefetch neighboring items when an item is fetched from main memory.

To enable the IF and MEM stages to operate in parallel, the five-stage pipeline is typically augmented with separate instruction and data caches, respectively. A *hit* in the instruction cache results in the IF stage taking only one cycle, whereas a *miss* causes the IF stage to stall for possibly many cycles until the instruction is retrieved from main memory. Likewise, a *hit* in the data cache results in the MEM stage taking only one cycle. A data *cache miss* stalls a load or store instruction for possibly many cycles, stalling instructions behind it in the pipeline too.

As the performance gap between processor and main memory grows, more levels of caching are added. The level-1 (L1) cache, to which the processor is directly attached, is the smallest and fastest. The L2 cache (if there is one) is larger and slower, but still much faster than main memory. The L1 instruction and data caches are almost always on the processor chip and closely integrated into the pipeline. Contemporary high-performance processors also have a unified (i.e., both instructions and data) L2 cache integrated on the chip [160,161]. Table 4-1 shows ballpark latencies, in cycles, for accessing caches and main memory at the time of this writing, for both a 3-GHz desktop processor and a 500-MHz

	L1 cache	L2 cache	Memory
3-GHz desktop processor	1–3 cycles	5–20 cycles	200–300 cycles
500-MHz embedded processor	1 cycle	*Not typically used*	50 cycles

4-1   Ballpark Cache/Memory Latencies.

TABLE

embedded processor. The table highlights the speed gap between processor and memory, which increases over time.

Like branch mispredictions, instruction and data cache misses are a major performance bottleneck. In the simple pipeline, a load that misses in the data cache causes later instructions not dependent on the load to stall. To alleviate this bottleneck, a means is required for allowing independent instructions to circumvent the load and instructions that depend on the load, i.e., out-of-order execution. This is the topic of the next section.

These days, caches are absolutely required for high performance on general-purpose programs. Unfortunately, their use in embedded processors is controversial (albeit increasingly common, especially in 32-bit MPUs). The problem is that caches introduce uncertainty in terms of latency for accessing memory, yet safe planning in real-time systems requires strict performance guarantees. Caching is only one of many high-performance pipelining techniques (e.g., dynamic branch prediction, out-of-order execution, and so on) whose adoption in embedded systems is possibly impeded by the need for performance guarantees. This theme is re-visited when we cover dual evolutionary paths of embedded processors in Section 4.4 and recent research that addresses these issues in Section 4.5.

4.3.4 Dynamic Scheduling

The simple five-stage pipeline executes instructions in strict sequential order. Although *in-order execution* is simple to implement, stalled instructions cause other independent instructions to stall needlessly. Dynamic scheduling provides a means for independent instructions to circumvent previous stalled instructions and, therefore, execute *out-of-order*. Out-of-order execution allows useful work to be overlapped with the load-use stalls described earlier, stalls caused by data cache misses, and stalls caused by long-latency arithmetic instructions (e.g., floating-point arithmetic).

The key mechanism for out-of-order execution is a pool of instruction buffers that decouples the instruction fetch/decode stages from later execution stages, as shown in Figure 4-12. These instruction buffers, collectively called the scheduling window or *instruction window*, provide a staging area in which instructions wait for their source operands to become available. An instruction may issue for execution as soon as its operands become available. Thus, prior stalled instructions do not stall later independent instructions and instructions issue from the window out-of-order. Dynamic scheduling adds another stage to the pipeline between decode (ID) and execute (EX), here called IS for *instruction schedule*.

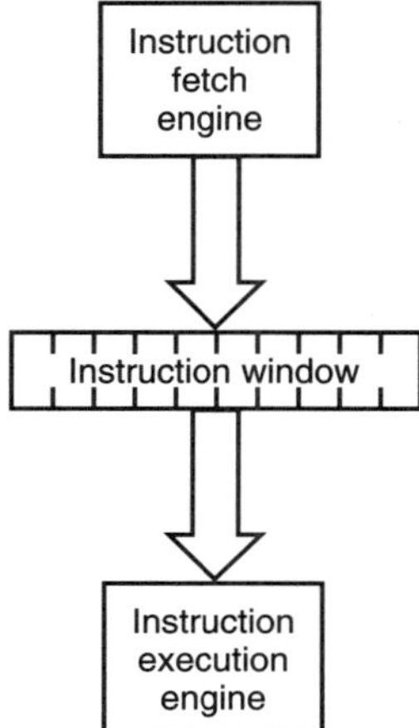

4-12 FIGURE   The instruction window enables out-of-order execution.

	cyc. 1	2	3	4	5	6	7	8	9	10	11	12	13	14	15	16	17
Instr. 1 *(load)*	IF	ID	EX	MEM				*cache miss penalty*						WB			
Instr. 2 *(load-use)*		IF	ID	(stall)	(stall)	(stall)	(stall)	(stall)	(stall)	(stall)	(stall)	(stall)	(stall)	EX	MEM	WB	
Instr. 3			IF	(stall)	(stall)	(stall)	(stall)	(stall)	(stall)	(stall)	(stall)	(stall)	(stall)	ID	EX	MEM	WB
Instr. 4														IF	ID	EX	MEM

4-13 FIGURE   Cache miss stall with in-order execution. (The arrow shows use of the MEM-EX bypass.)

The benefit of dynamic scheduling is evident by comparing in-order and out-of-order execution in the presence of a cache miss, as shown in Figures 4-13 and 4-14, respectively. Instruction 1 is a load that misses in the cache. Instruction 2 is dependent on the load. Instructions 3 and 4 are independent of instructions 1 and 2. With in-order execution, independent instructions 3 and 4 are not overlapped with the cache miss. In contrast, with out-of-order execution, instructions 3 and 4—and possibly many instructions after them—execute while the cache miss is being serviced.

	cyc. 1	2	3	4	5	6	7	8	9	10	11	12	13	14	15	16	17
Instr. 1 *(load)*	IF	ID	IS	EX	MEM	*cache miss penalty*									WB		
Instr. 2 *(load-use)*		IF	ID	IS	IS	IS	IS	IS	IS	IS	IS	IS	IS	IS	EX	MEM	WB
Instr. 3			IF	ID	IS	EX	MEM	WB									
Instr. 4				IF	ID	IS	EX	MEM	WB								

4-14

FIGURE

Cache miss stall with out-of-order execution.

4.3.5   Deeper Pipelining, Multiple-Instruction Issue, and Hardware Multithreading

The pipeline discussed so far is called a *scalar pipeline* because its peak throughput is limited to one instruction per cycle. Augmenting a scalar pipeline with branch prediction, caches, and out-of-order execution minimizes stalls due to control and data hazards, approaching—but never exceeding—1 instruction per cycle.

One way to improve performance further is to deepen the scalar pipeline. As shown in Figure 4-15, a pipeline twice as deep as the original pipeline is capable of twice the peak execution rate. Once the deep pipeline is filled, two new instructions are completed every *original* cycle. However, in practice, doubling pipeline depth does not double performance because there are relatively more stall cycles that need to be tolerated. A larger instruction window is needed to hide lengthier stalls due to data hazards (for example, data cache misses take more cycles to resolve since memory access time in nanoseconds is unchanged). A more accurate branch predictor is needed because the branch misprediction penalty is larger (more stages between IF and EX).

An alternative to deeper pipelining is to fetch, issue, and execute multiple instructions in parallel each cycle, exceeding the scalar bottleneck explicitly. This approach is called, appropriately, *superscalar* processing. A dual-issue superscalar processor doubles the peak execution rate with respect to the original scalar pipeline, as shown in Figure 4-16.

The most aggressive pipelines employ deep *speculation*, out-of-order execution, and *multiple-instruction issue*. Unfortunately, at some point, branch mispredictions and lengthy cache miss stalls limit the maximum achievable performance. Any further pipeline enhancements yield diminishing returns, and the pipeline *shallow scalar pipeline*

	cyc. 1	2	3	4	5	6	7
Instr. 1	IF	ID	EX	MEM	WB		
Instr. 2		IF	ID	EX	MEM	WB	
Instr. 3			IF	ID	EX	MEM	WB

deep scalar pipeline

	cyc. 1	2	3	4	5	6	7	8	9	10	11	12	13	14
Instr. 1	IF1	IF2	ID1	ID2	EX1	EX2	M1	M2	WB1	WB2				
Instr. 2		IF1	IF2	ID1	ID2	EX1	EX2	M1	M2	WB1	WB2			
Instr. 3			IF1	IF2	ID1	ID2	EX1	EX2	M1	M2	WB1	WB2		
Instr. 4				IF1	IF2	ID1	ID2	EX1	EX2	M1	M2	WB1	WB2	
Instr. 5					IF1	IF2	ID1	ID2	EX1	EX2	M1	M2	WB1	WB2

4-15

FIGURE

Scalar pipeline performance can be increased by deepening the pipeline.

scalar

	cyc. 1	2	3	4	5	6	7
Instr. 1	IF	ID	EX	MEM	WB		
Instr. 2		IF	ID	EX	MEM	WB	
Instr. 3			IF	ID	EX	MEM	WB

superscalar

	cyc. 1	2	3	4	5	6	7
Instr. 1	IF	ID	EX	MEM	WB		
Instr. 2	IF	ID	EX	MEM	WB		
Instr. 3		IF	ID	EX	MEM	WB	
Instr. 4		IF	ID	EX	MEM	WB	
Instr. 5			IF	ID	EX	MEM	WB
Instr. 6			IF	ID	EX	MEM	WB

4-16

FIGURE

Superscalar processing exceeds the scalar bottleneck explicitly.

becomes underutilized. For example, it is generally recognized that an eight-issue superscalar processor is significantly underutilized due to imperfect branch prediction and L2 cache misses. In terms of ballpark estimates, average utilization is around 20 to 30% for SPEC integer and floating-point benchmarks. (Tullsen et al. [171] report an average execution rate of 2.2 instructions-per-cycle for a superscalar processor with a peak execution rate of 8 instructions-per-cycle, for a subset of SPEC92 benchmarks.)

Hardware multithreading is a means for improving pipeline utilization when bottlenecks prevent further speedup of single-threaded programs. It enables the simultaneous execution of multiple independent programs on the same pipeline. Independent programs are an abundant source of *instruction-level parallelism*, since there are no dependences between instructions from different threads. If one thread incurs a lengthy cache miss, the pipeline can still be utilized by another thread. This is often referred to as thread-level parallelism. Hardware multithreading improves pipeline utilization by converting thread-level parallelism to instruction-level parallelism [172–174].

In fact, it can be argued that hardware multithreading provides an alternative to contemporary pipeline enhancements, especially branch prediction, caching, and out-of-order execution, since thread-level parallelism naturally enhances latency tolerance. However, this argument rests on the relative importance of single-threaded and multithreaded performance and the availability of independent threads. This trade-off may very well differ between high-performance processors (in which single-threaded performance is paramount and multiple threads are not always available) and embedded processors (embedded systems typically have many threads). These trade-offs are discussed in more depth when we survey the Ubicom embedded processor, in Section 4.4.3.

In early forms of hardware multithreading, instructions are fetched from a different thread each cycle [172]. An example of this form of hardware multithreading is shown in Figure 4-17. It shows a high-level diagram of a dual-issue superscalar pipeline (it fetches, decodes, executes, and so on, up to two instructions in parallel each cycle). Two instructions are fetched from thread 1 (T1), followed by two instructions from thread 2 (T2), and so on, until the current cycle, during which two instructions are fetched from thread 5 (T5), as shown. In this way, each pipeline stage contains two instructions from a different thread.

Simultaneous multithreading (SMT) is even more flexible: a single pipeline stage may mix instructions from multiple threads during the same cycle [173], as shown in Figure 4-18 (e.g., IF is simultaneously fetching one instruction from thread 1 and one instruction from thread 3).

Whatever form it takes, hardware multithreading requires replicating the logical register state (one set of registers for each hardware thread) and possibly other microarchitectural resources [171,175].

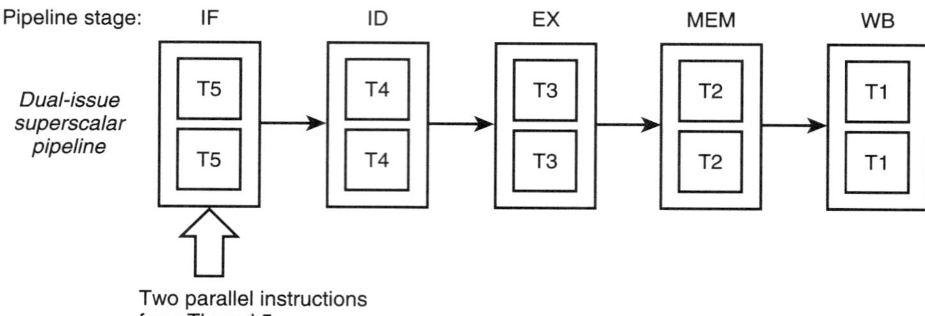

4-17 FIGURE   Early form of hardware multithreading [172] on a dual-issue superscalar pipeline.

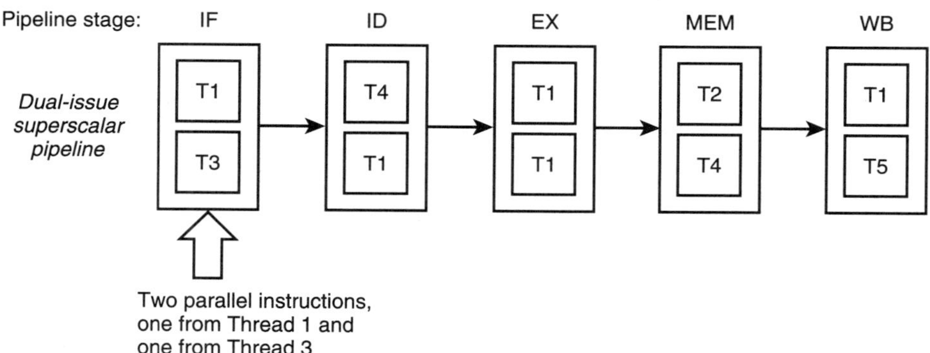

4-18 FIGURE   Simultaneous multithreading [171] on a dual-issue superscalar pipeline.

4.4   SURVEY OF GENERAL-PURPOSE 32-BIT EMBEDDED MICROPROCESSORS

In this section, we survey the architectures of several general-purpose 32-bit embedded microprocessors. Sections 4.4.1 (ARM MPUs), 4.4.2 (IBM, Intel, and MIPS MPUs), and 4.4.3 (Ubicom) are each intended to represent a point in the embedded application space. For example, ARM MPUs deliver moderate

	Automotive	Consumer electronics (e.g., set-top box, DVD)	Imaging (e.g., digital camera, printer)	Micro-controllers (e.g., hard real-time)	Networking	Storage (e.g., disk drive)	Wireless, mobile device (e.g., mobile phone, hand-held)	Broadband (e.g., cable modem, fire wall)
ARM	✓	✓	✓	✓	✓	✓	✓	No info
IBM	No info	✓	✓	No info	✓	✓	✓	✓
MIPS	No info	✓	No info	No info	No info	No info	✓	✓
Ubicom					✓			

4-2

TABLE

Application Spaces Penetrated by MPUs (based on available marketing literature).

performance with very low power and small packages. Accordingly, they are widely used in cell phone handsets and PDAs. IBM, Intel, and MIPS MPUs provide high-end embedded performance by integrating relatively complex memory hierarchies (e.g., larger caches and multiple pending memory requests) and moderate instruction-level parallelism (e.g., dual superscalar execution) for consumer electronics (e.g., set-top boxes, DVD), storage, and communications.

That said, it is difficult to tie these companies' MPUs to any particular embedded application space. Table 4-2 lists application spaces covered by these MPUs, based on on-line marketing literature that documents specific products using them [176]. (Note that unavailability of specific product information—marked "no info" in Table 4-2—does not imply lack of penetration in that application space.) For example, ARM MPUs penetrate many more segments than just handsets, including consumer electronics, storage, and so on. This trend somewhat confirms our earlier observations about the general-purpose nature of "instruction-set processors" and may even signal a trend toward consolidation, as we see in the desktop market.

Ubicom markets its MPUs as programmable network processor units (NPUs) [177]. NPUs represent an interesting segment of the embedded MPU market, deep enough architecturally (due to many levels of exploitable parallelism) that it deserves a wholly separate treatment beyond the scope of this chapter. NPUs are designed to process network packets efficiently. Although the Ubicom MPU has key features characteristic of network processors—such as multithreading for exploiting coarse-grain parallelism and special bit-level instructions—the core microarchitecture is not fundamentally tied to packet processing. In fact, Ubicom's scalar pipeline is simple in comparison with high-end NPUs, which typically employ many symmetric or packet-pipelined processing elements (each of which may be scalar or wide VLIW/superscalar), one or more specialized-coprocessors, and specialized function units and memories [152]. For the interested reader, Shah and Keutzer [152] provide an excellent tutorial of network processor architectures and neatly classify MPUs into five key dimensions: approaches to parallel processing, elements of special-purpose hardware, structure of memory architectures, types of on-chip communication, and use of peripherals.

Although we would like to correlate specific architectural features (e.g., pipelining, branch prediction, caches, and so on) with particular application spaces, this is not how things seem to work from an engineering standpoint. Different application spaces have different requirements in terms of *overall* performance, power, size, and cost. These high-level attributes are considerably influenced by architectural features in aggregate, i.e., the MPU as a whole. In some cases, it is possible to identify specific architectural features with particular application spaces, usually ISA enhancements such as DSP extensions and efficient instruction encodings (e.g., 16-bit Thumb instructions, which are explained later).

4.4.1   ARM

ARM holds the lion's share of the 32-bit embedded microprocessor market, according to one market analyst [146]. On-line ARM documentation [178–183] served as the primary source of the comparatively brief survey presented here. An excellent reference text for the ARM instruction-set architecture and extensions, ARM families of microprocessors, and fundamentals of computer architecture and organization is Furber's *ARM System-on-Chip Architecture* [184].

Overview

The ARM architecture continues to evolve over time and, currently, there are five 32-bit RISC MPU product families: ARM7, ARM9, ARM9E, ARM10, and ARM11. Successive families represent a significant shift in the microarchitecture (pipeline design). In addition, there are often corresponding ISA enhancements for new functionality. These functional enhancements include *Thumb*, a 16-bit instruction set for compact code; *DSP*, a set of arithmetic extensions for DSP applications; and *Jazelle*, an extension for direct execution of Java bytecode. Another ISA enhancement is *memory management unit* (MMU) support for open platforms.

ARM provides two to four implementations within each product family, tailored for certain domains. These are offered as *hard cores* and/or *soft cores* in 0.25-, 0.18-, and 0.13-μm technologies. Navigating the numerous implementations can be overwhelming. Suffice it to say that there are two evolutionary paths spanning all product families: cores for open platforms and cores for *real-time embedded operating systems* (RTOS). The first path features an MMU for virtual memory management and protection, targeting operating systems such as Windows CE, Linux, and Palm OS. Performance features span the full range, as high performance is required for many open platforms. The second path tailors implementations for closed embedded control systems that typically employ an RTOS. Memory protection varies from no protection to minimal protection. Unpredictable microarchitecture features are avoided (e.g., no cache), minimized (e.g., *cache locking*), or explicitly managed (e.g., tightly coupled memory instead of caches), for hard real-time applications.

ARM7 to 11 Microarchitectures

Pipeline stages for the ARM7 to 11 are shown in Figures 4-19 through 4-21. All the pipelines are scalar, i.e., limited to a peak execution rate of one instruction per cycle. Nonetheless, the ARM11 exploits limited parallelism in the back-end of the pipeline.

The ARM7 has a very shallow pipeline (Fig. 4-19), only three pipeline stages. As a result, it is exceedingly simple. It does not require even the most basic pipelining technique, bypassing, since a single pipeline stage fetches operands from the

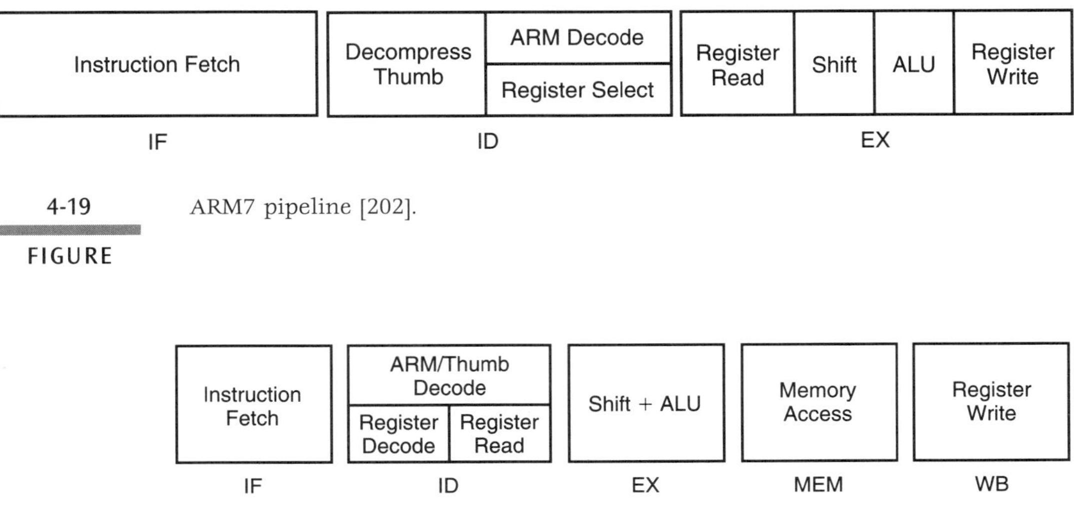

4-19   ARM7 pipeline [202].

FIGURE

4-20   ARM9 pipeline [202] resembles the classic five-stage pipeline.

FIGURE

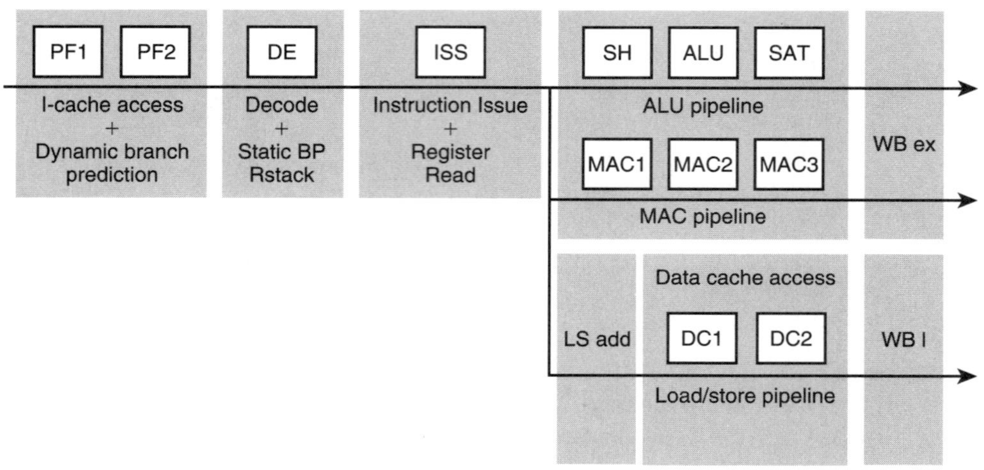

4-21

ARM11 pipeline [203].

FIGURE

register file, executes the instruction, and writes the result back to the register file, all in one cycle. Thus there are no data hazards between successive instructions. The first stage (IF) fetches an instruction from memory. The second stage (ID) decompresses 16-bit Thumb-encoded instructions to 32-bit ARM-encoded instructions (if operating in Thumb mode) and then decodes the ARM instruction. (We elaborate on Thumb ISA extensions in the next section, Thumb ISA Extensions) The third and final stage is the execute stage (EX), just described. No branch prediction is used.

There are four implementations in the ARM7 family. Only one of the four, at the high end, integrates an 8-Kb unified cache and an MMU, aimed at supporting the high-performance requirements and virtual memory demands of open platform applications such as Windows CE, Linux, Palm OS, and others.

To enhance performance, the pipeline depth is gradually increased with each new ARM family. For example, the ARM9 pipeline increases the number of stages from three to five. The ARM9 pipeline, shown in Figure 4-20, very closely resembles the generic five-stage pipeline used throughout this chapter. In spite of having a deeper pipeline than the ARM7, branch prediction is still not employed. All the ARM9 implementations have split instruction and data caches. The implementation targeting RTOS-based systems has 4-Kb caches and a basic memory protection unit, whereas implementations targeting open platforms have 8- or 16-Kb caches and full MMUs.

The ARM9E family has the same pipeline as the ARM9 but deploys new features. First, ARM9E provides DSP and Java extensions to the ISA. All the ARM9E implementations execute the DSP extensions (e.g., *multiply-accumulate instructions* [MACs]). One implementation can directly decode/execute 80% of Java bytecodes and trap to software for the remaining 20%. Second, the instruction and data caches are individually configurable. One implementation offers cache sizes from 0 (no cache) to 128 Kb. Another goes up to 1 Mb. Third, interfaces to *tightly coupled memory* (TCM) are provided. TCM is memory that operates at the same speed as the processor. TCM may be used as a substitute for caches in hard real-time systems, in which deterministic performance is paramount. One implementation supports TCM sizes ranging from 0 (no TCM) to 64 Mb. Finally, ARM9E implementations with data caches also feature write buffers to improve memory system performance.

The ARM10 and ARM11 families represent major steps forward in performance and functionality. We refrain from describing the ARM10 in detail since the ARM11 inherits and extends some of the ARM10's performance features (e.g., a separate load/store pipe with *hit-under-miss* capability). However, we do point out that the ARM10 has a six-stage pipeline (one more stage than the ARM9/ARM9E), and static branch prediction is introduced for the first time, to counteract the higher branch misprediction penalty.

The introduction of the ARM11 microarchitecture is a clear sign that embedded MPUs are under pressure to deliver ever-increasing performance (Fig. 4-21). The ARM11 has an eight-stage pipeline to achieve significantly higher frequencies, e.g., up to 1 GHz in 90-nm technology. Dynamic branch prediction is employed for the first time in an ARM processor, due to the now-significant branch misprediction penalty (six cycles). A simple form of hybrid branch prediction is employed. Each entry of a 64-entry BTB contains (1) a 2-bit counter for predicting the branch direction, and (2) a taken target for predicted-taken branches. If the BTB does not contain information for a given branch, the branch is predicted in the decode stage using the static *backward-taken/forward-not-taken* heuristic. In addition, a three-entry *return address stack* predicts targets of subroutine return instructions [185] in the decode stage.

The ARM11 also implements aggressive features for improved memory latency tolerance—flirting with aspects of out-of-order execution, although, strictly speaking, it is still an in-order pipeline. This is out of necessity, as the high-frequency target widens the processor-memory performance gap. Specifically, the pipeline continues issuing/executing instructions around a cache miss as long as the instructions are independent of the load. Coupled with this is *out-of-order completion* of instructions so that independent instructions can continue to make forward progress while a load miss is serviced. This middle-ground between in-order and out-of-order execution is enabled by decoupled ALU, MAC, and load/store pipelines. Furthermore, the data cache is *non-blocking* with up to two outstanding load misses: the load/store pipeline stalls only after a third load miss.

In addition to inheriting the DSP and Java ISA extensions introduced earlier, the ARM11 introduces new 8-bit and 16-bit *single-instruction, multiple-data* (SIMD) instructions. SIMD instructions perform the same operation on multiple bytes or half-words in parallel, a very cheap form of superscalar execution.

Thumb ISA Extensions

The ARM ISA (beginning with ARM7 and carrying forward through ARM11 for compatibility) is really two ISAs in one: "ARM" instructions are encoded with 32 bits, and "Thumb" instructions are encoded with only 16 bits. This empowers the programmer/compiler to balance a set of tradeoffs impacted by instruction width. Thumb instructions have the key advantage of reducing static code size and, therefore, the number of kilobytes of instruction storage. Code size is a crucial parameter in very small and deeply embedded systems because the size of the instruction memory (typically a ROM) significantly impacts on overall system size and cost. In such systems, the programmer/compiler may be able to identify program fragments (e.g., a function, loop, or basic block) that can be encoded with the fewest bytes using Thumb instructions. On the other hand, some program fragments are more compactly encoded using ARM instructions because they are able to exploit more powerful instructions (for example, complex addressing modes). A nice capability is that ARM and Thumb instructions can coexist in the same binary. Variable-length ISAs normally cause many implementation headaches. (The decode stage is generally more complex, and variable-length instructions complicate decoding multiple instructions in parallel, an important consideration for future superscalar implementations.) The ARM architects reconciled flexibility and implementation complexity in two ways: first, by using only two widths and, second, by using ARM/Thumb *modes*. By mode, we mean that the processor operates alternately in Thumb mode and ARM mode over extended program fragments. Certain branch instructions facilitate switching between ARM and Thumb modes, namely, the target of a branch may signal a mode switch.

Reducing code size (via Thumb instructions) must be carefully weighed against other factors. Although using Thumb instructions may reduce static code size, it may take more instructions at run-time to execute the program, degrading performance and increasing power consumption.

A more subtle point is that, from the perspective of the processor designer, there is potentially an intrinsic penalty for supporting Thumb instructions. As shown in Figure 4-19, the ARM7's decode stage must translate Thumb instructions to corresponding ARM instructions so that later pipeline stages only have to process ARM instructions. Depending on the translation complexity and how time-consuming the decode stage is relative to other pipeline stages, cycle time may be extended by the translation logic.

4.4.2 High-End Embedded MPUs

Embedded processors from IBM, MIPS, and Intel seem to be more oriented toward high-end embedded applications (e.g., set-top boxes) than ARM (e.g., cell phone handsets), although the ARM11 microarchitecture approaches the complexity/performance of these other MPUs. The focus on high-end embedded applications is not surprising, given these companies' history in the high-performance desktop processor market.

IBM PowerPC 440

IBM's PowerPC 440 [186] can issue two instructions per cycle (superscalar). However, it is still an in-order pipeline. There are seven pipeline stages that are conceptually similar to the ARM11 pipeline stages. The branch unit implements dynamic branch prediction via a 4K-entry *branch history table* (BHT) and a 16-entry *branch target address cache* (BTAC). The branch misprediction penalty is a minimum of four cycles. There are three execution pipelines: the complex integer pipe (where MACs are performed), the simple integer pipe, and the load/store pipe. Up to two instructions can be issued per cycle, one to the complex integer pipe and one to either of the simple integer or load/store pipes. Static scheduling by the compiler plays a key role in maximizing superscalar issue, in terms of both minimizing adjacent dependent instructions and observing dual-issue restrictions. The cache interface supports non-blocking caches of sizes ranging from 8 to 64 Kb. Support is provided for locking individual cache lines (both instruction and data) for hard real-time applications that require deterministic performance.

MIPS32 24K

The MIPS32 24K [187] is a scalar, in-order, eight-stage pipeline with a four-cycle branch misprediction penalty. A somewhat unique emphasis is placed on tolerating instruction fetch stalls, namely, the instruction cache can fetch up to two instructions each cycle into a six-entry instruction buffer that decouples the front-end and back-end of the pipeline. The instruction buffer provides a means of building up a reservoir of instructions, ensuring an uninterrupted supply of instructions to the execution pipelines in spite of potential instruction cache misses. Dual instruction fetching quickly builds up the reservoir after branch mispredictions.

Dynamic branch prediction is implemented via a 512-entry BHT (one-eighth the size of the PowerPC 440's BHT) comprised of 2-bit counters (bimodal predictor). The targets of subroutine returns are predicted via a four-entry return stack (one more entry than the ARM11's return address stack).

As with the ARM11 and PowerPC 440, separate execution and load/store pipelines allow independent instructions to execute around cache misses. The load/store pipeline implements hit-under-miss and only stalls after four outstanding cache misses.

Predictable performance for hard real-time applications in the presence of instruction and data caches is achieved in one of two ways. First, the caches support individual line locking. Second, there is robust support for software-controlled cache data movement.

The MIPS32 24K implements the MESI *cache coherence* protocol. Although it seems unusual to support coherence protocols in an embedded system, it makes complete sense in the context of a multiprocessor system-on-chip (MPSoC), in which multiple processors communicate via shared memory.

Intel Pentium

The embedded Pentium processor is a literal demonstration of our earlier observation regarding the parallel evolution of embedded and desktop processors: Intel recycles its aging desktop processor designs for its x86 embedded processors [188]. This makes good sense—whereas the original Pentium in a 0.35 to 0.8-μm technology was large and power-hungry, the same design in a 0.18-μm or 90-nm process exploits size and power scaling factors. Also, recycling previously verified designs is attractive.

Like the PowerPC 440, the embedded Pentium is an in-order dual-issue superscalar processor. The pipeline is rather shallow given its capabilities; only five stages. (Hence its peak frequency is somewhat limited.) There is one extra stage for variants that implement Intel's SIMD extensions called MMX.

Another advantage of recycling the Pentium design is that it already has an integrated floating-point pipeline. In contrast, many embedded processors, including the ARM11 and PowerPC 440, provide interfaces to separate-coprocessors. On the other hand, the integrated floating-point pipeline can sometimes be a liability—extra baggage for applications that do not require floating-point support.

There are two integer pipelines, the U-pipe and the V-pipe. The U-pipe can execute any integer instruction, whereas the V-pipe can execute only simple instructions. Up to two integer instructions (one to the U-pipe and the other to the V-pipe) or a single floating-point instruction can be issued in a single cycle. Because the pipeline is in-order and there are instruction-pairing restrictions, the compiler plays a key role in maximizing dual issue opportunities. Unlike the ARM11, instructions in the U-pipe and V-pipe may *not* complete out-of-order.

Instruction fetching is less ordinary in the Pentium than in the RISC processors. The fetch unit alternates fetching from the sequential path and the predicted path (the latter guided by the BTB), into two respective prefetch buffers (each has four 16-byte entries). We are unaware of the exact reason for multi-path fetching in the Pentium. We suspect it has to do with the complexity and latency of decoding the variable-length x86 instructions. This latency is normally exposed immediately following a branch misprediction, unless the alternate path is pre-decoded [189].

4.4.3   Ubicom IP3023: Deterministic High Performance via Multithreading

The Ubicom 8- and 32-bit embedded processors [177,190] represent an interesting design point and a glimpse of what may lie ahead for the evolution of general-purpose embedded processors. The overriding goal of Ubicom is achieving *deterministic high performance*. Normally, determinism and high performance are at odds. This trade-off is reconciled by hardware multithreading, introduced earlier in this chapter. Ubicom refers to this genre of embedded processors as multi-threaded architecture for software I/O (MASI), borrowing from the pioneering work by Burton Smith, who introduced multithreading in the HEP [172] and later the Tera supercomputer [191] and also from contemporary SMT research [171,173,174].

The architecture of the Ubicom IP3023 is the careful matching of ISA and pipeline features (for determinism) with multithreading and an awareness of embedded systems (for performance). Although there are many interesting aspects of the ISA (it is quite spartan—only 41 instructions—as it targets embedded communications), the aspect relevant to this discussion is the heavy preference toward *memory—memory instructions*, as opposed to separate load/store instructions. Memory—memory instructions use memory locations for their operands directly, instead of registers (a *complex instruction set computing* [CISC] trait, although the ISA is otherwise very RISC-like). Moreover, caches are explicitly avoided in favor of on-chip TCM. A picture begins to form, of an architecture with a very flat memory hierarchy. There are logical registers, but memory—memory instructions accessing TCM is the preferred model. A flat memory hierarchy is consistent with the goal of deterministic performance.

Although a flat memory hierarchy seems performance-limited, there are several factors that make it viable. First, TCMs are on-chip and (therefore) moderately sized—being close by and not too large allows for relatively fast access. Effectively, TCMs are like large caches, software-managed for determinism. Conventional wisdom is that software cache management is complex. However, the use of explicit memory—memory instructions eliminates this problem. (From this standpoint, it is more accurate to view TCMs as large register files instead of as large caches.) It is important to point out that the viability of on-chip memory is due to the often bounded (and transient) memory requirements of closed embedded systems (the first example of embedded systems awareness on the part of Ubicom), although this may not be universally true in the future.

The second factor that makes a flat memory hierarchy viable is hardware multithreading. Although accessing TCMs is not too cumbersome, they are comparatively slower than caches. Moreover, memory—memory instructions require more pipeline stages to support memory addressing modes. The higher latency for accessing operands extends the allowable distance between dependent instructions, i.e., data hazards cause lengthier stalls in the IP3023 than in other pipelines. This is reconciled via hardware multithreading, which enables the IP3023 to schedule independent instructions from other threads between dependent instructions. Here, we point out that embedded systems are highly concurrent—many independent threads are typically available (the second example of embedded systems awareness on the part of Ubicom).

The high-level block diagram of the IP3023 is shown in Figure 4-22. The processor provides register state for eight threads. (The ISA supports up to 32). Thus, there is zero-cycle context-switching for resident threads. The pipeline can simultaneously process instructions from all resident threads (so context-switching does not require draining the pipeline), in the same manner shown in Figure 4-17.

In keeping with the overriding goal of deterministic high-performance, static branch prediction is used in favor of dynamic branch prediction. Static predictions are conveyed by the compiler or assembler via a hint bit associated with each branch instruction. The IP3023's scalar in-order pipeline has 10 stages, as shown in Figure 4-23. In spite of a large branch misprediction penalty (seven cycles between I-Fetch 1 and Execute) and no dynamic branch prediction, high performance is maintained through moderately accurate static branch prediction

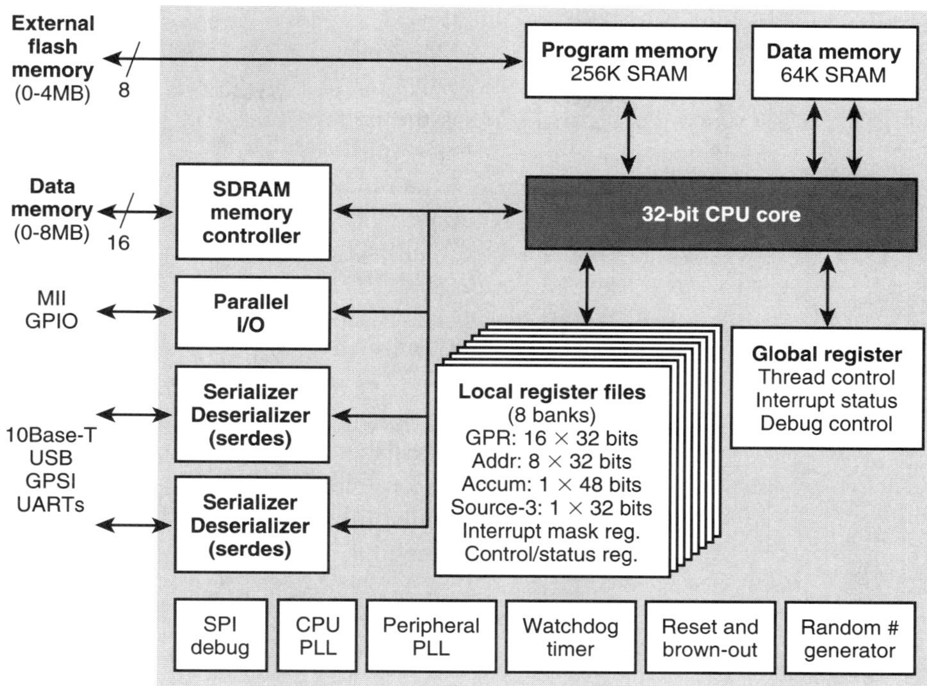

4-22   Ubicom IP3023 multithreaded embedded processor [177,190].

FIGURE

Schedule Threads	Instruction Fetch		Decode Instruction	Address Calculation	Read Memory		Execute Instruction	Store Result	
Schedule	I-Fetch 1	I-Fetch 2	Decode	Address	Read 1	Read 2	Execute	Write 1	Write 2

4-23   IP3023's 10-stage scalar in-order pipeline [177,190].

FIGURE

coupled with multithreading to reduce the apparent misprediction penalty. Mixing independent instructions from other threads with control-dependent instructions after a branch reduces the number of instructions that get squashed in the case of a misprediction.

The reason for the comparatively deep pipeline (10 stages) is evident from Figure 4-23. Instruction fetching is divided into two stages due to the access time of the 256-KB Program Memory (Fig. 4-22). Likewise, reading and writing memory operands from/to the 64-KB Data Memory each require two stages. Plus, there is an additional stage for calculating operand addresses that is not present in the prior RISC pipelines. Thus, there are five extra stages with respect to the canonical five-stage pipeline, due to thread scheduling (Schedule), a longer instruction fetch latency (I-Fetch 2), and memory—memory instructions combined with a longer memory access latency (Address, Read 2, and Writeback 2).

A potential drawback of the memory—memory ISA is the need for a highly-ported Data Memory. The architecture may not be scalable in the long-term, for example, if a dual-issue superscalar pipeline is considered. Multi-porting may partly explain why the Data Memory has fewer bytes of storage than the Program Memory. (Multi-ported memory is more area-intensive than single-ported memory.)

A potential advantage of hardware multithreading with respect to conventional multiprocessing is quick communication and synchronization among threads via the shared Data Memory. The ISA provides bit-set and bit-clear instructions for implementing synchronization, reminiscent of managing full/empty bits in the HEP machine.

The thread scheduler is guided by a static schedule contained in a 64-entry table, initialized by the programmer.

4.5   VIRTUAL SIMPLE ARCHITECTURE (VISA): INTEGRATING NONDETERMINISM WITHOUT UNDERMINING SAFETY

In the previous section, we showed how general-purpose embedded processors are evolving as the demand for performance grows. We also highlighted the tension between higher performance and the need for deterministic performance, in hard real-time systems. In this section, we frame this issue in terms of *static worst-case timing analysis* and describe how dual tracks for open and closed embedded systems can be bridged with a new microarchitecture substrate that combines deterministic and non-deterministic performance.

Worst-case execution times (WCETs) of hard real-time tasks are needed for safe planning and ensuring that the processor is never over-subscribed even under worst-case conditions. Static worst-case timing analysis tools provide a means for deriving safe (i.e., never exceeded) bounds for the WCETs of tasks on a particular pipeline. However, contemporary microarchitecture features (many of which have been covered in this chapter) exceed the capabilities of current static worst-case timing analysis tools. The most sophisticated tools to date are capable of analyzing scalar in-order pipelines with split instruction/data caches and static branch prediction [192–198]. Thus, even now, the ARM11 and PowerPC 440 (for example) are beyond the scope of worst-case timing analysis tools.

The complexity/safety trade-off in hard real-time systems is usually reconciled by either excluding contemporary processors altogether, or disabling features that cannot be analyzed for the full duration of hard real-time tasks [147,199,200].

The *virtual simple architecture* (VISA) framework offers a less conservative alternative [148]. A VISA is the pipeline timing specification for a hypothetical simple processor that can be analyzed by static worst-case timing analysis. The WCET of a task is determined using the VISA, i.e., it is assumed that the task is going to be executed on the hypothetical simple pipeline. However, this assumption is speculatively undermined at run-time by executing the task on a more complex pipeline. Since the task's execution time on the unsafe pipeline is not provably bounded by the WCET, its satisfactory progress must be continuously monitored. If progress is deemed insufficient for meeting the WCET, then the unsafe pipeline is re-configured to a simple mode of operation that directly implements the VISA, with enough time to spare for meeting the WCET. For example, dynamic branch prediction may be downgraded to static branch prediction, multiple-instruction issue may be downgraded to scalar execution, out-of-order execution may be downgraded to in-order execution, and so on. The key idea is that the decision to disable unsafe performance enhancements is deferred until they are problematic. In most cases, the WCET abstraction is enforceable without resorting to the simple mode.

Progress of the task on the unsafe pipeline is gauged by dividing the task into multiple smaller sub-tasks and assigning soft interim deadlines to each sub-task, called *checkpoints*. The checkpoint for a sub-task is based on the latest allowable completion time of the sub-task on the hypothetical simple pipeline. Thus, continued safe progress on the unsafe pipeline is confirmed for as long as sub-tasks meet their checkpoints, i.e., the unsafe pipeline is confirmed to be at least as timely as the hypothetical simple pipeline. If a sub-task misses its checkpoint, boundedness by the WCET cannot be enssured if execution continues on the unsafe pipeline. Switching to the simple mode at this point ensures that the WCET bound is met, in spite of missing the interim checkpoint.

Speculation incurs a small overhead. The WCET must be padded to accommodate a single missed checkpoint safely. When a sub-task misses its checkpoint, it is conservatively assumed that no fraction of the sub-task was completed on the unsafe pipeline, and hence the entire sub-task must be executed in the simple mode after switching (only from the standpoint of bounding WCET). Thus, the amount of speculation overhead depends on the granularity of sub-tasks (fine-grained is better than coarse-grained) and can be regulated accordingly.

In practice, non-deterministic performance enhancements accelerate periodic hard real-time tasks. However, this does not enable more periodic hard real-time tasks to be scheduled or the rates of existing ones to be increased. The reason is that the perceived *worst-case utilization* of the processor by periodic hard real-time tasks is not reduced. The VISA framework preserves the WCET abstraction, i.e., WCETs are not reduced with respect to the hypothetical simple pipeline.

Nonetheless, *actual utilization* of the processor is reduced, and the resulting dynamic slack in the schedule can be exploited for introducing other types of tasks (types that do not require advance planning, such as sporadic hard real-time tasks, periodic soft real-time tasks, and non-real-time tasks), permitting embedded systems with much more functionality. Alternatively, dynamic slack can be exploited for power and energy savings by lowering frequency/voltage. For six C-lab real-time benchmarks, a VISA-protected four-issue superscalar processor only needs to operate between 150 and 300 MHz, whereas an explicitly-safe in-order scalar processor needs to operate between 800 and 900 MHz [148]. This corresponds to power savings of 43 to 61 % without compromising overall timing safety [148].

A conventional embedded system and an embedded system based on the VISA framework are depicted side-by-side in Figure 4-24. The VISA framework allows the use of unsafe processors in safe real-time systems, and, by preserving the WCET abstraction, the framework is transparent to most aspects of embedded system design and software in particular, such as schedulability analysis, earliest deadline first (EDF) or other scheduling, dynamic voltage scaling (DVS) [201], and so on.

4.6   CONCLUSIONS

A close examination of contemporary general-purpose embedded processors reveals an underlying structure similar to (albeit simpler than) the structure of their high-performance desktop cousins. We believe this has to do with a common foundation grounded in general-purpose ISAs and the resulting confinement to generic instruction-level pipelining. Moreover, the evolution of embedded processors seems to parallel that of desktop processors, as more high-performance pipelining techniques are gradually deployed within successive generations of embedded processors.

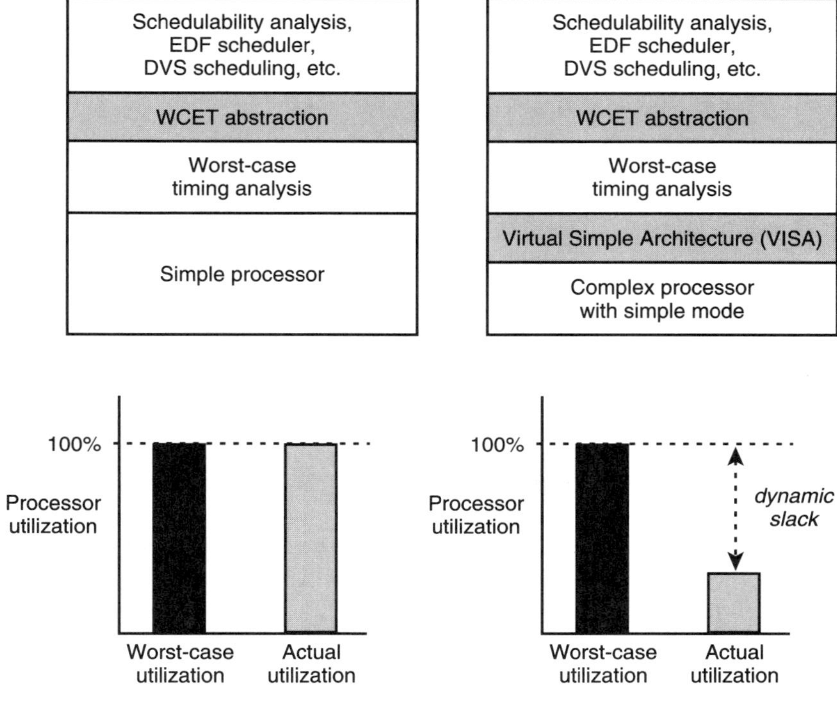

*Exploit dynamic slack for power/energy
savings, other functionality*

4-24

FIGURE

Conventional embedded system (left) and embedded system based on the VISA framework (right). The two systems have the same worst-case utilization (since tasks have the same perceived WCETs), but actual utilization is significantly reduced by the high-performance, complex processor in the VISA framework. The VISA framework enables the transparent use of unsafe processors in safe real-time systems, with key benefits such as significant power/energy savings and increased functionality.

This evolution has created a tension between higher performance and the need for deterministic performance. Currently, this tension is reconciled by dual offerings: on the one hand, unrestrained microarchitectures for open embedded systems; on the other, determinism-restrained microarchitectures for closed embedded systems. We have described an available commercial processor targeted for communication applications that revives key ISA principles and hardware multithreading to achieve deterministic high performance, within a constrained memory access model. Ironically, multithreading principles originally intended for supercomputers may ultimately find their optimal use in embedded systems, in which deterministic latency tolerance is most needed and the prerequisite abundance of threads is easily satisfied.

We concluded by describing a general framework for combining non-deterministic and deterministic performance through a VISA. The VISA framework shields worst-case timing analysis from the non-deterministic component, thereby extending the capabilities of real-time systems without undermining timing safety. VISA protection allows the use of unsafe processors in safe real-time systems, with key benefits such as major power/energy savings and increased functionality.

ACKNOWLEDGMENTS

This research was supported in part by NSF grants CCR-0207785, CCR-0208581, and CCR-0310860, NSF CAREER grant CCR-0092832, and generous funding and equipment donations from Intel.

Performance and Flexibility for Multiple-Processor SoC Design

Chris Rowen

5.1   INTRODUCTION

The rapid evolution of silicon technology is bringing a new crisis to system-on-chip (SoC) design. To be competitive, new communication, consumer, and computer product designs must exhibit rapid increases in functionality, reliability, and bandwidth and rapid declines in cost and power consumption. All these improvements dictate increasing use of high-integration silicon, for which many of the data-intensive capabilities are presently realized with register-transfer-level (RTL) hardware-design techniques. At the same time, the design productivity gap, the growing cost of deep-sub-micron semiconductor manufacturing, and the time-to-market pressures of a global electronics market all put intense pressure on chip designers to develop increasingly complex hardware in decreasing amounts of time.

One way to speed up the development of mega-gate SoCs is the use of multiple microprocessor cores to perform much of the processing currently relegated to RTL. Although general-purpose embedded processors can handle many tasks, they often lack the bandwidth needed to perform particularly complex processing tasks such as audio and video media processing. Hence the historic rise of RTL use in SoC design. A new class of processor, the extensible processor, can be configured to bring the required amount and type of processing bandwidth to bear on many embedded tasks. Because these processors employ firmware for their control algorithm instead of RTL-defined hardware, it is easier and faster to develop and verify processor-based task engines for many embedded SoC tasks than it is to develop and verify RTL-based hardware blocks to perform the same tasks.

A few characteristics of typical deep-sub-micron integrated circuit (IC) design illustrate the challenges facing SoC design teams:

✦ In a generic, 130-nm standard-cell foundry process, silicon density routinely exceeds 100 K usable gates per mm^2. Consequently, a low-cost chip (50 mm^2 of core area) can carry 5 M gates of logic today. Simply because it's possible, a system designer somewhere will find a way to exploit this potential in any given market.

✦ In the past, silicon capacity and design-automation tools limited the practical size of a block of RTL to smaller than 100-K gates. Improved synthesis, place-and-route, and verification tools are raising that ceiling. Blocks of 500-K gates are now within the capacity of these tools, but existing design and verification methods are not keeping pace with silicon fabrication capacity, which can now put millions of gates on an SoC.

✦ The design complexity of a typical logic block grows much more rapidly than does its gate count, and system complexity increases much more rapidly than the number of constituent blocks. Similarly, verification complexity has grown disproportionately. Many teams have reported that they now spend as much as 70% of their development effort on block- or system-level verification [205].

✦ The cost of a design bug is going up. Much is made of the rising cost of deep-sub-micron IC masks—the cost of a full 130-nm mask set is approaching $1M, and 90-nm masks may reach $2M [206]. However, mask costs are just the tip of the iceberg. The combination of larger teams required by complex SoC designs, higher staff costs, bigger non-recurring engineering (NRE) fees, and lost profitability and market share makes show-stopper design bugs intolerable. Design methods that reduce the occurrence of, or permit painless work-arounds for, such show stoppers pay for themselves rapidly.

✦ All embedded systems now contain significant amounts of software. Software integration is typically the last step in the system-development process and routinely gets blamed for overall program delays. Earlier and faster hardware/software validation is widely viewed as a critical risk reducer for new product-development projects.

✦ Standard communication protocols are growing rapidly in complexity. The need to conserve scarce communications spectra, plus the inventiveness of modern protocol designers, has resulted in the creation of complex new standards (for example, IPv6 Internet Protocol packet forwarding, G.729 voice coding, JPEG2000 image compression, MPEG4 video, Rjindael AES encryption). These new protocol standards demand much greater computational throughput than their predecessors. These new standards also require new, more flexible implementation methods compared with earlier protocols in equivalent roles (for example, IPv4 versus IPv6 packet forwarding; G.711

versus G.729 voice coding; JPEG, MPEG2, and DES coding and encryption versus no coding or encryption).

In most markets, competitive forces drive the ever-increasing need to embrace new technologies. Gordon Moore's observation and prophesy that silicon density would double roughly every 18 months sets a grueling pace for all chip developers in the electronics markets. The universal expectation for ever cheaper, faster transistors also invites customers to expect and demand constant improvements in functionality, battery life, throughput, and cost from the systems they buy.

Generally speaking, the moment a new function is technically feasible, the race is on to deliver it. The competition is literally that intense in many markets. In many volume markets, expectations for functionality are increasingly set by the economics of silicon. Capabilities expand to consume the entire available electronics budget, or perhaps even a little more than that. Just one CMOS process step, say from 180 to 130 nm roughly doubles the available number of gates for a given die size and cost.

In the last 5 years—the period in which the SoC has become a foundation technology for chip designers—we've seen silicon capacity increase roughly by a factor of 10. The International Technology Roadmap for Semiconductors forecasts a slight slowing in the pace of density increases, but exponential capacity increases are expected to continue for at least the next decade, as shown in Figure 5-1.

Competitive pressure has pushed us to mega-gate SoCs, characterized by dozens of functions working together. The trend toward the use of large numbers of RTL-based logic blocks and the mixing together of control processors and digital signal processors on the same chip is illustrated in Figure 5-2.

This ceaseless growth in IC complexity is a central dilemma for SoC design. If all these logic functions could be implemented with multiple, heterogeneous processor blocks that were cheap, fast, and efficient enough, a processor-centric design approach would be ideal. Using pre-designed and pre-verified processor cores for the individual functional blocks in an SoC would move the SoC design effort largely to software. This design-methodology change would permit bug fixes in minutes instead of months because software is much easier to change and verify than hardware. This change in design technique would also allow for the easy addition of new features at any time during product development, even after the product has been fielded because of the software's flexibility. Unfortunately, general-purpose processors fall far short of the mark with respect to application throughput, cost, and power efficiency for the most computationally demanding problems.

At the same time, designing custom RTL logic for these new, complex functions or emerging standards takes too long and produces designs that are too rigid to change easily. A closer look at the makeup of the typical RTL block in Figure 5-3 gives insights into this paradox.

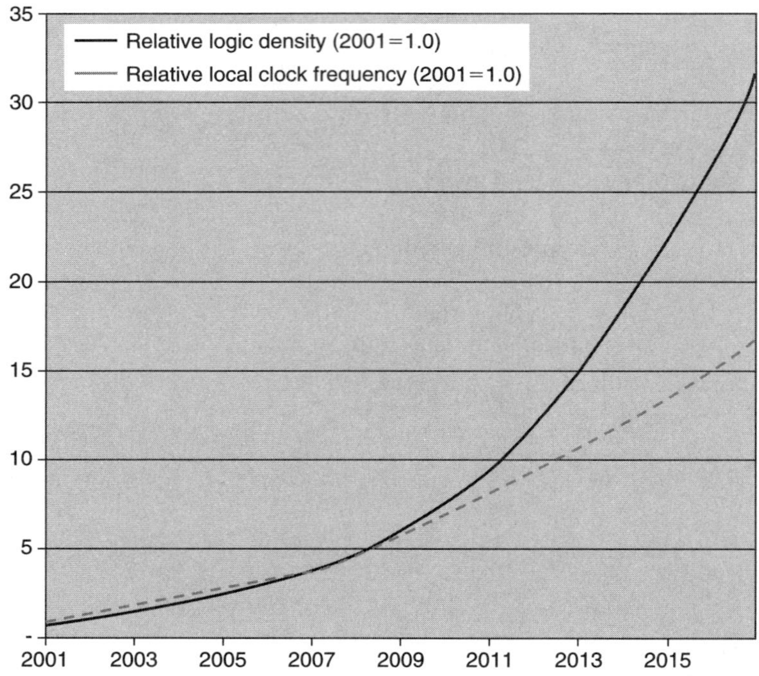

5-1

FIGURE

Silicon density and speed increase (ITRS'01).

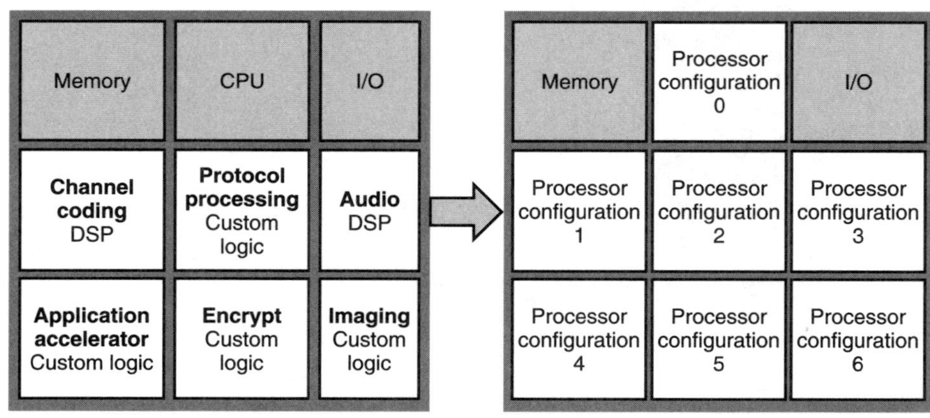

5-2

FIGURE

Complex SoC structure and transition.

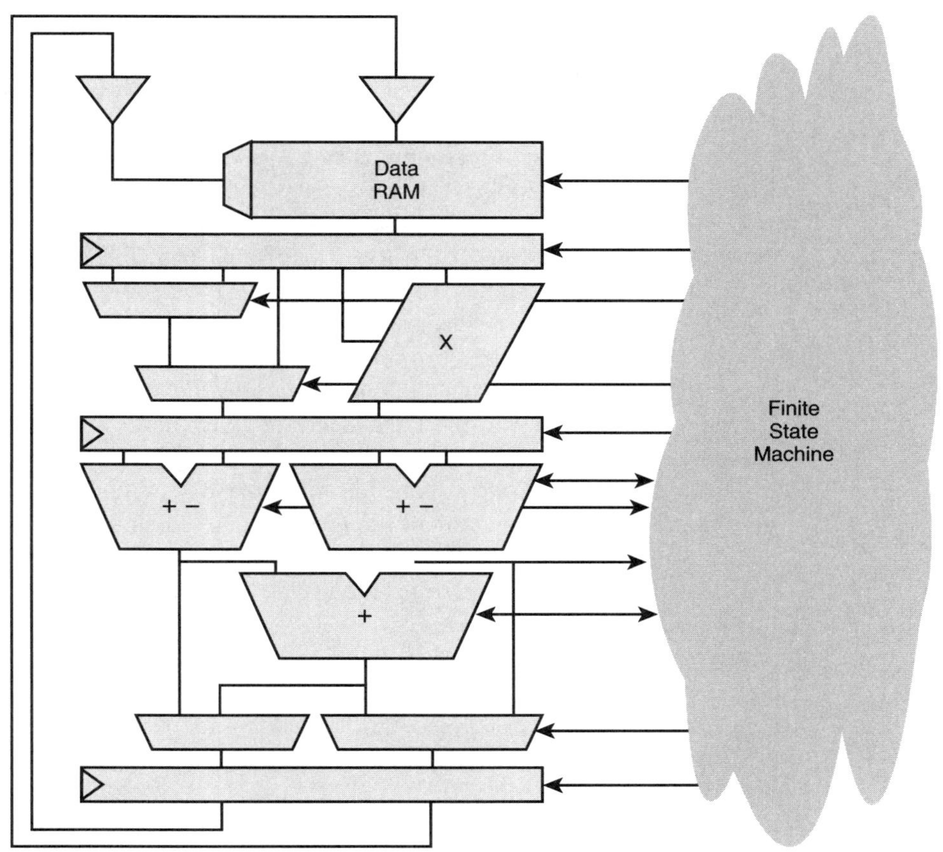

5-3   Hardwired RTL function: datapath + state machine.

FIGURE

In most RTL designs, the datapath consumes the vast majority of the gates in the logic block. A typical datapath may be as narrow as 16 or 32 bits or may be hundreds of bits wide. The datapath will typically contain many data registers, representing intermediate computational states, and will often have significant blocks of RAM or interfaces to RAM that are shared with other RTL blocks. The choice of datapath elements and the connections among these elements often directly reflect the fundamental data types on which the block operates. For example, a packet-processing block will probably employ a datapath that closely corresponds to the packet header's structure. An image-processing block might implement a datapath for a row or column of 8 pixels from an 8-by-8-pixel image block.

These basic datapath structures reflect the nature of the data and are largely independent of the finer details of the specific algorithm operating on that data. By contrast, the RTL logic block's finite state machine contains nothing but control details. All the nuances of the sequencing of data through the datapath, all the exception and error conditions, and all the handshakes with other blocks are captured in this subsystem of the RTL block. This state machine may consume only a few percent of the block's gate count, but it embodies most of the design and verification risk due to its complexity. If a late design change is made in an RTL block, the change is much more likely to affect the state machine than the structure of the datapath. This situation heightens the design risk.

One way to understand the risks associated with hardware state machines is to examine the combinatorial complexity of verification. A state machine with N states and I inputs may have as many as N^2 next-state equations, and each of these equations will be some function of the I inputs, or 2^I possible input combinations. Taken together, at least N^2*2^I input combinations must be tried to test all the state transitions of this state machine exhaustively.

Typically, it will take many cycles to test each such transition because a number of cycles may be required to reach a desired state and an additional number of cycles to determine whether the desired transition occurred. For a relatively simple state machine with 40 states and 20 inputs, truly exhaustive testing probably takes tens of billions of cycles requiring months of slow RTL simulations.

As a consequence, logic is rarely tested completely except with formal verification methods, but the basic problem remains: hardware-based state machines are hard to design and even harder to verify, yet they're required for RTL-based design. *Configurable, extensible processors*—a fundamentally new form of microprocessor—provide a way of reducing the risk of state-machine design by replacing hard-to-design, hard-to-verify state-machine logic blocks with pre-designed, pre-verified processor cores and application firmware.

5.2 THE LIMITATIONS OF TRADITIONAL ASIC DESIGN

As discussed, logic design and verification consume much of the SoC design effort. New chips are characterized by rapidly increasing logic complexity. Moore's-law scaling of silicon density makes multi-million-gate designs feasible. Fierce product competition in system features and capabilities makes these advanced silicon designs necessary. The well-recognized design gap widens every year [207].

Moreover, the market trend toward high-performance, low-power systems (long-battery-life cell-phones; 5-Mpixel digital cameras; fast, inexpensive color printers; high-definition digital televisions; and 3D video games) is also increasing the number of SoC designs. Unless something closes the design gap, it will become impossible to bring the enhanced, future versions of these system designs to market.

The low level of design expression (hardware data-path structures and finite-state machines) and the arduousness of verification limit the scalability of RTL-based design. If the task function and interface are absolutely stable, reuse is feasible. In addition, the widespread acceptance of synthesizability gives fixed functions easier fabrication process portability. When requirements change, however, especially when new modes and features must be added, RTL-level designs may not scale well, particularly if the original design and verification team is not available to do the redesign. RTL-based designs therefore have a limited shelf life, and a high upgrade cost, in the face of changing applications requirements.

The conventional SoC-design model closely follows the tradition of its predecessor: combining a standard microprocessor, standard memory, and RTL-built logic into an application-specific instruction set processor (ASIC). In fact, many SoC designs are philosophical descendants of earlier board-level designs. At the board level, chip-to-chip interconnect is expensive and slow, so shared busses and narrow datapaths (often only 32 bits wide) are typical of board-level designs. These relatively limited busses and narrow datapaths are often carried over to SoC designs simply because this organization is familiar to embedded systems architects, but they represent an unnecessary restriction on SoC designs.

Most commonly, the processors used for these board-level designs are general-purpose reduced instruction set computing (RISC) processors originally designed in the 1980s for general-purpose UNIX desktops and servers. Separate hardwired (non-programmable) ASICs implement all the logic functions that lie beyond the performance limit of the chosen general-purpose processor.

When all system components are combined on a single piece of silicon, clock frequency increases and power dissipation decreases relative to the equivalent board-level design. System reliability and cost often improve as well. These benefits alone can justify the investment in an SoC. However, the shift to SoC integration does not automatically change a design's organization or architecture. The architecture of these chips typically inherits the limitations and design trade-offs of board-level design. SoC architectures that are cloned from board-level designs are often organized around one or two 32-bit busses (often a fast memory bus, plus a slow peripheral bus) because this approach saves pins—an expensive commodity in a board-level design but much less relevant to an SoC's potential on-chip connections.

Architecturally cloned system designs often retain the rigid partitioning between a single microprocessor performing supervision tasks—running user-interfaces, real-time operating systems, and high-level user tasks—and a set of logic blocks, each dedicated to a single function such as data transformation, protocol processing, image manipulation, or other data-intensive tasks.

Architectures derived from board-level design concepts typically assume that the communications between the various logic blocks, and between logic blocks and the processor, are bottlenecks plagued with the long latency, low clock frequency, and narrow data-path width of board-level, chip-to-chip connections. The typical data bandwidth between ASIC-based logic chips and the control processor rarely reaches 100 MB per second in board-level designs, and the aggregate bandwidth among all logic functions rarely exceeds a few hundred MB per second.

5.2.1   The Impact of SoC Integration

Ironically, bus bottlenecks commonly disappear in SoC designs. Although a 64-bit bus might be prohibitively expensive for a board-level design, 128- and 256-bit connections running at hundreds of megahertz are easy to design into SoCs. Wide busses are efficient and appropriate to use between adjoining SoC logic blocks. The communications bandwidth between a processor and surrounding logic can exceed 1 GB per second on an SoC using these wider busses. Moreover, integrated circuits potentially offer much higher aggregate bandwidth. With deep-sub-micron line widths and six or more layers of metal interconnect, the theoretical cross-section bandwidth of a die 10 mm on a side can approach 10^{13} bits/sec (10 terabits/sec). Although few practical SoC designs will even approach this limit, wide on-chip busses create tremendous architectural headroom and invite a new, more effective approach to system architecture.

5.2.2   The Limitations of General-Purpose Processors

The traditional approach to SoC design is further constrained by the origins and evolution of microprocessors. Most popular embedded microprocessors, especially 32-bit architectures, are direct descendants of desktop computer architectures of the 1980s: ARM, MIPS, 68000/ColdFire, PowerPC, x86, and so forth. These processors were designed to serve general-purpose applications and were structured for implementation as stand-alone integrated circuits. Consequently, these processors typically support only the most generic data types (8-, 16-, and 32-bit integers) and operations (integer load, store, add, shift, compare, logical-AND, and so on).

The general-purpose nature of these processors makes them well suited to the extremely diverse mix of applications run on computer systems. These processor architectures are equally good (or equally bad) at running databases, spreadsheets, PC games, and desktop publishing. However, all these processors suffer from a common bottleneck: their need for complete generality dictates an ability to execute an arbitrary sequence of primitive instructions on an unknown range of data types. Even the most silicon-intensive, deeply pipelined, super-scalar, general-purpose processors can rarely sustain much more than two instructions per cycle (IPC), and the harder processor designers push against this IPC limit, the higher the cost and power per unit of useful performance extracted from the microprocessor architecture. Hence the rush to gigahertz clock rates to achieve performance improvements in this class of processors.

Compared with general-purpose computer systems, embedded systems are more diverse as a group but are individually more specialized. A digital camera may perform a variety of complex image processing but it never executes standard query language (SQL) database queries. A network switch must handle complex communications protocols at optical interconnect speeds but doesn't need to process 3D graphics. A consumer audio device may perform complex media processing but doesn't run payroll applications.

The specialized nature of individual embedded applications creates two issues for general-purpose processors in data-intensive embedded applications. First, there is a poor match between the critical functions of many embedded applications (e.g., image, audio, protocol processing) and a RISC processor's basic integer instruction set and register file. As a result of this mismatch, critical embedded applications require more computation cycles when run on general-purpose processors. Second, the more focused embedded devices cannot take full advantage of all of a general-purpose processor's broad capabilities, so expensive silicon resources built into the processor go to waste because they're not needed by the specific embedded task that's assigned to the processor.

A large number of embedded systems interact closely with high-speed, real-world events or communicate complex data at high rates. These data-intensive tasks could be performed by some hypothetical general-purpose microprocessor running at tremendous speed. For many tasks, however, no such processor exists today as a practical alternative. The fastest available processors typically cost orders of magnitude too much and dissipate orders of magnitude too much power to meet embedded system-design goals. Instead, designers have traditionally turned to hard-wired circuits to perform these data-intensive functions such as image manipulation, protocol processing, signal compression, encryption, and so on.

5.2.3 DSP as Application-Specific Processor

Digital signal processors (DSPs) have emerged as an important class of *application-specific processors*. Their instruction sets and memory systems are organized around a particular computational pattern: large numbers of multiply-add operations on uniform blocks of sampled data (most commonly 16-bit). DSPs are often used in tandem with RISC controllers on SoCs, especially when the end application calls for a mix of control and signal processing. Interestingly, the DSPs have evolved along similar paths as control processors, gaining more general-purpose processing features, better software tools, larger silicon area, and consequently higher power dissipation (adjusting for technology).

The emergence of complex very long instruction word (VLIW) DSPs such as Texas Instruments C6000 family and the StarCore architecture reflect this "quest for generality." Ironically, this evolution means that many of the most performance- and efficiency-intensive signal-processing tasks cannot be performed in DSPs. Most of the key channel-coding work in consumer communication devices (e.g., CDMA cell phone, DSL, and 802.11 modems) is performed by RTL logic, not general-purpose DSPs today. In many cases a programmable DSP would be attractive, but only if it could be sufficiently fast in the application to rival RTL performance.

In the past 10 years, the wide availability of logic synthesis and ASIC design tools has made RTL design the standard for hardware developers. RTL-based design is reasonably efficient (compared with custom, transistor-level circuit design) and can effectively exploit the intrinsic parallelism of many data-intensive problems. If similar operations are required on a whole block of data, RTL design methods—specialized operations, pipelining, replicated operators in a single data-path, replicated function units—can often achieve tens or hundreds of times the performance achieved by a general-purpose processor. Because they are not attempts to solve application-arbitrary sequential problems, RTL designs avoid the general-purpose, single-processor performance bottlenecks. The more computationally intensive the data manipulation, the greater the potential RTL speed-up over general-purpose processor-based design approaches.

5.3 EXTENSIBLE PROCESSORS AS AN ALTERNATIVE TO RTL

Hardwired RTL design has many attractive characteristics: small area, low power, and high throughput. However, the liabilities of RTL—difficult design, slow verification, and poor scalability to complex problems—are starting to dominate. A design methodology that retains most of the efficiency benefits of RTL but with reduced design time and risk has a natural appeal. Application-specific processors as a replacement for complex RTL fit this need.

5.3.1   The Origins of Configurable Processors

The concept of building new processors quickly to fit specific needs is almost as old as the microprocessor itself. Architects have long recognized that high development costs, particularly for design verification and software infrastructure, forces compromises in processor design. A processor had to be "a jack of all trades, master of none."

Early work in building flexible processor simulators—especially ISP [208] in 1970; in retargetable compilers, especially the GNU C compiler, in the late 1980s [209]; and in synthesizable processors in the 1990s—stimulated hope that the cost of developing new processor variants might be dramatically reduced. Research in application-specific instruction processors (ASIPs), especially in Europe (code generation at IMEC [210], processor specification at the University of Dortmund [211], micro-code engines ["transport-triggered architectures"] at the Technical University of Delft [212], and fast simulation at the University of Aachen [213]) all confirmed the possibility of developing a fully automated system for designing processors. However, none of these projects demonstrated complete microprocessor hardware design and software environment generation (compiler, simulator, debugger) from a common processor description.

5.3.2   Configurable, Extensible Processors

Like RTL-based design using logic synthesis, extensible-processor technology allows the design of high-speed logic blocks tailored to the assigned task. The key difference is that RTL state machines are realized in hardware, and logic blocks based on extensible processors realize their state machines with firmware. The tools for creating extensible processors allow the automatic generation of new application-specific processor hardware design and software tools from one high-level description. Automating hardware and software generation guarantees consistency among all the representations of the processor definition—the hardware design, test-benches, simulation models, compilers, assemblers, debuggers, and real-time operating systems. All these software-development tools are built for exactly the same architecture by the processor generator from the same definition used to build the processor itself.

This approach to configuring processors eliminates the "Tower of Babel" that surrounds many embedded processor environments today, whereby subtle differences among members of a processor family—differences in instruction set variations, memory system organization, debug facilities, and processor control structures—frustrate the system designer putting together the hardware, verification, and software tools. By generating the processor from a high-level description, the system designer controls all the relevant cost, performance, and functional attributes of the processor subsystem without having to become a microprocessor design expert.

This design approach effectively opens up processor design to a very broad population of system and application architects, just as the proliferation of ASICs and logic-synthesis tools democratized IC design in the previous decade. System developers may implement configurable and extensible processors using different hardware forms ranging from ASICs (with design cycles of many weeks) to field-programmable gate arrays (FPGAs; with implementation cycles measured in minutes).

The four key questions for the use of configurable and extensible processors in SoCs are these:

1. What target characteristics of the processor can be configured and extended?

2. How does the system designer capture the target characteristics?

3. What are the deliverables—the hardware and software components—to the system designer?

4. What are typical results for building new platforms to address emerging communications and consumer applications?

A fully featured configurable and extensible processor consists of a processor design and the design tool environment that permits significant adaptation of that base processor design by allowing a system designer to change major processor functions, thus tuning the processor to specific application requirements. Typical forms of *configurability* include additions, deletions, and modifications to memories, to external bus widths and handshake protocols, and to commonly used processor peripherals. To be useful for practical SoC development, configuration of the processor must meet two important criteria:

1. The configuration mechanism must accelerate and simplify the creation of useful configurations.

2. The generated processor must include complete hardware descriptions (typically synthesizable RTL descriptions in Verilog or VHDL), software

development tools (compilers, debuggers, assemblers, and real-time operating systems), and verification aids (simulation models, diagnostics, and test support).

An important superset of configurable processors is the extensible processor, a processor whose functions—especially its instruction set—can be extended by the application developer to include features never considered or imagined by designers of the original processor. The key to *extensibility* is the ability to expand instruction-set definitions, register sets, and execution pipelines. Automated software support for extended processor features is especially challenging, however.

A range of extensible or configurable processors is now widely available. Configurable products can be roughly categorized into five groups:

1. **Non-architectural processor configuration**: features of the processor that are invisible to software tools and applications are selectable or modifiable. Many synthesizable processors, such as ARM1026EJ-S, support variable cache sizes. This form of extension gives little leverage on application performance because the processor instruction set is untouched.

2. **Fixed menu of processor architecture configurations**: a preset range of features of the processor visible to software that can be enabled and disabled. Hardware design and software tools are configured in parallel, sometimes from the same user interface. ARC International's ARChitect 2 GUI allows interactive selection of instructions and groups of instructions [214].

3. **User-modifiable processor RTL**: the processor design supports a hardware interface for addition and/or modification of instructions. New operations on an existing processor state and new processor states can be added to the RTL manually. This approach generally precludes software support of the new instructions by the compiler, assembler, simulator, debugger, and/or real-time operating system because the hardware extensions are created by hand so there is no automatic support for adding knowledge of these extensions to the software tools. This approach also raises serious questions of processor correctness after manual RTL modification, especially for more complex, pipelined extensions. Consequently, such manual extensions must be carefully tested through additional verification. MIPS Technologies' "User Defined Instructions" and ARC International's "custom extensions" both fit into this group.

4. **Processor extension using an instruction-set description language**: an automated processor-generation tool that uses a high-level description of instructions and other processor extensions (processor state, operands, and

operation semantics are all expressed above the RTL level) to build automatically both the processor hardware and the corresponding software tools. Research at HP Labs, including the program-in, chip-out (PICO) [215] and Lx [216] projects, looked at application-directed configuration of non-programmable, VLIW-based processor accelerators. The *Tensilica Instruction Extension* (TIE) language and Tensilica's *Xtensa* Processor Generator also fit this model [217].

5. **Fully automated processor synthesis**: a compiler-based processor generator analyzes a set of profiled applications and creates an optimized instruction set to reach application performance and silicon cost goals. Tensilica's automated processor-development technology fits this category.

A configurable processor can implement data-path operations that closely match those of RTL functions. The logical equivalent of the RTL datapaths are implemented using the integer pipeline of the base processor and additional execution units, registers, and other functions added by the chip architect for a specific application. As an example of how such a processor is defined, the TIE language is optimized for high-level specification of data-path functions, in the form of instruction semantics and encoding. This form of description is much more concise than RTL because it omits all sequential logic, including state-machine descriptions, pipeline registers, and initialization sequences.

The new processor instructions and registers described in TIE are available to the firmware programmer via the same compiler and assembler that handle the processor's base instructions and register set. All sequencing of operations within the processor's datapaths is controlled by firmware, through the processor's existing instruction-fetch, decode, and execution mechanisms and are typically written in a high-level language such as C or C++.

Extended processors used as RTL-block replacements routinely use the same structures as traditional data-path-intensive RTL blocks: deep pipelines, parallel execution units, problem-specific state registers, and wide datapaths to local and global memories. These extended processors can sustain the same high computation throughput and support the same low-level data interfaces as typical RTL designs. The control of the extended-processor datapaths is very different, however. Cycle-by-cycle control of the processor's datapaths is not fixed in state transitions hardwired into the logic. Instead, the sequence of operations is explicit in the firmware executed by the processor. Control-flow decisions are made explicitly in branches, memory references are explicit in load and store operations, and sequences of computations are explicit sequences of general-purpose and application-specific computational operations. The vocabulary of the

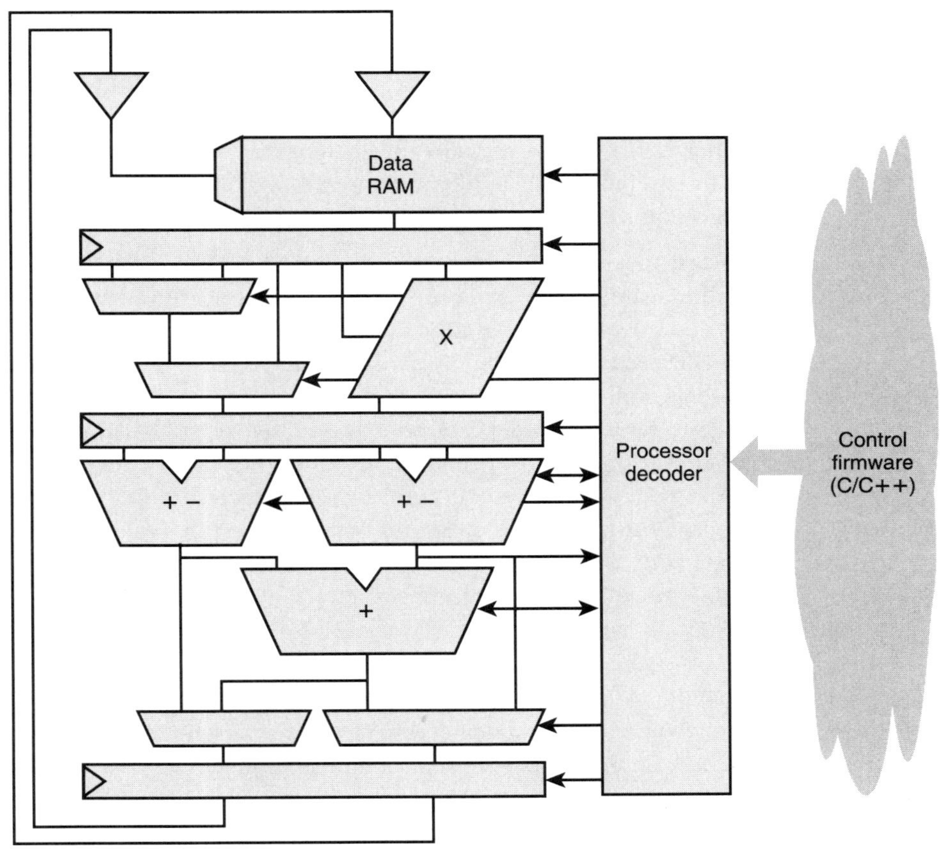

5-4   Programmable function: datapath + processor + software.

FIGURE

processor-based logic block is set by the processor's datapath and expressed as the processor instruction set, but the firmware program determines the overall function of the block, as shown in Figure 5-4.

This design migration from hardwired state machine to firmware program control has important implications:

1. **Flexibility.** Chip developers, system builders, and end-users (when appropriate) can change the block's function just by changing the firmware. New functions, bug fixes, and performance upgrades can potentially be made at any time, including the period after a product has been fielded.

2. **Software-based development.** Developers can use sophisticated, low-cost software development methods for implementing most chip features. PC-based native code development and source-level debug with graphical user interfaces make development, enhancement, and maintenance of functions significantly easier. Developers routinely report, for example, that software-style, real-time debug tools give much higher visibility and allow much faster bug fixes than typical hardware simulation and verification methods.

3. **Faster, more complete system modeling.** RTL simulation is slow. For a 10-million-gate design, even the fastest software-based logic simulator may not exceed a few cycles per second. Only the most basic system verification can be achieved. Simulation of realistic datasets is impossible. Even mixed-mode models suffer from RTL simulation bottlenecks. By contrast, firmware simulations for extended processors, including cycle- and bit-exact support for processor extensions, run at hundreds of thousands of cycles per second, per processor, so the simulation of even a complex multiple-processor SoC will typically run at least a thousand times faster than simulation of the RTL gate-level equivalent. Instruction-set simulators are often encapsulated in system-level modeling environments such as SystemC. These C-based modeling protocols can be used for transaction-level modeling for blocks destined for RTL-based implementation. Although high-level RTL block modeling may accelerate system-level simulation, development of the C model takes extra effort, both for initial development and for functional verification against the RTL. Even then, a block's transaction-level model rarely reproduces the cycle-exact behavior of the RTL. By contrast, fast processor simulators can fully reproduce the pin-level, cycle-exact behavior of RTL.

4. **Unification of control and data.** No modern system consists solely of hard-wired logic. There is always a processor and some software in the system—or on the chip for SoCs—although the processor may be restricted to simple initialization, the user-interface, or error handling. Moving functions previously handled by RTL into a processor removes the artificial distinction between control and data processing. This unification typically simplifies SoC development and eliminates unnecessary interface hardware, driver software, and communication bottlenecks.

5. **Time-to-market.** Moving critical functions from RTL to application-specific processors simplifies the SoC design, accelerates system modeling, and pulls in finalization of hardware. This design approach easily accommodates changes to standards, because implementation details do not get "cast in stone" but are instead implemented in run-time firmware. For the traditional RTL flow, IC prototypes cannot be built until every detail of every logic

function has been designed and verified. Almost any error, even in rare state transitions, mandates a new prototype run, incurring million-dollar costs and months of project delay. When an extended processor implements the critical data functions, the hardware prototypes can be safely built as soon as the basic data-path functions are fixed. All control functions can be developed in firmware using a simulator or hardware emulator while a hardware prototype is being fabricated or directly on the prototype hardware as soon as it is ready.

Most importantly, perhaps, migration from RTL-based design to the use of application-specific processors boosts the engineering team's productivity by reducing the engineering workforce needed both for RTL development and for verification. A processor-centric SoC design approach sharply cuts risks of fatal logic bugs and permits graceful recovery when (usually not if) a bug is discovered. This approach accelerates the start of firmware development and reduces the uncertainty typically surrounding hardware—software integration. It frees up resources for the more highly leveraged work of inventing new functions, developing improved system-level architectures, and supporting new customers and new markets instead of the low-level work of grinding out RTL code. The more complex the RTL task, the greater the impact of the switch to SoC design based on the use of application-specific processors.

A couple of caveats should be noted, however. For all the attractive properties of application-specific processors, they are not always the right choice. Two cases where application-specific processors are not the right choice for a block's design stand out:

✦ **Small, fixed-state machines:** some logic tasks are too trivial to warrant a processor. A state machine with just a handful of states, no data storage, no computation, and little risk of bugs or changes could, of course, be implemented as short sequence of processor instructions. However, if the function requires fewer than a few dozen states, or fewer than a few thousand total gates, a processor would plainly be inefficient. For example, bit-serial engines such as simple universal asynchronous receiver transmitters (UARTs) fall into this category.

✦ **Simple data buffering:** similarly, some logic tasks amount to no more than storage control. A first-in first-out (FIFO) controller built with a RAM and some wrapper logic can be emulated via memory operations within a processor, but a simple FIFO is faster and simpler, and FIFOs are commonly available in standard design libraries.

The migration of functions from software to hard-wired logic over time is well known. During early design exploration of pre-release protocol standards, processor-based implementations are common even for simple standards that eventually use efficient logic-only implementations. Some common protocol standards that have followed this path include MPEG2 video codecs, G.711 voice compression, and DES encryption. Data-encoding and -communication standards are becoming more complex, driven in part by demand for new modes and features, and driven in part by demand for denser encoding to preserve transmission bandwidth. These standards are increasingly complex, for example, MPEG4 is replacing MPEG2 for video encoding, G.729 is often replacing G.711 for voice compression, and AES is replacing DES for data encryption. Some of these newer standards may still be implementable as hardwired logic, but the initial development is more expensive and the risk of "feature-creep" is higher with the more complex algorithms.

The emergence of configurable and extensible application-specific processors creates a new design path that is quick and easy enough for the development and refinement of new protocols and standards yet efficient enough in silicon area and power to permit very high volume deployment.

5.3.3   Configurable and Extensible Processor Features

The end goals for processor configurability and extensibility are to allow features to be added or adapted in any form that optimizes the processor's cost, power, and application performance. In practice, configurability and extensibility can be broken into four categories, as shown in Figure 5-5.

As an example of a configurable processor, a block diagram for Tensilica's Xtensa processor is shown in Figure 5-6. This figure identifies the processor's baseline instruction-set architecture (ISA) features, scaleable memories and interfaces, optional and configurable processor peripherals, selectable DSP coprocessors, and facilities to integrate designer-defined instruction set extensions [218]. The basic flow for generating an Xtensa processor is shown in Figure 5-7.

The chip or system designer (the application expert) uses the web-based Xtensa generator interface to select or describe the instruction-set options, memory hierarchy, closely-coupled peripherals, and external interfaces required by the application. The Xtensa processor generator then produces both the complete synthesizable hardware design and the software environment in less than an hour. The processor hardware can be immediately integrated into the balance of the SoC or ASIC design. It is easily and legally portable to any semiconductor manufacturing process or fab, ensuring optimal performance and silicon leverage.

Instruction Set	Memory System	Interface	Processor Peripherals
• Extensions to ALU functions using general registers (e.g. population count instruction) • Coprocessors supporting application-specific data types (e.g. network packets, pixel blocks), including new registers and register files • Wide instruction formats with multiple independent operation slots per instruction • High-performance arithmetic and DSP (e.g. compound DSP instructions, vector/SIMD, floating point), often with wide execution units and registers • Selection among function unit implementations (e.g. small iterative multiplier vs. pipelined array multiplier)	• Instruction-cache size, associativity, and line size • Data-cache size, associativity, line size, and write policy • Memory protection and translation (by segment, by page) • Instruction and data RAM/ROM size and address range • Mapping of special-purpose memories (queues, multiported memories) into the address space of the processor • Slave access to local memories by external processors and DMA engines	• External bus interface width, protocol, and address decoding • Direct connection of system control registers to internal registers and data ports • Arbitrary-width wire interfaces mapped into instructions • Queue interfaces among processors or between processors and external logic functions • State-visibility trace ports and JTAG-based debug ports	• Timers • Interrupt controller: number, priority, type, fast switching registers • Exception vectors addresses • Remote debug and breakpoint controls

5-5

Types of processor configurability and extensibility.

FIGURE

Software development and tuning can also start immediately using the simultaneously created compiler, instruction-set simulator, and related firmware-development tools. With the integrated software profiler and 1-hour turnaround for the software tools and RTOS, the system designer can realistically tune a processor to fit a specific SoC application using this development approach.

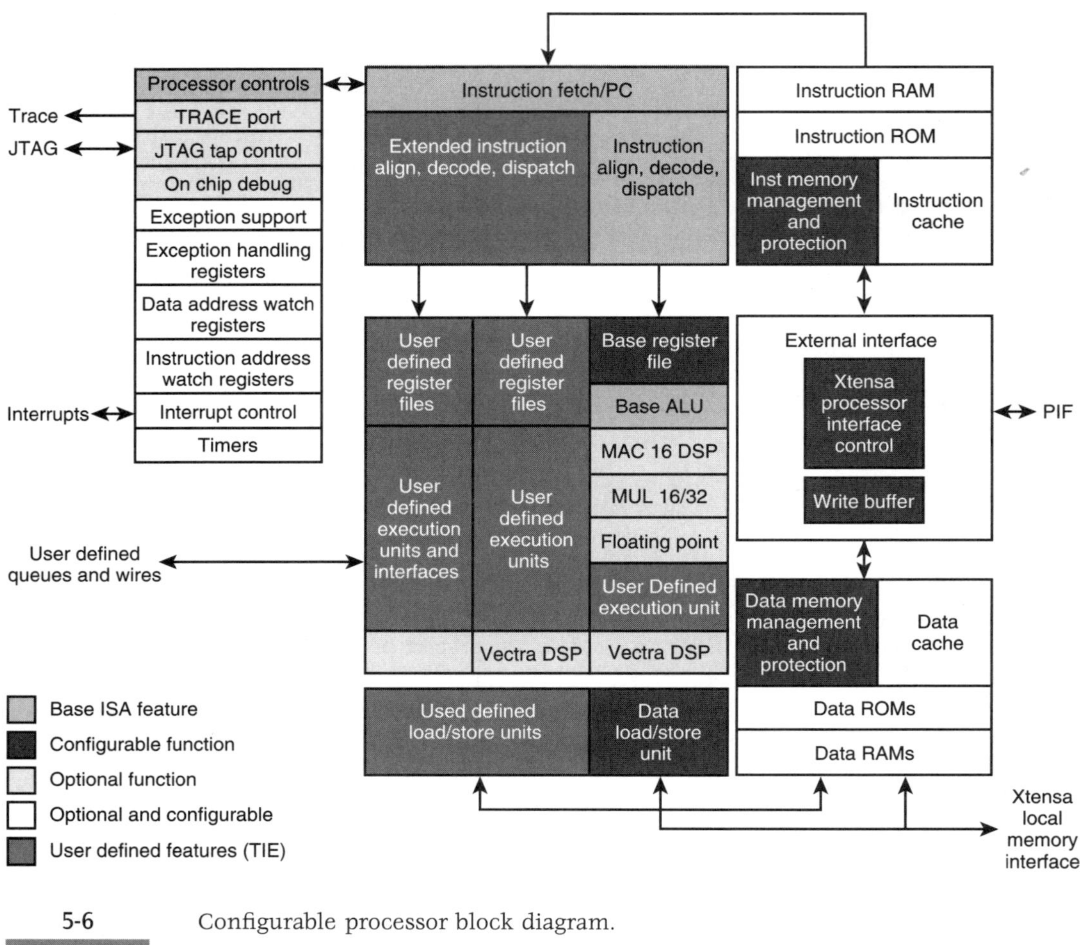

5-6   Configurable processor block diagram.

FIGURE

5.3.4   Extending a Processor

Fast and easy processor configuration depends on an appropriate means to describe and extend processor functions, especially the instruction set. The TIE language is the first widely used commercial instruction-set-extension description format, and the Xtensa generator technology is the first widely used processor-extension generator.

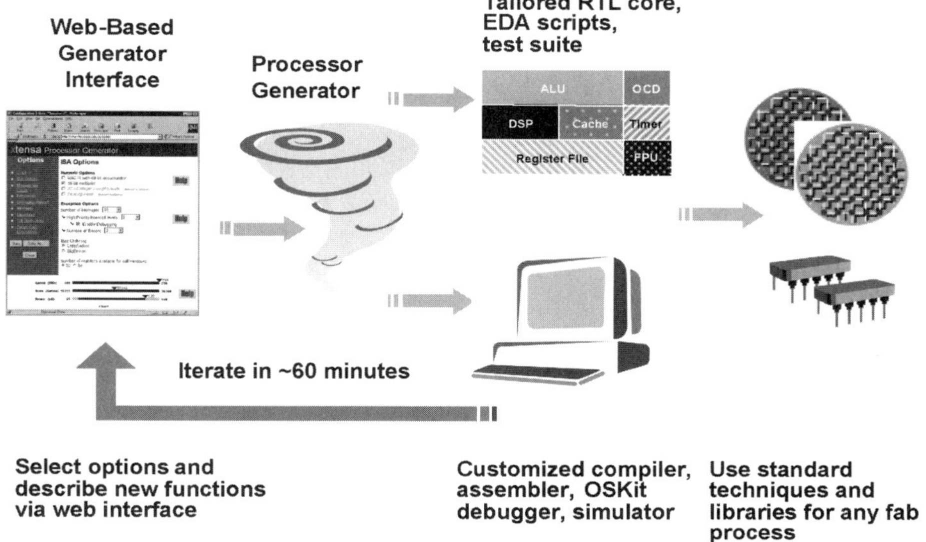

5-7   Configurable processor development cycle.

FIGURE

The TIE Language

A TIE description consists of two basic parts:

+ **State declarations:** designers can add state registers and register files of any arbitrary (non-binary) width and number.

+ **Instruction encodings and formats—operation descriptions:** designers can specify new instructions with as many as six source and destination operands, taken from base architecture register files, newly defined registers, and new states. TIE also supports encoded immediate fields to keep instruction encodings compact. For each new operation, designers specify the corresponding transformations from source registers to destination registers. The compiler, simulator, and hardware generator automatically stretch complex functions across multiple clock cycles using the pipeline-latency parameter.

A Simple TIE Example

Here is a simple but complete TIE example that adds a new register file and one new instruction to the Xtensa processor:

```
regfile LR 16 128 1
operation add128 {out LR sr,in LR ss,in LR st} {assign sr =
st + ss;}
```

This code defines a register file, LR, with 16 entries, each 128 bits wide. The processor generator creates a compiler that supports a new data type, LR, and defines new load and store instructions for that register file. Any scalar variable declared with the new data type will be automatically allocated to that register file, and any array or pointer that references that data type will use the new load and store operations. The operation statement defines a new instruction, add128, that takes two operands from the new register file, computes their sum, and places the result back into the new register file. A software developer might use the new 128-bit data type and operations directly as follows:

```
main() {
int i;
LR src1[256],src2[256],dest[256];
  for (i=0;i<256;i++} dest[i] = add128(src1[i],src2[i]);
}
```

With small additions to the TIE description, the designer could specify deep pipelining to the new instruction; add other instructions that operate on a combination of existing registers, data memory, and newly defined state; and define instructions sets with multiple operation slots per instruction.

5.3.5   Exploiting Extensibility

Application experts create TIE-based extensions, typically starting from the application kernel. They run code on a simulated version of the baseline Xtensa processor and use the associated simulation tools to profile the application and identify hot spots in the original code. The profiled code and identified hot spots lead to the creation of new data types and operations that fit the application problem more naturally. All hardware, verification infrastructure, and software tools are created together from a single Xtensa configuration file and delivered securely over the web. Using this process, a system designer can typically evolve an optimized processor configuration over just a few days. A typical process flow is shown schematically in Figure 5-8.

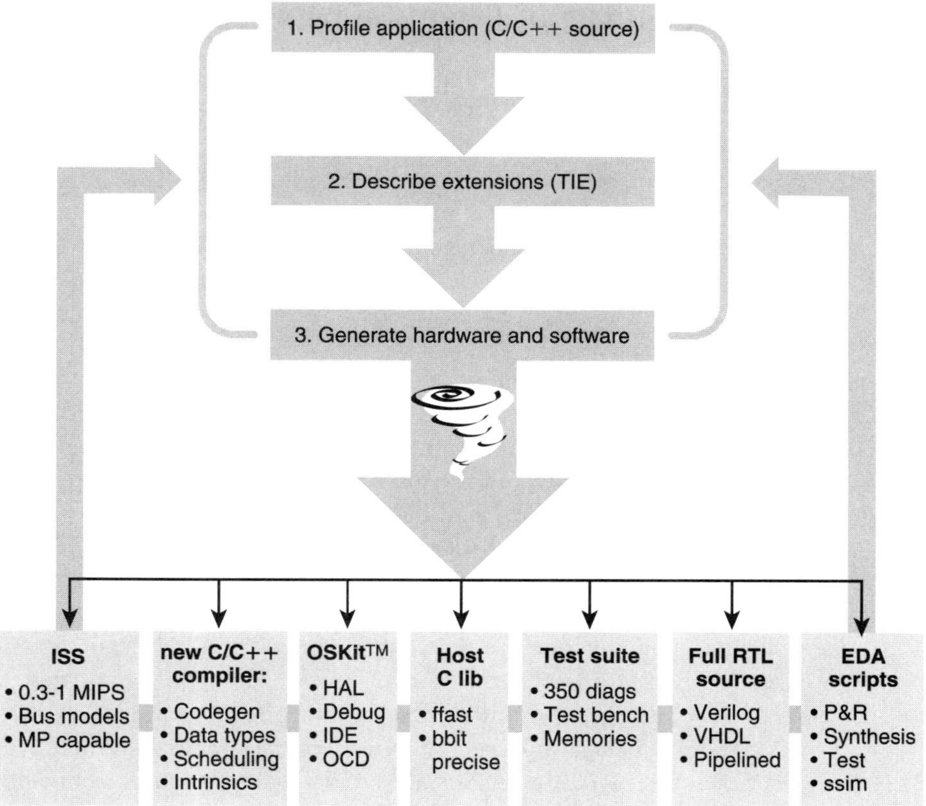

5-8 Xtensa processor design flow.

FIGURE

From a single configuration description, the Xtensa processor generator produces a fast, cycle-accurate instruction-set simulator and bus-functional model; complete C source libraries for the instruction-set extensions; GNU-derived C/C++ compilers (including a modified front-end, code generator, code scheduler, and intrinsic support); an *OSKit*—the support files required for kernel and debug extensions for third-party real-time operating systems; a hardware test-bench and complete hardware diagnostics (including coverage of designer-defined TIE extensions); scripts for synthesis, place and route, test generation, timing, and power optimization and simulation; and complete source code for the processor hardware in Verilog or VHDL.

5.3.6   The Impact of Extensibility on Performance

The promised benefits of automated processor generation are these: significant improvements in application efficiency (throughput, power, and silicon cost) combined with simplification of comprehensive platform development (completeness and immediacy of compilers, simulators, RTOS, and hardware). Examples in multimedia, telecommunications, and network processing illustrate how this process works. Although these application domains represent an enormous range of tasks, the Embedded Benchmark Consortium (EEMBC, pronounced "embassy") benchmark suites are the most widely used measures of performance in these areas [219].

For each of three suites—Networking, Telecommunications, and Consumer—we can compare the performance of several standard processors with Xtensa processors, before and after optimization. The processors include the ARM1026EJ-S [220] and two MIPS-based processors, the NEC VR5000 (which implements a 64-bit MIPS architecture), and the NEC VR4122 (which implements a 32-bit MIPS architecture [221]). To separate the effects of clock frequency from the effects of architectural features, certified results for the total benchmark performance and the performance per MHz are shown. Performance per MHz is a useful tool for comparing architectural efficiency, independent of fabrication technology, especially when processors with similar pipelines are being compared.

All these processors attempt to execute one integer instruction per clock using a five- or six-stage pipeline. Moreover, synthesizable ARM and Tensilica processors run at generally comparable clock rates in equivalent technology (for 130-nm technology, roughly 250 to 350 MHz). EEMBC certifies scores both for straight compilation of standard C code for a standard configuration and for hand coding of software and optimization of the configuration. The NEC VR5000 and VR4122 processors have roughly similar pipeline structures to MIPS Technologies' MIPS5K and MIPS4K synthesizable processor core families [222], but no EEMBC scores are available for these cores. The clock frequencies for the MIPS Technologies cores are also similar to the ARM and Tensilica cores' clock frequencies. For these benchmarks, the processors are generally all configured with large enough local caches that differences in off-chip memory speed are immaterial.

EEMBC Consumer Benchmark Results

Video processing lies at the heart of consumer electronics—in digital video cameras, in digital television, and in games. Common tasks include color-space conversion, 2D filters, and image compression. The EEMBC Consumer benchmarks

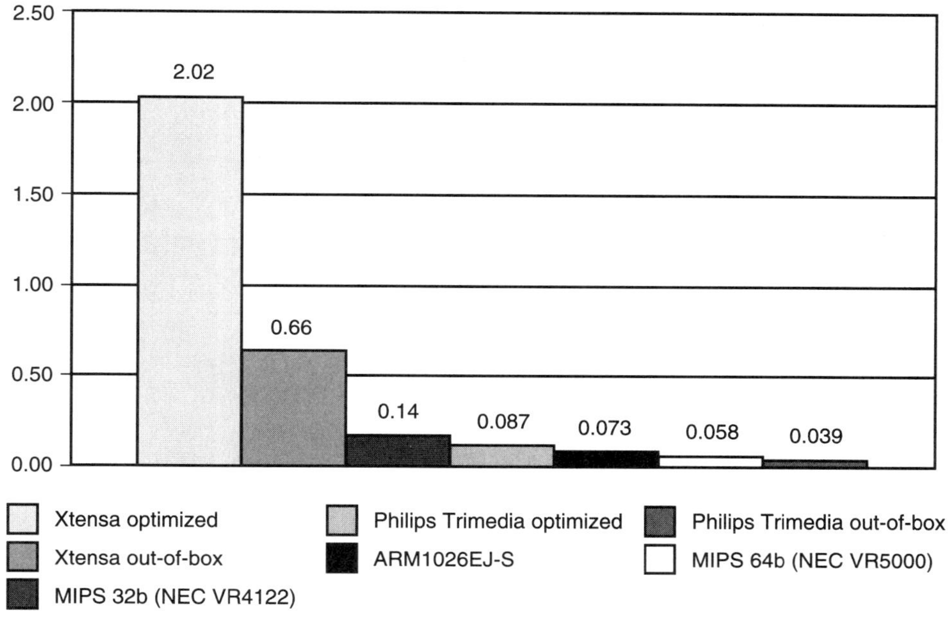

5-9 FIGURE EEMBC ConsumerMark comparison: performance/MHz.

include a representative sample of all these applications. Figure 5-9 shows performance per MHz for Tensilica's Xtensa V release and several other architectures. The other processors are the Philips Trimedia TM-1300 [223], the ARM1026 core, and the NEC VR5000 and NEC VR4122 MIPS architecture processors. The optimized scores can include both hand tuning of code and tuning of the processor configuration for this class of consumer applications. In the case of Xtensa, roughly 200 K gates of added consumer instruction set features have been implemented in TIE. The configured processor provides about three times the cycle efficiency of a good media processor (the Trimedia TM-1300), and about 14 times the cycle efficiency of a good 64-bit RISC processor (the NEC VR5000).

Clock frequency is an important component of processor performance. Although adding instructions can have a second-order impact on speed, other factors, especially process technology and effective processor pipelining, are typically more important. The full ConsumerMarks performance is shown in Figure 5-10, in which the baseline for performance (ConsumerMark = 1.0) scale is the ST Microelectronics ST20 at 50 MHz [224].

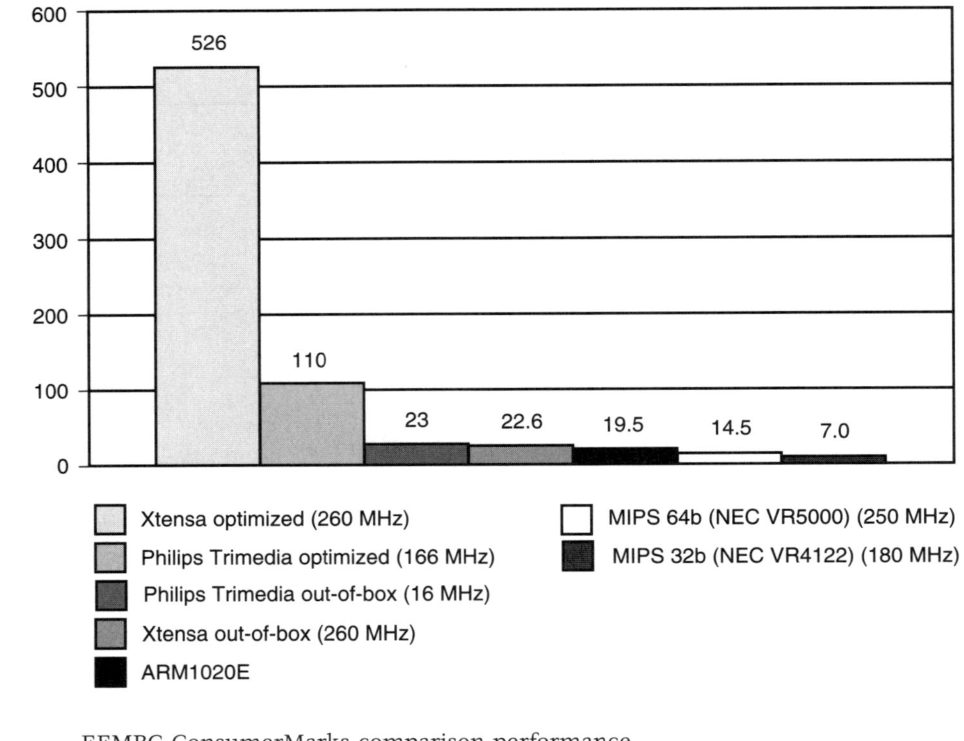

5-10

EEMBC ConsumerMarks comparison performance.

FIGURE

EEMBC Telecommunications Benchmark Results

Telecommunications applications present a different set of challenges. Here the data are often represented as 16-bit fixed-point values, as densely compacted bit-streams, or as redundantly encoded channel data. Over the past 10 years, standard DSP processors have evolved to address many filtering, error correction, and transform algorithms. The EEMBC Telecom benchmark includes many of these common tasks. The other architectures shown in these results include a high-end DSP (the Texas Instruments TMS320C6203 [225]), plus the ARM and MIPS processors.

The TI code has been hand-tuned, and the Xtensa processor has been optimized with instructions to aid convolutional coding and bit allocation. Even the out-of-box performance for Xtensa is strong, in part because the Vectra DSP instruction set (a configuration option for the Xtensa processor) is included in the

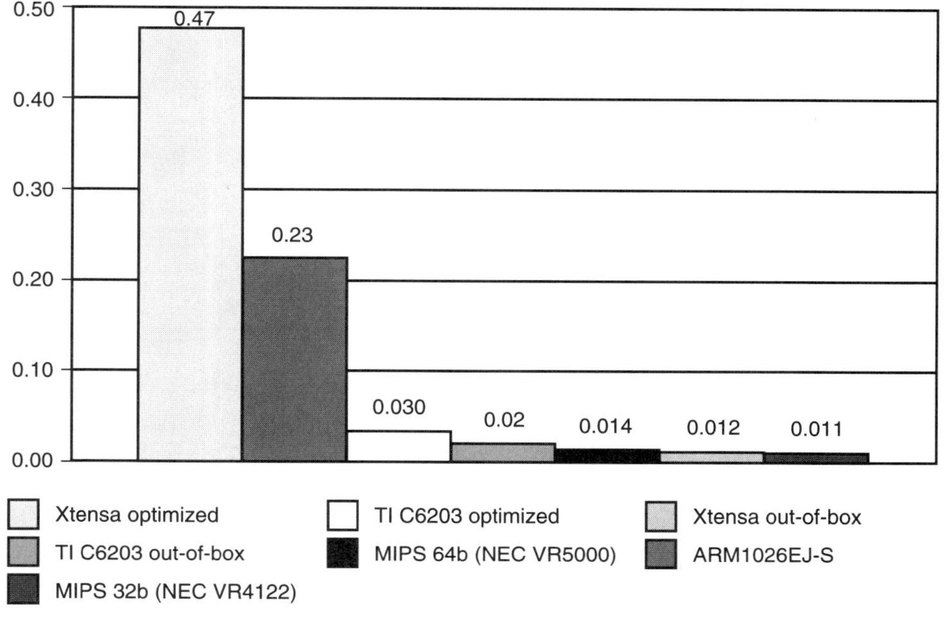

5-11   EEMBC Telemark comparison: performance/MHz.

FIGURE

baseline configuration and in part because of good DSP compiler performance. Even Xtensa's out-of-box performance is two to three times higher than that of the 32- and 64-bit RISC processors. The configured processor achieves about twice the cycle efficiency of a good DSP processor and almost 30 times the efficiency of a good 64-bit RISC processor (Fig. 5-11).

Once clock frequency is considered, the TI processors continue to show well, but slightly faster clock frequency cannot make up for the significantly more efficient architecture achieved through the Xtensa processor's extensibility. The Telemark scores are shown in Figure 5-12, in which it can be seen that the IDT 32334 (32-bit MIPS) at 100 MHz sets the performance reference of 1.0.

EEMBC Networking Benchmark Results

Networking applications have significantly different characteristics than consumer and telecommunications applications. They contain little computation, generally show less low-level parallelism, and frequently require rapid

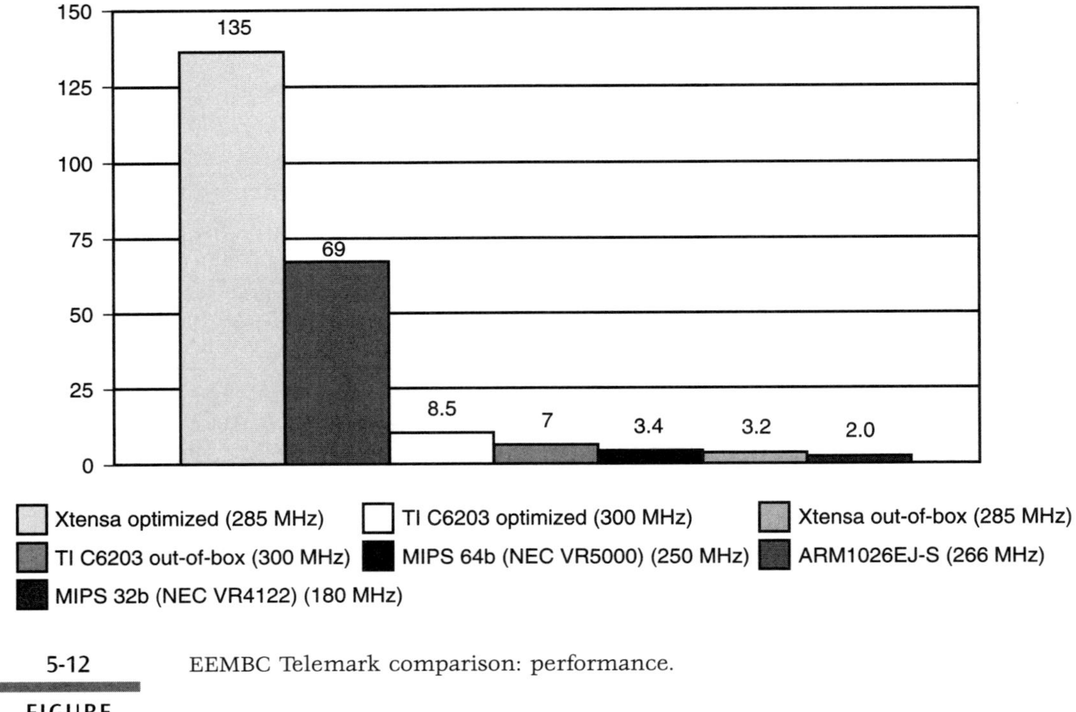

5-12   EEMBC Telemark comparison: performance.

FIGURE

control-flow decisions. We might therefore expect that application-specific instruction sets for network applications will not resemble those for image and signal processing.

The EEMBC Networking benchmark suite contains representative code for routing and packet analysis. Figure 5-13 compares EEMBC Netmark/MHz of out-of-box and optimized scores for an Xtensa processor and several other architectures; the IDT 32334 (32-bit MIPS) at 100 MHz is the performance reference of 1.0. The extended Xtensa processor achieves about seven times the cycle efficiency of a good 64-bit RISC processor (the NEC VR5000).

These processors achieve generally comparable clock frequencies, although the NEC4122 (32-bit MIPS) lags somewhat behind, giving the overall Netmark performance shown in Figure 5-14.

These EEMBC benchmark results suggest that extensible processors can achieve significant improvements in throughput across a wide range of embedded applications, relative to good 32- and 64-bit RISC, DSP, and media processor cores.

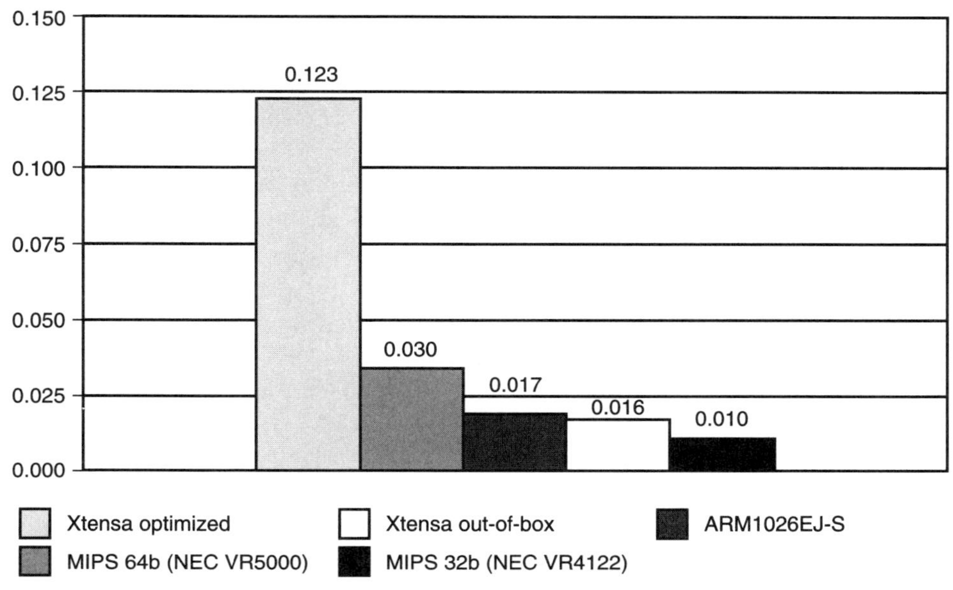

5-13   EEMBC Netmark comparison: performance/MHz.

FIGURE

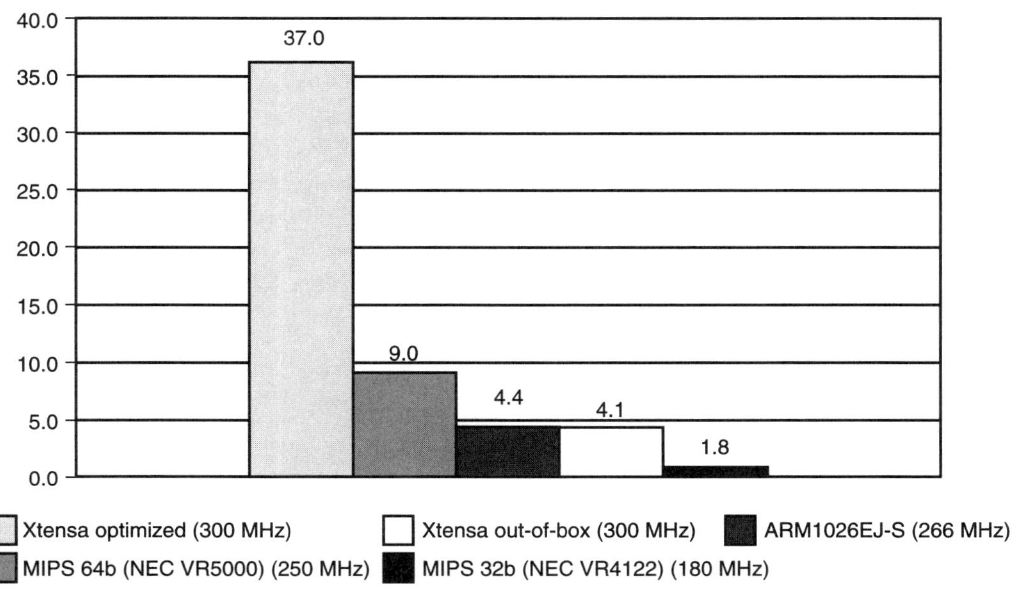

5-14   EEMBC Netmark comparison: performance.

FIGURE

5.3.7   Extensibility and Energy Efficiency

Designers often resort to hard-wired logic design rather than processors to improve power dissipation and energy efficiency. The power efficiency of RTL logic comes from the combination of small silicon area (low switched capacitance) and low cycle count (more useful work per cycle). Configurable processors benefit from the same effects. The base architectures are small, because features with uncertain utility can be omitted and configured on demand, and application-specific instructions and interfaces can be added as needed to reduce cycle count. The combination of effects should result in significantly better energy efficiency for complex algorithms compared with embedded RISC cores. Figure 5-15 compares the estimated typical power dissipation (including cache memories) and performance for baseline Xtensa processors, optimized Xtensa processors, and the ARM1026EJ-S running the EEMBC benchmark suites, all implemented with 130-nm technology.

The configured processors range from 12 times more energy efficient than the ARM1026 on the Network suite to 44 times more energy efficient on the Consumer suite.

5.4   TOWARD MULTIPLE-PROCESSOR SOCS

Although individual tasks and individual processors are important components in SoC solutions, the larger questions of system applications and SoC architecture are increasingly important. The aggregate complexity of SoC designs, particularly as the development cost grows into tens of millions of dollars and total gate counts reach toward 100 million gates, creates demand for faster initial design and greater post-fabrication flexibility. Broader use of processors as a basic building block for SoCs could address these demands. Two trends are working together to increase the number of processors used on a typical SoC:

1. The combining of functions traditionally implemented with discrete control processors and DSPs into a single SoC, and

2. The migration of functions traditionally implemented with RTL into application-specific processors, especially as the complexity of these functions grows.

As a result, concerns about the design requirements for multiple-processor SoCs become central concerns for all advanced integrated circuit and system

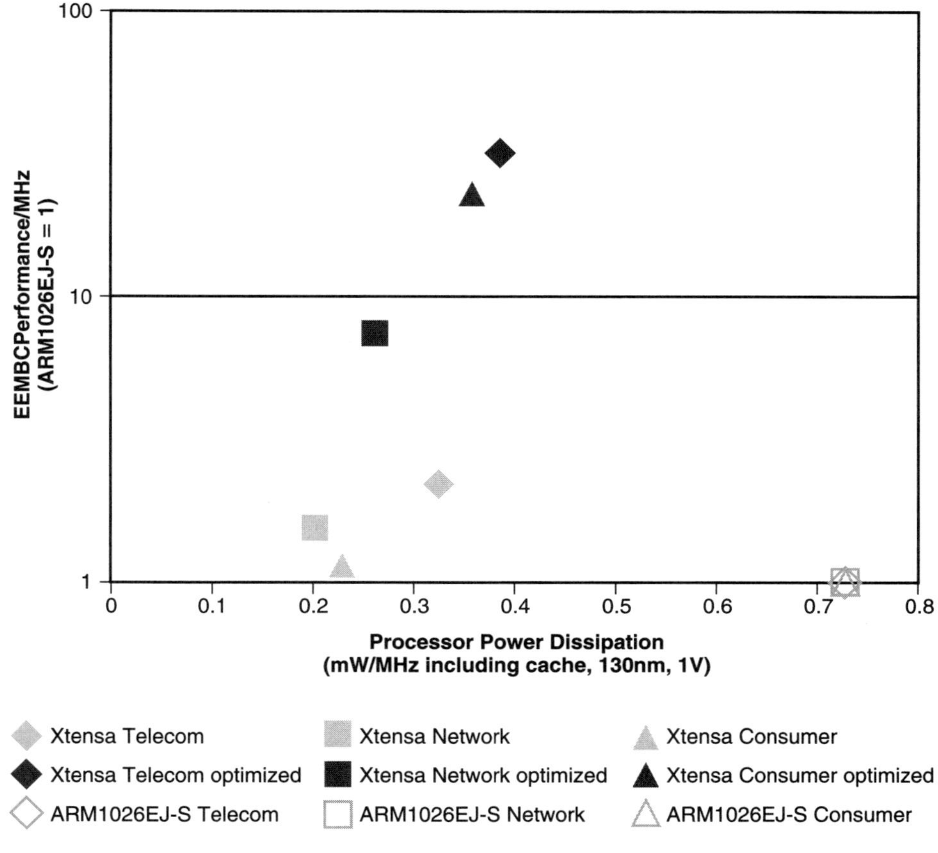

5-15

FIGURE

Performance and power for configurable versus RISC processors.

design. The topologies for interconnecting multiple processors range from simple bus structures, to dual-ported memories linking processors, to on-chip processor communication networks. The implementation of these topologies is relatively well understood. However, the selection criteria for choosing one connection approach over another are not as well understood and are system- and application-specific. Design-space exploration through the simulation of various multiple-processor system designs is essential to choosing the right connection topology in these systems. Let us therefore turn to a particularly pressing problem in the creation of multiple-processor SoCs, the development of fast, accurate multiple-processor system models.

5.4.1   Modeling Systems with Multiple Processors

A key element that is essential to the ability to replace multiple RTL logic blocks with multiple heterogeneous processors is the ability to simulate a system composed of these application-specific processors. The Tensilica development environment supports a sophisticated cycle-accurate instruction-set simulator (ISS), which is often used by itself or encapsulated in popular third-party hardware/software co-design environments such as Mentor's Seamless simulation environment [226]. In addition, Tensilica's instruction-set simulator API has been significantly extended to create the *Xtensa Modeling Protocol* (XTMP).

XTMP can be used for simulating homogeneous and heterogeneous multiple processor subsystems as well as complex uni-processor architectures. The XTMP API allows system designers to write customized, multi-threaded simulators to model complicated systems. Designers can instantiate multiple similar and dissimilar Xtensa cores and use the XTMP API to connect these simulated processor cores quickly to memories, peripherals, interconnects, and designer-defined subsystems. The environment supports synchronous and asynchronous debugging of complex sets of tasks distributed across large numbers of processors.

Because the Xtensa instruction-set simulator (a model of the application-specific processor created by the Xtensa processor generator) is written in C, it runs much faster than does simulation of a processor model coded in a hardware description language (HDL) such as Verilog or VHDL. It is also easier to modify and update a C simulation model than a hardware prototype or an RTL simulation. Thus, XTMP allows designers to create, debug, profile, and verify their combined hardware and firmware systems early in the design process, which encourages a more thorough investigation of the system's solution space. Like the stand-alone ISS, XTMP is a tool to help the designer quickly check the overall performance and correctness of firmware for the multiple-processor SoC design.

5.4.2   Developing an XTMP Model

A system simulation consists of a connected set of simulation components or models. The designer instantiates simulation components, connects them together to form a system, loads a program into each processor's memory, initializes a source debugger for each processor, and then starts the simulation.

In XTMP_main( ), the top level of the simulation program, the designer performs the following steps:

1. **Create the components to be simulated:** the designer can create Xtensa processors with XTMP_coreNew( ) or basic memory devices with XTMP_memoryNew( ) or write new models of more complicated devices and allow the rest of the simulator to access them with callback functions.

2. **Connect the components together:** after the designer has created the Xtensa processors and the devices that will be a part of the system simulation, the designer connects them to each other. The designer can connect a device directly to an Xtensa processor interface or create a connector, attach it to a processor interface, and then connect multiple devices to the connector with the XTMP_connect( ) function using either a single global address map or a different map for each processor.

3. **Control the simulation:** after creating and connecting all the simulation components, the designer can:

   + Load the program for each Xtensa processor core

   + Call XTMP_enableDebug( )

   + Start or stop the execution of all the simulated cores simultaneously

   + Disable individual processor cores while the others continue to run

   + Access the internal registers and memory spaces of any of the processor cores (allowing instrumentation of the simulation).

XTMP Components

XTMP has a powerful set of predefined components. Multiple independent instantiations of each processor type are fully supported. Some of the most common are:

XTMP_core: an Xtensa processor
XTMP_memory: a simple memory device
XTMP_connector: a device for connecting processor cores to other devices
XTMP_device: a designer-defined device that interacts with the rest of the simulator via callback functions
XTMP_procId and **XTMP_lock:** devices to help with multiprocessor synchronization.

XTMP Simulation Commands

The most common XTMP simulation commands include the following:

XTMP_loadProgram( ): loads target program into memories

XTMP_start( ): runs the simulator until finished or for a fixed number of cycles

XTMP_disable( ), XTMP_enable( ), and **XTMP_reset( ):** disables, enables and resets a particular core

XTMP_step( ): steps the program on one processor core, while other cores are disabled.

A Simple XTMP Example

The following example shows how to create a single-processor simulation using XTMP:

```c
#include "iss/mp.h"
int XTMP_main(int argc, char **argv)
{
  XTMP_params p;
XTMP_core core;
XTMP_singleAddressMapConnector singleMap;
XTMP_memory rom, ram;

  p = XTMP_paramsNew( "s1", NULL);
  core = XTMP_coreNew( "cpu", p, NULL);
  if ( core == 0 ) {
    fprintf ( stderr, "Failed to create core\n" );
    exit (1);
  }
rom = XTMP_sysRomNew( "rom", p );
ram = XTMP_sysRamNew( "ram", p );
  singleMap = XTMP_singleAddressMapConnectorNew ( "singleMap",
  XTMP_byteWidth (core) );
  XTMP_connect( singleMap, core );
XTMP_connect( singleMap, rom );
XTMP_connect( singleMap, ram );

  if ( !XTMP_loadProgram( core, "sieve.out", NULL ) ) {
    fprintf ( stderr, "Failed to load sieve.out\n" );
    exit (1);
   }
  XTMP_start (-1);
  return(0);
}
```

The impact of rapid, accurate multiple-processor modeling on system design is dramatic. System models can often be developed in a matter of hours, in contrast to the months typically required to stitch together large-scale RTL-based SoC simulations. C-based XTMP models run significantly faster than do simulations of RTL models—generally at least 100 times; compared with standard RTL logic simulators for the same level of system complexity.

For example, the Xtensa processor instruction set simulator, running in full cycle-accuracy mode runs about 350,000 cycles per second on a Linux PC (1 GB memory). RTL simulation of the same processor on the same PC achieves a simulation rate of about 1500 cycles per second. Even if the RTL implementation of a function were simpler than the application-specific processor implementation, it could not close this 200–fold simulation performance gap. A system with 30 blocks, of complexity comparable to an application-specific processor (30 to 100-K gates), might reach only 50 cycles per second, whereas the 30-processor system would simulate more than 5000 cycles per second, making fully cycle-accurate modeling of long application scenarios feasible.

5.5   PROCESSORS AND DISRUPTIVE TECHNOLOGY

In his book, *The Innovator's Dilemma*, Harvard professor and author Clayton Christensen defines a "disruptive technology" as an innovation that establishes a new standard in product cost and accessibility [227]. Disruptive technologies gradually displace even successful market and technology leaders. Extensible processors, a disruptive technology for SoC-based design, combine the dramatic productivity of complete software programmability with the exceptional performance and efficiency of optimized RTL logic.

Measured by conventional criteria applied to PC processors, the configurable processor is pretty weak. In its un-extended configuration, it delivers only a few hundred MIPS of performance and runs only a limited range of well-known operating systems. For example, Tensilica's Xtensa processor runs Linux and Wind River's VxWorks, but not Microsoft Windows [228]. When configured for a specific application domain, the application-specific processor can be as fast or faster than the Pentium, but with less generality. On the other hand, power dissipation for the configurable processor, including its small memories, is well below $100\,\mu\text{W/MHz}$, and it consumes less than $1\,\text{mm}^2$ of 130-nm silicon (the equivalent of a few cents of cost).

For an increasing range of applications, the performance of these application-specific processors is fast enough, and the cost, power, and focused application performance are compelling enough to enable embedding in many places where no other processor could suffice. The penetration of these application-specific processors into roles previously reserved for hardwired logic fulfills the "disruptive technology" criteria of enabling new uses. These processors compete more with non-consumption of processors than with consumption of some more established processor families.

Intel's pursuit of the most demanding stand-alone processor customers is echoed in the evolution of embedded RISC CPU cores. ARM Limited and MIPS Technologies are the established leaders in embedded RISC cores. ARM, in particular, originally focused on low-end cores, cores that lacked the performance and features common in stand-alone processors of the day. They were "lousy CPUs" by general-purpose computing criteria. Over time, however, these cores have grown in performance and features, such that they can satisfy the needs of even very demanding embedded applications. The progression, for example, from ARM6 to ARM7 to ARM9 to ARM10 to ARM11, and soon to ARM12, demonstrates remarkable sustained improvements in absolute performance. As a result of these improvements, however, the ARM11 has now accumulated a wide range of sophisticated features. It requires roughly 10 times the silicon area of a base-line configurable processor and dissipates almost 10 times the power per MHz.

Processors like the ARM11 are clearly able to displace stand-alone processors in a wide range of personal computing devices such as personal digital assistants (PDAs). On the other hand, the "improved" RISC core can no longer compete for lowest-cost, highest-efficiency embedded computing functions. Consequently, many designers must satisfy their data-intensive computing in hardwired logic, not ARM processors. Figure 5-16 shows how the disruptive technology model might be applied to two classes of customer needs, embedded control tasks and personal computing—and how three types of processors (stand-alone microprocessor chips, RISC CPU cores, and configurable processors) compare. For this comparison, the performance axis is taken to be general-purpose performance and availability of a wide variety of applications.

The impact of the configurable processor is not just to lower silicon size and cost for embedded processing, but to change the basis of competition. Automatically generated processors serve in roles that embedded RISC cores cannot easily play. The adaptability to new applications, the high performance in data-intensive applications, and the more complete support for multiple-processor SoC design all open new markets for processors in which customers value criteria besides general-purpose performance and breadth of application availability.

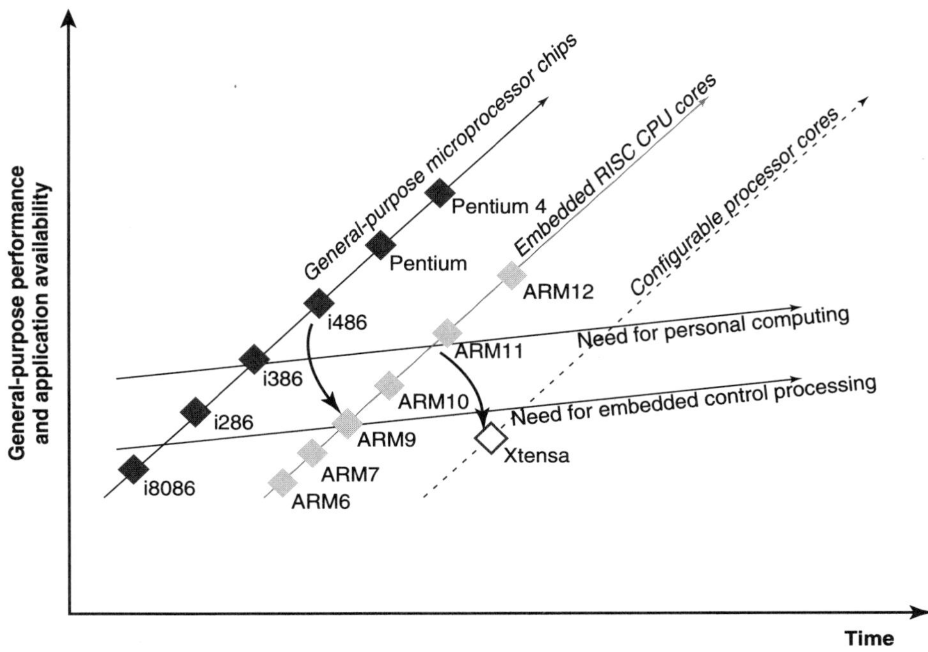

5-16   Applying the disruptive technology model to embedded processors.

FIGURE

5.6   CONCLUSIONS

The steady, rapid improvement in SoC gate density opens the door to a wide spectrum of new digital products, from optical-rate routers to multi-mega-pixel digital cameras. However, the difficulty of designing millions of gates of new logic for each SoC threatens to frustrate this potential. A new design methodology—configuration, extension, and programming of application-specific processors in place of RTL logic design—offers important benefits and allows system designers to exploit the gate counts of mega-gate SoCs.

The net result of introducing the configurable, extensible processor core as a design component and methodology is a fundamental disruption of the traditional decade-old methodology for RTL-based chip design. This new design methodology allows each of the increasing number of functional blocks in an SoC to be replaced by an analogous application-specific processor with similar throughput and cost but much greater flexibility. Extensible processors reduce SoC design time and design risk, because of simpler specification, much faster verification, and seamless functional changes at any point in the product development cycle. The extensible processor core liberates engineers from routine and repetitive low-level RTL design tasks so they can develop new functions and explore higher-level system optimizations.

HDL languages such as Verilog and VHDL will remain important, but only to design simple logic blocks and as intermediate data formats used for process portability and low-level validation of design integration. Design reuse in large organizations and market transactions for silicon intellectual property can now shift from RTL representations to formal descriptions of processor configurations plus the source firmware that exploits those configurations.

SystemC and System Verilog may become important as system simulation frameworks. System Verilog is likely to be used most widely by hardware designers building test-benches for RTL-based subsystems, whereas SystemC is likely to be used most widely by SoC architects to combine hardware blocks, allocate software to embedded processors, and embed other more abstract or unallocated tasks into a system-level simulation model. If the configurable processor premise is correct—that application-specific processors will be used in large numbers and take over many of the roles once played by hardwired logic—then system-modeling languages must become ever more effective at integrating large numbers of different processor configurations. Moreover, the system-modeling environment will be closely tied to the system-programming environment, in which software-centric methods for code development, task-to-task communication, and parallel debug will become the focal point for tools, methodologies, and SoC building blocks.

The transition to a design methodology based on application-specific processors even holds some promise to change the fundamental economics of silicon development. As this new methodology takes hold (as did RTL design over 10 years ago), the design of very complex chips will become easier and less expensive. Pervasive programmability will also expand the number of applications each design can support. This is a path that leads to greater silicon product longevity and better return on the development investment.

Broad use of extensible processors may also fundamentally disrupt the embedded microprocessor market. These very small, very fast processors will be built in such large numbers, and used by such a wide range of engineers, that the popularity of traditional processors will be seriously undermined. The lower cost, easier integration, and more-than-adequate general-purpose performance all tend to make the extensible processor an increasingly attractive alternative to general-purpose processors for SoC design. As extensible processors become broadly understood as basic building blocks, they may displace legacy processors in most general-purpose embedded-processing tasks. Extensible processors promise to simplify significantly the whole SoC design and programming challenge, and they may very well ultimately displace current leading processor architectures. Christensen's model of "disruptive innovation" is clearly at work in microprocessors.

ACKNOWLEDGMENTS

Credit is due to the whole Tensilica engineering team, whose work is represented by the Xtensa architecture and processors, the processor generator, and key tools such as the Xtensa Modeling Protocol. Moreover, 5 years of tight interaction between Tensilica and countless customers have honed the capabilities of the processor and associated tools to fit the demands of multiple-processor systems-on-chips. Special thanks go to Steve Leibson for his work in organizing and editing this chapter.

MPSoC Performance Modeling and Analysis

Rolf Ernst

6.1   INTRODUCTION

The productivity requirements for multiprocessor system-on-chip (MPSoC) design can only be met by systematic reuse of components and subsystems. MPSoC components are specialized to specific tasks, such as a digital signal processor (DSP), a reduced instruction set computing (RISC) processor, or a coprocessor and are connected via an on-chip communication infrastructure ranging from simple point-to-point interconnects to busses to complex on-chip networks. An increasing share of an MPSoC dice is occupied by memories of different sizes and types. Reuse and specialization result in increasingly heterogeneous MPSoC structures.

6.1.1   Complex Heterogeneous Architectures

Architecture templates, called platforms [229], have been introduced to improve component reusability and to simplify the design process. An example is the Philips Nexperia platform [230,231] for multimedia applications. There are many types of processors, coprocessors, memories, and busses available as library elements. Figure 6-1 shows an instantiation, the PNX8500 VIPER processor [232]. Two types of processors, a TriMedia very long instruction word (VLIW) for signal processing and a MIPS RISC processor core for control and general purpose tasks, both equipped with caches (not shown), are complemented by coprocessors, many of them used in several places. These coprocessors have fixed functionality, such as a peripheral interface or a video format converter, or they are weakly programmable, i.e., have a small set of specialized functions for stream transport or

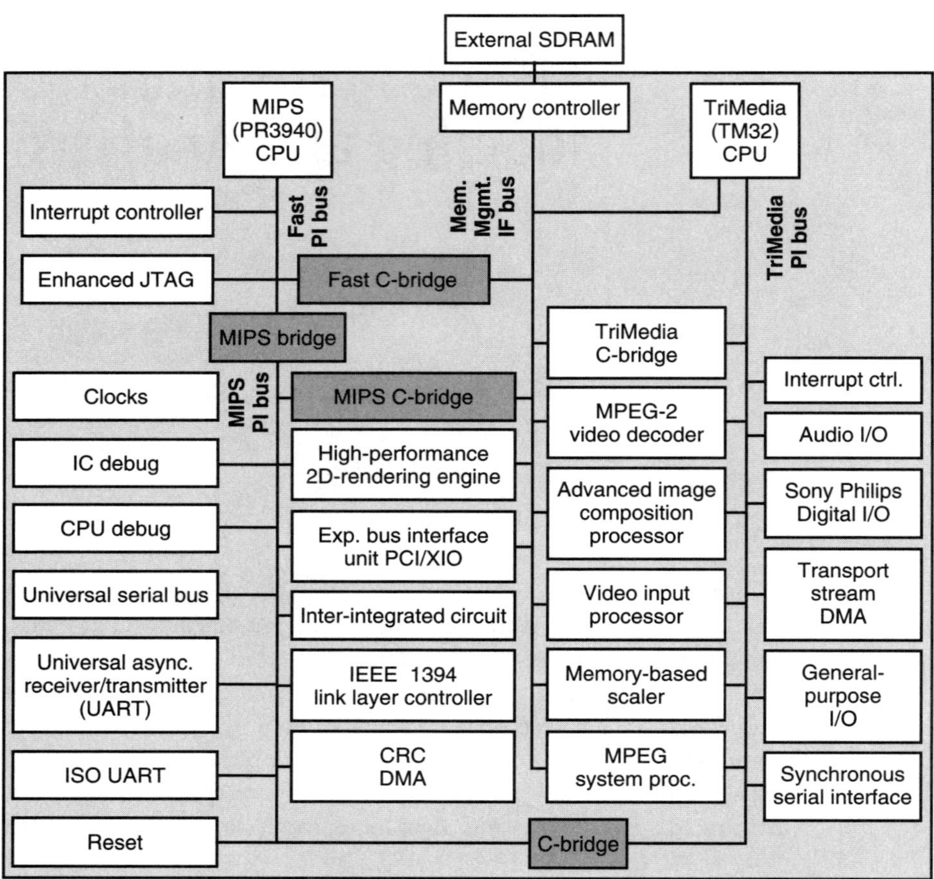

6-1   Platform example: Philips VIPER [232].

FIGURE

as drawing engines. There are several types of busses to interconnect these components and to attach an external larger SDRAM memory. The layered Nexperia software architecture includes support of an OS running on the MIPS processor (e.g., VxWorks), a real-time kernel running on the TriMedia VLIW (pSOS+), and an application programmer interface (API) containing streaming software components and software codecs.

Figure 6-2 gives an example of how the software of a programmable component can be structured on top of the hardware. The processor core is

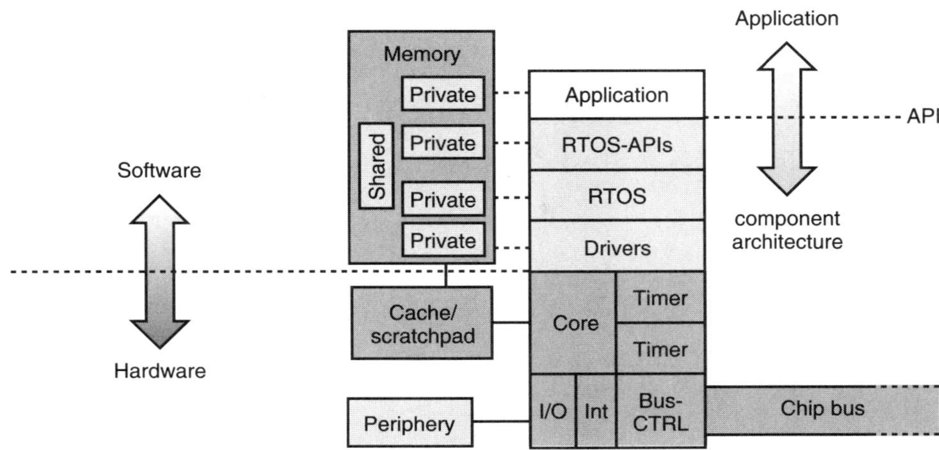

6-2   Software architecture: example.

FIGURE

complemented by interface units to the periphery and to the communication infrastructure as well as timing and interrupt resources. Drivers control access to these hardware resources. The real-time operating system (RTOS) includes a process scheduler, whereas the RTOS API implements an interface to the application program, e.g., process communication primitives and functions for resource access. The memory space not only consists of several different memories but is also shared by different levels of the component hardware/software hierarchy requiring a memory protection mechanism. Obviously, the layers of this rather general structure are not required for all components. A DSP with a fixed process execution order and a manually optimized memory communication will need fewer if any RTOS functions. On the other hand, in most major applications there are some parts that require such an operating system support, such as a mobile phone or a multimedia home platform running user software. Moreover, operating systems support portability and interoperability, such as the OSEK operating system for automotive applications.

Hence, in any sufficiently complex MPSoC, we will find different component hardware/software architectures and different scheduling strategies, such as RISC processors with static or dynamic priority scheduling, hardware components or DSPs with a fixed schedule, or communication devices with time-triggered resource sharing. In other words, the heterogeneous hardware architecture leads to a similarly heterogeneous software architecture. Component hardware and software form a unit to be integrated with other hardware/software components.

6.1.2   Design Challenges

As a result, component integration becomes the major challenge in MPSoC design including component and subsystem interfacing, design space exploration and optimization, integration verification, and last, but not least, design process integration.

In this chapter, we will focus on the verification aspect. MPSoC verification can, again, be separated into

- Function verification, i.e., verification that the specified function has correctly been implemented, and

- Target architecture performance verification, validating that the target architecture

    - Provides sufficient processor and communication performance to run the application

    - Meets the timing requirements

    - Avoids memory underflow or overflow

    - Does not introduce run time-dependent behavior such as deadlocks.

This separation is a fundamental assumption in real-time analysis (e.g., ref. 233). It is the basis for the many performance verification approaches that have found their way into the embedded system design practice. We will discuss some of these approaches later.

What is new in MPSoC design is the heterogeneity of scheduling algorithms, leading to almost arbitrary combinations. Some of the advanced analysis techniques cover more than one scheduling approach, such as the work by Tindell and Clark [234], but are limited to few combinations and, therefore, do not scale to large heterogeneous systems. Again, the problem is found in the integration of different scheduling strategies.

At this point, we want to clarify the terms component and subsystem. A component shall be defined as a computer architecture component, i.e., processor, memory, or communication link. A subsystem shall be a set of components that is controlled by a common scheduling strategy and a common layered software architecture in case there is such an architecture.

6.1.3   State of the Practice

Today, simulation is state of the art in MPSoC performance verification. Tools, such as Mentor Seamless CVE [235] or Axys MaxSim [236], support hardware and software co-simulation of the complete system. The co-simulation times are extensive, but there are several obvious advantages to this approach. Performance verification and function verification can use the same simulation environment. The same simulation patterns can be used for performance and function verification. Second, the system designer can reuse components or subsystem simulation patterns for system integration. Application benchmarks can be reused for simulation-based performance verification. Finally, simulation provides use cases for a given set of input patterns that support system understanding and debugging.

However, there are critical disadvantages of simulation-based performance verification. First, performance verification is much more time-consuming than function verification. Performance verification can use the same simulation pattern but needs timed models, whereas untimed models are sufficient for function verification. The higher model and simulation complexity of timed simulation results in much higher run times for timed model simulation given the same number of simulation patterns. Therefore, simulation for performance verification is a bottleneck in the design process. This can be particularly painful if the detailed timed models used for verification shall also be used for performance evaluation in an iterative design space exploration process. Performance simulation with abstract performance models, such as in Cadence VCC [237] is, therefore, a reasonable simplification step but at the cost of introducing yet another set of simulation models.

However, there are even more serious limitations to simulation. Other than in function verification, in which integration verification checks the correct function of known component interactions, the component interactions in performance verification are only partially visible in the system function.

Figure 6-3 illustrates a component dependency introduced by resource sharing. Suppose a subsystem consisting of processor CPU_1 and memory M_1 communicates with subsystem CPU_2-M_2 via the system bus B. Concurrently, the fixed function component HW_3 reads data from memory M_3 using the same bus. Bus B arbitration introduces a nonfunctional performance dependency that is not reflected in the system function. This dependency can be sophisticated, turning component or subsystem best-case performance into system worst-case performance. An example is shown in Figure 6-4. Suppose process P_1 running on CPU_1 sends one data packet per process execution. P_1 has a best-case execution time t_{bc1} and a worst-case execution time t_{wc1}. Reading data from a buffer in M_1, P_1 is

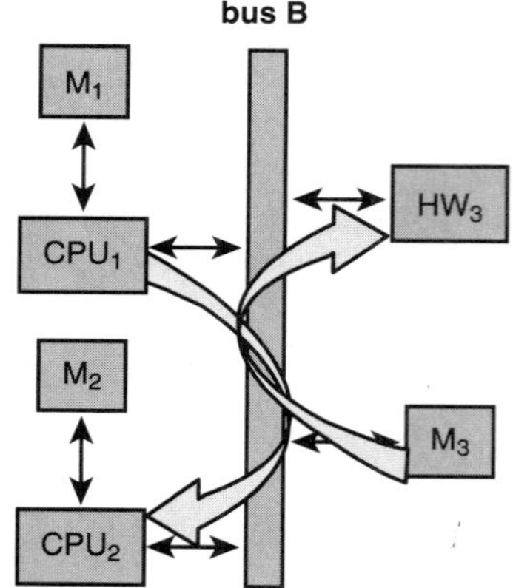

6-3

FIGURE

Complex run-time interdependencies.

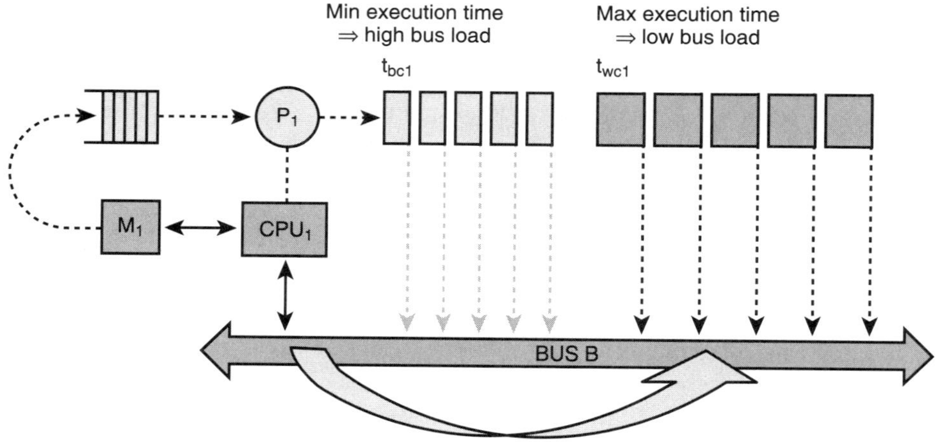

6-4

FIGURE

Interdependency: example.

iterated with maximum frequency. P_1 could, e.g., be a driver process. The shorter the process execution time is, the shorter will be the distance of packets sent over the bus and the higher will be the transient bus load. So, the minimum execution time corresponds to the maximum transient bus load, slowing down communication of other components. The "burst" behavior of CPU_1 depends on the input queue state, which in turn depends on previous executions of other processes.

Resource sharing introduces a confusing variety of run-time interdependencies. They lead to data-dependent "transient" run-time effects that are difficult to find and to debug. System corner cases, i.e., worst-case system states, can be quite complex and difficult to reach in simulation.

So, where do we get the stimuli for performance verification? Function verification patterns are not sufficient since function corner cases do not cover nonfunctional dependencies introduced by resource sharing. The situation here is very similar to the functional hardware test: testing the hardware by testing the system function only does not reach high fault coverage since the hardware-independent function test does not cover all hardware faults. It could at best do so if function testing could be guaranteed to cover all possible input sequences—an impractical requirement. To stress this comparison a little more, a situation like the one in Figure 6-3 is comparable to a short circuit or crosstalk between wires introducing new signal dependencies. This is a known hard problem.

The example in Figure 6-3 also shows that the system corner cases are different from component or subsystem corner cases. So, the system designer cannot rely on the component simulation patterns, nor can the system designer easily add new corner cases due to a lack of detailed component knowledge. To make the situation even more complicated, the corner cases of a software process often depend on the component on which it is executed. As an example, take two processors, one with an integer multiplier and the other using shift-and-add. If there are several execution paths in a process, then the path with multiplication can take longer time on the one processor whereas it is shorter on the other. The difference in execution time can even depend on the input data to be multiplied. More complex situations are introduced when caches are used. There is no hope that a software developer could provide all the corner cases, except with complete execution path coverage which, from software testing, is known to be impractical. We will come back to this problem later.

6.1.4   Chapter Objectives

This book chapter has two objectives. It aims to give a better understanding of the target architecture run-time effects and to review more efficient and reliable alternatives to co-simulation based performance verification based on formal performance analysis. We assume that the following is given:

+ the system function and the environment are modeled by a set of communicating processes (e.g., described in SystemC and/or VHDL),

+ the target architecture,

+ an implementation of the processes and their communication on the target architecture.

We want to model and analyze

+ the target architecture information flow,

+ the system performance.

6.1.5   Structuring Performance Analysis

Formal software performance analysis differentiates two problems, formal process performance analysis, usually in the form of process execution time analysis, and schedulability analysis. The first one analyzes the execution time of a process running on a single processor without resource sharing, and the second one analyzes the effects of resource sharing on one or multiple processors. In the following discussion, we will demonstrate that these problems can be seen as orthogonal and treated independently, as is usually done in real-time computing. However, the classical approaches to schedulability analysis are limited to subsystems, as defined above. As a novelty, we will introduce another level of performance analysis whereby individual subsystem analysis is composed to investigate complete heterogeneous systems.

The approaches to software performance analysis can easily be extended to hardware/software systems, since the respective principles of execution and communication are strongly related. Instead of process execution time analysis, we will look at a more general architecture component performance analysis; schedulability analysis will be extended to a more general subsystem performance analysis complemented by a global system performance analysis.

Figure 6-5 gives an example illustrating the different performance analysis problems throughout the chapter. The target architecture is already known from Figure 6-3. As a subsystem example, we use CP_1 and M_1 running two processes P_1 and P_2 on a software architecture. We introduce activation models that define the activation of the processes. In Figure 6-3, this is shown for P_1.

After this introduction, Section 6.2 reviews architecture component modeling and analysis techniques. The results are the execution time and the communication volume of a process or the timing of a communication element. To analyze the total load, the frequency of process activation and communication must be determined. This is investigated in Section 6.3. Up to this point, the individual process or communication is regarded. In Section 6.4, resource sharing is introduced, including standard scheduling and performance analysis approaches. Current resource sharing approaches are focused on homogeneously scheduled

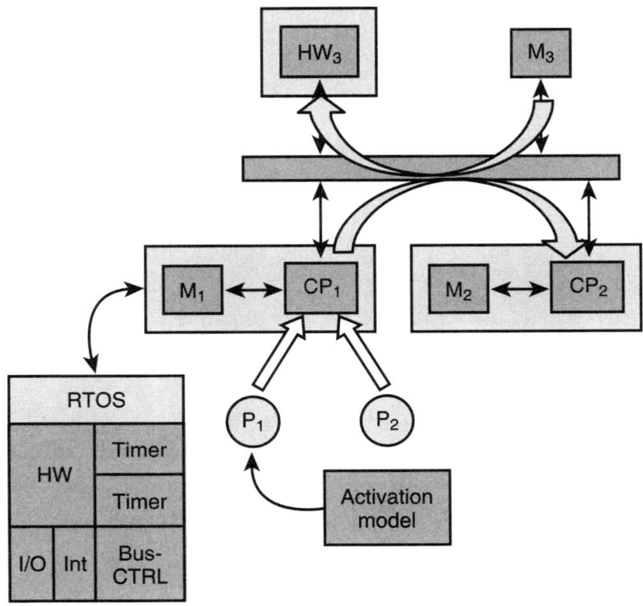

6-5　　Performance model structure.

FIGURE

systems. Section 6.5 presents three approaches to the analysis of heterogeneous systems consisting of several differently scheduled subsystems. Section 6.6 draws a conclusion.

6.2　ARCHITECTURE COMPONENT PERFORMANCE MODELING AND ANALYSIS

For modeling and analysis at the component level, we distinguish the three architecture component types, processing elements, communication elements, and memories. We start with processing element modeling and analysis.

6.2.1　Processing Element Modeling and Analysis

To enable subsystem analysis, processing element modeling and analysis must not only provide the execution time but also the communication activity. The latter is needed to determine communication load and, hence, communication component timing.

The execution time is determined by the execution path and the function implementation. To define execution path, we first assume that a component performs a sequence of actions when it is activated. This sequence can be defined by a software program, or by a state machine (FSM) in case of a hardware component. Then, the unique sequence of actions that is executed for a given initial state and for given input data is an execution path.

The execution time of an execution path is the time needed for the execution of that path. An execution path does not necessarily cover a whole program, but how much time is spent between two communication or synchronization statements can be of equal interest.

Figure 6-6 shows a small example to explain the relation of execution path timing and component timing. The execution path is determined by control statements in the program or FSM, i.e., by branches and loops. The conditions that control branches and loops depend on input data or on internal control variables, such as in the case of a fixed loop. In between control statements are basic blocks, in our case b1 to b4. Basic blocks have a single entry and a single exit point for all execution paths. Since there are no internal branches, a basic block has a single

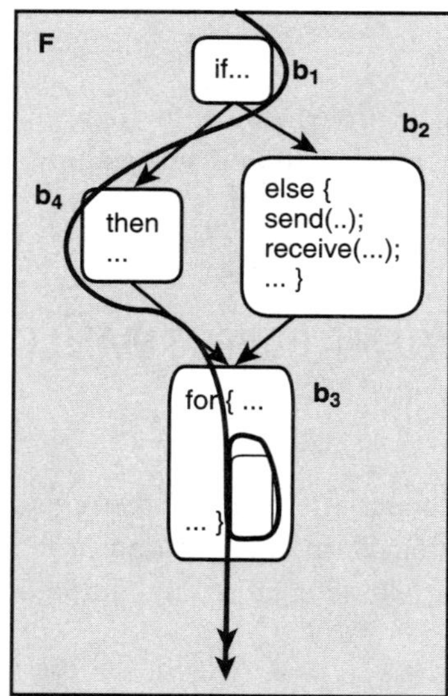

6-6   Execution path.

FIGURE

execution path. That property has been exploited for optimization in compilers and high-level synthesis tools.

If the execution path of a component depends on the input data, then that component will have data-dependent execution times. The execution paths are hardware architecture-independent. An accurate approach to execution time analysis must conservatively cover all possible execution paths.

The second parameter, function implementation, captures the target architecture influence. It includes the component hardware architecture, the software architecture, and, if used, the compiler or synthesis tool.

In practice, processing element analysis today is based on process simulation. Process execution times are measured or simulated using, e.g., breakpoints. The simulation pattern are, e.g., taken from component design. Data-dependent execution paths and data-dependent instruction execution lead to observed upper and lower timing bounds. Even when subsystem performance analysis uses formal methods, the basic component data are derived via simulation.

6.2.2   Formal Processing Element Analysis

With increasing component and function complexity, processing element simulation becomes less reliable due to the increasing number and complexity of worst-case situations that must be covered. On the other hand, formal analysis has for a long time been too inefficient, providing very conservative timing bounds and requiring user support. There has been significant progress in recent years. Commercial formal tools at the process level are just about to become available but still lack library support [238].

To explain formal analysis, we return to Figure 6-6. In principle, the execution time of a program path can be determined by summing up the execution times of all basic blocks along an execution path. In case of pipelined architectures, there will be overlap between the execution of subsequent basic blocks that must be taken into account. To determine the architecture-dependent timing interval, basic blocks can be analyzed for best- and worst-case execution times, and the overlap can be considered for the best case (e.g., perfect pipelining) and worst case (pipeline empty). This leads to the following simple equation:

$$t_{\mathrm{pe}}(F, pe_j) = \sum_I t_{\mathrm{pe}}(b_i, pe_j) \cdot x(b_i) \tag{1}$$

where F is the process, b_i are the basic blocks of F, $x(b_i)$ is the execution path-dependent number of basic block executions, and $t_{\mathrm{pe}}(b_i, pe_j)$ is the execution time (also called the running time) of b_i when it is executed on processing element pe_j. We would now have to solve the equations for all possible paths. Li and Malik [239] had the idea to consider Equation 1 as an integer linear programming (ILP) optimization problem to be solved for the minimum, corresponding to the best-case execution time (BCET), and for the maximum, corresponding to the worst-case execution time (WCET). The program structure is given as a constraint. This "implicit path enumeration" technique was an important step, but it still required user interaction to define the dependencies between control statements. Many of these dependencies can be found by a symbolic evaluation of data dependencies, called abstract interpretation [240]. Another improvement is the clustering of basic blocks to longer program segments with a single execution path (single feasible path [SFP]), such as a fixed loop structure implementing a digital filter or an FFT. This reduces the problem size, and analysis can provide tighter upper and lower performance bounds, as shown in Wolf [241] and Wolf et al. [242].

In more complex component architectures, execution paths and function implementation cannot always be treated independently. Caches have a significant performance impact, but cache conflicts introduce non-functional dependencies between data variables and between program segments. For direct mapped instruction caches, Li and Malik [239] have introduced a directed cache conflict graph. The graph nodes represent program segments corresponding to cache lines, and the edges represent possible conflicts derived from the control flow graph. For set associative caches, a more complicated cache state transition graph is used.

Once the cache architecture-dependent graph is generated, implicit path enumeration can be applied to cache analysis, leading to the cost equation:

$$t_{\text{cache}} = \sum_{i=1}^{N} \sum_{j=1}^{n_i} \left(t_{i,j}^{\text{hit}} \cdot x_{i,j}^{\text{hit}} + t_{i,j}^{\text{miss}} \cdot x_{i,j}^{\text{miss}} \right) \tag{2}$$

where N is the number of basic blocks and n_i is the number of cache lines in block b_i. $x_{i,j}^{\text{hit}}$ and $x_{i,j}^{\text{miss}}$ are the numbers of cache hits and misses, respectively, on line j in block b_i. The time t_{cache} of Equation 2 is added to Equation 1 assuming non-overlapping cache operation. Overlapping cache operation such as cache line prefetch or write buffering can, e.g., be treated by taking the larger of the minimum values of Equations 2 and 1 as BCET. Hit time $t_{i,j}^{\text{hit}}$ and miss time $t_{i,j}^{\text{miss}}$ are memory-dependent and will be discussed later.

Clustering of adjacent basic blocks to cache lines [243] reduces the node count of the graph. Clustering of basic blocks to SFP segments further reduces the node count and, combined with data dependency analysis, increases analysis precision [241]. Analysis of data caches is harder, in particular if pointer analysis shall be involved to increase precision. Several approaches have been proposed with different limitations, abstract interpretation [244], data flow analysis [245], and single data sequence (SDS) analysis [241]. The commercial tool Abslnt [238] uses abstract interpretation.

Determination of $t_{pe}(b_i, pe_j)$ is the second challenge of formal analysis. For simple component architectures, it is sufficient to add the instruction execution times of basic block b_i [239], whereas more complex architecture models include pipelining effects [246]. In Ferdinand et al. [244], the reader finds a modeling case study of the Motorola Coldfire processor. Since there is only a single execution path in b_i, it is also possible in principle to simulate b_i or, even better, the larger SFP [241], using standard cycle true simulation models.

Formal analysis can also provide the communication activity of a process. Figure 6-6 shows a send and a receive function in one of the basic blocks. Communication is in general data-dependent, just like execution time. The data volume communicated can be expressed by:

$$s(F) = \sum_I s(b_i) \cdot c(b_i); \quad r(F) = \sum_I r(b_i) \cdot c(b_i) \tag{3}$$

with $s(b_i)$ being the data volume sent in b_i, and $r(b_i)$ being the data volume received. Given the target architecture-dependent data encoding and protocol, the ILP optimization approach used for timing analysis can be adopted to analyze the minimum and maximum data to be communicated, as shown in Wolf et al. [242].

6.2.3 Communication Element Modeling and Analysis

Communication element behavior is typically far less complex than processing element behavior. More complexity is found in communication links with protocols that are used to share the communication resource or to ensure correct transmission. Resource sharing protocols will be deferred to a later section, whereas transmission protocols can either be added to the communication latency or modeled as running on the processing elements on which they are implemented. This is a matter of modeling granularity. In effect, the communication time will be an interval, just like the execution time of a process.

Communication element modeling and analysis follow the same principles as processing element analysis. In addition to simulation, profiling and statistical load analysis are popular performance validation approaches. This appears useful if the source of a communication is statistically characterized and/or if communication latency and memory cost are not critical. None of these assumptions hold on an MPSoC running an embedded system. On the other hand, communication element behavior is not statistical in nature. For some important resource sharing protocols, there are simple analysis algorithms if the individual transmission time is known. Therefore, we want to include formal analysis mechanisms for communication elements.

6.2.4   Formal Communication Element Analysis

We start with a simple point-to-point communication link, as in Figure 6-7. Suppose processor CP_i sends data over communication element CE_j to processor CP_k. Suppose CE_j transfers data word by word. Then the communication time to transfer the data volume x, $t_{com}(x, ce_j)$ grows linear with the transmitted data, as given in Equation 4:

$$t_{com}(x, ce_j) = x \cdot t_{com}(1, ce_j) \tag{4}$$

$$t_{com}(x, ce_j) = \left\lceil \frac{x}{pl_{ce_j}} \right\rceil \cdot t_{com}(pl_{ce_j}, ce_j) \tag{5}$$

For packet transmission with a fixed packet length pl, the formula is given in Equation 5. This simple yet accurate model has, e.g., been used for optimization in hardware/software co-design (e.g., ref. 247). We might need to extend Equation 5 by packet header cost, which corresponds to a simple additive term.

6-7   Communication link: example.

FIGURE

A one-to-many link can logically be split into many one-to-one links with possibly different timing. Many-to-many links, i.e., busses, originate in resource sharing and are treated in the respective chapters. Finally, t_{com} ( ) can be an interval to include transmission error handling, for example. Hence, the communication time can be a time interval.

6.2.5   Memory Element Modeling and Analysis

The different types of memories shall be classified according to their timing properties. In SRAMs, access timing is independent of the memory access sequence. FLASH and EEPROM have different read and write times but are otherwise similar to SRAM memory. Dynamic RAM has an internal multiple bank structure to reduce the effect of different row and column access times by interleaving. Since the memory words are much wider than the communication links, words are best communicated via burst transmission, which is exploited in SDRAM, DDRAM, or Rambus architectures. Dynamic memories are inevitable for many MPSoC applications with large memory requirements, e.g., in multimedia applications, regardless of whether they are implemented on chip or externally.

On MPSoC, instruction and data memories are typically FLASH or SRAM. Their latency time is usually included in the instruction fetch and memory access times and is thus included in process element analysis.

DRAM timing analysis depends on the memory architecture. If caches are used for DRAM access, the access times are part of the cache miss costs. A cache miss leads to a cache line transfer, which corresponds to a fixed-length sequence of DRAM accesses. Since the communication link between cache and memory is typically smaller than the cache line size, the cache line transfer translates to a burst transfer to or from the DRAM. Modern DRAM architectures, such as SDRAM or DDRAM, use a burst transfer protocol with a fixed total time. This is the required $t_{i,j}^{miss}$.

If there is no cache, then the burst access is reduced to a single word transmission. Modern DRAMs treat such memory accesses as bursts of length 1. The same holds for caches with write-through mechanism.

If the communication link between memory and processing element or the memory itself are shared resources, then the access time is nonfunctionally dependent on other processes. This makes accurate single process simulation hard and raises an interesting analysis problem which we will discuss in the context of resource sharing.

6.2.6   Architecture Component Modeling and Analysis: Summary

We summarized approaches to modeling and analyzing the execution times and communication volumes of processing elements, communication elements, and memories. As an alternative to simulation and measurement, we outlined formal approaches that can be combined to analyze single process execution. The models included memories and communication.

6.3   PROCESS EXECUTION MODELING

So far, we have focused on the timing of a single process execution or communication activity. To determine the load of a component, the number of such executions per time is needed.

6.3.1   Activation Modeling

To model the execution of a process, we must know how the process is activated. We can distinguish two very different types of processes. A process can be started once and then run forever, following an internal loop. The execution time of such a process is infinite, and the communication volume is, in principle, infinite, as well. In this case, the process and communication take whatever resources are available. If such a process uses blocking read then the timing depends on other processes and, eventually, on the environment as primary sources of data. Hence, the behavior of systems of infinitely looping processes is only limited by the environment and by the available processing and communication performance. It is hard to analyze delay times and buffer requirements of systems with such processes, and there are few general resource sharing algorithms applicable, mainly the time-driven algorithms that assign time slices to processes and communication (see Section 6.4.3).

Here, we would like to focus on the second type of processes, i.e., processes with finite execution time. Such processes can be time- or event-activated. Time activation means that a process is executed when a given time has elapsed. Typically, a time-activated process is activated periodically. The process reads input data, but activation is input data-independent. Event-activated processes are activated upon arrival of an event and/or because a certain input data condition holds. The implementation of event-activated processes often uses interrupts caused by events, whereas time-activated processes often use timer-generated interrupts and poll the input data.

In the following, we subsume both activation types as activation models. Together with the execution time and communication volume of a process, the activation model defines the system load caused by a process.

Events are generated by the environment or by other processes. Often, several events are needed to activate a process, such as a packet router, which only routes a packet when the packet has completely been received. On the other hand, a dense sequence of events may lead to overlapping executions of the same process. Last but not least, a sequence of events may follow a certain pattern that can be exploited to better utilize a resource.

Other than simulation that considers each event individually (even though there are first proposals to bundle events in simulation as "event signatures" [248]), formal analysis uses abstract event models defining event sequences. In the following, we will introduce some of the important event model classes used in real-time processing:

* Events with periods

  * Periodic events

$$t_{e_{i+1}} - t_{e_i} = t_p$$

  * Periodic events with jitter

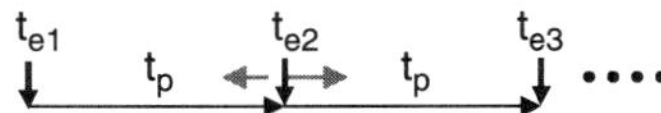

$$t_p - j \leq t_{e_{i+1}} - t_{e_i} \leq t_p + j; \quad j < t_p$$

  * Periodic events with bursts

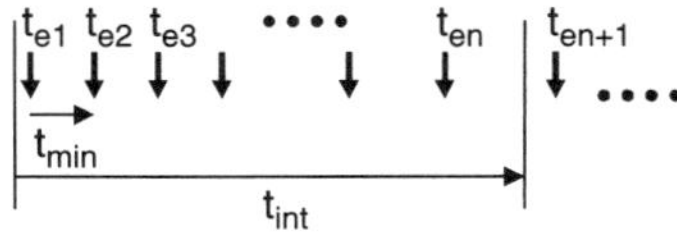

$$t_p - j \leq t_{e_{i+1}} - t_{e_i} \leq t_p + j;$$

$$t_{e_{i+n}} - t_{e_i} \geq t_{\text{int}}; \quad t_{e_{i+1}} - t_{e_i} \geq t_{\min}$$

* Events with minimum interarrival time

  * Sporadic events with minimum interarrival times

$$t_{e_{i+1}} - t_{e_i} \geq t_{\min}$$

◆ Sporadic events with burst

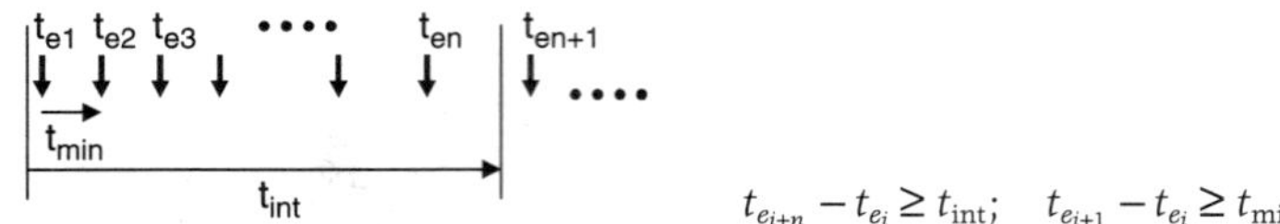

$$t_{e_{i+n}} - t_{e_i} \geq t_{int}; \quad t_{e_{i+1}} - t_{e_i} \geq t_{min}$$

Periodic event sequences are typical for signal processing and control applications. The events are generated by periodic sampling of an A/D converter. Periodic events with jitter usually originate from a periodic source, but the timing has been distorted by preceding processes or communication. If the jitter becomes large enough, then the sequence turns into a burst sequence, which is then bounded by the minimum interarrival time and an "outer" period t_{int} limiting the event frequency over a longer time.

If the source is not periodic, then we need at least some knowledge about the distance between activations and, hence, between event arrivals, to determine a maximum system load. So the second group comprises event sequences with minimum interarrival times. The most general model is the sporadic event model with minimum interarrival times. As in the case of periodic signals with bursts, there is a second model, sporadic events with bursts, which allows higher transient event frequencies.

6.3.2   Software Architecture

For a complete picture, we want to discuss briefly the component hardware/software architecture impact. Figure 6-8 shows a simple example of a two-processor architecture with a hard maximum delay time between the sensor and the actuator. The dashed arrows show the event path from the sensor signal to the actuator. The example illustrates that the application processes are only a small part of the chain of activations necessary to trigger the actuator. All hardware and software components must be included in path delay calculation. For that purpose, software libraries, hardware components, and communication must be fully characterized, whether by formal analysis, by simulation, or by in-system measurement.

6.3.3   Process Execution Modeling: Summary

We summarize that the activation model defines the frequency and context of process activation. In the discussion, we have focused on processes with a finite

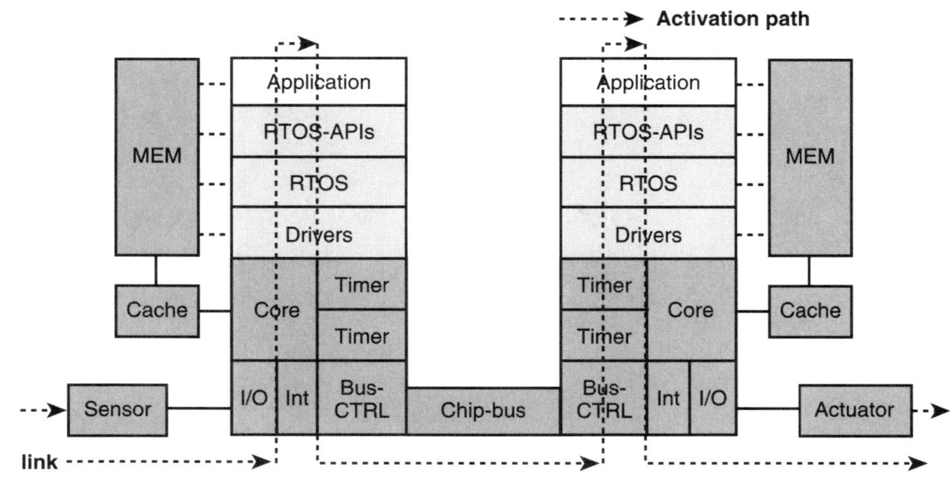

6-8

Software event path: example.

FIGURE

execution time. Periodic time activation is a simple model whereby the process is activated independently of the system state. Event activation is more flexible but needs upper bounds of the activation frequency to limit the system load.

The same activation models discussed for process execution apply to communication.

6.4   MODELING SHARED RESOURCES

Processes and communication activities share the MPSoC resources to which they are mapped.

6.4.1   Resource Sharing Principle and Impact

Resource sharing requires resource arbitration, i.e., scheduling, and context switching. Scheduling strategies can be divided into

◆ Static order versus dynamic order scheduling and

◆ Preemptive scheduling (interrupt) or non-preemptive ("run-to-completion") scheduling.

Context switching implies overhead. On a processing element, pipeline states and register contents must be saved and restored, and caches must be (partially) flushed. On a communication element, context switch includes all bus arbitration overhead. Context switch overhead in memory access is usually negligible since modern DRAM types operate using fixed transactions, independent of the process issuing a memory access (except the processes reprogram the DRAM interface), and all other memory types have no context defining internal states. Context switching time is mostly constant and can, therefore, usually be determined at design time. Scheduling effects, however, are highly execution time dependent.

In this section, we will look at both process scheduling and communication scheduling. There are three main classes of scheduling strategies:

+ Static execution order scheduling

+ Time-driven scheduling with subclasses of

    + Fixed time slot assignment

    + Dynamic time slot assignment

+ Priority driven scheduling with subclasses of

    + Static priority assignment

    + Dynamic priority assignment.

The efficiency of these models depends heavily on the activation model, as we will see in the following.

6.4.2   Static Execution Order Scheduling

Figure 6-9a gives an example of two processors CP_1 and CP_2 communicating over a shared communication element CE_1. CP_1 runs processes P_1, P_2, and P_5, CP_2 runs P_3 and P_4. Figure 6-9b shows a Gantt chart of the scheduling sequence.

CP_1 and CP_2 start processing P_1 and P_3, respectively. When P_1 has finished, it sends data to the dependent process P_4, followed by a context switch of CP_2 to P_2. P_4 sends data back to CP_1 which enables P_5 to execute. Now, all processes have been executed and the scheduling sequence can repeat. As seen in the figure, static execution order scheduling is applicable to both process element and communication scheduling.

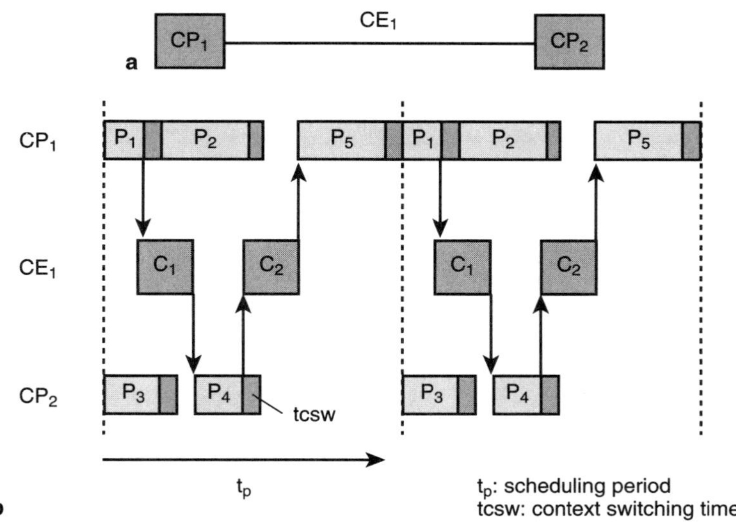

6-9

Static execution order scheduling. (a) Example architecture. (b) Schedule.

FIGURE

Static process execution has a number of important advantages. It supports interleaved utilization of processing and communication elements minimizing idle times, since there is full control on the execution order (e.g., refs. 249 and 250). For the same reason, buffer sizes can be efficiently optimized. Sequences of processes on one processing element can be clustered into one process and then compiled. The compiler will implement an optimized context switch and might be able to find more optimizations across processes. The scheduler is easily implemented as a state machine.

6.4.3 Time-Driven Scheduling

Time-driven scheduling is a very flexible scheduling strategy. It assigns time slices to processes or communication links independent of activation, execution times, or data dependencies.

TDMA

The time division multiple access (TDMA) strategy keeps a fixed assignment of time slices to processes or communication links. This assignment is periodically repeated. Figure 6-10 shows an example. Process P_1, P_2, P_3, and P_4 are assigned 12, 10, 5, and 13 ms, respectively. This results in a total period t_{pTDMA} of 40 ms. The total execution time of P_1 is 45 ms, such that it ends at $t_r = 129$ ms. t_r is the response time of P_1. After that time, the P_1 slot remains idle until P_1 is activated again. For simplicity, we have omitted the context switching times in the figure. P_2 has an execution time of 23 ms and a response time of $t_r = 95$ ms. It is again activated at $t = 150$ ms and continues execution at $t = 172$ ms. P_3 with an execution time of 54 ms has a response time of $t_r = 426$ ms.

The greatest advantages of TDMA are predictability and simplicity [251]. Processes or communications with arbitrary behavior and activation can be merged on one resource without influencing each other. In effect, the available performance is scaled down according to Equation 6. (For readability, we use this simple upper bound.) This is an excellent property for integration that has been exploited in many integration applications, such as in automotive design (TTP bus) or by the on-chip MicroNetwork offered by Sonics as communication IP [252,253]. (The MicroNetwork has more options–see below.) The main limitations are efficiency and long total response times. There is some flexibility since the time slots can be adapted at system startup time.

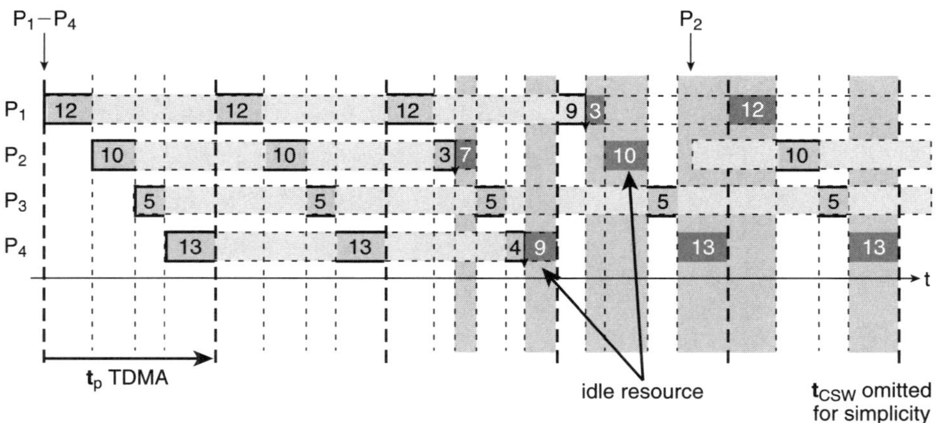

6-10   Scheduling and idle times in TDMA.

FIGURE

$$t_{\mathrm{peTDMA}}(P_i,\, pe_i) = \left\lceil \frac{t_{\mathrm{pe}}(P_i,\, pe_i) - t_{\mathrm{csw}}}{t_{\mathrm{Pi}}} \right\rceil \cdot t_{\mathrm{pTDMA}} \tag{6}$$

Round Robin

Round-robin scheduling departs from the fixed time slot assignment and terminates a slot if the corresponding process ends. Therefore, slots are omitted or shortened, and the cycle time, t_{RR}, of the round-robin schedule is time-variant. Figure 6-11 shows the example of Figure 6-10, this time for a round-robin strategy. P_1 now ends at $t_r = 113\,\mathrm{ms}$, but, more impressively, P_3 has a response time of $t_r = 179\,\mathrm{ms}$.

Round robin avoids the idle times of TDMA and reaches maximum resource utilization. On the other hand, process execution is no longer independent, losing the most important integration property of TDMA. P_3 only finished so quickly because the other processes were not executing. However, round robin guarantees a minimum resource assignment per process, since under full load conditions it falls back to a TDMA schedule. This is suitable for applications with soft deadlines and quality of service requirements in which a given resource level must be guaranteed.

Again, round robin is applicable to communication and processing. It is found in many applications, such as in the Sonics MicroNetwork for on-chip interconnect [252,253], or in standard operating systems.

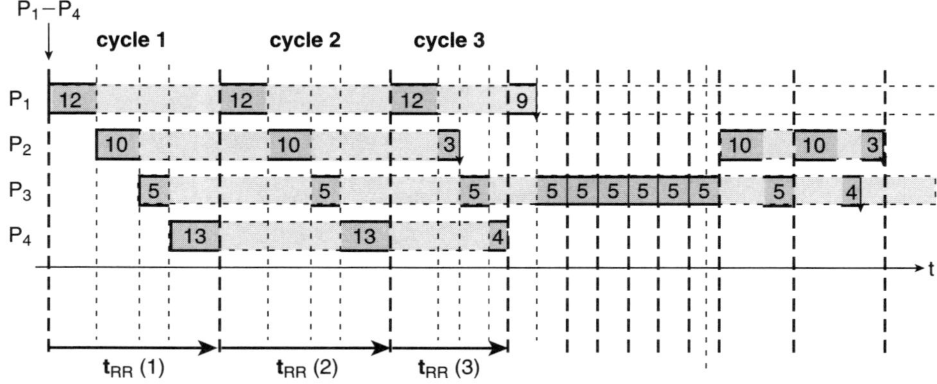

6-11   Round-robin scheduling.

FIGURE

6.4.4   Priority-Driven Scheduling

Static Priority Assignment

The third class of scheduling strategies uses process or communication priorities. Static priority assignment allows one to offload the scheduling problem to a simple interrupt unit. Vectorized interrupt units reducing interrupt latency and context switching overhead are found even in small 8-bit microcontrollers, such as the 8051. A scheduler process or control unit is not needed. Finally, there is efficient analysis and optimization algorithms are available.

We discuss three different static priority assignment strategies that differ in their activation model. Again, we will look at processing elements, but the same discussion applies to communication.

Model 1. Processes are activated by the arrival of an input event. Input events are periodic with jitter. The process deadline is at the end of the period. Therefore, process execution must be periodic with jitter, as well. The input event and, hence, the process execution rates may have different periods.

This classical and widely used model was first investigated by Liu and Layland [254]. They proved that the optimal solution for single processors is to order the process priorities according to increasing process execution rates, i.e., the process with the shortest period is assigned the highest priority. This rate monotonic scheduling (RMS) is very popular in embedded system design due to its simplicity and ease of analysis. It has, e.g., been extended to cover synchronization for mutual exclusive resource access and multiprocessing [255]. Deadline monotonic scheduling (DMS) is a straightforward extension of RMS for deadlines smaller than a period. A nice property of RMS and DMS in the context of more complex systems is that the processes finish periodically with jitter. This property allows control of buffer sizes and the load on other system parts that use the output events of these processes as input.

Model 2. Like model 1, except that a subset of the processes is dependent such that a process P_2 that depends on P_1 can only be executed if process P_1 has been finished in this period. Obviously P_1 and P_2 must have the same period. This relation can be represented in a task graph.

This model has been investigated by Yen and Wolf [256]. The dependencies constrain the possible event jitter. They also imply a static process order for each period. Both can be exploited for better utilization of processors and communication links, for single processors as well as multiprocessors. In effect, the schedule within a period appears like a loosely coupled version of static execution order scheduling.

Model 3. Like model 1, but with arbitrary deadlines.

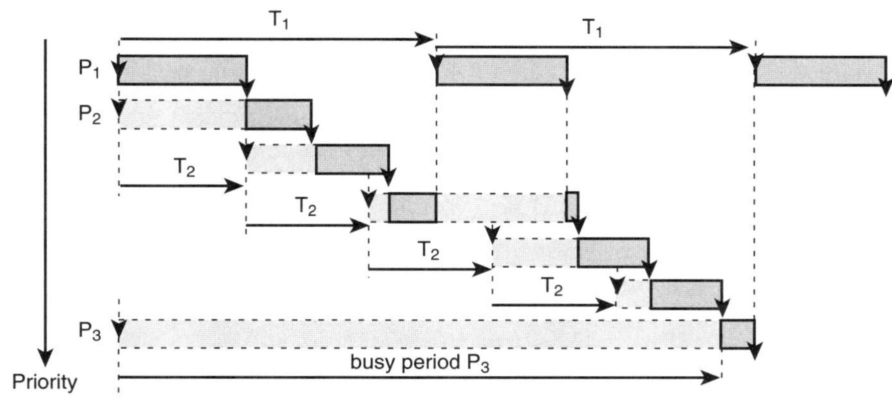

6-12

Static priority scheduling with arbitrary deadlines.

FIGURE

This seemingly small change has a major impact on optimization, analysis, and system load. On the other hand, this model is frequently found in more complex systems, in which a deadline covers multiple components and subsystems. Take the example in Figure 6-12. P_1 with execution period T_1 has the shortest deadline and is assigned the highest priority. P_2 with execution period T_2 is next in priority and P_3 is last. The first input events are assumed to be available at $t = 0$, a known worst-case situation [254]. Execution starts with the highest priority task P_1. If it has terminated, P_2 will be executed. Before P_2 has even started, the next input event for P_2 has arrived at $t = T_2$. As soon as P_2 ends, it is started again to process the buffered second input event. The third input event arrives at $t = 2T_2$, before the second P_2 execution has ended. The third execution of P_2 is interrupted because the second event for P_1 has arrived at $t = T_1$. P_2 can only resume at the end of T_1. P_3 can only be processed shortly before $t = 2T_1$.

The execution sequence is rather complex. With its input buffers filled, P_2 execution runs in a burst mode with an execution frequency that is only limited by the available processing element performance. This burst execution will lead to a transient P_2 output burst that is modulated by P_1 execution. A larger system with more priority levels generates complicated event burst sequences. Using another processing element (or a faster bus if we apply this scheduling strategy to communication) will increase the P_2 burst frequency and, consequently, the transient load caused by P_2 output. This is another illustration of why MPSoC integration is that complicated (see Fig. 6-3).

On the other hand, we can see in the example that P_2 execution frequency can be easily bounded over a larger time interval. This observation justifies the introduction of event burst models and corresponding analysis approaches.

There are many solutions to the analysis and optimization of model 3 systems. Lehoczky [257] provided a solution to timing analysis for arbitrary deadlines, Audsley et al. proposed an iterative heuristic algorithm for priority assignment [258]. Finally, Tindell and Clark presented an algorithm [234] that extends model 3 to handle periodic input events with jitter and bursts, effectively introducing input—output event model compatibility and, thus, interoperability for static priority systems with arbitrary deadlines. However, this approach applies to single processing element systems only.

Dynamic Priority Assignment

Static priority algorithms cannot reach maximum resource utilization, not even for the simple model 1 [254]. To reach higher resource utilization (or shorter deadlines), the priority must be assigned dynamically at run time. The best dynamic priority assignment strategy is the one that gives the process with the earliest deadline the highest priority. This is shown in Liu and Layland [254] as well. The advantage of this "earliest deadline first" (EDF) scheduling is its flexible response to input event timing and process execution times. However, it depends on the availability of hard deadlines for each process. In MPSoC, these deadlines must be derived from global system requirements. If the process execution times are known, then such deadlines can be derived from task graphs, as already shown in Blazewicz [259] and later extended to multirate periodic process systems with circular dependencies in Ziegenbeine et al. [260]. There is a host of work in the domain of dynamic priority assignment and the respective timing analysis for different input models (see, e.g., Gresser [261]), which cannot be discussed here.

In any case, dynamic priority assignment requires a scheduler process running the assignment strategy and thereby observing the (local) system state. Therefore, in MPSoCs, it is practically restricted to run on microcontrollers. There, it adds scheduling overhead to the context switching overhead and eventually also increases power consumption.

6.4.5   Resource Sharing: Summary

The efficiency of resource sharing strategies is largely dependent on the activation model. Time driven scheduling is very robust but reaches less efficiency than other strategies or is only suitable for best effort applications. Static order scheduling reaches highest efficiency but also imposes the tightest constraints on the input event stream and on narrow execution time intervals to reach this efficiency. Static priority scheduling provides good adaptation to a wide range of input event model parameters but can create burst event sequences for the general case of arbitrary deadlines. Static priority scheduling is well supported by algorithms for analysis and optimization. Dynamic priority scheduling provides the highest flexibility but incurs significant scheduling overhead.

For all resource sharing strategies presented in this section, formal analysis techniques have been proposed. Many of these can even be applied manually. Only a small selection has been presented here, which is, however, sufficient background to understand the next level of global performance modeling and analysis.

6.5   GLOBAL PERFORMANCE ANALYSIS

None of the resource sharing strategies is sufficiently optimal to be used for all functions of an MPSoC. In any sufficiently complex MPSoC, there will always be parts using different scheduling strategies, i.e., subsystems in our terminology. In this last section, we want to look at solutions to analyze such subsystem combinations. Figure 6-13 gives an example. There is a RISC processor using dynamic scheduling, a VLIW processor with static priorities, a DSP with static execution order scheduling, e.g., running a filter, and several simpler schedules of coprocessors. The system bus uses TDMA scheduling.

The approaches to analysis of such compound systems can be divided into three classes:

+ coherent analysis for several subsystems.

+ event model generalization for a set of scheduling strategies.

+ event model adaptation.

Due to lack of space, we will briefly introduce the first two approaches and then focus on the latter one.

An obvious approach is to extend the scope of analysis to cover all scheduling strategies of a system coherently. A look at Figure 6-13 tells us that such a "holistic" approach will be very complex in general, since there is a huge variety of possible combinations. But there are some frequent configurations for which coherent analysis proved feasible and efficient. Tindell and Clark [234] and Pop

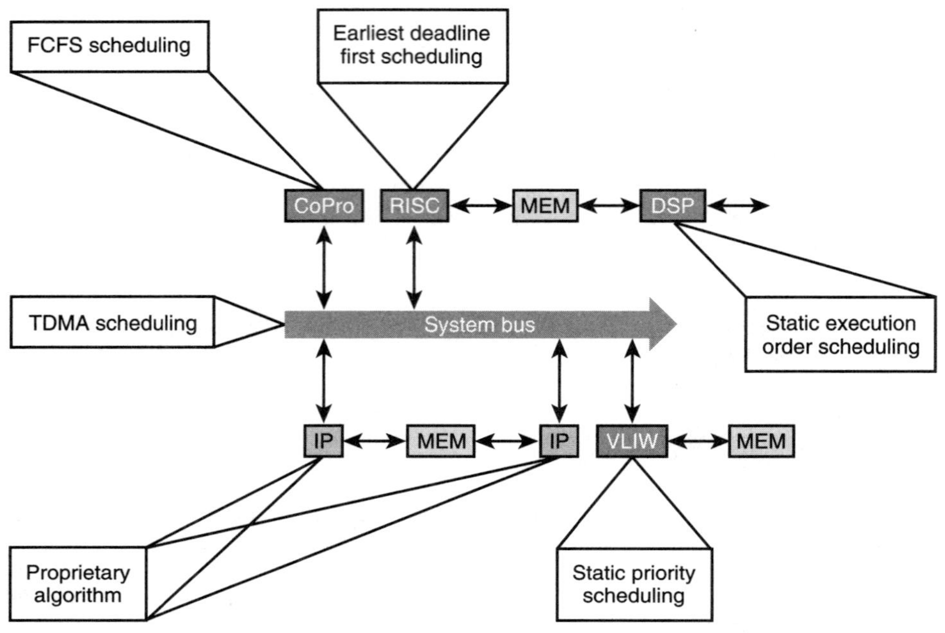

6-13   MPSoC with multiple scheduling strategies.

FIGURE

et al. [262,263] have investigated static priority process scheduling combined with TDMA communications. Such a combination is, e.g., found in automotive electronics, in which a standardized TDMA bus (e.g., with a TTP protocol) connects control units running a standard RTOS (here OSEK) using static priorities. TDMA busses alone do not appear sufficiently efficient for MPSoC on-chip communication, however. There is rather a tendency toward complex bridged busses, as we have already seen in the VIPER example in Figure 6-1.

The other two approaches, event model generalization and event model adaptation, regard a system as a set of independent subsystems communicating via event streams. Communication, then, corresponds to event propagation. Figure 6-14 shows two subsystems S_1 and S_2 communicating via one or several event streams. The figure distinguishes an input event model that is the type of event model that a subsystem assumes as its input, and an output event model that the subsystem generates as a result of scheduled process execution (or communication). So far, we have mainly been interested in the input model, but given an input model and a scheduling strategy, we can derive an appropriate output event

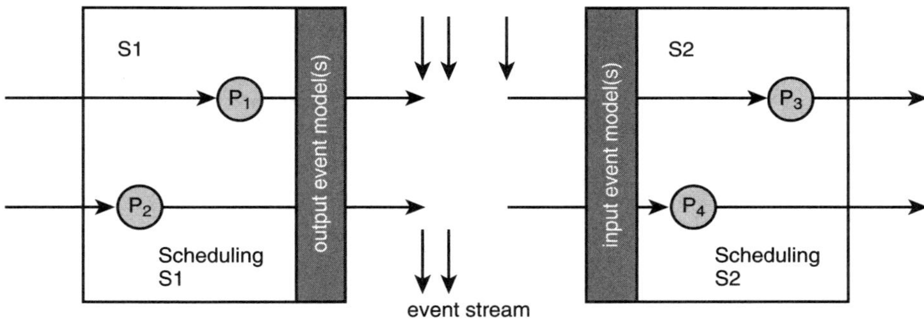

6-14   Subsystem coupling via event streams.

FIGURE

model. A static process order scheduling with periodic input generates a periodic output event stream with jitter, a TDMA scheduling adds jitter to whatever input model it gets, and a static priority scheduling with arbitrary deadlines takes periodic input events with jitter and generates periodic output events with bursts, just to name a few examples. Each subsystem can have different input and output models.

Event model generalization tries to cover all possible event models between subsystems by a single, general event model. Then, all global analysis and optimization techniques are based on this single "interface" event model. An early approach was introduced by Gresser [261]. His generalized model uses a set of vectors to declare the number of possible events in a given time interval. In Thiele et al. [264], a more intuitive model consisting of minimum and maximum event arrival curves is defined. Event processing is defined as service curve, again given as minimum and maximum service curve. The difference between both curves determines buffer contents and delay times. Given the process communication volume [cf. s() and r() functions in Section 6.2.2], output event curves can be calculated for a given service curve, again resulting in a minimum and a maximum curve. These curves are then used as input event arrival curve for the next subsystem. As in Gresser's approach, the curves can take an almost arbitrary shape (they must at least be monotonous) which is then linearly (and conservatively) approximated and analyzed with a generalized analysis approach. The approach has been applied to network processor analysis [265].

The event generalization approach trades generality for analysis complexity. The definition using arrival and service curves is intentionally flexible to cover as many event models as possible.

A third approach is to start with the specialized and highly efficient analysis approaches available today and adapt their event models to employ these approaches in a hierarchical analysis. This is the basis for the event model adaptation approach, as presented in Richter et al. [266,267].

Event model adaptation transforms an output event model to an input event model. This transformation is called event model interface (EMIF). We distinguish lossless transformation, whereby the input model can use all information the output model provides, and lossy transformation, whereby the input event model contains less information than available in the output event model. As a simple example, a periodic output event model with period t_p can be transformed losslessly to an input event model that is periodic with jitter, with period t_p and jitter $j = 0$. An output event model with period t_p with jitter j can be transformed to a sporadic signal with minimum distance t_p-j. This transformation is lossy, since the periodic character of the output event stream is lost in the sporadic model.

These transformations are just mathematical functions enabling event propagation for analysis algorithms but do not appear in the target architecture and, therefore, do not cause any overhead. Unfortunately, subsystem interfaces are not always so elegant. A periodic event stream with burst can be transformed to a periodic event stream, but at the cost of an interface function with an internal state, effectively buffering and resynchronizing the event flow. Such a function, called event adaptation function (EAF), must be implemented as a buffer in hardware or software. This buffer overhead is not introduced as transformation overhead, but it is always needed to enable correct systems integration. In Richter and Ernst [266], the event model adaptation approach is explained in detail and proofs are given for the transformation functions.

Figure 6-15 depicts the possible event model transformations for the four commonly used event models. Transformations from sporadic to periodic event flows are missing since such a transformation requires event "stuffing" or nonblocking read, i.e., polling, in the subsystem that expects periodic input events. This is a requirement of the target architecture. Otherwise, such subsystem combinations are not feasible, leading to potential buffer underflows. Underflow hazards would also be detected by the coherent or general event model approaches.

Once the event models are adapted, the global system can be analyzed by performing the local subsystem analysis, which provides a mapping from the input event stream to an output event stream. Then the output event model parameters (t_p, t_{min}, t_{int}, and so on) are transformed to the respective input model parameters. This process iterates until convergence (i.e., a fix point) is reached or until a contradiction occurs. The global analysis approach is presented in Richter et al. [267]. There, we also give an example of how nonfunctional

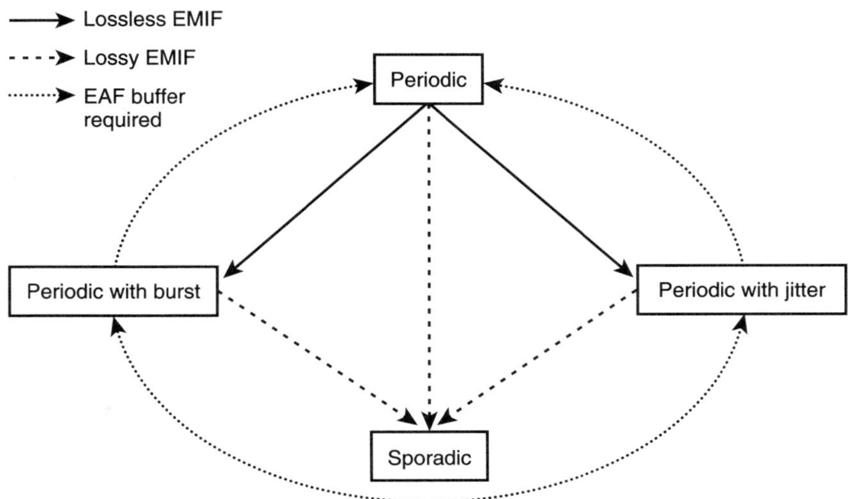

6-15   Event model transformations.

FIGURE

circular dependencies can occur as a consequence of priority assignment and how they can be analyzed.

We close the chapter with an example that we took from Richter et al. [268,269], shown in Figure 6-16. Two subsystems are integrated on one MPSoC and share the communication network-on-chip, NoC. Subsystem 1 consists of a sensor, Sens, which sporadically sends blocks of 8-KB data to process P_1 that runs on a CPU. The sporadic data have a minimum distance of 590 µs, corresponding to a maximum event frequency of 1.7 kHz. Process P_3 runs on the same CPU, is executed with a period of 50 µs, and produces output events of 3 KB at the same frequency. These data are sent to a hardware accelerator module "HW" that can handle sporadic input at a minimum distance of 20 µs. Subsystem 2 consists of several IP modules that periodically send 1-KB data blocks to a DSP. The DSP is statically scheduled to work with this periodic input signal sequence. If taken alone, both subsystems work correctly.

Next, these two subsystems are integrated using a bus (NoC) that employs different packet sizes of 128, 256, or 1024 B with a 6-byte header. The bus protocol uses static priority. Subsystem communication is modeled as logical channels C_1 to C_3, which share the bus. We observe a distortion of the periodic stream over C_3

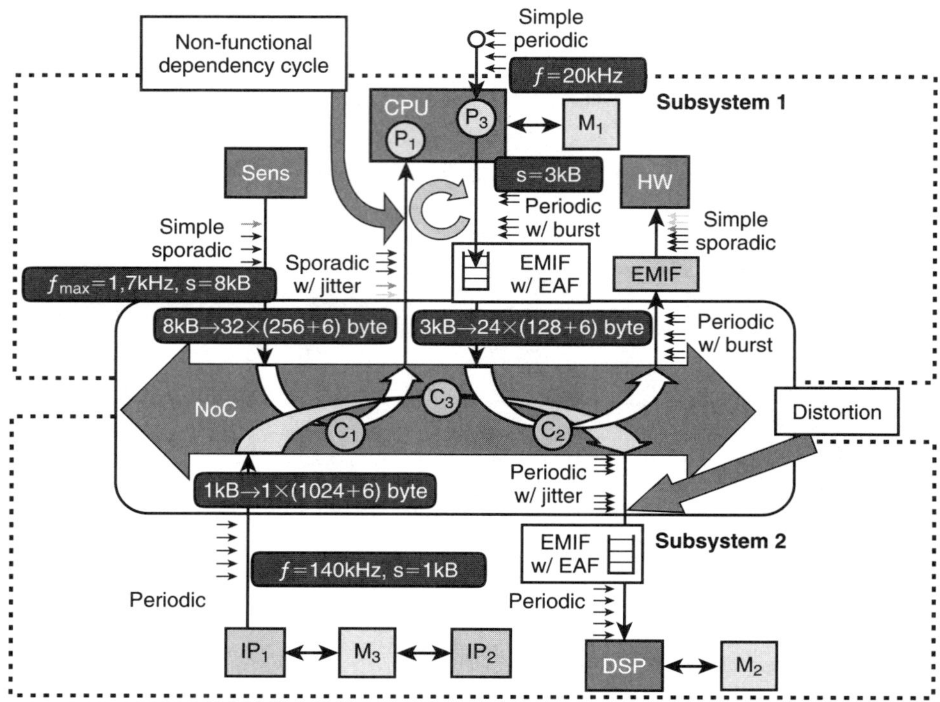

6-16

Example system.

FIGURE

as a consequence of interleaving with the other streams. This leads to a signal jitter, which must be eliminated with a buffer (EAF) to match the periodic input requirement. In a more sophisticated fashion, the input and output streams of the CPU are coupled via the bus, leading to a nonfunctional dependency cycle. That cycle leads to a large jitter, which requires a large subsystem 2 buffer and a large buffer to keep the required $20\,\mu s$ minimum distance at HW. Large buffers are not only costly but lead to long end-to-end latency times of the overall system. To avoid that problem, the designer might decide to insert a traffic "shaper" in the C_2 stream. That traffic shaper effectively breaks the nonfunctional dependency and controls the jitter, leading to a much better system with smaller buffers and lower latency.

Figure 6-17 shows results for different bus speeds. Experiments 1 through 3 have no buffers in C_2, whereas 4 through 6 use a buffer C_2 that reduces jitter and, hence, worst-case latency. In experiment 3, the jitter grows over all bounds. The

	Experiment	Network speed [MByte/s]	Network util [%]	C_1 maximum response [us]	Jitter C_1, out [us]	Buffer size [KB]
	1	480	46	102.6	81.5	19
	2	300	74	466.2	438.4	43
	3	250	89	> 600	-	Deadline violation
	4	480	46	34.7	17.2	18
	5	300	74	96.8	69.0	18
	6	250	89	233.5	199.8	21

6-17

FIGURE

Example system results.

experiments were done with the analysis tool, SYMTA/S [270–273], in less than a second.

6.6 CONCLUSIONS

MPSoCs show complex run-time behavior due to the implementation of multiple hardware, software, and communication scheduling strategies in one system. Resource sharing creates transient dependencies, which are not reflected in the system function. This event "crosstalk" is a hard problem in systems integration. Finding and quantifying the impact on performance and correctness by simulation is increasingly hard. Systematic formal MPSoC performance analysis is possible with a hierarchical approach exploiting knowledge from real-time systems analysis and formal execution time analysis combined with novel approaches to heterogeneous systems analysis.

ACKNOWLEDGMENTS

I would like to thank Kai Richter for helpful comments and support in developing some of the figures.

Design of Communication Architectures for High-Performance and Energy-Efficient Systems-on-Chips

Kanishka Lahiri, Sujit Dey, and Anand Raghunathan

7.1 INTRODUCTION

On-chip communication is increasingly being regarded as one of the major hurdles for complex system-on-chip (SoC) designs [274]. As technology scales into the nanometer era, chip-level wiring presents numerous challenges, including large signal propagation delays, high-power dissipation, and increased susceptibility to errors [275,276]. On the other hand, at the system-level, the integration of an increasing number and variety of complex components is resulting in rapid growth in the volume and diversity of on-chip communication traffic, imposing stringent requirements on the underlying on-chip communication resources. As a result, on-chip communication has started to play a significant role in determining several system-wide metrics, including overall system performance, power consumption, and reliability.

The growing importance of on-chip communication adds a new facet to the process of system design, as illustrated in Figure 7-1. Under the traditional notion of system design, a system's functional requirements are refined and mapped onto a well optimized set of computational resources and storage elements. Today, system designers are required to pay increasing attention to the relatively less understood process of mapping a system's communication requirements onto a well-optimized on-chip communication architecture. We have reached a point at which system design practices need to evolve from being computation-centric to being increasingly communication-aware (Fig. 7-1). Recognizing this, new architectures and system design techniques have started to emerge that address

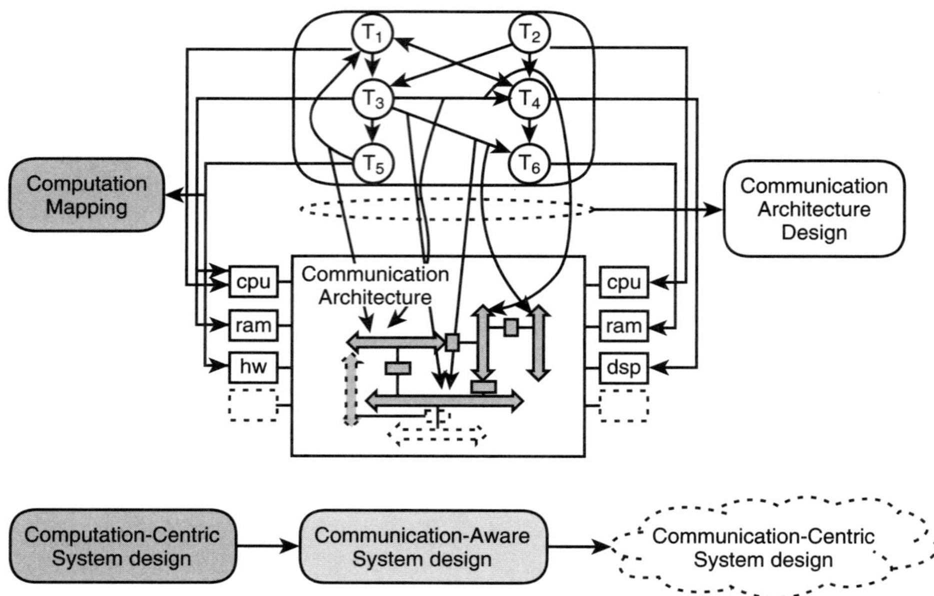

7-1

FIGURE

The two dimensions of system design and the evolution of system design methodologies.

on-chip communication. In this chapter, we describe such recent advances, focusing in particular on techniques for designing on-chip communication architectures for high-performance and energy-efficient SoCs.

We address the problem of communication architecture design for SoCs in terms of the following steps: (1) design or selection of an appropriate network topology, which defines the physical sructure of the communication architecture; (2) design of the on-chip communication protocols, which define the mechanisms by which system components exchange data; and (3) communication mapping, which specifies a mapping of system communications (dataflows) to physical paths in the communication architecture.

This chapter describes advances in tools and design methodologies for communication architecture design that are concerned with the above problems. We point out the significant advantages (performance, energy efficiency) that can be obtained by customizing the communication architecture to characteristics of the communication traffic generated by the SoC components. An important requirement for any communication architecture design and customization environment is the availability of automatic techniques for analyzing the impact of the communication architecture (including the effects of the topology, protocols, and communication mapping) on system performance and power consumption. In this chapter, we provide an overview of alternative approaches for such automatic analysis, describing one such technique in detail. We describe techniques for customizing the design of the communication architecture to exploit the characteristics of the on-chip communication traffic generated by an application.

Several systems exist, in which components may impose time-varying communication requirements on the communication architecture. For such systems, a one-time customization of the communication architecture may often prove ineffective. Hence, techniques are required that can dynamically customize the communication architecture to time-varying requirements. We describe techniques for dynamic (run-time) adaptation of the on-chip communication protocols for achieving high performance in such systems. In this chapter, we also describe techniques for designing communication architectures for energy-efficient systems. We describe two classes of techniques: (1) those that target reducing the energy consumption of the communication architecture itself, and (2) techniques that exploit the communication architecture to improve the energy-efficiency of the entire system. We provide an example of the latter, in which "power aware" on-chip communication protocols are used for system-level power management that aim at improving battery life.

The rest of this chapter is organized as follows. Section 7.2 presents background material on communication architectures, including a survey of typical topologies and communication protocols in use today. Section 7.3 describes automatic system-level analysis techniques for driving communication architecture design. Section 7.4 describes techniques for design space exploration and customization of the communication architecture. Section 7.5 describes techniques that allow adaptive on-chip communication, via dynamic customization of the communication protocols. Section 7.6 describes techniques for communication architecture design for energy-efficient and battery-efficient systems.

7.2   ON-CHIP COMMUNICATION ARCHITECTURES

The on-chip communication architecture refers to a fabric that integrates SoC components and provides a mechanism for the exchange of data and control information between them. The first basis for classifying communication architectures is the network topology, which defines the physical structure of the communication architecture. In general, the topology could range in complexity from a single shared bus to an arbitrarily complex, regular or irregular interconnection of channels. The second basis for classification is the communication protocols employed by the communication architecture, which specify the exact conventions and logical mechanisms according to which system components communicate over the communication architecture. In addition, these protocols define resource management and arbitration mechanisms for accesses to shared portions of the communication architecture. As mentioned in the previous section, communication mapping refers to the process of mapping the system-level communication events to physical paths in the topology.

In the rest of this section, we survey existing and emerging communication architecture topologies and communication protocols. We also describe recent advances in communication interface design. First, we define some basic terminology that is used in the rest of this chapter.

7.2.1   Terminology

Two kinds of components may be connected to the communication architecture. Masters are system components that can initiate communications (reads/writes). Examples include CPUs, DSPs, DMA controllers, and so on. Slaves are system components (e.g., on-chip memories and passive peripherals) that do not initiate communication transactions by themselves but merely respond to transactions initiated by a master. Bridges or routers allow communication between component pairs that are connected to different communication channels. Bridges may support one-way or two-way communications and may themselves consist of multiple master and slave interfaces. Communication resource management and arbitration algorithms are implemented in centralized or distributed arbiters or controllers. Numerous parameters help define properties of the communication channels and their associated protocols, such as bus widths, burst transfer size, and master priorities.

7.2.2   Communication Architecture Topologies

In this subsection, we describe existing and emerging communication architecture topologies, in order of increasing sophistication and complexity.

Shared Bus

The system bus is the simplest example of a shared communication architecture topology and is commonly found in many commercial SoCs (e.g., PI-bus [277]).

The bus consists of a set of address, data, and control lines shared by a set of masters, that contend among themselves for access to one or more slaves. In its simplest form, a bus arbiter periodically examines accumulated requests from the multiple master interfaces, and grants access to a master using arbitration mechanisms specified by the bus protocol. However, limitations on bus bandwidth (due to increasing load on global bus lines) are forcing system designers to use more advanced topologies.

Hierarchical Bus

In recognition of the problems associated with topologies based on a flat shared bus, topologies consisting of multiple busses have recently begun to appear. This architecture usually consists of several shared busses interconnected by bridges to form a hierarchy. Different levels of the hierarchy correspond to the varying communication bandwidth requirements of SoC components. Commercial examples of such architectures include the AMBA bus architecture (ARM) [278] and the CoreConnect Architecture (IBM) [279]. Different levels of the hierarchy are connected by bridges. Transactions across the bridge involves additional overhead, and, during the transfer, both busses remain inaccessible to other components. However, multiple word communications can proceed across the bridge in a pipelined manner. Such topologies typically offer large throughput improvements over the shared topology, due to decreased load per bus, and the potential for transactions to proceed in parallel on different busses.

Rings

In certain application areas, ring-based communication architectures have been used, such as high-speed ATM switches [280]. In such architectures, each component (master/slave) communicates using a ring interface, which typically implements a token-passing protocol. The advantage of ring-based architectures is that each communication channel is of shorter length (point to point between neighboring components) and therefore can potentially support higher clock speeds.

Packet Switched Fabrics

Recently, the use of more complex topologies has been proposed for SoCs consisting of larger numbers of processing elements. These architectures avoid the use of globally shared busses but rely on switching mechanisms to multiplex communication resources among different master and slave pairs. Although these architectures have been studied in detail in the context of general-purpose systems such as multiprocessors and network routers, examples of such emerging architectures in the SoC domain include architectures based on crossbars [280], fat-trees [281], octagons [282], and two-dimensional meshes [274,283]. The advantages such architectures are expected to provide include higher on-chip communication bandwidth and predictable electrical properties, due to regularity in the communication architecture topology.

7.2.3 On-Chip Communication Protocols

In this section, we describe existing and emerging communication protocols, focusing on the different types of resource management algorithms employed for determining access rights to shared communication channels.[1] In this regard, these on-chip communication protocols are analogous to those used in the "datalink" layer of wide area networks, which determine access rights to shared LAN resources (e.g., CSMA/CA) [284]. As systems continue to grow in complexity, it is expected that other, higher layer protocol concepts from the domain of large scale networking will become increasingly applicable to communication architectures, such as routing, flow control, and quality of service [285].

Static Priority

This is a commonly employed arbitration technique used for shared bus-based communication architectures [277–279]. In this protocol, a centralized arbiter examines accumulated requests from each master, and grants access to the requesting master that is of highest priority. Transactions may be non-preemptive (i.e., once a transaction of multiple bus words begins, it runs to completion, during which time, all other components requiring access to the bus are forced to wait) or preemptive (lower priority components are forced to relinquish the bus if higher priority components are waiting). Although this protocol can be implemented efficiently, it is not suited to providing fine-grained control over the allocation of on-chip communication bandwidth to different system components.

TDMA

In this type of architecture, the arbitration mechanism is based on a timing wheel with each slot statically reserved for a unique master. Techniques are typically

---

[1] On-chip communication protocols also feature a variety of other functions, such as support for burst mode transfers, split transactions, multithreaded transactions, and so on.

employed to alleviate the problem of wasted slots (inherent in time-division multiple access [TDMA]-based approaches). One approach is to support a second level of arbitration. For example, the second level can keep track of the last master interface to be granted access via the second level and issue a grant to the next requesting master in a round-robin fashion. A commercial example of a TDMA-based protocol is one offered by Sonics [286,287].

Lottery

In this architecture, a centralized lottery manager accumulates requests for ownership of shared communication resources from one or more masters, each of which is (statically or dynamically) assigned a number of "lottery tickets" [288]. The lottery manager probabilistically chooses one of the contending masters to be the winner of the lottery and grants access to the winner for one or more bus cycles, favoring masters that hold a larger fraction of lottery tickets. This architecture provides fine-grained control over the allocation of communication bandwidth to system components and fast execution (low latencies) for high-priority communications, at the cost of more complex protocol hardware.

CDMA

A code division multiple access (CDMA)-based protocol has been proposed for sharing on-chip communication channels [289], to exploit the same advantages that CDMA has been known to provide in sharing the air medium in wireless networks, namely, resilience to noise/interference, and the ability to support multiple, simultaneous data streams [290]. The proposed architecture features synchronous pseudorandom code generators (a code is assigned to a communicating pair of components), modulation and demodulation circuits at the component bus interfaces, and differential signaling.

Token Passing

Token passing protocols have been used in ring-based architectures [280]. In such protocols, a special data word circulates on the ring, which each interface can recognize as a token. An interface that receives a token is allowed to initiate a transaction. If the interface has no pending request, it forwards the token to its neighbor. If it does have a pending request, it captures the token and reads/writes data from/to the ring, one word per cycle, for a fixed number of cycles. When the transaction completes, the interface releases the token.

7.2.4 Communication Interfaces

An important issue in designing components for use in SoC designs is the use of a consistent communication interface (across different on-chip busses and communication architectures) to facilitate plug-and-play design methodologies. Using such interfaces can provide the additional advantage of freeing the SoC component designer from having to be aware of the details of the communication architecture to which the component will be connected. Hence, this approach facilitates the development of innovative communication architectures that are not constrained by the interfacing requirements of system components that it may potentially need to serve. According to an analogy drawn in Zhu and Malik [291], communication interfaces provide an abstraction that is similar to the instruction set architecture of a microprocessor, whose purpose is to make microarchitectural details transparent to the programmer and, at the same time, facilitate advances in microarchitecture design. Currently, the standards being proposed by various industry consortia in an effort to help realize this goal include interfaces based on VSIA's Virtual Component Interface [292], and the Open Core Protocol [293].

In addition to such standardization initiatives, a large body of research has examined various issues in the design and implementation of on-chip communication interfaces, including techniques for interface modeling, simulation, and refinement (or synthesis). These techniques aim at providing ways of exploring different interfaces, and converting high-level protocol descriptions into efficient hardware implementations. Techniques for model refinement of communication interfaces are described in refs. 294 and 295, and techniques for synthesizing interfaces between components that feature incompatible protocols are described in ref. 296. Generic module interfaces ("wrappers") are described in ref. 297 that facilitate simulation of a complex SoC, while providing for the flexibility of mixing multiple levels of abstraction during simulation. Interface synthesis using these wrappers is described in ref. 298.

7.3 SYSTEM-LEVEL ANALYSIS FOR DESIGNING COMMUNICATION ARCHITECTURES

In recognition of the growing role of on-chip communication, recent work has addressed the development of system-level analysis techniques for estimating the impact of the communication architecture on overall system performance and power consumption. In this section, we describe such techniques, which aim at providing automatic support to drive the process of communication architecture selection, design, or optimization. These techniques can be broadly divided into the following categories.

1. System simulation-based techniques. In this class of techniques, the effects of the communication architecture are incorporated by developing suitable simulation models of the communication architecture topology and protocols [291,294,299–303]. Techniques that rely on simulation of the complete system are typically not feasible for exploring large design spaces, such as those offered by existing and emerging communication architectures. Simulation speed up is typically achieved by using abstract models of system components and the communication architecture. However, the use of such models typically trade off accuracy for efficiency.

2. Static estimation-based techniques. This class of techniques makes use of "static" models of the communication latency between system components [304–308] or static characterizations of the power consumption of system tasks, using simple power models for alternate implementations, and inter-task communication (e.g., refs. 309 and 310). These techniques often assume systems in which the computations and communications can be statically scheduled. For many systems, using such techniques could result in inaccurate performance and power estimates, since they usually ignore or make simplifying assumptions regarding the occurrence of dynamic effects (e.g., waiting due to bus contention).

3. Trace-based techniques. In this class of techniques, the effects of the communication architecture are incorporated into system-level analysis using a trace-based approach. These techniques derive accuracy from an initial detailed simulation and computational efficiency from fast processing of execution traces. In the next section, we describe one of these techniques in detail. A similar approach has been taken by Lieverse et al. [311,312], in which trace transformation techniques are used for communication refinement of signal processing applications modeled using Kahn process networks. Givargis et al. [313] also use trace-based analysis: although they do consider the effects of the communication architecture, their work is aimed mainly at SoC parameter tuning (e.g., cache parameters, buffer sizes, bus widths), rather than analyzing and exploring different communication architectures.

7.3.1   Trace-Based Analysis of Communication Architectures

In this section we describe a trace-based technique for fast and accurate system-level performance analysis and power profiling, to drive the design of the communication architecture [314,315]. The analysis technique enables evaluation of general communication architecture for use in complex SoC designs. The techniques provide the system designer with the freedom to select a network topology, communication protocols, and communication mapping. The topology could range from a single shared bus to an arbitrary interconnection of dedicated and shared communication channels. For each channel, the designer can select and configure the communication protocols and specify values of channel related parameters. The designer can also specify a communication mapping: i.e., an arbitrary mapping of system communications to paths in the specified topology. For a given specification of the communication architecture, the designer can evaluate system performance, the system power consumption profile, and various statistics regarding the usage of communication architecture resources.

The complete methodology is shown in Figure 7-2. The first phase of this methodology constitutes a preprocessing step in which simulation of the entire system is performed, without considering the detailed effects of the communication architecture. In this phase, communication is modeled in an abstract manner, through the exchange of events or tokens. The output of this step is a timing-inaccurate system execution trace. The second phase (enclosed within the dotted box in Fig. 7-2) is responsible for analyzing system performance and generating a system power consumption profile over time, including the effects of a selected communication architecture. The second phase operates on a symbolic execution graph (SEG), which is constructed from the traces obtained in the first phase, and captures the computations, communications, and synchronizations observed during simulation of the entire system. The system designer specifies the communication architecture, in terms of the network topology, communication protocols, and communication mappings. Analysis algorithms manipulate the SEG to incorporate effects of the selected communication architecture. The outputs, all of which account for the effects of the communication architecture, include (1) a transformed version of the SEG, (2) system performance estimates, (3) communication architecture usage statistics, such as channel utilization, resource conflicts, and so on, (4) system critical path (or paths), and (5) a profile of the system power consumption over time, as well as power profiles of individual components.

The advantage of this approach is that the time-consuming simulation step need not be performed iteratively, and alternative communication architectures

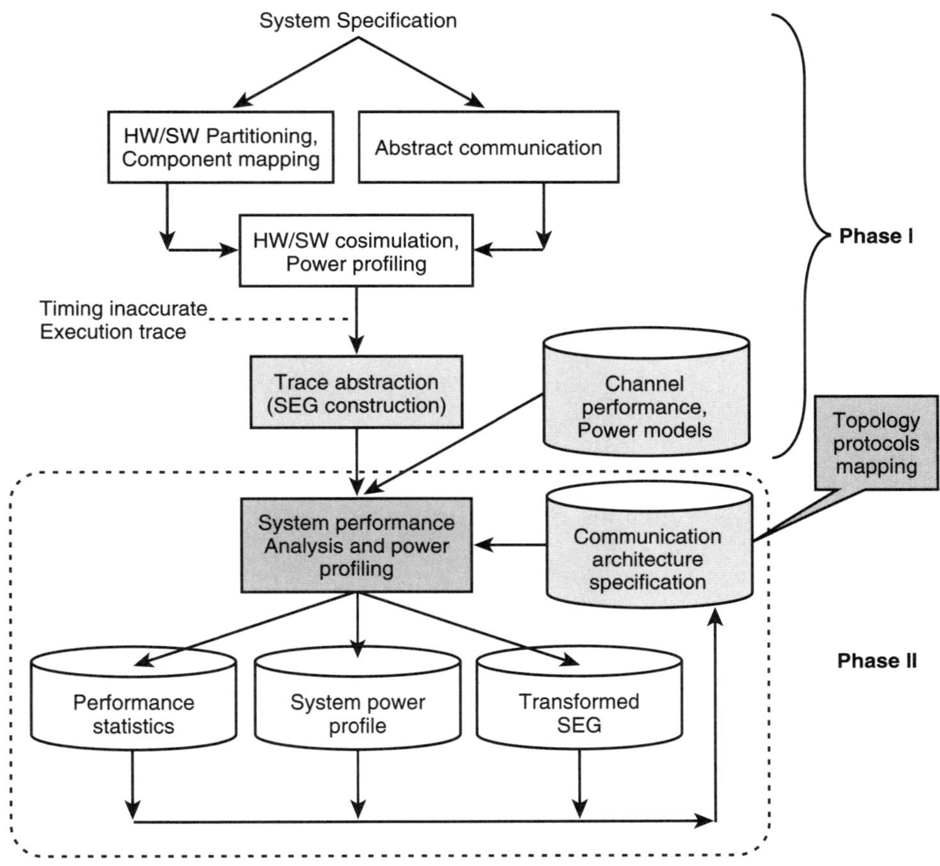

7-2

Trace-based analysis methodology for communication architecture design.

FIGURE

can be evaluated using the second phase of the methodology alone. The second phase, being fast and accurate (for reasons discussed later in this section), can provide the designer with important feedback regarding the impact of the communication architecture within very short analysis times, allowing efficient evaluation of many alternative architectures.

In the following subsections, we illustrate each of the shaded steps of Figure 7-2. These include (1) abstracting information from the simulation trace and

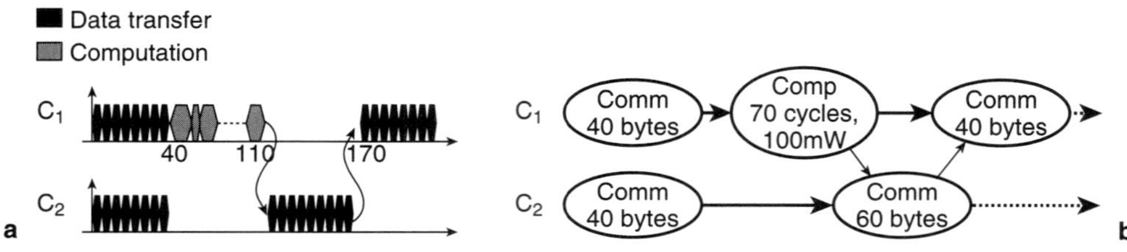

7-3 FIGURE Trace abstraction via SEG construction. (a) Traces generated by HW/SW co-simulation. (b) Corresponding SEG.

constructing the SEG, and (2) analyzing the system performance under the selected communication architecture.

Trace Abstraction

Simulation traces can be highly complex and detailed, making them undesirable as a starting point for analysis. Hence, only information that is necessary and sufficient to perform accurate analysis and to incorporate the effects of the communication architecture is extracted from the traces. The extraction step selectively omits unnecessary details (such as values of internal variables, communicated data, and so on). In addition, the extraction procedure groups contiguous computation and communication events into clusters, to construct an SEG. The SEG, being an abstract and compact representation of the traces, can be efficiently analyzed to obtain a variety of useful information.

Example 1. *Figure 7-3a shows a set of execution traces as two components, C_1 and C_2; each executes computations, communication with slaves, and synchronization between each other. Figure 7-3b illustrates the SEG that corresponds to the traces of Figure 7-3a. The graph has two types of vertices—computation and communication— and a set of directed edges representing timing dependencies, which arise due to the sequential component control-flow, or due to intercomponent synchronization and communication. Note that the SEG is acyclic because it is constructed from simulation traces in which all dependencies have been unrolled in time.*

The example illustrates that the SEG accurately captures the dynamic control-flow behavior of the HW/SW partitioned and mapped system. However, it should be noted that the SEG is "timing-inaccurate," since communication between system components was modeled in an abstract manner during simulation for trace collection.

We next describe how the analysis techniques incorporate the effects of a communication architecture to provide the required accuracy.

Analyzing Abstract Traces

The analysis technique is based on manipulating the SEG to incorporate the effects of the selected communication architecture, which includes a specification of the network topology, protocols, and communication mapping. The analysis technique makes use of designer-provided information and executes a pass of the SEG, adding new vertices and edges, splitting existing vertices, and calculating time stamps and power estimates for each vertex. Next, we illustrate the execution of the analysis algorithm with an example. We consider the processing of the SEG of Figure 7-3b, in order to incorporate the effects of a communication architecture consisting of a static priority-based shared bus. Note that although we use a shared bus for ease of explanation, the technique is applicable to arbitrary topologies and protocols.

Example 2. *Figure 7-4 illustrates the SEG that results from the execution of the analysis algorithms. We indicate example transformations that the algorithms introduce*

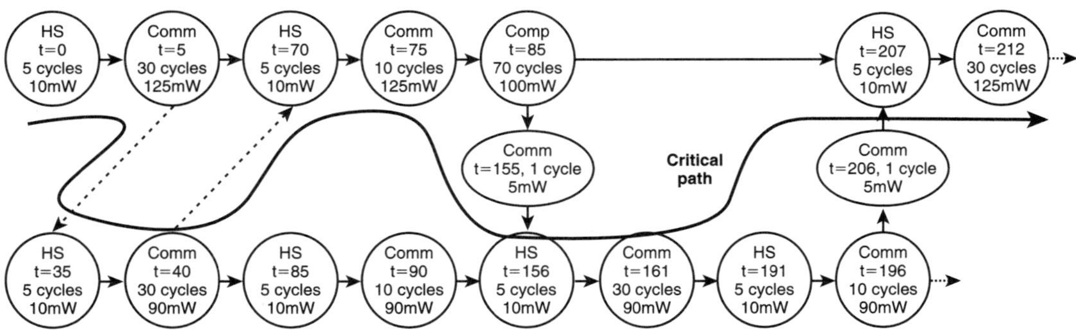

7-4 FIGURE Transformed SEG after incorporating effects of a priority-based shared bus.

while incorporating effects of the communication architecture. Intrinsic protocol over-heads, such as a connection setup, or handshaking, precede every data transfer. This is incorporated by the insertion of new vertices of type HS in the SEG, prior to each communication vertex. The protocol specifies a maximum permissible burst size. Hence, some communication vertices are split into multiple vertices, none of which exceed the burst size (30 cycles, in this example). The SEG of Figure 7-3b contains communication vertices from different masters that overlap. Since the communication architecture in this example is a shared bus, this represents a conflict, and one of the components must wait. The transformation that incorporates this conflict is shown in Figure 7-4, in which the analysis tool "grants" access to C_1, in accordance with designer-specified priorities ($C_1 > C_2$). As a result, C_2 waits, and an additional dependency (dotted arrow) is intro-duced to enforce the mutually exclusive nature of access to the shared bus. SEG vertices are annotated with time stamps, which are obtained by examining time stamps of pred-ecessor vertices and calculating, or looking up, vertex execution times. For communi-cation vertices, the execution times are calculated using provided raw channel bandwidth and latency information. Lookup techniques are used for HS vertices. For the computation vertices, the execution times are preserved from the original SEG. The analysis algorithms also annotate each vertex in the transformed SEG with average power estimates. For the communication and HS vertices, analytical power models of the communication channels, master and slave interfaces, are used, whereas for the computation vertices, again, the estimates are obtained from the original SEG.

From the resulting SEG, various useful statistics can be derived, to provide feed-back to the designer of the communication architecture. For example, by examining time stamps on specific SEG vertices, system performance metrics can be calculated (time taken to complete a specific task, number of missed deadlines, and so on). As shown in Figure 7-4, the technique allows generation of the system critical path(s). Also, since each vertex is annotated with a power estimate, the technique can efficiently generate the system power consumption profile versus time under the selected communication architecture. Statistics regarding the usage of communication architecture resources, such as channel utilization, conflicts, and so on, are provided to the designer and can be used to guide automatic design space exploration techniques.

Case Study: Exploration of Communication Architecture Parameters for a TCP/IP Subsystem

The trace-based analysis technique has been applied to a range of system designs and communication architectures. To illustrate the utility of the analysis tech-nique, we present an example in which an exhaustive search of the design space of a particular communication architecture is conducted to obtain a configuration of the communication architecture that maximizes system performance.

Example 3. *Consider a system consisting of three components and a shared memory, which together implement a subsystem of the TCP/IP protocol, relating to computation and verification of the TCP checksum. The parameter space of the communication architecture consists of all possible priority assignments and available burst sizes. Results of the analysis for each configuration are presented in Figure 7-5, which shows the performance of the TCP/IP system when one is processing 10 packets of size 512 bytes under 36 different design points. The best performance is obtained when the burst size is 128 and priorities are assigned so the three components Ether_Driver, IP_Chk, and Chksum are in descending order of priority. The curves in the x-yplane are isoperformance contours. The system performance is observed to vary between extremes of 2077 cycles and 3570 cycles. The entire space exploration experiment took less than 1 second of CPU time.*

The above example illustrates that (1) it is possible to perform thorough and fast exploration of the communication architecture design space using the described trace-based techniques, and (2) obtaining the optimum communication architecture parameters is a nontrivial problem, calling for design space exploration algorithms for customization of the communication architecture. In addition to demonstrating over two orders of magnitude improvements in efficiency over simulation-based approaches, this technique results in imperceptible loss of accuracy, with respect to both system performance (on average 2%) and generated power profiles (on average 4%).

Summary

In this section, we discussed techniques that help to analyze the impact of communication architecture design choices on overall system performance and power consumption. We described a technique providing designers with accuracy that is comparable to that of complete system simulation and yet is orders of magnitude faster. The technique derives its efficiency from the fact that it operates on a highly abstract representation of the system execution traces. The technique derives its accuracy from two factors. First, the traces are derived from simulation, which ensures that control flow (loop iterations, branches taken) is fully determined and is available during analysis. Second, since communications are not isolated from the rest of the system (computations and synchronizations), it is possible to account for the indirect effects that the communication architecture has on the timing and power profile of the system components. These two features make the analysis technique a useful addition to communication architecture design environments.

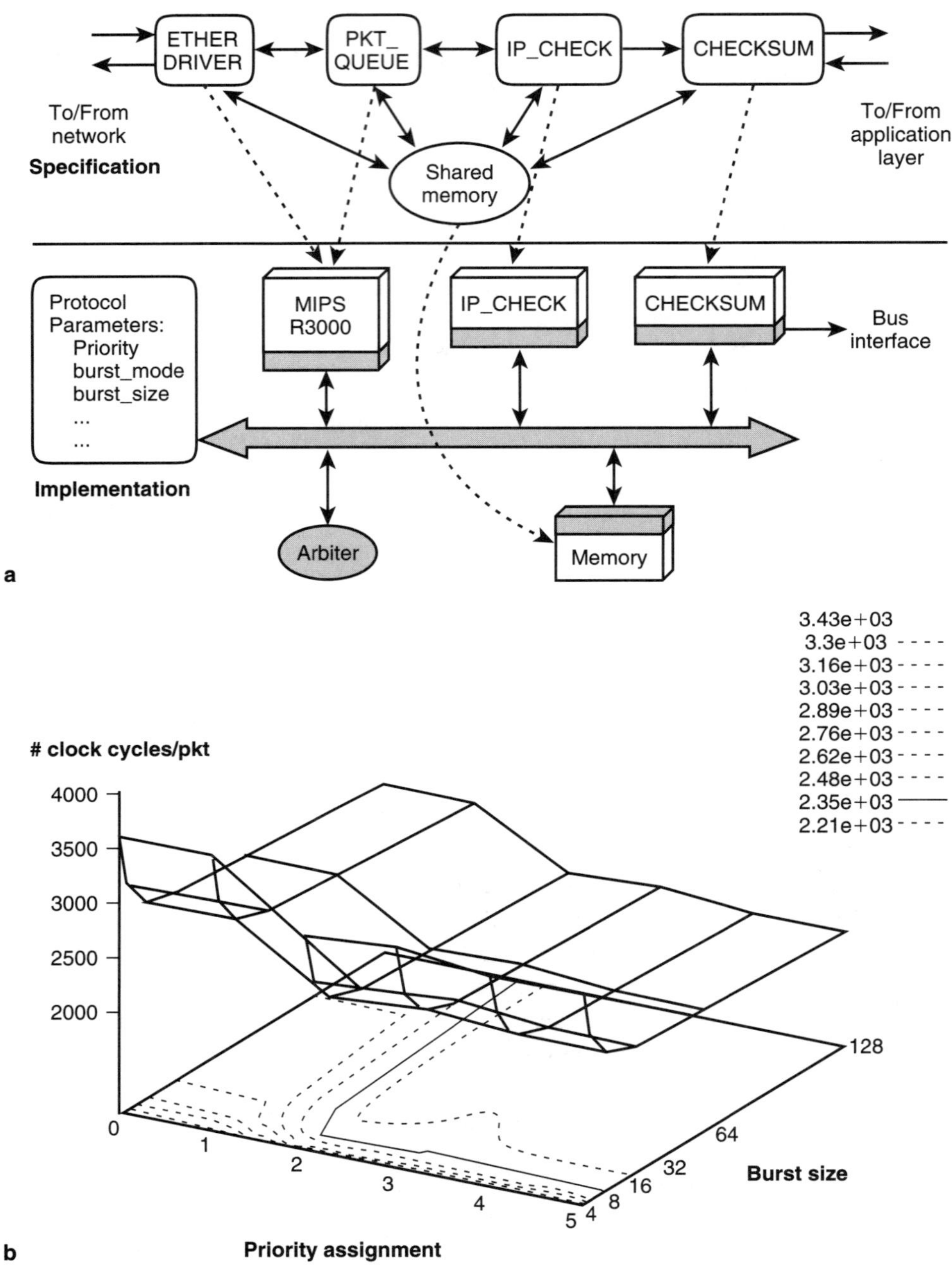

7-5 FIGURE Design space exploration of a bus-based communication architecture. (a) TCP/IP subsystem. (b) Performance versus priority and burst size for TCP/IP subsystem.

7.4   DESIGN SPACE EXPLORATION FOR CUSTOMIZING COMMUNICATION ARCHITECTURES

The design of a communication architecture calls for careful attention to several issues in order to meet all the desired design goals. These include (1) selection of an appropriate network topology; (2) selection of appropriate communication protocols, along with careful configuration of protocol parameters; and (3) optimization of the mapping of the system communications to physical paths in the topology. As communication architectures grow in complexity, each of these steps potentially presents system designers with a large number of alternatives. In addition (as we shall see later in this section) they may interact, leading to tradeoffs that are difficult to identify or evaluate without automatic design space exploration tools. In this section, we describe techniques for exploring this design space, focusing on techniques for customizing the communication architecture to best suit the characteristics of the communication traffic generated by an application [316].

7.4.1   Communication Architecture Templates

Existing techniques for design space exploration and customization of communication architectures mostly deal with generation of network topologies [304,305,308,317,318]. Although topology selection is a critical step in the design of a communication architecture, we believe that future communication architecture designs will be based on one of several architectural "templates" that can be customized for the needs of an application. A few such templates have started to become available from interconnect IP providers, examples of which include hierarchical busses [278], shared micronetworks [286], and mesh-based networks [283]. Modern templates often feature standard topologies, while providing designer configurable parameters for the communication channels (e.g., bus widths) and protocols (e.g., priorities, split transactions, and so on). As available templates increase in complexity and configurability, the parameter space of the communication architecture, together with the number of alternative ways in which the system's communications can be mapped to the template, will make exhaustive design space exploration computationally intractable. As shown in this section, even for relatively simple communication architectures (e.g., consisting of just two busses), naive use of the communication architecture could result in poor system performance. Hence, it is crucial that the underlying communication architecture template be optimized for, or customized to, the specific characteristics of the on-chip communication traffic generated by the application. Since the design space offered by such templates is complex, efficient methodologies are required for template customization. In the next subsections, details of such a methodology are presented.

7.4.2 Communication Architecture Template Customization

Figure 7-6 presents a design space exploration and optimization technique for communication architecture customization. The methodology takes as inputs (1) a partitioned and mapped system, and (2) a communication architecture template, consisting of a fixed network topology, configurable protocol parameters, and support for arbitrary mapping of system communications to paths in the topology. The methodology uses efficient algorithms to generate (1) an optimized mapping of the system's communications onto the communication architecture topology, and (2) optimized values for communication protocol/channel parameters. The methodology consists of two interacting parts: (1) a fast and accurate performance analysis tool, and (2) a design space exploration framework. For a given point in the communication architecture design space, the analysis tool efficiently estimates system performance, profiles intercomponent communication

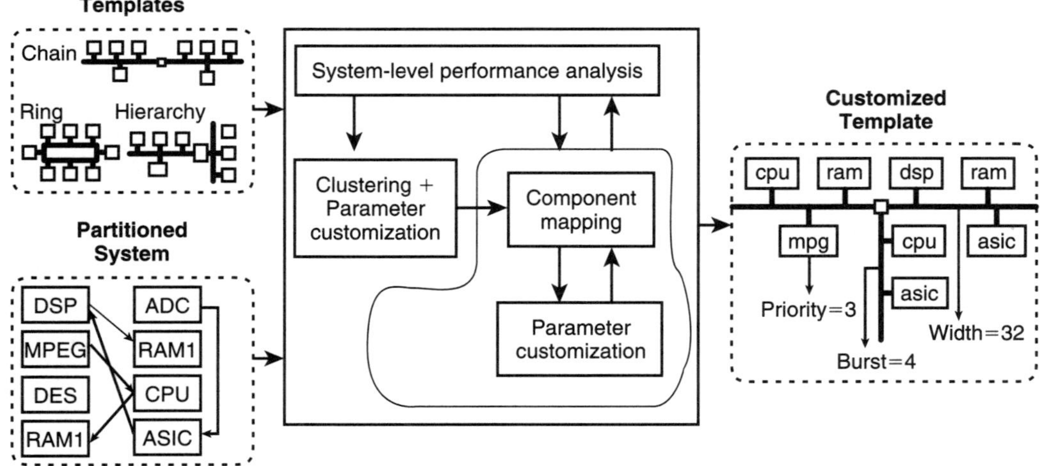

7-6   Design space exploration and customization of communication architectures.

traffic, and provides accurate estimates of contentions for shared resources. This information is used to guide the design space exploration algorithms for communication architecture customization.

In the next subsections, we present details of two techniques used by the methodology: a clustering-based initial solution construction and a subsequent design space exploration procedure. We use examples to illustrate some of the issues that are taken into consideration by the different algorithms.

Clustering-Based Initial Solution Construction

The first step toward customizing the communication architecture for the application uses the principle of clustering frequently communicating components closer in the topology. The intuition is that clustering reduces the number of communications that undergo high-latency, multi-hop communications (e.g., those spanning multiple busses connected by bridges). A list of components is constructed in decreasing order of communication bandwidth requirements, based on statistics obtained via performance analysis. Each component from the list is considered in turn and is mapped to a channel in the template topology. Metrics are used to estimate the "affinity" of the component with different channels, based on measurements of the volume of traffic it exchanges with other (already assigned) components. Wherever possible, components are assigned to channels with which they have maximum affinity, subject to constraints on the maximum number of components per channel.

Example 4. *Figure 7-7a shows a system consisting of a set of masters that execute computation tasks and exchange data and control as indicated. (Solid lines indicate data transfer, and dotted lines indicate synchronization.) Application of the clustering-based mapping strategy to a template consisting of two shared busses connected by a bridge results in Arch1 (Fig. 7-7b). In this mapping, communications between C_1 and C_2 are mapped to Bus1, and those between C_5 and C_6 are mapped to Bus2. This mapping minimizes the volume of transactions that go across the bridge, by clustering frequently communicating components on the same bus. Figure 7-8a symbolically represents Arch2, a naive mapping to the template (chosen at random). In Arch2, communications between C_1 and C_4 are mapped to a path of busses: Bus1 $\rightarrow$ Bus2. Table 7-1 reports system performance under each of these architectures. From the table, we observe that Arch1 is 24% faster than Arch2.*

The above example illustrates that better system performance can be obtained by customizing the communication mapping to the traffic characteristics of the application. In this case, simple clustering of frequently communicating components in the communication architecture helped improve system performance

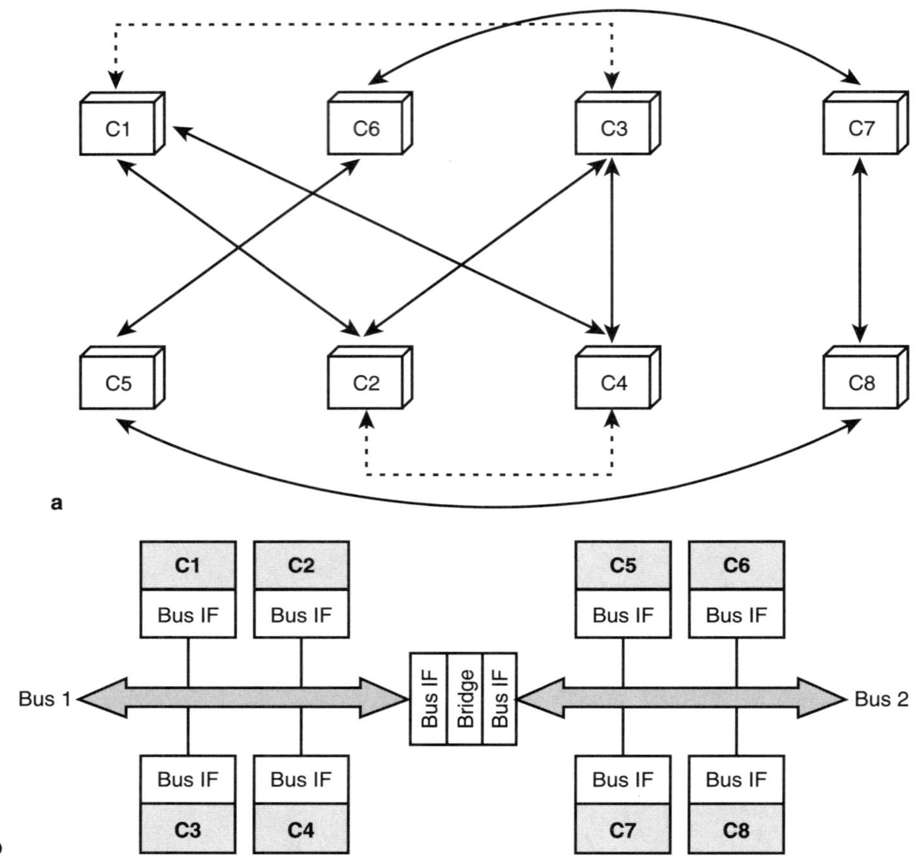

7-7

FIGURE

Example system of communicating components. (a) Logical view of intercomponent communication. (b) Candidate communication mapping to the template topology–Arch1.

compared with a random assignment. However, as we shall see next, this approach does not necessarily yield the best solution.

Analysis-Guided Iterative Refinement

The second step of the exploration framework aims at improving system performance beyond what can be achieved via simple clustering. This step starts with an initial solution (consisting of a communication mapping and a configuration of the protocol parameters for each channel) and uses efficient exploration

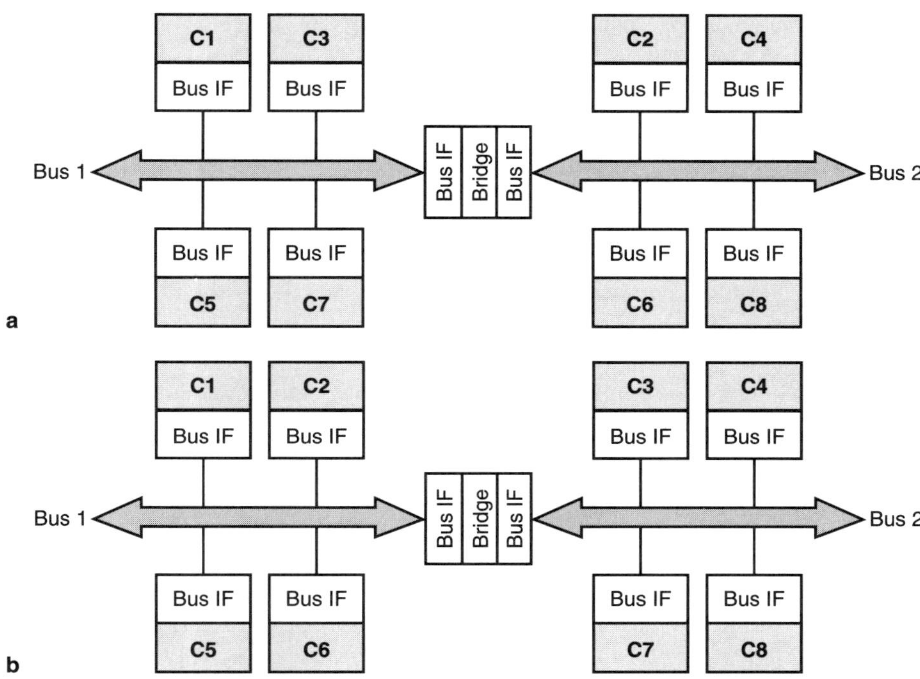

7-8

FIGURE

Alternative mappings for the example system. (a) Arch2. (b) Arch3.

Cases	Bus 1	Bus 2	Performance (clock cycles)
Arch1	C1 C2	C5 C6	11723
	C3 C4	C7 C8	
Arch2	C1 C3	C2 C4	15314
	C5 C7	C6 C8	
Arch3	C1 C2	C3 C4	9242
	C5 C6	C7 C8	

7-1

TABLE

Performance for Different Communication Mapping Alternatives.

algorithms that consider the benefits of moving components from currently mapped channels to other channels, while simultaneously reconfiguring the protocol parameters. The exploration technique is based on a well-known iterative improvement strategy for partitioning that maximizes the cumulative gain of a sequence of moves [319]. The procedure employs efficient analytical techniques to estimate the potential gain of moving a component from its currently mapped channel to another, which takes into account communication characteristics of the component, properties of the channels, and their utilization levels, as well as resource conflicts. To illustrate the advantage of the iterative refinement step, we return to the previous example.

Example 5. *Consider again the example system of Figure 7-7a. Applying the exploration procedure with Arch1 as a starting point results in Arch3 (shown in Fig. 7-8b). Analysis of Arch3 indicates that the system completes its execution in 9242 cycles, an improvement of 27% over Arch1 (which was deemed optimal from a clustering viewpoint). Arch3 consists of a mapping that is superior to Arch1 because it is conscious not only of the volume of data communicated between components but also the timing and concurrency of different communications. In particular, it separates components that have largely overlapping communication lifetimes (C_1 from C_3 and C_5 from C_7), resulting in fewer resource conflicts and hence enhanced concurrency at the system level. The price paid is increased traffic that spans multiple busses, since component C_1 communicates with component C_4 and C_3 with C_2 (Fig. 7-7a).*

The example illustrates that in order to derive the best communication mapping, it is important to consider all the characteristics of the application's communication traffic, including its temporal properties. The exploration procedure described above is capable of accounting for such characteristics and outperforms techniques that are based solely on the volume of data exchanged by components.

Interdependency Between Communication Mapping and Protocol Customization

We next illustrate how the problems of optimizing the communication mapping and optimizing the parameters of the communication protocols are interdependent. In other words, a strategy that first optimizes the communication mapping, and then optimizes the protocol parameters could result in significantly suboptimal system performance. The methodology of Figure 7-6 addresses this interdependence by interleaving refinement of protocol parameters with refinements to the communication mapping. To illustrate the importance of considering this interdependence, we return to the example of Figure 7-7 and evaluate an exploration strategy whereby the communication mapping is first optimized, assuming

Case	Bus 1 Protocol	Bus 2 Protocol	Performance (cycles)
Arch3-subopt	C1 > C2 > C5 > C6 DMA = 5	C3 > C4 > C7 > C8 DMA = 10	12504
Arch3-opt	C1 > C6 > C2 > C5 DMA = 10	C3 > C7 > C4 > C8 DMA = 10	9242

7-2

Effect of Protocol Parameters on Design Space Exploration.

TABLE

a fixed set of protocol parameters, and the protocol parameters are subsequently tuned.

Example 6. *Consider again the example system of Figure 7-7. Each bus in the template is a static priority based shared bus supporting burst mode transfers. The protocol parameters (the maximum burst transfer size, bus access priorities) were fixed during exploration of alternative communication mappings. After obtaining a solution for the communication mapping, the protocol parameters were optimized as a separate step. This yields Arch1, the same solution as that obtained via clustering. This solution, as reported earlier, is 27% worse than the best solution. In this approach, Arch3, the best solution, gets overlooked. This is because when the exploration algorithms consider the communication mapping of Arch3 with the a priori fixed selection of protocol parameters, performance estimates show that the system takes 12,504 cycles to complete (indicated in Table 7-2 by Arch3-subopt). Since this represents a performance degradation of of 6.6% compared with Arch1, the communication mapping of Arch3 is discarded in favor of that of Arch1. In contrast, when communication mapping and protocol customization are performed simultaneously (using the methodology of Fig. 7-6), the optimal architecture (Arch3) is obtained.*

The above example demonstrates that it is important to consider and exploit the interdependence between the protocol parameters and the communication mapping while customizing a communication architecture template.

Summary

In this section, we described a design space exploration strategy for customizing a communication architecture template to best suit the communication traffic characteristics of an application. The techniques described demonstrate the advantages of customizing the template architecture while taking into account spatial locality of the communication traffic, temporal properties of the communication traffic, and the interdependence between the problems of protocol customization and communication mapping. These techniques lead to the design of customized communication architectures that are better suited to handling the requirements imposed by an application, resulting in significant gains in overall performance.

7.5 ADAPTIVE COMMUNICATION ARCHITECTURES

The previous section described techniques for statically customizing the communication architecture toward specific characteristics of the on-chip communication traffic generated by an application. In several systems, the traffic generated by SoC components could exhibit significant temporal variation in the requirements they impose on the communication architecture. This may occur due to variability in the exact functions being executed by the system components or the properties of the data being processed. For such systems, a mere static customization of the communication architecture (using approaches such as the ones described in the previous section) may prove insufficient. In other words, the communication architecture should be capable of detecting variations in the communication requirements of system components over time and of adapting the underlying communication resources to best meet those requirements. In recognition of this need, commercial bus architectures have recently started provisioning for the ability to configure communication protocol parameters dynamically. Examples include programmable arbiters in the AMBA bus [278] and dynamically variable slot reservations in the Sonics micronetwork [286]. Although hardware support for dynamic configurability is becoming available, there is a lack of established methodologies that enable designers to take advantage of such configurability. In this section, we describe the use of communication architecture tuners, a technique that enables the on-chip communication protocols to recognize and adapt to varying requirements of the application's traffic [320].

7.5.1 Communication Architecture Tuners

Communication architecture tuners (CATs) constitute a layer of circuitry that surrounds a communication architecture topology and provides mechanisms for communication protocols to adapt to run-time variations in the communication needs of system components. The CATs monitor the internal state of each component, analyze the generated communication transactions, and "predict" the relative importance of communication transactions in terms of their impact on system-level performance metrics. The results of the prediction are used to configure available communication protocol parameters to best suit each component's changing communication needs[2]. For example, more performance-critical data transfers may be handled differently, leading to lower communication latencies. The CATs approach yields improved utilization of the on-chip communication bandwidth and consequently significant improvements in overall system performance (e.g., the ability to meet real-time deadlines).

In this section, we first describe the overall architecture of a CAT-based system, along with basics of the tuner circuitry that needs to be added to SoC components. We then illustrate the functioning of a CAT with an example and finally provide an overview of a methodology for designing CAT-based communication architectures.

CAT-Based Communication Architectures

A typical system with a CAT-based communication architecture is shown in Figure 7-9a. The system contains several components, including a processor core, memories, and peripherals. The topology (consisting of a combination of shared and dedicated channels) is enclosed in the dotted boundary. The portions of the system that are added or modified as a result of the CATs methodology are shown shaded in Figure 7-9a. A detailed view of a component with a CAT is shown in Figure 7-9b. The CAT consists of a partition detector circuit and parameter lookup tables (LUTs).

1. Partition detector. A communication partition is a subset of the communication transaction instances generated by the component during system execution. For each component, the methodology defines a set of partitions[3] and the conditions that need to be satisfied by a communication instance for it to be classified as belonging to a specific partition. The partition detector circuits monitor and analyze the following information generated by the component:

   ✦ Identifiers: identifiers t_1, t_2, ... are emitted by a component when the history of its control flow satisfies associated expressions defined over a

---

[2] Note that the addition of CATs does not change the topology of the communication architecture but allows dynamic variation of the parameters governing the on-chip communication protocols.

[3] The term "partition" is often used to refer to the decomposition of a set into a set of disjoint subsets, as well as to one of the individual subsets. In this section, we used the term "partition" with the latter connotation.

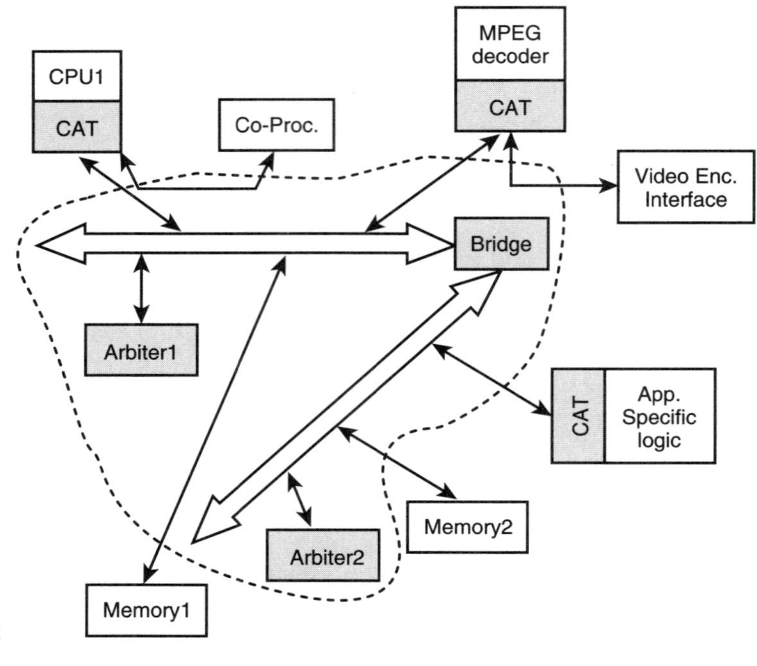

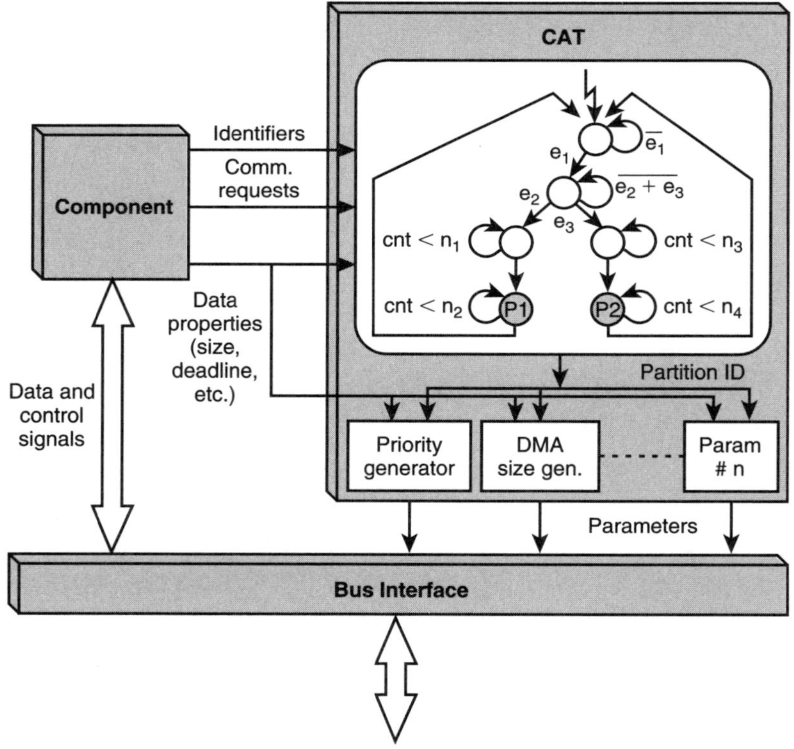

<table>
<tr><td>7-9</td><td>(a) An example SoC with a CAT-based communication architecture. (b) Detailed view of a master component enhanced with a CAT.</td></tr>
</table>

FIGURE

set of control flow events $e_1 \ldots e_n$. The component specification is enhanced to generate these identifiers purely for the purpose of the CAT.

+ The communication requests that are generated by the component.

+ Any other application-specific properties of the communicated data (e.g., fields in the data that indicate its relative importance).

The partition detector uses heuristic techniques based on observing identifiers (t_1, t_2, . . .), communication instances (c_1, c_2, . . .), and data properties to classify communication instances into partitions. These heuristics take the form of regular expressions. For example, the expressions $t_1 \cdot c^4$ and $t_1 \cdot c^8$ when associated with a specific partition indicate that the fourth to seventh communication instances that are generated following the identifier event t_1 will be classified as belonging to that partition.

2. Parameter LUTs. These LUTs contain precomputed values for communication protocol parameters (e.g., priorities, burst sizes, and so on), which are indexed by the output of the partition detector, and other application-specific data properties specified by the system designer. These parameter values are sent to the communication architecture, which may result in a change in the performance offered by the communication architecture to the SoC component.

Example 7. *The functioning of a CAT-based communication architecture is illustrated using symbolic waveforms in Figure 7-10. The first two waveforms represent identifiers generated by the component. The third waveform represents the communication instances generated by the component. The fourth waveform shows the state of the partition detector circuit. The state of the partition detector circuit changes first from S0 to S1, and later from S1 to S2, in reaction to the identifiers generated by the component. The fourth communication instance generated by the component after the partition detector reaches state S2 causes it to transition into state S3. All communication instances that occur when the partition detector FSM is in state S3 are classified as belonging to partition CP. The fifth waveform shows the output of the priority generation circuit, which assigns a priority level of 2 to all communication instances that belong to partition CP. This increase in priority leads to a decrease in the latency associated with the communication instances that belong to partition CP.*

The above example illustrates how CATs provide the mechanisms for run-time detection of time-varying communication requirements and appropriate adaptation of the communication protocols.

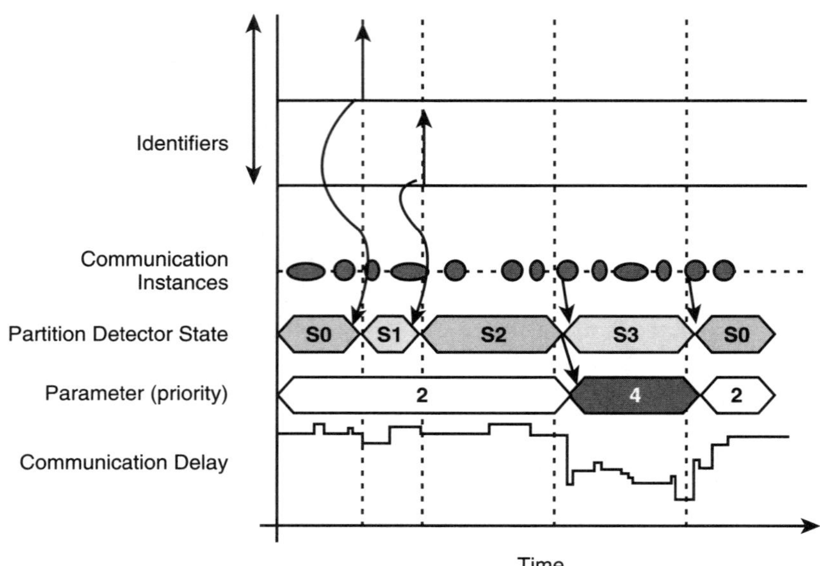

<table>
<tr><td>7-10</td><td>Symbolic illustration of the execution of a CAT-enhanced component.</td></tr>
<tr><td>FIGURE</td><td></td></tr>
</table>

Design Methodology for CAT-Based Systems

The overall methodology for designing CAT-based communication architectures is shown in Figure 7-11. The methodology takes as inputs a simulatable partitioned/mapped system description, a selected network topology and protocols, typical input stimuli, and a set of performance objectives (e.g., time taken to complete a specific task, number of real-time deadlines met or missed). The output is a set of optimized communication protocols for the target system. From a hardware point of view, the system is enhanced through the addition of CATs wherever necessary and through the modification of the controllers/arbiters for the various channels in the communication architecture.

In step 1, simulation of the partitioned/mapped system is carried out, with a statically configured communication architecture. The resulting traces are converted into a compact representation (SEG) that facilitates fast analysis. (Details of the SEG and the techniques used to analyze it have been described in Section 7.3.) In step 2, the SEG communications are classified into a set of distinct "communication partitions," each partition consisting of communications that impose similar performance requirements on the communication architecture. Communications belonging to a particular partition may benefit from a unique

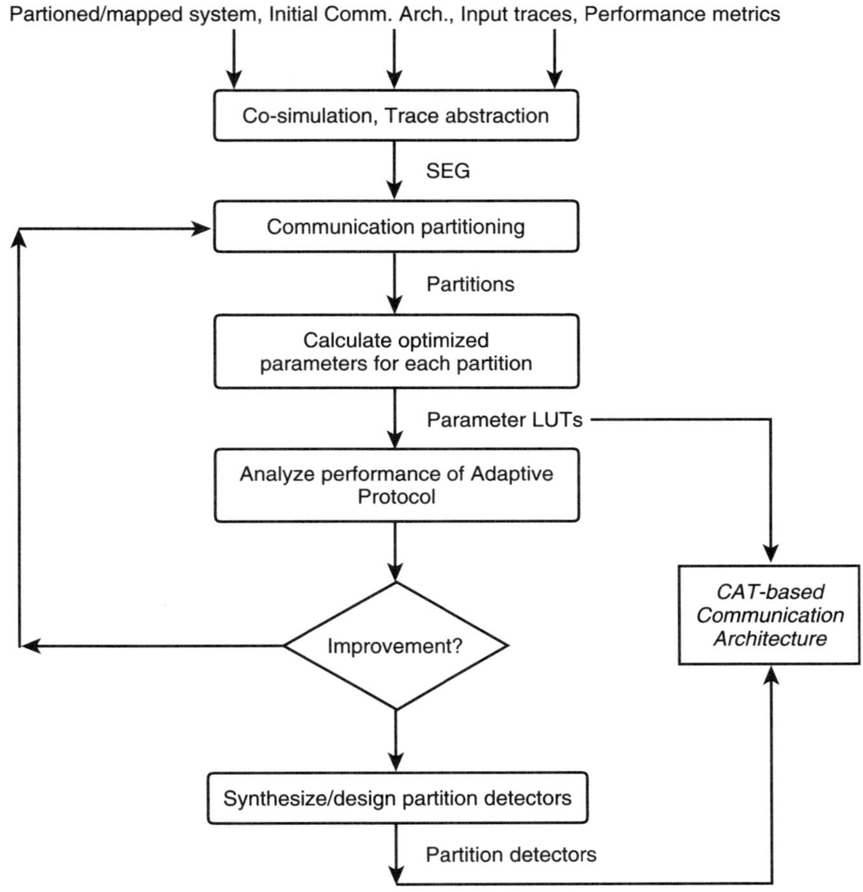

7-11   Methodology for CAT-based communication architecture design.

FIGURE

specification of the communication protocol parameters. In step 3, the communication partitions are analyzed to determine an optimized set of parameters for each communication partition. The output of step 3 is a set of lookup tables that map partitions to parameter values. Step 4 reevaluates system performance, taking into account the parameter assignments derived in step 3. Steps 1 to 4 are repeated until no further performance improvement is obtained.

Note that the off-line determination of communication partitions (step 2) is relatively easy, due to the availability of the complete execution trace. However, since knowledge of future events is not available at run time, dynamic partition detection is performed using heuristic strategies, based on component control-flow history and data properties. Step 5 deals with generation of such partition detector circuits. Parameters for the detected partitions are obtained from LUTs generated in Step 3. These LUTs together with the partition detector circuits constitute the CAT hardware that is used to enhance the communication architecture.

It bears mentioning that the notion of associating protocol parameters to communication partitions is quite general. For example, a static priority-based protocol represents a special case of such an approach, wherein all communications generated by a component belong to a single partition, and there are as many partitions as components.

7.6   COMMUNICATION ARCHITECTURES FOR ENERGY/BATTERY-EFFICIENT SYSTEMS

In this section, we turn our attention to the design of communication architectures for energy-efficient and battery-efficient systems. For battery-efficient systems, it is important not only to reduce the total energy consumption of the system but also to tailor the manner in which energy is drawn to specific characteristics of the battery.

We consider two categories of techniques that address the design of the communication architecture, keeping energy/battery-efficiency in mind. The first category consists of techniques that attempt to reduce the energy consumption of the communication architecture itself, by reducing the power consumed by the wires constituting the communication architecture topology. The second category consists of techniques that are concerned with the impact of the communication architecture on the energy consumption of the rest of the system. We describe the details of an approach that uses "battery-aware" on-chip communication protocols to regulate the total system power consumption dynamically, as opposed to merely reducing energy consumption of the communication architecture. This communication-based power management technique is battery-driven and hence can be used to maximize battery life.

7.6.1   Minimizing Energy Consumed by the Communication Architecture

We first examine techniques that help reduce the energy consumption of the communication architecture itself. We describe techniques for reducing the power consumed while transferring data across busses in deep submicron designs and then describe recent work on optimizing the topology for energy minimization.

Energy-Efficient DSM Interconnects

Low-power design of DSM interconnects is an active area of research that aims at coping with the new sources of power consumption associated with global interconnects in deep submicron (DSM) technologies. These effects (e.g., cross-coupling between neighboring wires) play a significant role in determining the power consumption of the communication architecture, due to the combined effects of large wire lengths, high aspect ratios, and small interwire spacing. Analysis of technology trends suggests that the power consumed by such DSM communication architectures could constitute a significant fraction of total system power, in some cases, reaching 40 to 50% [276,321]. Although in the past, numerous techniques have been proposed for minimizing the power consumed on bus lines (e.g., refs. 322 and 323), these techniques were largely concerned with off-chip busses, for which it was sufficient to minimize the total switching activity on the bus. However, it is increasingly clear that such approaches provide limited advantage for on-chip busses, in which DSM effects further complicate the problem.

Several power models have recently been developed for DSM interconnects [324–327]. These models, being more accurate than traditional approaches (based on signal transition counts), have been applied to developing design techniques for low-energy data communication on DSM busses. These techniques are largely based on encoding techniques that transform the communicated data so as to reduce the impact of DSM effects [326,328–331]. Specialized techniques are used for address busses to exploit the high correlation between successive bus words. Another approach that avoids power consumption due to deep submicron effects includes modifying the organization of bus lines. Shin and Sakurai [332] describe an approach whereby the ordering of parallel bus lines is optimized (based on a profile of the data values transferred) so as to minimize power consumption. Machiarullo et al. [333] describe a technique for optimizing the spacing between adjacent address bus wires (also using value profiling), to help reduce power consumption, while satisfying design rules. Recently, low swing signaling techniques have been proposed for on-chip busses [334] that make use of dynamic voltage scaling techniques to reduce energy consumption.

Energy-Efficient Communication Architecture Topologies

Techniques described in refs. 335 and 336 illustrate the potential savings in energy consumption that can be achieved via optimizations in the communication architecture topology. In each of these approaches, shared bus-based architectures are considered. These techniques evaluate the benefit of splitting globally shared bus lines into smaller, segmented busses (e.g., by separating them with tristate buffers, or pass transistors). It has been demonstrated that when the communications between system components exhibit high spatial locality, splitting the shared medium into segments can result in large energy savings. This is because, in the split architecture, local communications can be performed by charging/discharging shorter, segmented busses, which have lower parasitic load and hence require less energy.

7.6.2 Improving System Battery Efficiency Through Communication Architecture Design

In this section, we describe the importance of the communication architecture in affecting the power consumption characteristics of the entire system, rather than merely considering the energy consumed by the communication architecture itself. In addition, the technique we describe in this section helps improve system battery efficiency, as opposed to just reducing energy consumption. Battery-efficient design aims at maximizing battery life by tailoring the run-time power consumption profile of a system to battery discharge characteristics. A large body of work has shown that such techniques can yield significant improvements in battery life, over and beyond what can be accomplished by merely minimizing total energy consumption (see ref. 337 for a survey of the area).

Communication-based power management (CBPM) is a system-level power management methodology that uses new "battery-aware" on-chip communication protocols to regulate the execution of system components dynamically, in order to achieve battery-friendly system-level power profiles. CBPM differs significantly from conventional power management techniques, which aim at minimizing power drawn during idle periods, or voltage scaling techniques, which mainly target reducing average power. In contrast, CBPM exercises proactive control over system components, delaying components that may be executing less critical operations, even if they are not idle. There are several advantages to using the communication architecture for battery-driven, system-level power management. These include:

- ✦ Low hardware overhead: the communication architecture physically and logically integrates all the components of the system. Embedding power management functionality in the communication architecture saves the cost of dedicated hardware and wiring for power management units or controllers.

◆ System-wide visibility: to exercise dynamic, regulatory control over the system power profile, it is important to gain access to system-wide information (e.g., current execution states of components). The communication architecture provides system-wide connectivity, allowing (1) effective monitoring of component execution states, and (2) efficient delivery of power management decisions to system components.

◆ Control over system power: the communication architecture directly controls the timing of intercomponent communications and hence, if properly exploited, can regulate component execution as well. CBPM features communication protocols that exercise this control, regulating the occurrence of component idle/active states, which allows shaping of the overall system power profile.

We illustrate how CBPM works by considering its application to an IEEE 802.11 MAC processor design. Next, we provide a brief description of a methodology for designing CBPM-based systems, and finally we describe the operation and implementation of a CBPM-based communication protocol.

CBPM: IEEE 802.11 MAC Processor Example

The IEEE 802.11 MAC processor architecture that we consider consists of two embedded processors, three coprocessors, and a hierarchical bus-based communication architecture. It implements the MAC layer functionality of a device connected to an IEEE 802.11b wireless LAN. Although for this example we consider only the system power profiles and execution traces of the processor, details of the design are available in ref. 338.

Example 8. *We compare the power profiles and battery efficiency of (1) an original, performance-optimized MAC processor and (2) the same design, with a CBPM-enhanced communication architecture. Figure 7-12a shows a snapshot of the current profile for the original MAC processor as it processes three MAC frames.[4] This snapshot is part of a longer profile, consisting of several thousand MAC frames processed while one is streaming video over an IEEE 802.11b wireless LAN. For the profile of Figure 7-12a, the energy delivered by a 250-mA/h lithium-ion battery was estimated at 154.3mA/h over a lifetime of 3209.2s [339]. The poor efficiency (62%) is attributable to regions of Figure 7-12a, in which it can be seen that the current drawn is significantly higher than the battery's rated current.*

---

[4] Current profiles are obtained by HW/SW cosimulation-based power estimation followed by division by the supply voltage (assumed constant).

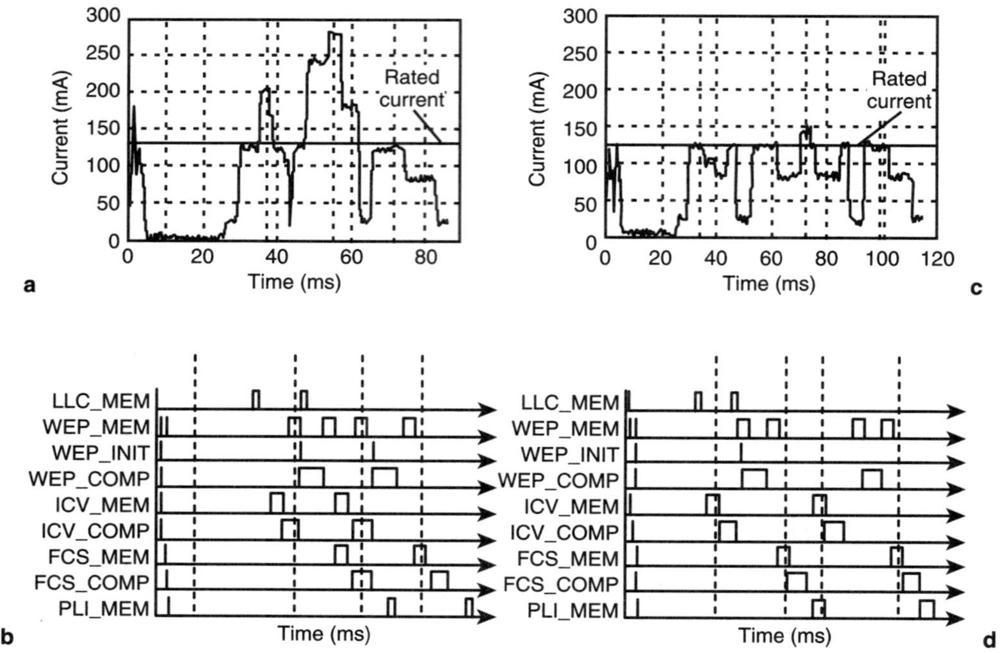

7-12 FIGURE   Current discharge profiles for an IEEE 802.11 MAC processor (a) without CBPM and (b) with CBPM. Symbolic execution traces of component macrostates (c) without CBPM and (d) with CBPM.

To understand how CBPM improves battery efficiency, we consider a macrostate execution trace (Fig. 7-12c), which corresponds to the profile of Fig. 7-12a. Macrostates correspond to specific subtasks that are performed by system components at different times. For example, the component wep could be in one of three macrostates: wep_init, which corresponds to the subtask of initializing a state array for subsequent use during encryption; wep_mem, which corresponds to the subtask of reading/writing MAC frame data from/to memory; and wep_comp, which corresponds to the subtask of encrypting frame data while accessing the state array. From Figure 7-12a and c, it is clear that the current drawn can vary significantly with time, depending on the macrostate combination. For example, Figure 7-12a indicates that, at t = 55 ms, while two hardware components are engaged in lengthy iterative computations (icv_comp and fcs_comp), and a third component accesses memory over a shared bus (wep_mem), the total drawn current is 280 mA, which is well in excess of the battery's rated current [340].

Next, we consider the same sequence of frames being processed by a CBPM-enhanced MAC processor. Figure 7-12b and d shows the current profile and the corresponding macrostate execution trace for the optimized design. A comparison of Figure 7-12c and d shows that CBPM delays the execution of certain macrostates (shaded) to keep the system within the rated current. For example, in Figure 7-12d CBPM delays the occurrence of macrostate wep_mem to t = 42 ms, to avoid overlap with macrostate icv_comp. From the current profiles, it is clear that the occurrence of macrostate combinations that violate the rated current have been greatly reduced. The battery life of the new system was observed to be 10,199 s (a 3.2 × improvement), and battery capacity was 225.4 mA/h (a 1.5 × improvement).

The above example illustrates that the communication architecture can play a significant role in affecting the power profiles of components that are connected to it and can thereby regulate the power consumption of the entire system, improving battery efficiency.

CBPM Design Methodology

In order to design a CBPM-based system, the specification of each component is decomposed into a set of macrostates whose execution is under the control of the communication architecture. Efficient performance analysis techniques are used to prioritize identified macrostates, and power profiling tools are used to estimate the average power consumption of each macrostate. Once the macrostates have been identified, the specification of each component is enhanced to generate special requests at each macrostate transition. These requests convey the component's destination macrostate to the communication architecture. A component with a pending macrostate transition request must wait until the (CBPM-enhanced) communication architecture issues it a grant. These grants are issued in accordance with a predetermined but configurable CBPM policy that is implemented in the on-chip communication protocols.

CBPM-Enhanced Communication Protocol

A CBPM-based system consists of a communication architecture that is enhanced with new battery-aware communication protocols. The CBPM protocols implement policies that regulate macro state execution. The CBPM protocol periodically polls for accumulated pending macrostate requests; it then grants as many pending macrostates as possible, starting with the macrostate of the highest priority, while making sure that a specified constraint on the total power consumption is not violated. This constraint helps control the extent to which the run-time power profile is tailored towards the battery, by regulating the occurrence of macrostate combinations that violate the battery's rated current. These policies are implemented as extensions to the existing communication protocol hardware. CBPM-enhanced communication architectures can be configured through specification of parameters (including the power constraint), t trade off performance versus battery efficiency statically or dynamically.

7.7   CONCLUSIONS

As SoCs become increasingly complex, the on-chip communication architecture is expected to become more and more critical from the point of view of system performance, energy consumption, and battery life. In this chapter, we provided an overview of techniques for designing communication architectures for high-performance, energy-efficient system-on-chips. We described techniques for fast and accurate system-level performance/power analysis to drive communication architecture design and automatic techniques for customizing the communication architecture to the characteristics of an application's communication traffic, through both static customization (of communication architecture templates) and dynamic customization (using CATs). We also described a system power management methodology (CBPM) that utilizes new, battery-aware on-chip communication protocols to enhance the overall battery efficiency of a system. We believe that in the near future, in order to combat the increasing challenges posed by on-chip communication, such communication-aware design methodologies will be widely integrated into design practice.

ACKNOWLEDGMENTS

The text of this chapter is partly based on material that has been published (or is pending publication) in the following conference proceedings and journals: the International Conference on VLSI Design (2000), the IEEE/ACM Design Automation Conference (2000, 2001, 2002), the Intl. Conf. on Computer-Aided Design (1999, 2000), the International Symposium on HW/SW Co-Design (2002), the IEEE Transactions on Computer-Aided Design of Circuits and Systems (vol. 20, no. 6), and the IEEE Design and Test of Computers (vol. 19, no. 4).

Design Space Exploration of On-Chip Networks: A Case Study[1]

Bishnupriya Bhattacharya, Luciano Lavagno,
and Laura Vanzago

8.1 INTRODUCTION

The design of wireless protocols and their implementations on heterogeneous architectures, including dedicated IP, programmable logic, and one or more embedded processors, is a difficult task. One challenge is to provide validation and performance estimation of a given mapping of the protocol functionalities on the architecture without doing cycle-accurate simulation at the register-transfer level (RTL). This allows a pragmatic approach to architecture and application performance optimization, in an attempt to achieve both flexibility and efficiency, by running simulations with several test benches. Despite the high level of abstraction used in the modeling of both the application and the architecture components (programmable and nonprogrammable), one must be able to evaluate performance indices, such as throughput, latency, and resource usage, which are essential to optimize the configurable platform parameters for the application at hand. The refinement of the initial specification onto the architecture must be performed in a unified framework that also addresses the problem of IP reuse for both functional and architecture IPs.

Our case study consists of the design of part of a standard protocol stack defined by the European Telecommunications Standards Institute (ETSI) and called Hiperlan/2 [341], specifically focusing on the data-dominated physical layer and its implementation onto a real heterogeneous architecture, aimed at

<hr>

[1] Based on "Design space exploration for a wireless protocol on a reconfigurable platform" by L. Vanzago, B. Bhattacharya, J. Cambonie, and L. Lavagno, which appeared in *Proceedings of the Conference on Design Automation and Test.* © 2003 IEEE.

low-power transceivers for wireless applications. In particular, we analyze the requirements that the application imposes on a parameterized on-chip communication network, a crossbar switch allowing the designer to choose the number of possible parallel communications.

The design methodology adopted has the fundamental goal of quickly evaluating several architectural solutions that are available due to the reconfigurable nature of the target architecture. We address this problem by orthogonalizing the models of function, architecture, communication, computation, and timing and by using a design framework that supports the above methodology and facilitates the model integration with the aid of a graphical user interface and design flow management support. This methodology, as proposed in refs. 342 and 343, permits an efficient design space exploration based on IP reuse since each orthogonal concern can be optimized and refined separately from the others.

The application functionality is described as a network of sequential processes, finite state machines, and dataflow actors, on top of a modeling and simulation infrastructure based on the C++ language. The communication between functional blocks is heterogeneous, mixing the co-design finite state machine (CFSM) [342] and dataflow (DF) models of computations [344], as dictated by the mixed control/data processing characteristic of the application. For this case study, we used the VCC™ tool from Cadence Design Systems [345], extended to model dataflow, but the method could be adapted to other application domains and tools.

For executing performance simulations, in which the timing effects of a function/architecture mapping [342] are taken into account, we also used an original technique to model the mapping of an application on a complex abstract pipelined architectural model of the datapath. Latency and throughput of the pipeline can be customized through the usage of appropriate parameters.

The behavioral communication arcs, both control (CFSM) and dataflow, are separately refined by describing the effect of the schedulers and communication protocols used in the architecture [346]. These communication behaviors are organized in libraries of C++ classes known as pattern and architecture services [347] that can be customized using various parameters (e.g., CPU clock speed, bus transfer speed, arbitration latency, interrupt response latency, and so on) and that describe the communication at the transaction level of abstraction.

The communication refinement methodology is based, as in SpecC [346] and SystemC [348], on the separation between interface and its implementation. In VCC, this programming paradigm, coupled with a clear separation between architecture, functionality and timing, enables a more general reuse methodology, supported by graphical editors, configuration management, and so on.

Similar methodologies and design frameworks have been evaluated in the automotive [349], telecommunications [350], and multimedia [351] domains.

However, some of these case studies (e.g., refs. 349 and 350) mainly focus on control and decision-dominated applications that lend themselves well to the CFSM model of computation, based on lossy communication. Others, such as ref. 351, require manual introduction of explicit queues among processes in order to specify the application using an extension of Kahn process networks [352].

Our case study is interesting because it shows the results that can be obtained on a real application, from the wireless networking domain, by using abstract functional and performance models to explore several design alternatives rapidly on a reconfigurable field-programmable gate array (FPGA)-based datapath with a parameterized on-chip communication network. Due to the data-intensive nature of the application, we extended to the dataflow model of computation some performance analysis and communication refinement techniques that were previously applied to more reactive models. In addition, we defined a technique that allows successful modeling of timing delays of complex pipelined architecture components by preserving the original structure of both the functional model and the hardware implementation.

The rest of the paper is organized as follows. Section 8.2 provides some background information on modeling in VCC. Section 8.3 describes the mixed control/dataflow model of computation, and the functional modeling of the application is described in Section 8.4. Section 8.5 describes the architecture modeling and refinement of the transmitter onto the architecture. Results are given in Section 8.6. Section 8.7 concludes with the insights gained from this project and charts some directions for future work.

8.2 BACKGROUND

Although the case study described in this chapter uses tool-specific terminology, the methodology and flow are much more general and have been implemented using a variety of academic, internal, and commercial tools that were mentioned in the previous section (e.g., refs. 342, 343, and 346). In particular, the SystemC communication modeling methodology [348] allows one to use the same communication refinement mechanisms described in this chapter for architectural exploration.

8.2.1 Function/Architecture Co-Design Methodology

The design methodology followed in this project and supported by VCC is represented in Figure 8-1.

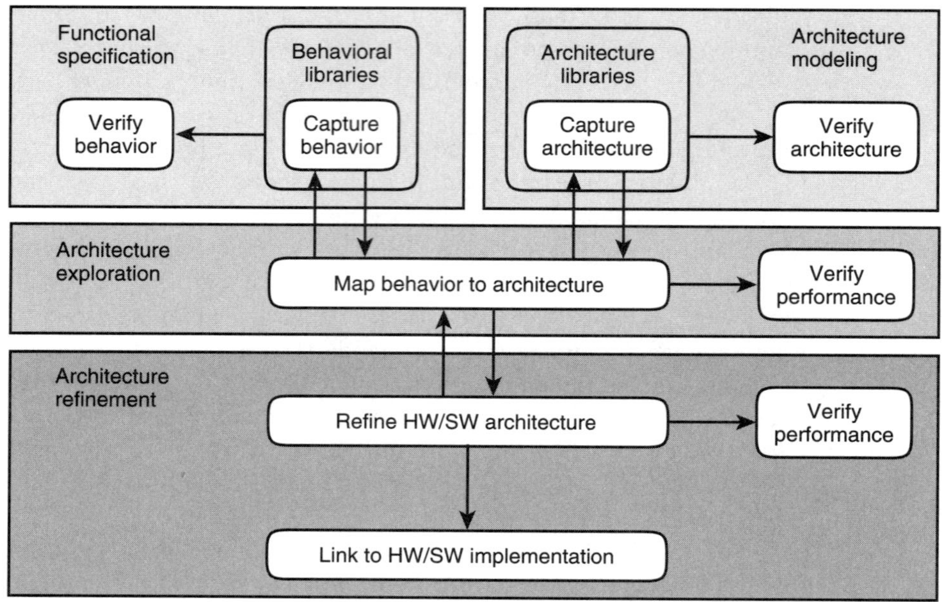

8-1

Design methodology.

FIGURE

Functional Specification

The specification of the system is supported in functional executable terms, meaning that it is simulatable and it is independent of the implementation, without any measure of cost or performance.

Blocks can be modeled in C++ (ANSI C and hierarchical FSMs are also available) and communicate by sending messages from the output port of one block to the input port of another block via the Post( ) application programming interface (API) call. A block is activated (run) if it receives a token on any one of its inputs. When activated, a block can use the Enabled( ) call to determine which input activated it, and it can read tokens from that input through the Value() call. Every input port has a one-place buffer, and the Post( ) call is nonblocking. This implies a lossy communication between blocks, in the sense that it is possible for the sender to overwrite a token on the receiver's input port before the receiver has had a chance to read the previous token.

VCC also supports communication via shared variables that do not activate the receiver block and are called behavioral memories.

Architecture Modeling

A set of parameterized alternative architectures that can be used to implement the specification is modeled. Architectures are described in terms of programmable units (CPUs), ASICs, interconnect networks, and real-time operating systems (RTOS; schedulers). They capture the computational capabilities, the degree of parallelism, the sequentialization of blocks sharing the same resources, and the capacity of the communication channels.

Architecture Exploration

The next step in the design phase is to map the behavior to the architecture. The mapping captures the partitioning of the behaviors onto the computational resources (processors and ASICs). For example, a given behavior can be mapped to an RTOS, implying that it is to be implemented as a software task on a processor, under control of the specified RTOS. As a refinement, the communication between the behaviors can be mapped to communication patterns (e.g., interrupt or polling), as explained in Section 8.2.2.

The mapped design can then be simulated to determine whether the performance of the architecture meets the system specifications. For example, performance analysis might indicate that the bus needs higher bandwidth or the processor needs a higher clock speed.

Architecture Refinement

Upon completion of the mapping step, the architecture with the implemented behavior is successively refined to microarchitectural levels at which detailed instruction sets, RTL models, and programming languages are selected and tested for compliance with the higher level requirements. It is at this step that ideal components described at an abstract level are actually mapped into "real" components, albeit still at a rather high-level description stage, using either automatic synthesis procedures or manual procedures. Since the ideal components are better characterized, it may be necessary to back-annotate the higher level models with more accurate parameters, so that performance evaluation can be assessed with enhanced confidence regarding the final outcome.

8.2.2   Performance Modeling with Architecture Services

Architecture services are a component of VCC that supports performance modeling of complex architectures. In addition to modeling the communication as bus

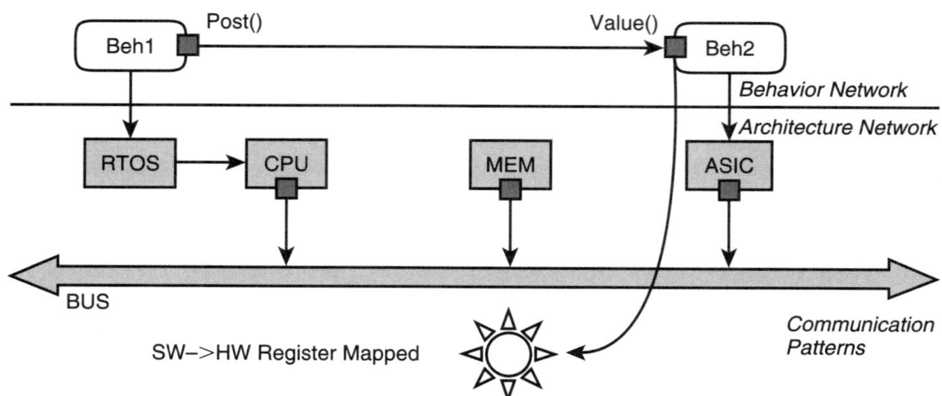

8-2

Behaviors mapped to architecture.

FIGURE

traffic and delay, it supports the specification of communication patterns and performance models distributed on each of the architecture resources. As a refinement step, the system designer maps each communication between the behaviors to a communication pattern (represented by a star in the mapped diagram of Fig. 8-2).

In Figure 8-2, the Beh1 functional block is mapped to the RTOS (implying that it will be implemented in software), and Beh2 is mapped to the ASIC (implying that it will be implemented in hardware). Therefore, the communication between Beh1 and Beh2 will be an SW → HW communication. The system designer then selects an appropriate communication pattern. In this example, the designer may choose to write a register on the ASIC directly or may use the shared memory resource MEM. In the former case, the two behaviors are more tightly synchronized for every communication than in the latter, since values must be read from the register immediately after being written, but require fewer resources (memory space and bus transactions). The pattern and each architecture component in the diagram specify a performance model. These models cooperate with each other to reflect the overall impact of mapping choices on system level performance. Similar mechanisms are used also to model the performance of each behavior on its architectural resource. We do not focus on this aspect in this paper.

Architecture services are a set of distributed models, each component specifying the impact it has on the system performance. Thus this methodology is flexible enough to handle new and diverse hardware/software platforms. The architecture services and communication patterns may also be written at different levels of accuracy, allowing the user to refine the design gradually. One analysis might explore the loading of the data bus, and then a refinement might include the impact of interrupts. The system designer might want to refine the analysis further to include the impact of various memory hierarchies, including caches and DMAs, instruction, and data fetching.

8.2.3 Mechanics of Architecture Services

Architecture services associated with an architecture element, e.g., the bus, the RTOS, or the CPU, conceptually declare the API for that element. For system modeling and performance analysis, the implementation of this API can be fairly abstract and only needs to describe that which has significant performance impact. Multiple implementations of the service can be provided at different levels of precision. Eventually, each service can be modeled at the hardware or software implementation level. Examples of services associated with an RTOS may include scheduling, Standard C Library, and software timers. Services supported by a CPU may include instruction fetching, interrupt handling, and bus adapters.

Each service declares a set of methods (API). The performance view of the architecture element implements those functions. Separating the declaration from the implementation permits multiple implementations for the same element. The different implementations may model the service at different points on the accuracy versus simulation-time tradeoff curve, or may represent different performance abstractions (i.e., delay, power). By simply binding a different view in the architecture diagram, the system performance can be refined from a high level to a more accurate level or may analyze different aspects of the system in terms of power or delay.

The service definitions are C++ classes that derive from the service declaration. These models "implement" the service declaration and can "use" other service declarations. For example, an RTOS service might implement the software timer service, which in turn "uses" the interrupt service on the processor.

A service definition can use parameters to specify implementation-dependent attributes, such as clock frequencies, access delays, scheduling overheads, I/O bit size, and so on. This allows reuse when one is describing different technology implementations of the same component, as well as design space exploration by executing performance simulations with various combinations of values of those parameters.

8.2.4   Architecture Topology Binds Services

The architecture diagram shows the topology of the architecture elements. As shown in Figure 8-3, the RTOS is running on the processor, and the processor is connected to a particular data bus. The architecture diagram does not need to define the individual signal connections between the processor and the data bus. It only means that they are physically connected and that data tokens can be transferred between elements connected to this bus.

The services of an architecture element can be associated with the element itself or with a particular port of the element. As mentioned above, these ports are not real signal connections but simply identify connection points to other architecture elements on a diagram.

The service definition can declare that it "uses" other service declarations. The architecture topology is automatically explored, at simulation setup time, to search for an element that implements the needed service declaration. Following the example above, the RTOS Standard C Library service uses a Memory Access service declaration. Since the RTOS is running on (and connected to in the architectural diagram) a given processor, the search should find a matching Memory

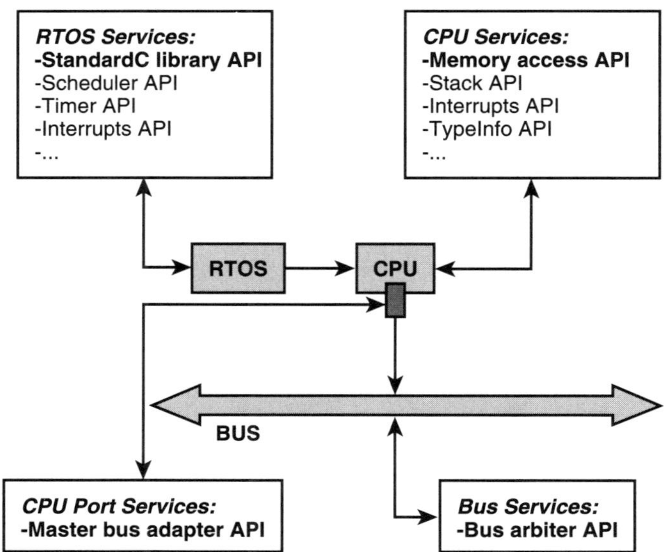

8-3   Architecture diagram annotated with sample architecture services.

FIGURE

Access service on the processor. Architecture services define a clean interface between the architecture elements. The designer of the architecture platform can mix and match architecture elements if they support compatible service declarations.

8.2.5   Communication Patterns

In addition to modeling the communication between architecture elements, services are used to model the full communication path between behaviors. For example, Figure 8-4 shows a behavior running in software, which sends data to a behavior running on an ASIC. This communication may involve the RTOS "memcpy" call from the Standard C Library, which in turn uses the memory access service of the CPU, which in turn uses the master bus adapter of the CPU to send tokens across the bus to a register on the ASIC. In order to send tokens on a bus, the master bus adapter requests ownership of the bus using the arbitration service on the bus. Finally the token reaches the ASIC register, and the ASIC polls the presence register before processing the data.

The communication patterns can be represented compactly by a sender service and a receiver service. The sender source implements the "output" port service declaration ("Post" function) and the receiver service implements the

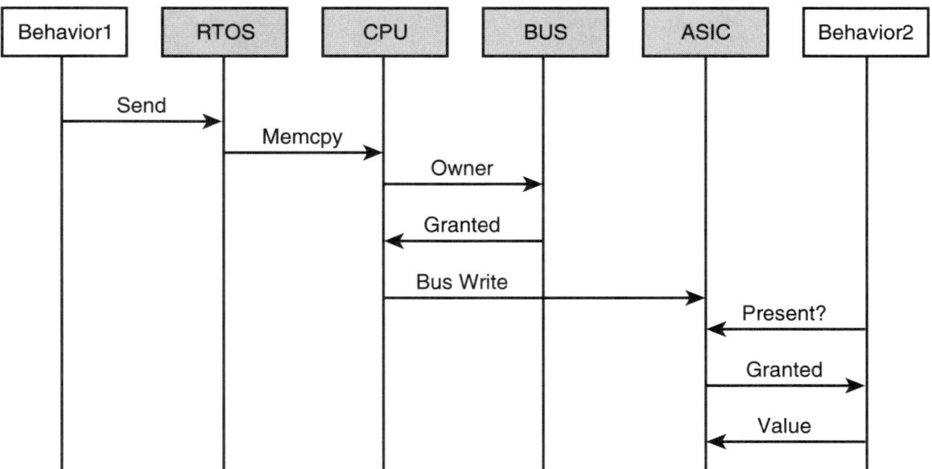

8-4   Messages in a register-mapped communication pattern.

FIGURE

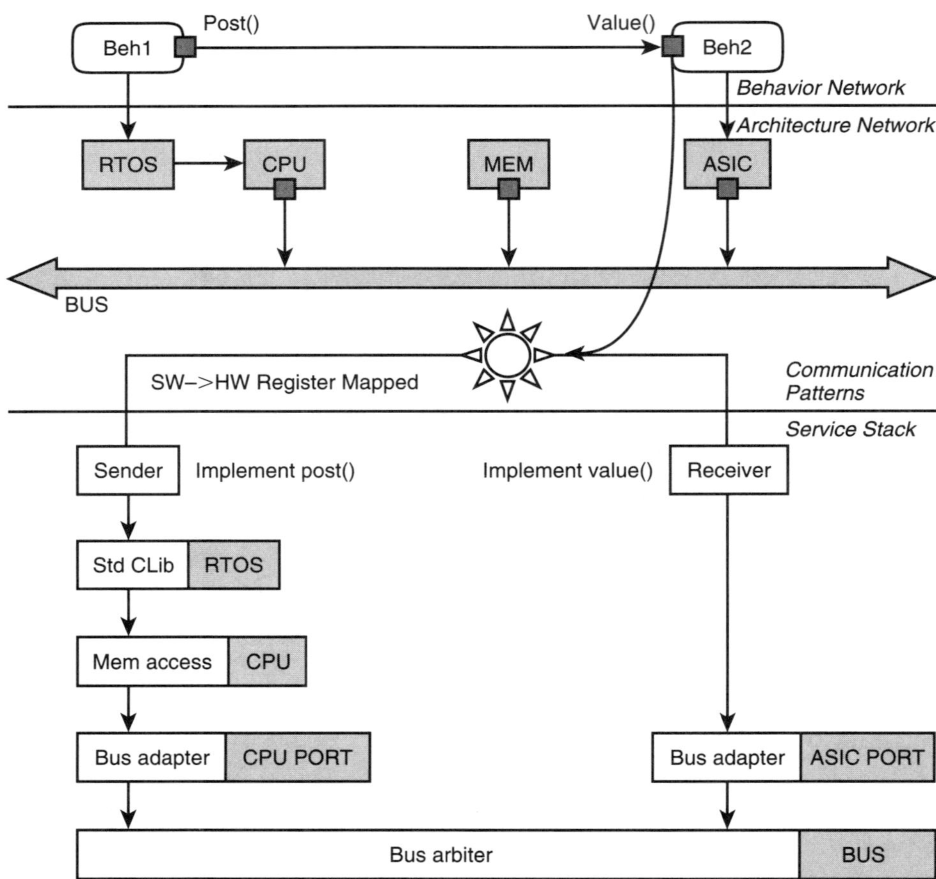

8-5   Service binding for register-mapped pattern.

FIGURE

"input" port service declaration ("Enabled" and "Value" functions). The sender and receiver services in turn can "use" services implemented by the architecture elements to which the behavior is mapped. For example, the register-mapped pattern would have a "sender" pattern that "uses" services supported by the RTOS/processor, and the "receiver" pattern "uses" services supported by the ASIC.

Figure 8-5 is an example of the services used in a Register Mapped pattern. The Post function is implemented by the RegisterMappedSWSender service. This sender pattern uses the Standard C Library API, which is implemented on the RTOS by the Standard C Library Service. The Standard C Library service uses the

CPU Memory API, which is implemented on the processor by the CPU Memory service. The CPU Memory service uses the Master Adapter API that is implemented by the FCFS Bus Adapter on the CPU port that writes the data to the Bus Slave Adapter on the ASIC port. In addition, the ASIC supports an ASIC Memory API, which provides the RegisterMappedReceiver service a way to read the data when the Value function is called.

The "sender" and "receiver" services only need to get the communication started and let the binding of services across the architecture topology transfer the data to its appropriate destination. Thus the "sender" and "receiver" patterns are reusable across different architecture platforms. For example, the Register Mapped sender and receiver services do not need to know the type of data bus that is used. In Figure 8-4, the data bus is modeled as first come, first served. The bus adapters can be replaced with different bus models, and the pattern will still work.

8.3   MODELING OF DATAFLOW NETWORKS

In this project, we extended the default CFSM semantics in VCC [342], which activates a behavior every time any one of its input ports has a token (the Enabled( ) function returns true), in order to support the dataflow [344] model of computation.

In dataflow, each block, when activated, consumes a certain number of tokens at each input port (called the data rate at that port) and produces a certain number of tokens at each output port. In static (also called synchronous) dataflow (SDF), the data rate at each port of a block is fixed and is statically known at compile time. In dynamic dataflow (DDF), a block can change the data rate at its ports at run time, as long as the block lets the scheduler know its firing rule for the next block activation. SDF restricts expressive power but provides important compile-time benefits like static scheduling, whereas DDF increases expressivity at the cost of decreased compile-time predictability.

The dataflow semantics primarily differ from CFSM semantics in terms of:

+ lossless communication between blocks via FIFO channels that queue up tokens between the sender and receiver.

+ firing rule of a block, whereby a block is activated only when all input ports have received sufficient tokens as determined by the data rate at that port.

◆ initial tokens (also called delay) on a channel, which are initially present in the FIFOs and correspond to delays in discrete control algorithms.

The DF model of computation is implemented in VCC by both changing the semantics and extending its standard communication API of Post()/Value()/Enabled(). Two additional API routines SetDataRate() and GetDataRate() are introduced for dataflow input/output ports, thus allowing the application designer to control the data rate at a port both at initialization time (SDF) and at run time (DDF). Strictly speaking, the Enabled() function (returning whether there are tokens or not on a given port) is not needed for dataflow input ports, since they are guaranteed to have sufficient tokens, according to the firing rule, when the block is activated. However, implementing it helps in reusing the same functional blocks in models of computation (such as CFSMs [342] and YAPI [351]) that allow a block to probe its input channels. The Value() call on a DF input port de-queues and returns the first token present on the FIFO channel at that input port. Post() on a DF output port writes a token onto the output FIFO channel. The number of initial tokens can be specified as a parameter on a dataflow input port. The implementation guarantees lossless communication in functional simulation, assuming infinite-length FIFO channels. For practical implementations onto an architecture, the FIFO length for a channel can be specified as a parameter on the receiving input port.

The examples in Figure 8-6 show the usage of the I/O API for a static multi-rate dataflow block (a; here the rate is specified via a block parameter called LengthPar) and for a dynamic dataflow block (b). Init() is a function executed at VCC simulation initialization. Run() is a function executed every time the block firing rule is satisfied. In Figure 8-6a, Real and Imag are the block input ports. SetDataRate and GetDataRate are used to set and retrieve the token threshold in the firing rule. Data are written at the output ports OutReal and OutImag by means of the Post() API interface function.

The dataflow capability described above is provided in the form of a simple communication pattern service in the VCC library that models the sender and receiver sides of the DF channel. Firing rule checks and block activation are implemented as a part of the receiver pattern service. This pattern has to be instantiated on each communication arc that follows DF semantics. In this prototype, no attempt is made to perform static scheduling for SDF blocks; instead, the block firing rule is checked dynamically for each DF block. Externally (e.g., to the VCC simulator) a DF block appears just as a CFSM block, and all the DF-specific activities are performed by the DF pattern. Hence interfacing with CFSM blocks is naturally supported, resulting in an unconstrained mix and match of CFSM and DF blocks, which is valuable for design space exploration. In our application, we extensively used this mix-and-match feature to describe the heterogeneous aspects

```
void Init() {
//Read the value of block parameter
LengthPar.
  Length=LengthPar.Value();
  Real.SetDataRate(Length);
  Imag.SetDataRate(Length);
}
void Run() {
for (i=0; i<Real. GetDataRate();i++)
  {
  data[i] [0]=Real. Value();
    data[i] [1]=Imag.Value();
  }
  //The block executes the FFT procedure..
  fft_cns_rot_bfp(data,Length....);
  for(i=0; i<Length; i++) {
    OutReal.Post(data[i][0]);
    OutImag.Post(data[i][1]);
  }
}
```
a

```
void Init() {
  // "Margin" is a block parameter
  In.SetDataRate(Margin.Value());
}
void Run() {
  if (RunCount==0) {
  //At the first execution the threshold of the
  //Input port In is changed; it will
  //affect only successive runs.
    In.SetDataRate(2); RunCount++;
  //At its first execution, this block just
  //consumes the input tokens without
  //producing output data.
    for (int i=0; i<In. GetDataRate();i++)
    In. Value();
  } else
  //At the successive executions the block
  //just copies to the output the input data.
    Out.Post(In. Value());
}
```
b

8-6

(a) FFT block. (b) Margin block.

FIGURE

of the transceiver, composing together both control and data processing blocks that are naturally modeled by CFSM and DF semantics, respectively. Such composition of CFSM and DF blocks for functional simulation is also possible in other existing tools like Ptolemy [353] and CoCentric System Studio [354]. The additional value that VCC provides is performance simulation of a mixed-model of computation (MoC) design for implementation onto a real architecture, as described in Section 8.4. El Greco moreover is based on a statically scheduled MoC, which may be difficult to implement efficiently on a distributed mixed hardware/software architecture.

8.4   CASE STUDY: HIPERLAN/2 APPLICATION

Our target application is the physical layer specification for high-performance radio local area network type 2 (Hiperlan/2), which targets professional and wireless local area network (WLAN) applications [355].

The Hiperlan/2 physical layer is based on the orthogonal frequency division multiplexing (OFDM) modulation scheme, which has been selected due to its improved performance on highly dispersive channels.

A key feature of the Hiperlan/2 physical layer is that it provides several physical layer modes with different coding rates and modulation schemes, as selected by the link adaptation mechanism. As a result, the data rate ranges from 6 to 54 Mbit/s, thus improving the radio link capabilities under different conditions of interference and of distance between mobile terminals and the access point. BPSK, QPSK, and 16QAM are used for subcarrier modulation, and 64QAM can be used as an optional mode.

The multirate feature introduces an element of dynamic configuration that matches well the use of reprogrammable components in the architecture, as discussed below. The Hiperlan/2 transmitter is compliant with the specification in ref. 341, including the functions of scrambling, forward error correction (FEC) encoding, interleaving, mapping, OFDM modulation by fast Fourier transform inverse (IFFT), guard interval insertion, and oversampling by a finite impulse response (FIR) filter. A significant part of the Hiperlan/2 receiver has also been modeled. It includes decimation by a FIR, preamble sequence recognition with autocorrelation of incoming complex samples, carrier frequency offset (CFO) acquisition and correction, timing synchronization, FFT, equalization, de-mapping, and de-interleaving.

8.4.1   Modeling the Hiperlan/2 Physical Layer

A careful definition of the block granularity at the functional level is a key aspect of the mapping-based system-level design methodology that we used, since functional blocks are the unit of partitioning over architectural resources. Moreover, they directly expose the parallelism in both computation and communication that can be exploited by the resources instantiated in the architecture.

We manually imported in our architecture exploration environment the transceiver functional model, which had been initially developed and validated using a combination of C code and purely functional dataflow models developed in Cossap™ (currently available as CoCentric System Studio, from Synopsys) [356]. Although we kept the granularity of Cossap (with a negligible effort in translating the block interfaces between environments), we partitioned the original C procedures into multiple blocks. This essentially involved copying and pasting the data processing code into the template that VCC requires for each behavioral block and defining some control functions, requiring about 150 additional lines out of approximately 2500 lines of untimed C++ behavioral code.

In particular, we separated the control from the data processing functions because they require different models of computation (CFSM and dataflow, respectively) at their interface. In addition, such separation allowed us to evaluate

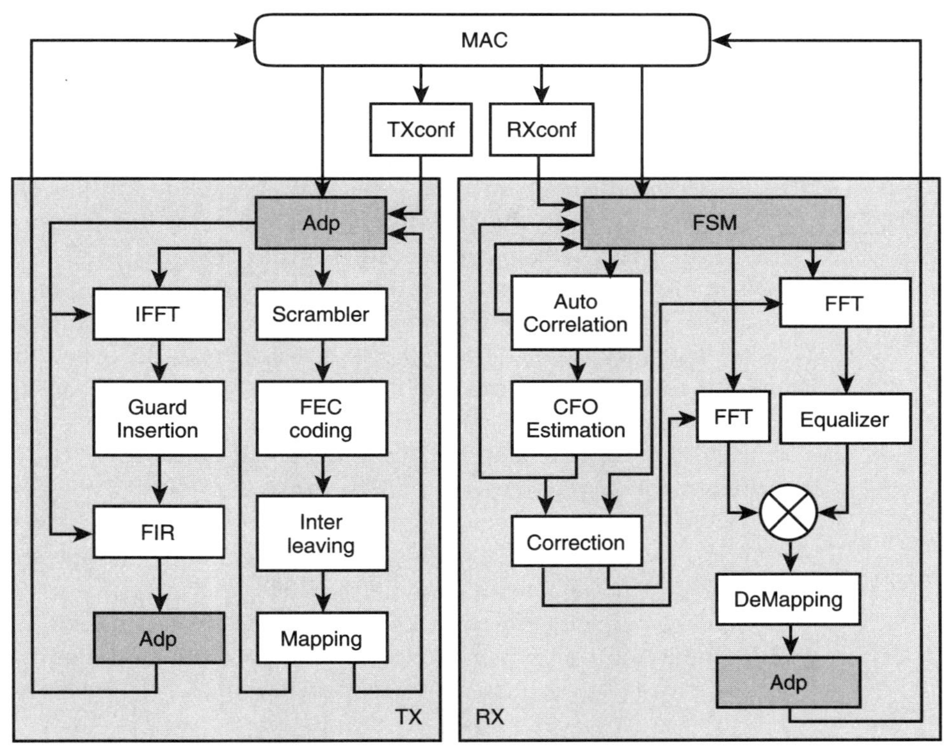

8-7   Hiperlan/2 application: behavioral diagram.

FIGURE

different mappings onto specialized cores and onto the FPGA, as discussed in Section 8.5.2.

Figure 8-7 shows the transmitter and receiver networks organized as separate hierarchical blocks, together with a simplified model of the medium access control (MAC). The model includes the blocks called TXconf and RXconf that control changes of modulation scheme or switch between transmission and reception modes. At the functional level they are not needed, but they act as placeholders for delays that will be added during the refinement step. The function of the MAC model is to trigger the transmitter and receiver executions alternately and to configure the modulation scheme of the TX network, according to the Hiperlan/2 specification. The MAC uses the CFSM method to communicate with TX, RX, and configuration blocks. TX and RX are modeled as dynamic dataflow networks.

The RX or TX quantum of computation defines the point at which the mode of the system can be switched by the MAC between RX and TX. It is signaled to the MAC by the presence of a token at their outputs, thus indicating the end of an iteration multicycle. The quantum can be defined both in TX and RX by using the relation between the number of OFDM symbols transmitted or received, and the number of times each process acting on the payload must be executed.

In the transmitter, the quantum of computation is one frame and the iteration cycle corresponds to the execution of the process chain shown in the TX network of Figure 8-7 to generate one of the OFDM symbols of the frame that is being transmitted. For example, a broadcast frame in normal traffic conditions can correspond to about 70 OFDM symbols. Consequently, the quantum of computation of the transmitter corresponds to 70 executions of the processes it includes.

The definition of the quantum of computation for the RX is slightly more complex, due to the initial synchronization stage that analyzes a sequence of complex numbers whose length is not known a priori. Nevertheless one can define the condition of termination of this stage as the evaluation of the correlation angle. After the synchronization, the number of complex numbers requested by the evaluation of equalization coefficients and that of the bits generated by the payload de-mapping and de-interleaving is also related to the number of OFDM symbols of the frame received. Then one can define the RX quantum of computation as in the TX case, by using an FSM that switches among the synchronization, preamble recognition, and payload de-mapping states.

The second function of the MAC module is to configure the TX network to execute one of the following modulation and demodulation schemes: QPSK, BPSK, QAM16, and QAM64, as requested by the Hiperlan/2 specification. The MAC sends this information to the TX and configuration blocks through dynamic parameters implemented by means of behavioral memories. They model a communication mechanism based on shared variables that does not activate the block.

The model described above allows the designer to execute a functional (untimed) simulation of the transmitter and receiver by driving their execution with the simplified MAC FSM and using data read from a file. The simulation results are collected as textual files that can be analyzed directly by the designer or with external visualization tools.

8.5   THE ARCHITECTURAL PLATFORM

In this project we mapped the Hiperlan/2 model onto a real reconfigurable and heterogeneous platform for low-power transceivers used in wireless applications.

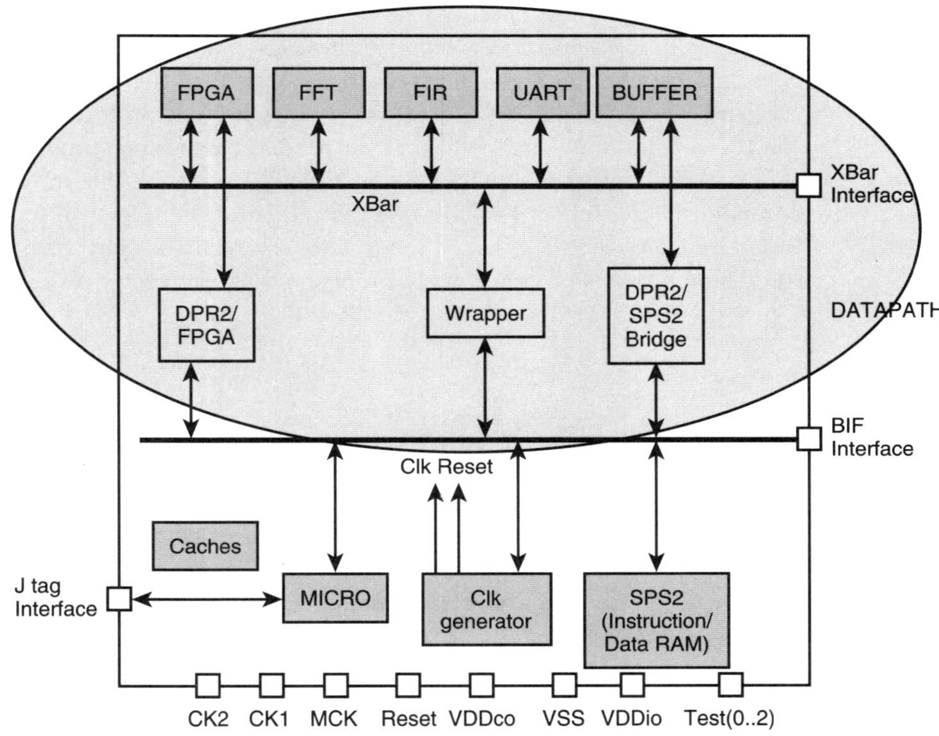

8-8   The architecture diagram of the wireless platform.

FIGURE

It is specialized for an OFDM-based physical layer but also supports the implementation of high-level protocol tasks on an embedded processor.

Several cores are connected through a flexible communication resource, a crossbar bus called XBar in Figure 8-8. Some of the cores, i.e., FFT and FIR, implement computation-intensive functions as highly optimized IP with limited range of programmability. Other cores are very flexible. An embedded low-power FPGA [357] provides bit-level programmability, and a RISC microcontroller provides resources for dataflow management control functions, as well as for MAC layer protocol tasks.

Each data item sent via the crossbar is associated with an attribute that describes which target it has to reach and to which thread it belongs. The crossbar is a parameterized block with a flexible tradeoff between area and performance. The number of parallel slave accesses can be selected by the designer at platform configuration time. The crossbar arbiter uses a "first-come, first-served" scheme with fixed priorities whenever there are multiple access requests to some slave or whenever there are more access requests than the designer-chosen crossbar parallelism. Finally, a Request/Grant/Acknowledge protocol is used between the IPs to adapt the dataflow to their respective computing speeds.

The datapath is reconfigured dynamically between the transmission and reception phases, as well as between the transmission of several frames, when a different modulation scheme is requested. The configuration mainly affects the FPGA and consists of either overwriting some internal registers or downloading a new configuration stream from a dedicated memory.

8.5.1   Architectural Modeling

We modeled the datapath coprocessor of the architecture, including FFT, FIR, FPGA, memories, and crossbar. Each component is described by an abstract API defining the services that it offers to the other architecture components and that impact on the overall architecture behavior and performance. Those services describe, for example:

+ the I/O transmission protocol and associated delays (sending or receiving a word through the crossbar or via the bus, including arbitration effects)

+ the storage delay access and capabilities (reading or writing a word from memory, including cache effects).

In this project we reused several services provided by the standard VCC library to describe memories, registers, schedulers, and data formatters. In addition, we designed new service definitions to model the crossbar interconnection resource supporting a number of concurrent communications. The overall behavior of the various peripheral interfaces, arbiter, and connection wires is modeled as the cooperation of three architecture services associated with the crossbar instance and with the connection ports of masters and slaves, as shown in Figure 8-9.

The arbiter service uses a prioritized FIFO for each slave, to order the requests coming from several masters. A data transmission involving several bus transactions may be suspended if during its execution the arbiter receives a request with higher priority.

The overall crossbar model is flexible in terms of the number of supported masters and slaves. For example, the number of FIFOs handled by the arbiter is defined when the simulation model is built, based on information derived from

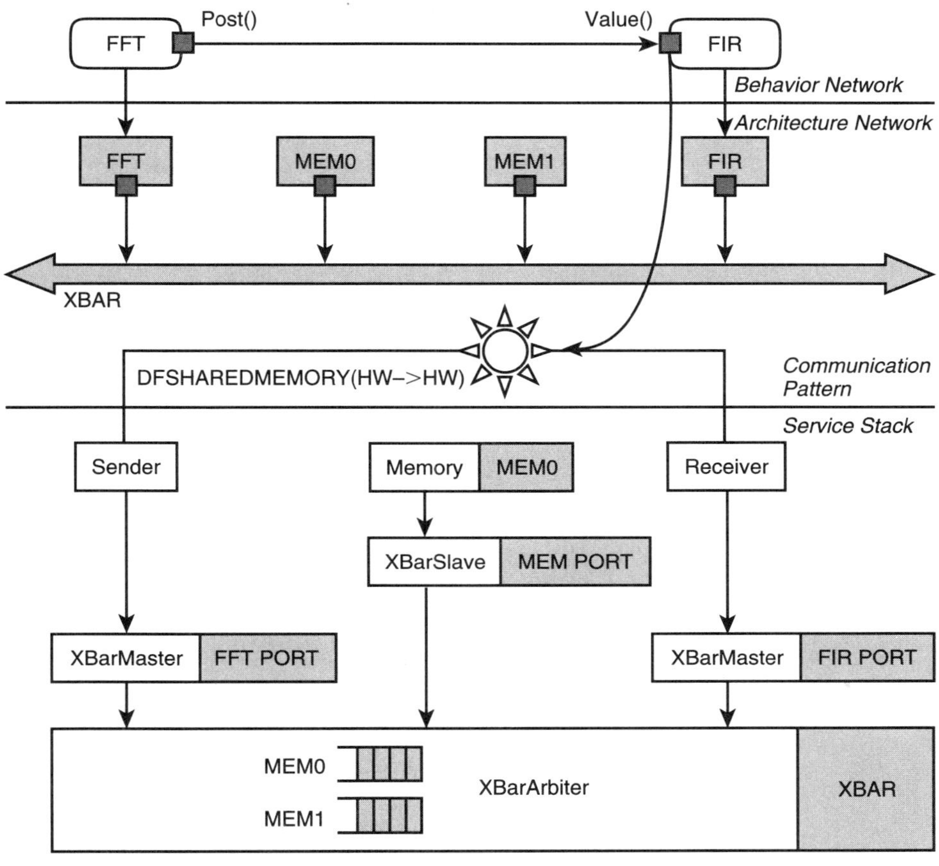

8-9   XBAR service stack.

FIGURE

the layout of the architecture diagram and on the set of services instantiated for each resource. The delay that characterizes the arbitration overhead, the slave access delay, and the number of simultaneous parallel accesses for each slave port are some of the parameters that are defined for these services.

8.5.2   Mapping and Communication Refinement

The mapping of a function onto an architecture resource specifies a possible implementation, e.g., as hardware or software, and its performance cost in terms

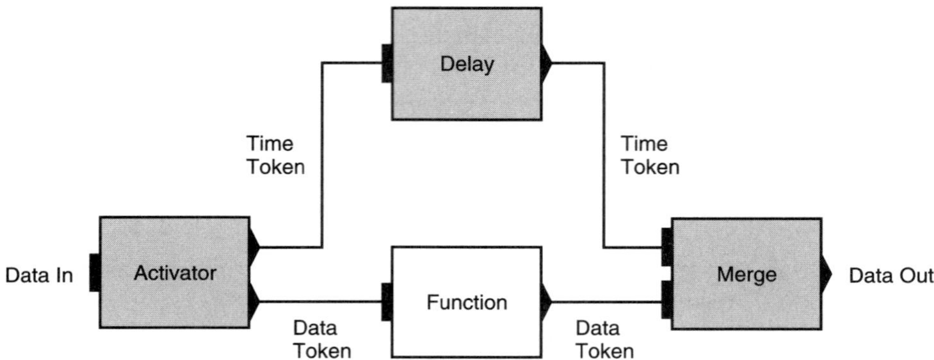

8-10

FIGURE

The delay wrapping technique used to model performance of a behavioral block.

of estimated delays. Clearly, the quality of these delay estimates affects the precision of the overall architecture performance simulation by the tool. In our case, the delays of the FFT and FIR peripherals are not data-dependent, because they are statically pipelined IP. Thus the timing estimates are quite precise and can be derived from documented IP specifications and RTL simulations.

The performance models for the FPGA are derived from the RTL code of the functions mapped to it. On the other hand, modeling the performance of a complex functional block, whose netlist structure is quite different from that of the pipelined shared datapath, required the use of the mechanism shown (conceptually) in Figure 8-10.

The modeling scheme allowed us to describe separately the timing and the functionality of each block. The Activator and Merge blocks split and remerge each token, carrying both a data value and a time stamp. The Function block is a "normal" VCC netlist, modeling the untimed functionality of an FIR filter or an FFT. On the other hand, the Delay block models a skeleton of the pipeline architecture, for example, that of the FFT core represented in Figure 8-11. It is used to experiment with various architectural options, such as the number of stages, the amount of buffering and the related stalls, and so on.

We also defined some new patterns to represent Shared Memory and Register Mapped communications, as required by the various possible architectural implementations of the dataflow pattern used in the functional simulation. The refinement consists of the redefinition of Post( ) and Value( ). A transaction record (for debugging and analysis) and bus access requests (for performance modeling) are generated with a rate that depends on the value specified by the SetDataRate( )

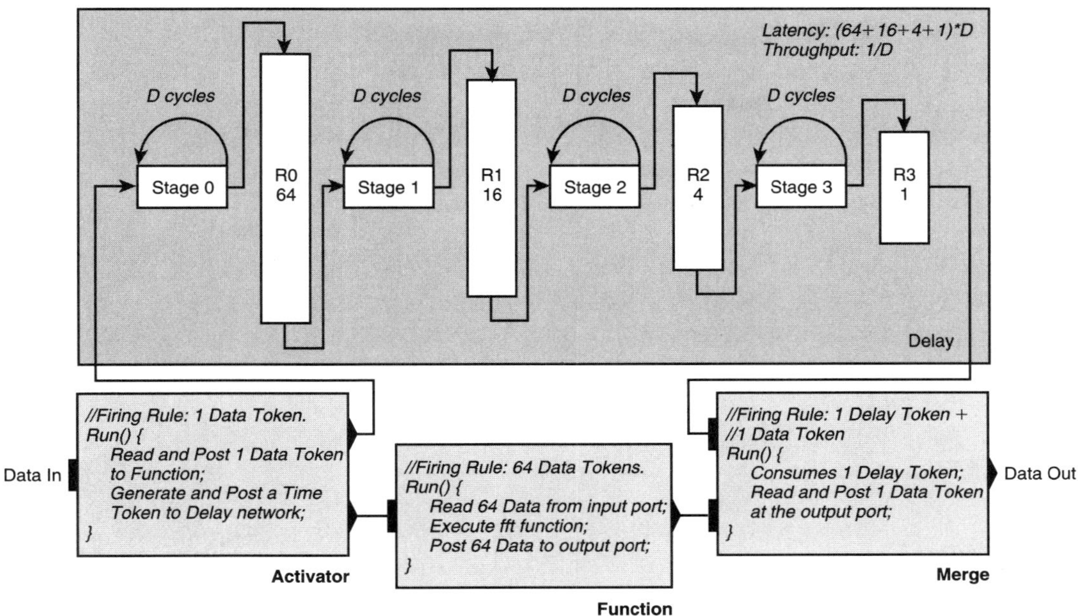

8-11   FFT skeleton pipeline (delay block).

FIGURE

function. If tokens are smaller than the bus width, the pattern collects them together in a buffer before issuing a bus request, as shown in Figure 8-12, in which the pattern service refines the Post() output function call so that it uses a shared Memory communication.

8.6   RESULTS

This section presents the results of some explorations that we performed, by evaluating the different mapping scenarios listed in Figure 8-13 for the transmitter application shown in Figure 8-7, using the architectural model of the datapath shown in Figure 8-8.

We report the global latency and throughput effects of the architectural choices, such as memory configuration (single port versus multiple ports),

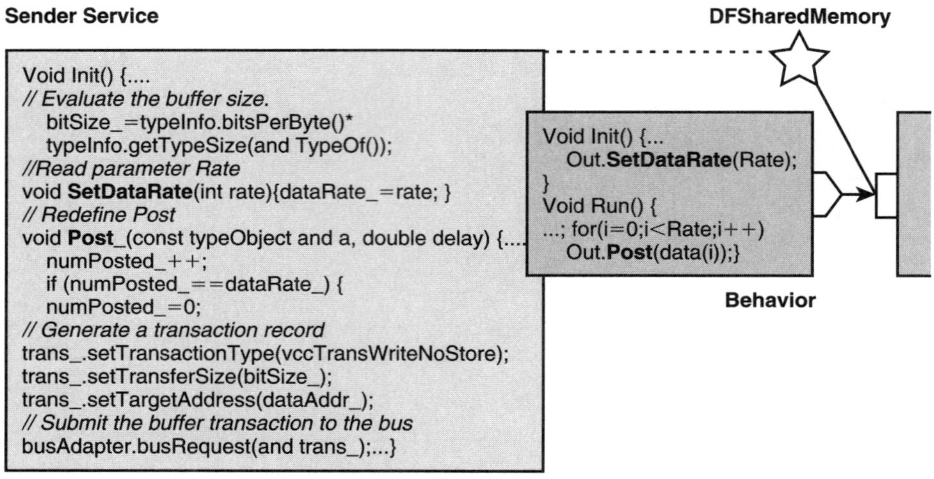

8-12

FIGURE

Example of communication refinement for a dataflow FIFO.

FFT Throughput: $\frac{1}{4}$clk cycle (i=1) 1/1 clk cycle (i=2)		Buffer: Num. of Ports on the crossbar	
		One	Two
FPGA–>FFT Comm. Refinement	Shared Memory	SimAi	SimBi
	Direct Comm.	SimCi	SimDi

8-13

FIGURE

Explored design space for transmitter application.

communication refinement (direct connect versus shared memory), and FPGA configuration strategies. A similar exploration would also be possible, in theory, with an RTL simulation, but the cost, in terms of model development, modification, and simulation time, would be unacceptable. The methodology and tools used in this case study, on the other hand, required only a few hours to define and simulate each design space point.

The mean error of the VCC model, in terms of number of clock cycles with respect to the "golden" RTL model available for one of the analyzed scenarios (SimA1), is better than ±10%. The difference originates mainly from the fact that the RTL model uses the microprocessor for feeding the FPGA, whereas our VCC model does not include the microprocessor.

The simulation is also reasonably fast, comparable to other cycle-accurate system-on-chip modeling methodologies. Each test case described below requires between 4 and 26 s (depending on how many probes are added to the design for instrumentation and analysis) to simulate the transmission of 6 OFDM symbols, representing from 4000 to 2000 simulated clock cycles (depending on the architecture instance), on a Pentium III 600-MHz machine running Windows NT.

However, the main advantage of the methodology is not the speed of the simulation but the speed of mapping and refinement changes in order to explore the design space. This dramatically reduces the precious time devoted by a designer to this task with respect to RTL modeling.

8.6.1 Communication Refinement

Figure 8-13 shows the design space covered by our first exploration. In this case, we varied the communication refinement between FPGA and FFT by using shared memory or a direct connection through the bus. For each case we also changed the access mode to the memory (one or two access ports from the crossbar). All cases are then evaluated by using two different FFT throughput estimates, which correspond to two different FFT architectures. The results are given in Figure 8-14. The chart shows the number of clock cycles (y-axis) for which each core is busy in each mapping scenario (x-axis). A balanced load, such as in SimC1, is generally better than an unbalanced one, such as in SimA1.

We verified that increasing the number of access ports to the memory through the crossbar (SimAi/SimBi or SimCi/SimDi) changes the bit rate for our application by no more than 25%.

From these charts, assuming a clock rate of 70 MHz, we can evaluate the data-path bit rate for a given mapping (Table 8-1). Those rates show that only some mappings are compliant with the physical layer Hiperlan/2 specification at full stream speed, which requires a bit rate in excess of 12 Mbit/s. However, each one shows different characteristics that may make it more or less desirable for other application families, such as low-bit-rate ad hoc radio. For instance, the scenarios involving the fastest FFT architecture result in the highest bit rate, at the price of a larger area and power consumption estimate.

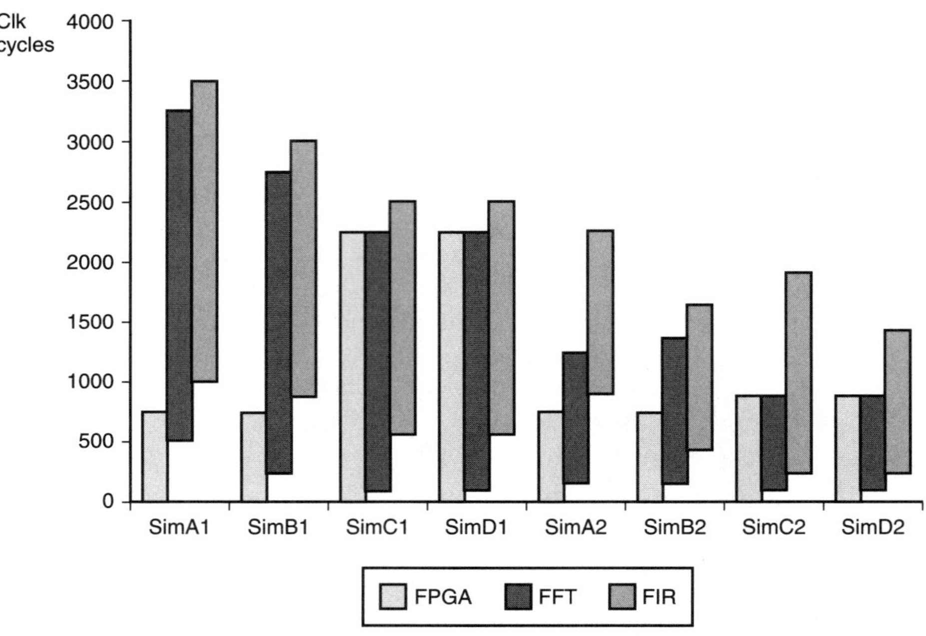

8-14 FIGURE   Results of design space exploration.

Sim run	A1	B1	C1	D1	A2	B2	C2	D2
Bit Rate	5.8	7.2	8.2	8.2	9.6	13.7	12.5	15.6

8-1 TABLE   Datapath Bit Rates (Mb/s) Evaluated for Different Mappings.

The current system model provides only latency/throughput information, but the architectural services could be extended to provide also area and activity (power) information.

8.6.2   FPGA Alternatives

The next exploration shows that it is possible to evaluate the cost of adding a new function to the algorithm and explore two implementation scenarios involving dynamic embedded FPGA reconfiguration.

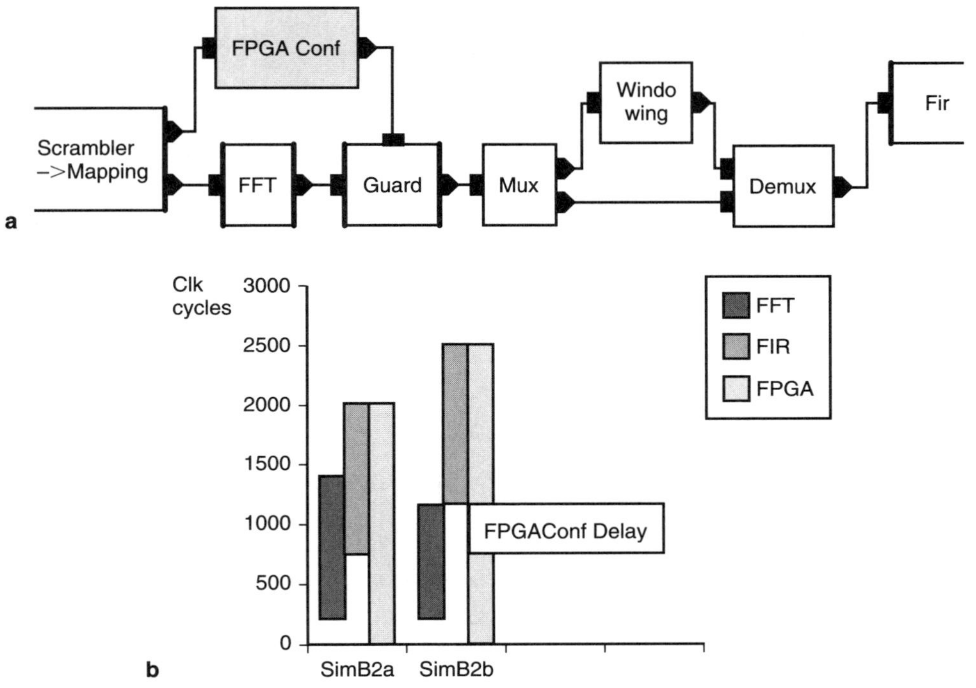

8-15

FIGURE

Design space exploration with windowing. (a) Functional network. (b) Simulation results.

Starting from case SimB2, we expanded the design space by inserting a new function, called windowing, in the behavioral network between the cyclic guard and the FIR functions. This function is used to decrease the dynamic range of the OFDM signal significantly before transmission.

We modified the functional network as shown in Figure 8-15a. Windowing must act only on the first and last 10 elements of each OFDM symbol. The dataflow between Guard and FIR is thus controlled by the dynamic dataflow blocks Mux and Demux, whereas the FPGAConf block is added as a placeholder for the FPGA reconfiguration delay.

As shown in Figure 8-15b, we executed two different simulations. In the first case (SimB2a), we assumed an embedded FPGA that was large enough to support a configuration including both mapping and windowing, which provides maximum parallelism, but little area efficiency.

In the second case (SimB2b), we assumed only one smaller embedded FPGA, which must be reconfigured on the fly to switch from the mapping function to the windowing function. This creates a configuration cost (number of cycles to reconfigure the FPGA) in the performance evaluation that results in a measurable bit rate decrease. In Figure 8-15b, we show the performance obtained when the reconfiguration cost is 500 clock cycles. Thus, in this exploration, we traded off the size of the FPGA against the performance of the system.

8.7   CONCLUSIONS

We applied a design methodology, based on separating functionality from architecture, and communication from computation, to a real application and architectural platform from the wireless networking domain. We performed architectural exploration considering both on-chip communication aspects and various pipelining and reconfiguration options for computation. During this experiment, we also extended the capabilities of the tool that we used, in order to describe the mixed control/dataflow nature of the application. We applied a methodology that allows one to refine behavioral dataflow communications onto bus transactions in the architecture. We also used an intuitive method to describe timing views for functions mapped to pipelined hardware. The flow was applied to a single-processor architecture, but it is completely general and, due to its high level of abstraction, it is much more scaleable than traditional, HDL-based design space exploration.

Further project developments will include a detailed functional model of the MAC layer and the synchronization stage of the receiver, followed by their mapping on the microprocessor.

ACKNOWLEDGMENTS

We acknowledge support from the Italian Ministry of Education, under a 40% grant Mixed Hardware/Software Analog/Digital Co-Design Techniques. We also thank the Central R&I team and Joel Cambonie of STMicrolectronics Berkeley Laboratory for their help in creating the application and architecture models and the VCC team of Cadence Design Systems for their support throughout the project and for some of the material in Section 8.2.

PART II

SOFTWARE

*T*his section concentrates on software for MPSoCs. Embedded software covers a broad range of topics. Not only must we design programs to perform individual tasks but we must also design the system of processes that will jointly execute on the MPSoC.

When designing embedded software, we are concerned not only with functionality but also with metrics such as performance and energy consumption. We measure those metrics through the lens of the hardware platform on which the software will operate. For example, the performance of an individual program is influenced by the size of the cache on its processor. The performance is also affected by other programs that share the processor, for example, tasks may interfere in the cache, resulting in worse performance than we would expect by analyzing each program alone. When designing software for MPSoCs, we need a hierarchy of models that let us refine the characteristics of the software as we refine its functionality.

Chapter 9, by Mahmut Kandemir and Nikil Dutt, looks at memory systems. Memory system performance is critical to both performance and power consumption. Some memory system optimizations, such as array techniques, have been developed for general-purpose computing systems, but many new techniques have been introduced to handle the special characteristics of embedded computing systems. Specialized techniques include aggressive optimizations for standard memory hierarchies as well as software that utilizes special features such as scratch-pad memories.

Chapter 10, by Jan Madsen, Kashif Virk, and Mercury Jair Gonzalez, describes a SystemC-based model for real-time operating systems. This model allows the designer to experiment with various characteristics of real-time operating systems (RTOSs), such as scheduling, before committing to a particular implementation.

This model takes advantage of the features of SystemC, such as multiple abstraction levels, synchronous and asynchronous communication, and refinement, in order to provide a flexible modeling environment for the choice of RTOS characteristics and parameters.

Chapter 11, by Peng Yang, Paul Marchal, Chun Wong, Stefaan Himpe, Francky Catthoor, Patrick David, Johan Vounckx, and Rudy Lauwereins, describes methods for task mapping, particularly for multimedia systems. Multimedia systems have highly complex system specifications that often require complex dynamic behavior with tasks created and destroyed at run time. The quality of the mapping can greatly affect both performance and energy consumption. The authors propose a task concurrency management (TCM) methodology and a two-phase methodology for task management. This methodology optimizes for both energy consumption and real-time performance.

The final chapter, Chapter 12, by Jens Palsberg and Mayur Naik, describes methods for software compilation taking resource constraints into account. Given the tight constraints in many embedded computing systems, strongly optimizing compilation methods can make the difference between feasibility and infeasibility. The authors use integer linear programming (ILP) as a fundamental tool for resource-constrained compilation.

Memory Systems and Compiler Support for MPSoC Architectures

Mahmut Kandemir and Nikil Dutt

9.1 INTRODUCTION AND MOTIVATION

System-on-chip (SoC) architectures are being increasingly employed to solve a diverse spectrum of problems in the embedded and mobile systems domain. The resulting increase in the complexity of applications ported into SoC architectures places a tremendous burden on the computational resources required to deliver the required functionality. An emerging architectural solution places multiple processor cores on a single chip to manage the computational requirements. Such a multiprocessor system-on-chip (MPSoC) architecture has several advantages over a conventional strategy that employs a single, more powerful (but complex) processor on the chip. First, the design of an on-chip multiprocessor composed of multiple simple processor cores is simpler than that of a complex single-processor system [358, 359]. This simplicity also helps reduce the time spent in verification and validation [359]. Second, an on-chip multiprocessor is expected to result in better utilization of the silicon space. The extra logic that would be spent on register renaming, instruction wake-up, speculation/predication, and register bypass on a complex single processor can be spent on providing higher bandwidth on an on-chip multiprocessor. Third, an MPSoC architecture can exploit loop-level parallelism at the software level in array-intensive embedded applications. In contrast, a complex single-processor architecture needs to convert loop-level parallelism to instruction-level parallelism at run time (that is, dynamically) using sophisticated (and power-hungry) strategies. During this process, some loss in parallelism is inevitable. Finally, a multiprocessor configuration provides an opportunity for energy savings through careful and selective management of individual processors. Overall, an on-chip multiprocessor is a suitable platform for executing array-intensive computations commonly found in embedded image and video processing applications.

One of the most critical components that determine the success of an MPSoC based architecture is its memory system. This is because many applications spend a significant portion of their cycles in the memory hierarchy. In fact, one can expect this to be even more so in the future, considering the ever-increasing dataset sizes, coupled with the widening processor-memory gap. In addition, from an energy consumption angle, the memory system can contribute up to 90% of the overall system power [360, 361]. In fact, one can expect that a significant portion of the transistors in an MPSoC-based architecture will be devoted to the memory hierarchy.

There are at least two major (and complementary) ways of optimizing the memory performance of an MPSoC-based system: (1) constructing a suitable memory organization/hierarchy and (2) optimizing the software (application) for it. This chapter focuses on these two issues and discusses different potential solutions for them. On the architecture side, one can employ a traditional cache-based hierarchy or can opt to build a customized memory hierarchy, which can consist of caches, scratch pad memories, stream buffers, LIFOs, or a combination of these. It is also possible to make some architectural features reconfigurable and tune their parameters at run time according to the needs of the application being executed. Different memory architectures for MPSoCs are detailed in Section 9.2.

Traditional compilation techniques for multiprocessor architectures focus only on performance (execution cycles). However, in an MPSoC-based environment, one might want to include other metrics of interest as well, such as energy/power consumption and memory space usage. Therefore, the compiler's job is much more difficult in our context compared with the case of traditional high-end multiprocessors. In Section 9.3, we investigate different MPSoC compilation issues in more detail and make a case for constraint-based compilation, which compiles a given application code for an objective function and under several constraints based on performance, energy/power, and memory usage. We summarize the chapter in Section 9.4.

9.2   MEMORY ARCHITECTURES

The application-specific nature of embedded systems presents new opportunities for aggressive customization and exploration of architectural issues. Since embedded systems typically implement a fixed application or problem in a particular domain, knowledge of the applications can be used to tailor the system architecture to suit the needs of the given application. Such an architectural exploration scheme is quite different from the development of general-purpose computer systems that are designed for good average performance over a set of typical benchmark programs that cover a wide range of applications with different behaviors. However, in the case of embedded systems, the features of the given application can be used to determine the architectural parameters. This is particularly true for MPSoC-based systems, in which we have numerous power-hungry components. For example, if an application does not use floating point arithmetic, then the floating point unit can be removed from the MPSoC, thereby saving area and power in the implementation.

Since the memory subsystem will dominate the cost (area), performance, and power of an MPSoC, we have to pay special attention to how the memory subsystem can benefit from customization. Unlike a general-purpose processor, in which a standard cache hierarchy is employed, the memory hierarchy—indeed the overall memory organization—of an MPSoC-based system can be tailored in various ways. The memory can be selectively cached; the cache line size can be determined by the application; the designer can opt to discard the cache completely and employ specialized memory configurations such as FIFOs and stream buffers; and so on. The exploration space of different possible memory architectures is vast, and there have been attempts to automate or semiautomate this exploration process [362, 363].

Traditionally, memory issues have been separately addressed by disparate research groups: computer architects, compiler writers, and the CAD/embedded systems community. Memory architectures have been studied extensively by computer architects. Memory hierarchy, implemented with cache structures, has received considerable attention from researchers. Cache parameters such as line size, associativity, and write policy, and their impact on typical applications have been studied in detail [364]. Recent studies have also quantified the impact of dynamic memory (DRAM) architectures [365]. Since architectures are closely associated with compilation issues, compiler researchers have addressed the problem of generating efficient code for a given memory architecture by appropriately transforming the program and data. Compiler transformations such as blocking/tiling are examples of such optimizations [366, 367]. Note that many of these designs/optimizations need a fresh look when an MPSoC-based system is under consideration.

Finally, researchers in the area of computer-assisted design (CAD)/embedded systems have typically employed memory structures such as register files, static memory (SRAMs), and DRAMs in generating application-specific designs. Although the optimizations identified by the architecture and compiler community are still applicable in MPSoC design, the architectural flexibility available in the new context adds a new exploration dimension. To be really effective, these optimizations need to be integrated into the design process as well as enhanced with new optimization and estimation techniques. In this section, we first present an overview of different memory architectures and then survey some of the ways in which these architectures have been customized.

9.2.1   Types of Architectures

In this section, we outline some common architectures that can be used in memories of MPSoC-based systems, with particular emphasis on the more unconventional ones. Note that MPSoC-based systems can contain both hardware-managed and software-managed memory components.

Cache

As application-specific systems became large enough to use a processor core as a building block, the natural extension in terms of memory architecture was the addition of instruction and data caches. Since the organization of typical caches is well known [364], we omit the basic explanation. Caches have many parameters (e.g., line size, associativity) that can be customized for a given application. Some of these customizations are described later in this chapter.

Scratch Pad Memory

An MPSoC designer is not restricted to using only a traditional cached memory architecture. S/he can use unconventional architectural variations that suit the specific application under consideration. One such design alternative is scratch pad memory (SPM) [362, 368].

SPM refers to data memory residing on-chip that is mapped into an address space disjoint from the off-chip memory but connected to the same address and data busses. Both the cache and SPM (usually SRAM) allow fast access to their residing data, whereas an access to the off-chip memory requires relatively longer access times. The main difference between the scratch pad SRAM and a conventional data cache is that the SRAM guarantees a single-cycle access time, whereas an access to the cache is subject to cache misses. The concept of SPM is an important architectural consideration in modern embedded systems, in which advances in embedded DRAM technology have made it possible to combine DRAM and logic on the same chip. Since data stored in embedded DRAM can be accessed much faster and in a more power-efficient manner than that in off-chip DRAM, a related

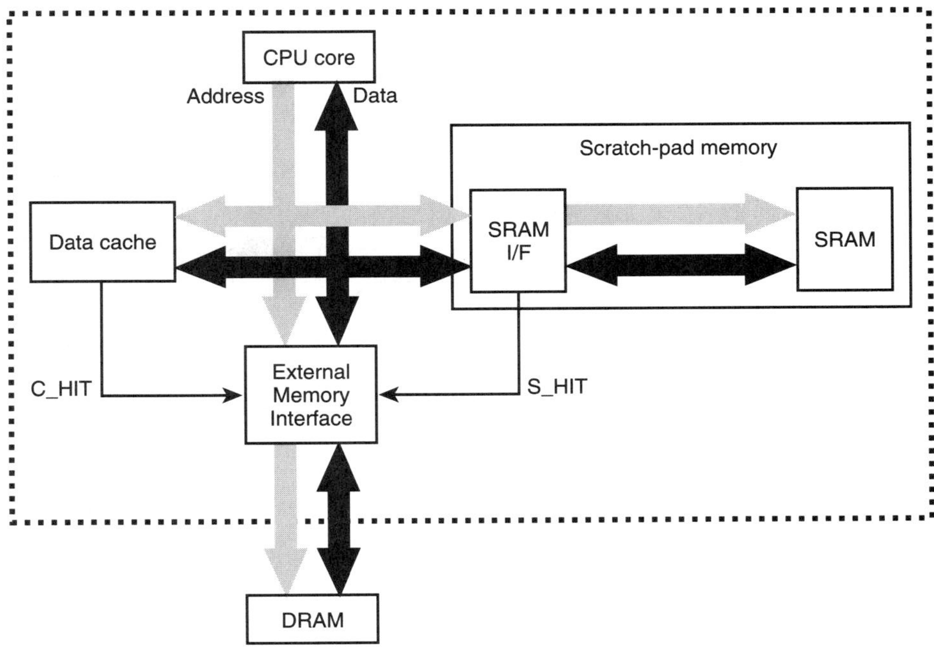

9-1

FIGURE

Block diagram of a core with SPM (from ref. 368, with permission).

optimization problem that arises in this context is how to identify critical data in an application, for storage in on-chip memory.

Figure 9-1 shows an SPM from the perspective of a single processor, with the parts enclosed in the dotted rectangle implemented in one chip, interfacing with an off-chip memory, usually realized with DRAM. The address and data busses from the CPU core connect to the data cache, SPM, and external memory interface (EMI) blocks. On a memory access request from the CPU, the data cache indicates a cache hit to the EMI block through the C_HIT signal. Similarly, if the SRAM interface circuitry in the SPM determines that the referenced memory address maps into the on-chip SRAM, it assumes control of the data bus and indicates this status to the EMI through the signal S_HIT. If both the cache and SRAM report miss, the EMI transfers a block of data of the appropriate size (equal to the cache line size) between the cache and the DRAM.

One possible data address space mapping for this memory configuration is shown in Figure 9-2, for a sample addressable memory of size N data words.

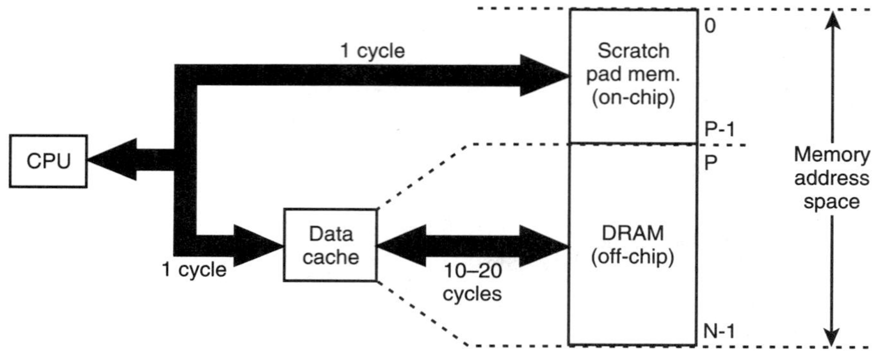

9-2

FIGURE

Dividing data address space between SPM and off-chip memory (from ref. 368, with permission).

Memory addresses $0 \ldots (P - 1)$ map into the on-chip SPM and have a single cycle access time. Memory addresses $P \ldots (N - 1)$ map into the off-chip DRAM and are accessed by the CPU through the data cache. A cache hit for an address in the range $P \ldots N - 1$ results in a single-cycle delay, whereas a cache miss, which leads to a block transfer between off-chip and cache memory, may result in a delay of, say, 50 to 100 processor cycles for an embedded processor operating in the range of 100 to 400 MHz. We illustrate the use of this SPM with the following example from [368].

Example 1. *A small (4 × 4) matrix of coefficients (mask) slides over the input image (source) covering a different 4 × 4 region in each iteration of y, as shown in Figure 9-3. In each iteration, the coefficients of the mask are combined with the region of the image currently covered, to obtain a weighted average, and the result (acc) is assigned to the pixel of the output array (dest) in the center of the covered region. If the two arrays source and mask were to be accessed through the data cache, the performance would be affected by cache conflicts. This problem can be solved by storing the small mask array in the SPM. This assignment eliminates all data conflicts in the data cache—the data cache is now used for memory accesses to source, which are very regular. Storing mask on-chip ensures that frequently accessed data are never ejected off-chip, thereby significantly improving the memory performance and energy dissipation.*

The memory assignment described in ref. 368 exploits this architecture by first determining a ttal conflict factor (TCF) for each array based on the access frequency and possibility of conflict with other arrays and then considering the arrays for assignment to SPM in the order of TCF/(array size), giving priority to high-conflict/small-size arrays.

```
# define N 128
# define M 4
# define NORM 16
int source[N][N], dest [N][N];
int mask [M][M];
int acc, i, j, x, y;
.
.
.
for (x=0; x<N−M; x++)
for (y=0; y<N−M; y++) {
acc=0;
for (i=0; i<M; i++)
for (j=0; j<M; j++)
acc=acc+source[x+i][y+j]* mask[i][j];
dest[x+M/2][y+M/2]=acc/NORM;
}
```

a

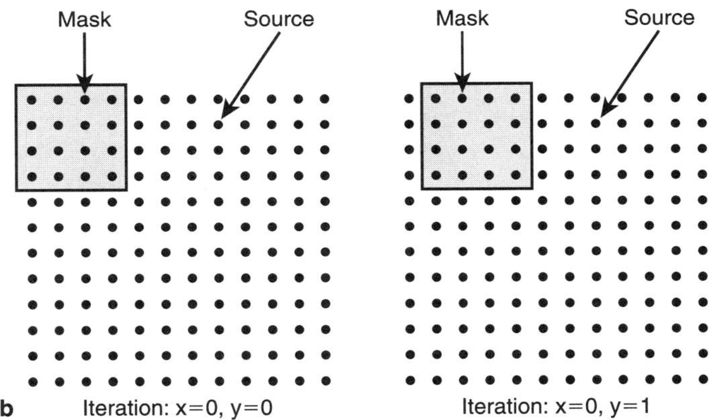

b

9-3
FIGURE

(a) Procedure CONV. (b) Memory access pattern in CONV.

Dynamic data transfers. In the above formulation, the data stored in the SPM were statically determined. This idea can be extended to the case of dynamic data storage. However, since there is no automatic hardware-controlled mechanism to transfer data between the SPM and the main memory, such transfers have to be explicitly managed by the compiler. In the technique proposed in ref. 369, the compiler uses a tiling-like transformation, moves the data tiles (blocks) into SPM (for processing), and then moves it back to main memory after the computation is complete. Section 9.3 presents a more detailed discussion of these compilation issues.

Storing instructions in SPM. An SPM storing a small amount of frequently accessed data on-chip has an equivalent in the instruction cache. The idea of using a small buffer to store blocks of frequently used instructions was first introduced by Jouppi [370]. Recent extensions of this strategy are the decoded instruction buffer [371] and the L-cache [372].

Researchers have also examined the possibility of storing both instructions and data in the SPM. In the formulation proposed in ref. 373, the frequency of access for both data and program blocks is analyzed and the most frequently occurring ones among them are assigned to the SPM. Chen et al. [374] present a compiler-directed management strategy for an instruction SPM.

DRAM

DRAMs have been used in a processor-based environment for quite some time, but the context of their use in embedded systems—both from a hardware synthesis viewpoint and from an embedded compiler viewpoint—have been investigated relatively recently. DRAMs offer better memory performance through the use of specialized access modes that exploit the internal structure and steering/buffering/banking of data within these memories. Explicit modeling of these specialized access modes allows the incorporation of such high-performance access modes into synthesis and compilation frameworks. New synthesis and compilation techniques have been developed that employ detailed knowledge of the DRAM access modes and exploit advance knowledge of an embedded system's application to improve system performance and power.

A typical DRAM memory address is internally split into a row address consisting of the most significant bits and a column address consisting of the least significant bits. The row address selects a page from the core storage, and the column address selects an offset within the page to arrive at the desired word. When an address is presented to the memory during a READ operation, the entire page addressed by the row address is read into the page buffer, in anticipation of spatial locality. If future accesses are to the same page, then there is no need to access the main storage area since it can just be read off the page buffer, which acts like a cache. Thus, subsequent accesses to the same page are very fast.

A scheme for modeling the various memory access modes and using them to perform useful optimizations in the context of behavioral synthesis is described in ref. 375. The main observation is that the input behavior's memory access patterns can potentially exploit the page mode (or other specialized access mode) features of the DRAM. The key idea is the representation of these specialized access modes as graph primitives that model individual DRAM access modes such as row decode, column decode, precharge, and so on; each DRAM family's specialized access modes are then represented using a composition of these graph primitives to fit the desired access mode protocol. These composite graphs can then be scheduled together with the rest of the application behavior, both in the context of synthesis and for code compilation. For instance, some additional DRAM-specific optimizations discussed in ref. 375 include:

1. Read-modify-write (R-M-W) optimization that takes advantage of the R-M-W mode in modern DRAMs, which provides support for a more efficient realization of the common case in which a specific address is read, the data are involved in some computation, and then the output is written back to the same location.

2. Hoisting, whereby the row-decode node is scheduled ahead of a conditional node if the first memory access in both branches is on the same page.

3. Unrolling optimization in the context of supporting the page mode accesses.

A good overview of the performance implications of the architectural features of modern DRAMs can be found in ref. 365.

Synchronous DRAM. As DRAM architectures evolve, new challenges are presented to the automatic synthesis of embedded systems based on these memories. Synchronous DRAM represents an architectural advance that presents another optimization opportunity: multiple memory banks. The core memory storage is divided into multiple banks, each with its own independent page buffer, so that two separate memory pages can be simultaneously active in the multiple page buffers.

The problem of modeling the access modes of synchronous DRAMs is addressed in ref. 376. The modes include:

+ burst mode read/write: fast successive accesses to data in the same page.

+ interleaved row read/write modes: alternating burst accesses between banks.

+ interleaved column access: alternating burst accesses between two chosen rows in different banks.

Memory bank assignment can be performed by creating an interference graph between arrays and partitioning it into subgraphs so that data in each part are assigned to a different memory bank. The bank assignment algorithm is related to techniques such as the one in ref. 377 that address memory assignment for DSP processors such as the Motorola 56000, which has a dual-bank internal memory/register file [378, 379]. The bank assignment problem in ref. 377 is targeted at scalar variables and is solved in conjunction with register allocation by building a constraint graph that models the data transfer possibilities between registers and memories followed by a simulated annealing step. Note that such techniques can be particularly suitable for MPSoCs in which a single reduced instruction set computing (RISC)-like core manages multiple DSP-like slave processors.

Chang and Lin [380] approached the SDRAM bank assignment problem by first constructing an array distance table. This table stores the distance in the DFG (dataflow graph) between each pair of arrays in the specification. A short distance indicates a strong correlation, possibly indicating that they might be, for instance, two inputs of the same operation and hence would benefit from being assigned to separate banks. The bank assignment is finally performed by considering array pairs in increasing order of their array distance information.

Whereas the previous discussion has focused primarily on the context of hardware synthesis, similar ideas have been employed to exploit aggressively the memory access protocols for compilers [381, 382]. In the traditional approach of compiler/architecture co-design, the memory subsystem was separated from the microarchitecture; the compiler typically dealt with memory operations using the abstractions of memory loads and stores, with the architecture (e.g., the memory controller) providing the interface to the (typically yet unknown) family of DRAMs and other memory devices that would deliver the desired data. However, in an embedded system, the system architect has advance knowledge of the specific memories (e.g., DRAMs) used; thus we can employ memory-aware compilation techniques [381] that exploit the specific access modes in the DRAM protocol to perform better code scheduling. In a similar manner, it is possible for the code scheduler to employ global scheduling techniques to hide potential memory latencies using knowledge of the memory access protocols and in effect, improve the ability of the memory controller to boost system performance [382].

Special Purpose Memories

In addition to the general memories such as caches, and memories specific to embedded systems, such as SPMs, there exist various other types of custom memories that implement specific access protocols. Such memories include memory implementing last-in, first-out protocol (LIFO), memory implementing queue or first-in, first-out protocol (FIFO), and content-addressable memory (CAM). Typically, CAMs are used in search applications, LIFOs are used in microcontrollers, and FIFOs are used in network chips.

9.2.2   Customization of Memory Architectures

We now survey some recent research efforts that address the exploration space involving on-chip memories. A number of distinct memory architectures could be devised to exploit different application-specific memory access patterns efficiently. Even if we restrict the scope of the architecture to those involving on-chip memory only, the exploration space of different possible configurations is too large, making it infeasible to simulate exhaustively the performance and energy characteristics of the application for each configuration. Thus, exploration tools are necessary for rapidly evaluating the impact of several candidate architectures. Such tools can be of great utility to a system designer by giving fast initial feedback on a wide range of memory architectures [362, 363].

Cache

Two of the most important aspects of data caches that can be customized for an application are: (1) the cache line size and (2) the cache size. The customization of cache line size for an application is performed in the study presented in ref. 362 using an estimation technique for predicting the memory access performance, that is, the total number of processor cycles required for all the memory accesses in the application. There is a tradeoff in sizing the cache line. If the memory accesses are very regular and consecutive, i.e., exhibit spatial locality, a longer cache line is desirable, since it minimizes the number of off-chip accesses and exploits the locality by prefetching elements that will be needed in the immediate future. On the other hand, if the memory accesses are irregular, or have large strides, a shorter cache line is desirable, as this reduces off-chip memory traffic by not bringing unnecessary data into the cache. The maximum size of a cache line is the DRAM page size. The estimation technique uses data reuse analysis to predict the total number of cache hits and misses inside loop nests so that spatial locality is incorporated into the estimation. An estimate of the impact of conflict misses is also incorporated. The estimation is carried out for the different candidate line sizes, and the best line size is selected for the cache. The customization of the total cache size is integrated into the SPM customization described in the next section.

Scratch Pad Memory

MemExplore [362], an exploration framework for optimizing the on-chip data memory organization, addresses the following problem: given a certain amount of on-chip memory space, partition this into data cache and SPM so that the total access time and energy dissipation is minimized, i.e., the number of accesses to off-chip memory is minimized. In this formulation, an on-chip memory architecture is defined as a combination of the total size of on-chip memory used for data storage and the partitioning of this on-chip memory into: scratch memory, characterized by its size; data cache, characterized by the cache size; and the cache line size. For each candidate on-chip memory size T, the technique considers different divisions of T into cache (size C) and SPM (size S = T − C), selecting only powers of 2 for C. Among the data assigned to be stored in off-chip memory (and hence accessed through the cache3), an estimation of the memory access performance is performed by combining an analysis of the array access patterns in the application and an approximate model of the cache behavior. The result of the estimation is the expected number of processor cycles required for all the memory accesses in the application. For each T, the (C, L) pair that is estimated to maximize performance is selected (L denotes the line size of the cache).

Example 2. *Typical exploration curves of the MemExplore algorithm are shown in Figure 9-4. Figure 9-4a shows that the ideal division of a 2-K on-chip space is 1K SPM and 1K data cache. Figure 9-4b shows that very little performance improvement is observed beyond a total on-chip memory size of 2KB. The exploration curves of Figure 9-4 are generated from fast analytical estimates, which are three orders of magnitude faster than actual simulations and are independent of data size. This estimation capability is important in the initial stages of memory design, in which the number of possible architectures is large, and simulation of each architecture is prohibitively expensive.*

DRAM

The presence of embedded DRAMs adds several new dimensions to traditional architecture exploration. One interesting aspect of DRAM architecture that can be customized for an application is the banking structure.

Figure 9-5a illustrates a common problem with the single-bank DRAM architecture. If we have a loop that accesses in succession data from three large arrays A, B, and C, each of which is much larger than a page, then each memory access leads to a fresh page being read from the storage, effectively canceling the benefits of the page buffer. This page buffer interference problem cannot be avoided if a fixed-architecture DRAM is used. However, an elegant solution to the problem is available if the banking configuration of the DRAM can be customized for the application [383]. Thus, in the example of Figure 9-5, the arrays can be assigned to separate banks, as shown in Figure 9-5b. Since each bank has its own private page buffer, there is no interference between the arrays, and the memory accesses do not represent a bottleneck.

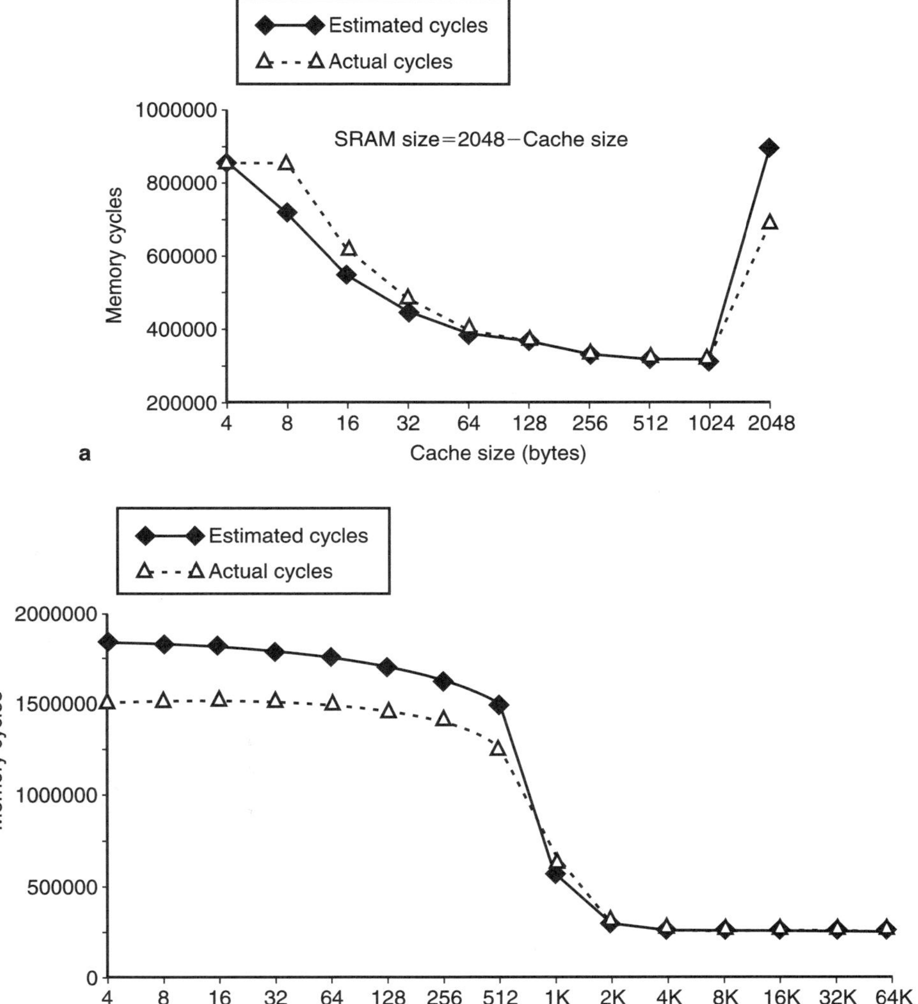

9-4

FIGURE

Histogram example. (a) Variation of memory performance with different mixes of cache and SPM, for total on-chip memory of 2 KB. (b) Variation of memory performance with total on-chip memory space (from ref. 368, with permission).

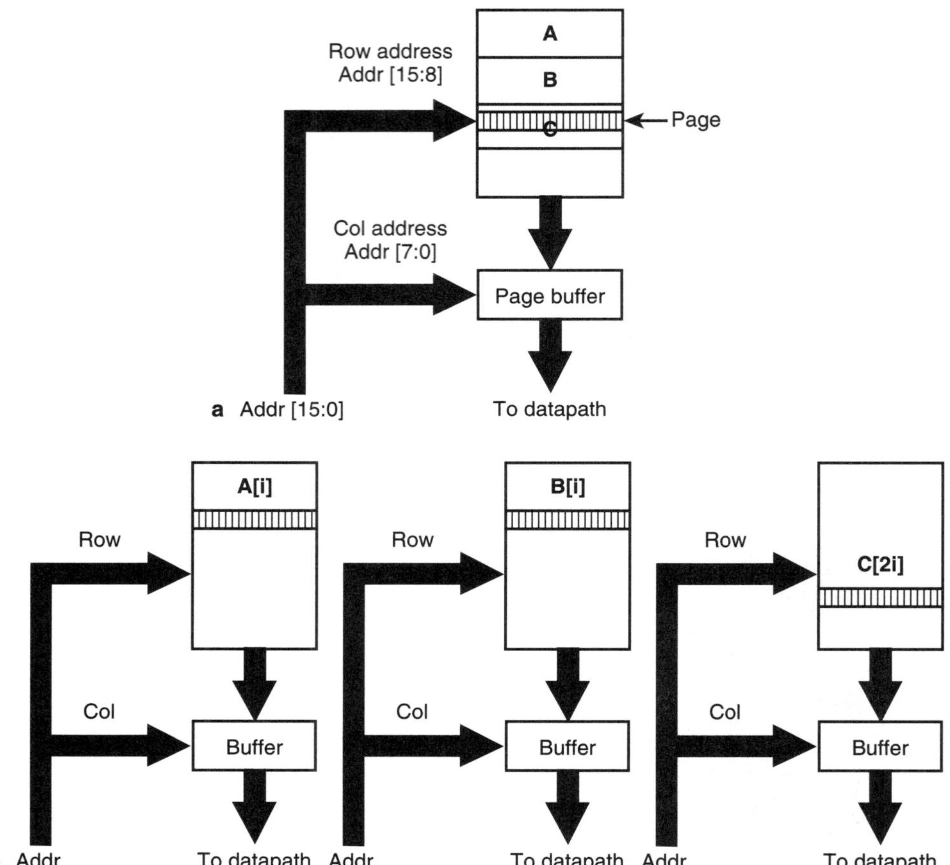

9-5

FIGURE

(a) Arrays mapped to a single-bank memory. (b) A three-bank memory architecture (from ref. 383, with permission).

In order to customize the banking structure for an application, we need to solve the memory bank assignment problem—determine an optimal banking structure (number of banks) and determine the assignment of each array variable into the banks such that the number of page misses is minimized. This objective optimizes both the performance and the energy dissipation of the memory subsystem. The memory bank customization problem is solved in ref. 383 by modeling the assignment as a partitioning problem, i.e., partition a given set of nodes into a given number of groups such that a given criterion (bank misses in this case) is optimized. The partitioning proceeds by associating a cost of assigning two arrays into the same bank, determined by the number of accesses to the arrays and the loop count. If the arrays are accessed in the same loop, then the cost is high, thereby discouraging the partitioning algorithm from assigning them to the same bank. On the other hand, if two arrays are never accessed in the same loop, then they are candidates for assignment into the same bank. This pairing is associated with a low cost, guiding the partitioner to assign the arrays together.

Multiple SRAMs

In a custom memory architecture, the designer can choose memory parameters such as the number of memories and the size and number of ports on each memory. The number of memory modules used in a design has a significant impact on the access times and power consumption. A single large monolithic memory to hold all the data is expensive in terms of both access time and energy dissipation than multiple memories of smaller size. However, the other extreme, in which all array data are stored in distinct memory modules, is also expensive, and the optimal allocation lies somewhere in between. The memory allocation problem is closely linked to the problem of assigning array data to the individual memory modules. Arrays need to be clustered into memories based on their accesses [384]. The clustering can be vertical (different arrays occupy different memory words) or horizontal (different arrays occupy different bit positions within the same word) [384]. Parameters such as bit width, word count, and number of ports can be included in this analysis [385]. The required memory bandwidth (number of ports allowing simultaneous access) can be formally determined by first building a conflict graph of the array accesses and then storing in the same memory module the arrays that do not conflict [386].

Special Purpose Memories

Special purpose memories such as stacks (LIFO), queues (FIFO), frame buffers, streaming buffers, and so on can be utilized when one is customizing the memory architecture for an application. Indeed, analysis of many large applications shows that a significant number of the memory references in data-intensive applications are made by a surprisingly small number of lines of code. Thus it is possible to customize the memory subsystem by tuning the memories for these segments of code, with the goal of improving performance, and also for reducing the power dissipation. In the approach described in ref. 387, the application is first analyzed and then different access patterns identified. Data for the most critical access patterns are assigned to memory modules that best fit the access pattern profiles. The system designer can then evaluate different cost/performance/power profiles for different realizations of the memory subsystem.

Processor-Memory Co-Exploration

In many embedded applications, it is also critical to explore both processor and memory architectures simultaneously, to capture the synergy between them.

Datapath width and memory size. The CPU's bit width is an additional parameter that can be tuned during architectural exploration of customizable processors. Shackleford et al. [388] studied the relationship between the width of the processor data path and the memory subsystem. This relationship is important when different data types with different sizes are used in the application. The key observation made is that as datapath width is decreased, the data memory size decreases because of less wasted space. For example, storing 3-bit data in a 4-bit word (instead of 8-bit word) reduces memory space demand, but the required instruction memory capacity might increase. For example, storing 7-bit data in an 8-bit word requires only one instruction to access it but requires two instructions if a 4-bit datapath is used. The authors use a RAM and ROM cost model to evaluate the cost of candidate bit widths in a combined CPU-memory exploration.

Architectural description language-driven co-exploration. Processor architecture description languages (ADLs) have been developed to allow for a language-driven co-exploration and software toolkit generation approach [389, 390]. Currently most ADLs assume an implicit/default memory organization or are limited to specifying the characteristics of a traditional memory hierarchy. Since embedded systems may contain nontraditional memory organizations, there is a great need to model explicitly the memory subsystem for an ADL-driven exploration approach. A recent approach [391] describes the use of the EXPRESSION ADL [392] to drive memory architecture exploration. The EXPRESSION ADL description of the processor-memory architecture is used to capture the memory architecture explicitly, including the characteristics of the memory modules (such as caches, DRAMs, SRAMs, and DMAs), the parallelism and pipelining present in the memory architecture (e.g., resources used, timings, access modes). Each such explicit memory architecture description is then used to generate automatically the information needed by the compiler [381, 382] to utilize efficiently the features in the memory architecture and to generate a memory simulator, allowing feedback to the designer on the match among the application, the compiler, and the memory architecture [363].

Split Spatial and Temporal Caches

Various specialized memory structures proposed over the years could be candidates for MPSoC-based embedded systems. One such concept is split spatial/temporal caches.

Variables in real life applications present a wide variety of access patterns and locality types (for instance scalars, such as indexes, usually present high temporal and moderate spatial locality, whereas vectors with small stride present high spatial locality, and vectors with large stride present low spatial locality and may or may not have temporal locality). Several approaches including the one presented in ref. 393 have proposed splitting a cache into a spatial cache and a temporal cache that store data structures with high temporal and high spatial locality, respectively. These approaches rely on a dynamic prediction mechanism to route the data to either the spatial or the temporal caches, based on a history buffer. In an embedded system context, the approach of Grun et al. [387] uses similar split-cache architecture but allocates the variables statically to the different local memory modules, avoiding the power and area overhead of the dynamic prediction mechanism. Thus, by targeting the specific locality types of the different variables, better utilization of the main memory bandwidth is achieved. The useless fetches due to locality mismatch are thus avoided. For instance, if a variable with low spatial locality is serviced by a cache with a large line size, a large number of the values read from the main memory will never be used. The approach described by Grun et al. [387] shows that the memory bandwidth and memory power consumption could be reduced significantly. Note that, in an MPSoC-based architecture, each processor may demand a customized cache (or SPM) for the best behavior.

9.2.3   Reconfigurability and Challenges

In MPSoC-based embedded systems, modifying a given code to improve data locality is one way of enhancing performance. An alternative approach is to reconfigure the cache (or SPM) architecture dynamically according to the application at hand. That is, it might be useful to have a morphable (reconfigurable) memory/cache system that adapts itself to the application's requirements (from both performance and energy/power consumption angles) dynamically. In fact, an optimizing compiler can analyze a given application, divide its code into regions, and, for each region, select an optimum cache configuration for each processor. However, there are several key issues that need to be addressed in translating the promise of reconfigurable cache architectures into practice:

+ architectural and circuit mechanisms for efficient and fast reconfiguration are essential. This issue has been the focus of several recent efforts that provide architectural mechanisms to support dynamic reconfiguration of the cache and memory parameters. It is important to note that the support for reconfiguration can also increase the access and energy costs of a reconfigurable cache, in contrast to that of a non-reconfigurable cache with identical cache parameters such as cache size (capacity) and associativity. This cost depends on the flexibility supported by the reconfigurable cache and the implementation mechanism.

+ control mechanisms for deciding when to reconfigure these caches are required. The reconfiguration can be performed at various levels of code granularity such as entire application, a single subroutine, a nested loop, or some specific segment of the application. The chosen level of granularity will depend on the overhead associated with reconfiguration and the potential benefits due to better customization with more frequent cache reconfiguration.

+ mechanisms to determine the optimal configuration of the cache are required. The desired cache configuration is a function of the application behavior and can be determined either statically using compile-time estimates or dynamically using run-time behavior. For example, the cache miss rates can be used as a metric to resize the caches dynamically.

+ techniques for minimizing the overhead of data invalidation across different reconfiguration phases are essential. For example, if the associativity of a cache is changed, the function for mapping the memory location onto the cache changes. Consequently, the data that can be reused across reconfiguration need to be invalidated after reconfiguration.

Kadayif et al. [394] focus on a morphable cache architecture and array-dominated embedded codes (which are suitable for an MPSoC-based environment) and show the potential benefits that can be obtained from such a system. They conduct a limit study for potential benefits (from energy and performance perspectives) for going from one configuration to another. The granularity that they focus on is a nested loop, which is the natural computation/access pattern boundary for array-dominated applications from a scientific domain and an image/video processing domain. Using a set of array-dominated codes, they investigate what the best cache configuration is for each nested loop under different objective functions (optimization criteria) such as cache energy, memory energy, cache misses, performance (execution time), overall energy, and energy-delay (energy-execution time) product. In addition to morphing conventional cache parameters, they also consider reconfigurability of energy-aware features found in some cache architectures such as block buffering [395]. Their results indicate that there are potential performance and energy benefits in adopting a morphable cache subsystem. The results also show that, depending on the optimization objective targeted, one may select an entirely different cache configuration. For example, minimizing cache memory energy requires a cache configuration for each nest that is different from an objective criterion that tries to minimize the overall memory system energy under a performance constraint.

9.3 COMPILER SUPPORT

An optimizing compiler that targets MPSoC environments should tackle a number of critical issues. In the following discussion, we first explain these issues and then study potential solutions.

9.3.1 Problems

From the performance viewpoint, perhaps the two most important memory-related tasks to be performed in an MPSoC environment are optimizing parallelism and locality.

Parallelism

Optimizing parallelism is obviously important, since parallelism is the main reason to employ multiple processors in a single unit. In fact, a parallelization strategy determines how memory is utilized by multiple on-chip processors and can be an important factor for achieving an acceptable performance. However, maximum parallelism may not always be easy to achieve because of several factors. For example, intrinsic data dependences [396] in the code may not allow full utilization of all on-chip processors. Similarly, in some cases, interprocessor communication costs can be overwhelming as one increases the number of processors used. Finally, performance benefits due to increased interprocessor parallelism may not be sufficient when one considers the increase in power consumption. Because of all these, it may be preferable to avoid increasing the number of processors arbitrarily. In addition, the possibility of different parts of the same application demanding different number of processors can make the problem much harder.

Instruction and Data Locality

An equally important problem is ensuring locality of data/instruction accesses. Although achieving acceptable instruction cache performance is not very difficult (since instructions are read-only and exhibit perfect spatial locality), the same cannot be said for data locality. This is because straightforward coding of many applications can lead to poor data cache utilization. In addition, in an MPSoC environment, interprocessor communication can lead to frequent cache line invalidations/updates (due to interprocessor data sharing), which in turn increases overall latency. This last issue becomes particularly problematic when false sharing occurs (i.e., the multiple processors share a cache line but not the same data in it). Therefore, an important task for the compiler is to minimize false sharing as much as possible.

Power/Energy Consumption

Considering the environments for which MPSoC based systems can be employed, one can see that performance is only one part of the big picture. Minimizing power/energy consumption can become vital for system availability. In fact, power/energy consideration is one of the most important factors that distinguish an MPSoC-based system from its high-end counterparts [397]. In other words, one cannot directly transfer pure performance-oriented compilation policies from a high-end parallel computing domain [398]. This is because these techniques in general favor increasing the number of processors (to be used for execution of the application) as long as there are additional performance benefits in doing so. However, such a strategy may not be preferable in our context since increasing the number of processors means powering up more processors along with their local caches (and/or SPMs). Obviously, this can lead to higher power consumption and a larger overall energy-delay product value. Therefore, an optimizing compiler should be able to balance the increase in power consumption and decrease in execution cycles. In addition, it is well known from high-end parallel computing that parallelism and data locality can sometimes impose conflicting demands and require different program/data optimizations. Adding power/energy minimization as an additional metric makes the overall optimization problem extremely hard and justifies, in our opinion, the use of heuristic solutions.

Memory Space

Yet another memory-related issue is the memory space required by the application. Since most prior existing locality-oriented studies have focused on improving memory access patterns, there is no guarantee that such techniques will reduce space demands. However, reducing space consumption might be of critical importance as doing so increases the effectiveness of on-chip memory utilization and can reduce the number of off-chip references.

9.3.2   Solutions

In this section, we discuss several techniques that can be used for enhancing memory behavior of MPSoC architectures. We focus on both performance and energy aspects.

Optimizing Parallelism

Parallelism can either be expressed by the programmer at the source level or be automatically derived by an optimizing compiler from the sequential code [397]. In the former case (which is called the user-assisted method), the compiler's job is relatively easy, and consists of distributing the computations across the processors (as indicated by the user) and minimizing interprocessor communication. This may involve program transformations that reduce the number of communications (messages) as well as the volume of data communicated at each message. In the latter case (i.e., when the compiler is to generate parallel code from a sequential one), the compiler needs to analyze data dependences and extract available parallelism. After that, by assessing the volume and intensity of interprocessor communication, the compiler can come up with a suitable parallelization strategy (e.g., in a loop nest-based code, it determines which loops should be run in parallel). After this step, the compilation process proceeds as in the user-directed (user-assisted) parallelism case.

Although in the high-end computing arena it is generally accepted that user-assisted parallelism is a more viable approach than compiler-initiated parallelism [397], the choice is not clear in the context of MPSoC due to additional constraints of low power and limited memory space. That is, it can be extremely difficult for the programmer to strike a balance among performance, power, and memory space consumption for a given application. What makes it even more difficult is that not all MPSoC applications are array/loop nest-based, making the task of identifying parallelism very difficult for the user.

In order for the compiler to parallelize a code given performance (execution cycles), energy, and memory space consumption constraints, it needs to carry out two major tasks: (1) estimating performance, power consumption, and space requirements of a given piece of code; and (2) optimizing the objective function under multiple constraints (which might be of performance, power, memory space, or a combination of these). Based on this observation, there are at least two different ways of dealing with this MPSoC parallelization problem:

+ static approaches: These techniques are generally applicable when the code is statically analyzable. Many applications from embedded image and video processing areas fall into this category. In this case, the compiler can analyze the application code and estimate its execution cycles, power/energy consumption, and memory space demands. Then it can conduct a similar estimation process for each potential optimized version, and it selects the best candidate for the output code. Obviously, an important question here is how to generate different optimized versions. Among the techniques proposed so far in literature are heuristic schemes and integer linear programming (ILP)-based techniques that make use of profiling [399]. It is to be noted that the static analyzability of the code also makes it possible for the compiler to come up with a small set of candidate optimized versions.

✦ dynamic approaches: The schemes in this category also work with a candidate set of optimized versions; but it is less clear (compared with the previous case) at compile time how these candidates would rank with respect to each other. In other words, the selection of the best optimized version is postponed until run time. One approach recently proposed in ref. 400 is first to execute a portion of the code fragment to be optimized and then, based on the run-time statistics collected, select the most suitable parallelization strategy at run time. Clearly, the time (and energy) spent in determining the best optimized version at run time is an overhead and needs to be optimized away as much as possible.

It should also be emphasized that whether the memory hierarchy in question is constructed using caches or SPMs can make a difference in the success of the parallelization strategy chosen. More specifically, it is easier to estimate the performance and energy consumption with SPMs (as opposed to caches) since the compiler is in full control of memory transfers in the former (whereas the second one is managed by the hardware).

Optimizing Locality

Locality optimization can be performed for both code and data. As mentioned earlier, since code locality is in general very good to begin with, most recent studies on the topic concentrate on data locality. Broadly speaking, there are two facets to the data locality problem: (1) single processor perspective, and (2) multiprocessor perspective. From the viewpoint of a single processor in an MPSoC, the locality optimization problem is not very different from one in a single processor architecture. The main goal is to maximize the number of data accesses satisfied from the highest level of memory in the hierarchy.

If the hierarchy has private L1 caches (one per processor), then one can make use of the theory developed in the optimizing compiler community (for a comprehensive discussion of this theory, we refer the interested reader to refs. 396 and 398). Compiler optimizations that target improving data locality can be broadly divided into two major categories: (1) optimizations that target the reduction of memory latency, and (2) optimizations that target the hiding of memory latency. In the first group are the techniques that modify program access pattern, memory layout of variables, or both. Loop transformations change the data access pattern exhibited by the loop nests in the application to make them more cache-friendly [401]. Their major drawback is that data dependences in the code may prevent the compiler from using the best transformation from the locality perspective. Among the most popular loop transformations are loop interchange (which changes the order of two loops in the nest), loop fusion (which combines two neighboring loop nests into a single nest), and tiling (which implements a block matrix style of computation automatically [398]).

The drawbacks of loop transformations motivated researchers to investigate alternative optimizations. One of these is data space (memory layout) transformations, or data transformations for short [402–404]. In applying a data transformation, instead of the program access pattern, the compiler modifies the underlying memory layouts of the data structures involved. For multidimensional array structures, implementing a data transformation involves transforming the array subscript expressions and modifying the corresponding array declarations. Recent work in this area focuses on pointer-intensive codes [405] and is based on the idea of co-locating data items that are accessed in close proximity in time. It should be noted that it is possible to apply data transformations statically or dynamically. In static transformations, the best layout for a given data structure is determined statically (at compile time), and all references to the old data structure are replaced by the references to the new data structure. Many layout transformations that target temporary data structure variables fall into this category.

In the dynamic transformation case, on the other hand, the layout transformations take place during the course of execution. Specifically, the compiler statically determines the layouts of the data structures of interest at compile time at each program point and inserts code for explicit layout transformation. However, the actual layout transformations themselves take place at run time. For example, in ref. 406, a dynamic layout optimization approach is presented. This technique extends the static layout optimization techniques to select dynamically changing layouts to further improve the locality of data accesses. It divides the job of optimizing locality between compile time and run time. More specifically, the layouts of arrays for different regions of code are determined at compile time (e.g., running a static layout optimizer for each loop nest in the code), and the book-keeping code that is necessary to transform memory layouts dynamically between different program regions is inserted in the code (again at compile time). These bookkeeping codes, however, are executed at run time.

The techniques discussed so far try to improve data locality, thereby reducing the number of accesses to slower levels in the memory hierarchy. Software data prefetching, in contrast, tries to bring data to the fast memory ahead of time (i.e., before the data are actually needed). If this can be achieved, the data can be accessed from cache memory when they are needed. In order to issue the prefetches ahead of time, a compiler transformation known as software pipelining needs to be used. Mowry [407] presents a compiler-directed software prefetching algorithm that consists of the following three steps: (1) determine data references that are likely to be cache misses and therefore need to be prefetched; (2) isolate the predicted cache miss instances through loop splitting [398]; and (3) apply software pipelining and insert explicit prefetch instructions in the code.

However, if the first level of memory in the hierarchy is an SPM (one per processor), the compiler needs to employ a different set of optimizations. This is because, as mentioned earlier, in an SPM-based system, the compiler is in full control of data transfers between SPM and the lower levels of the memory hierarchy. Therefore, it needs to insert in the code explicit data movement instructions at compile time. These instructions, when they are executed at run time, cause data transfers between SPM and other memory components. Obviously, this is a desired compilation strategy only if one wants to manage SPMs dynamically (as defended in refs. 369 and 408). On the other hand, it is also possible to fix the contents of the SPMs statically, i.e., at compile time. In this case, the compiler does not insert explicit data movement calls in the code since there are no run-time data movements. As discussed earlier in this chapter, Panda et al. [362] present such a strategy for a single processor embedded architecture. Although it is possible to extend the approach to the MPSoC case, one needs to be careful in selecting the frequently shared data items to be placed in the SPM.

It is also possible to allow multiple processors to share a single (or a set of) on-chip SPM. A solution along this direction, called the virtually shared SPM (VS-SPM), has been discussed in ref. 408. The execution model in a VS-SPM-based architecture is as follows. The system takes as input a parallelized application (user-assisted or through compiler-based dependence analysis). In this model, each loop nest is parallelized as much as possible. In processing a parallel loop, all processors in the system participate in computation, and each executes a subset of loop iterations. When the execution of a parallel loop has completed, the processors synchronize using a special construct called a barrier before starting the next loop. The synchronization and communication between processors are maintained using fast on-chip communication links. Based on the parallelization strategy, each processor works on a portion of each array in the code. Since its local SPM space is typically much smaller than the portion of the array it is currently operating on, it divides its portion into chunks (also called data tiles) and operates on one chunk at a time. When a data tile has been used, it is either discarded or written back into off-chip memory (if modified).

In order to improve the reuse of data in the VS-SPM, one can consider intraprocessor data reuse and interprocessor data reuse. Intraprocessor data reuse corresponds to optimizing data reuse when considering the access pattern of each processor in isolation. Previous work presented in refs. 369 and 362 and discussed briefly above addresses this problem. It should be noted, however, that exploiting intraprocessor reuse only may not be very effective in a VS-SPM-based environment. This is because intraprocessor data reuse has a local processor-centric perspective and does not take interprocessor data sharing (communication) effects into account. Such effects are particularly important for applications in which data regions touched by different processors overlap. This is very common in many array-intensive embedded image processing applications. Interprocessor data reuse, on the other hand, focuses on the problem of optimizing data accesses considering access patterns of all processors in the system. In other words, it has an application-centric view of data accesses. The compilation approach proposed in ref. 408 maximizes interprocessor data reuse for SPM-based data. More specifically, the data tile accesses of individual processors are restructured (coordinated) in such a way that whenever a nonlocal array element is required (to perform some computation), it can be found in some remote SPM (instead of going to the off-chip memory). If this can be achieved for all nonlocal accesses, one can eliminate all extra off-chip memory accesses due to interprocessor communication.

From a multiple processor perspective, the locality problem is more challenging to tackle than the single processor case. The SPM approach is simple and has been summarized in the previous two paragraphs. If the processors have conventional L1 caches instead of SPMs, the situation becomes much more difficult to handle. This is because on-chip processors can share cache lines in many different ways. In true-sharing, processors share the same data item, and the integrity of the item should be maintained using an on-chip coherence protocol. In false-sharing, they just share the cache line, not the data item itself. Note that this case also causes a coherence activity (which costs extra execution cycles as well as energy), and in the false-sharing case, this activity is pure overhead that needs to be minimized. Although we expect that the approaches used in high-end computing [409] may be very useful in this context, one also needs to consider the energy consumption problem. Therefore, we see energy-aware coherence protocols/algorithms as an important potential research direction.

Optimizing Memory Space Utilization

Another important issue in an MPSoC-based environment is the management of limited memory space. It should be observed that optimizing for data locality does not mean that the optimized application will execute using fewer data items. To reduce data space requirements of a given application, one may need to consider lifetimes of program data structures/variables. In particular, if the lifetimes of two different data items do not overlap, they can potentially share the same memory location, thereby reducing overall memory space demand.

For array-based applications, in a given loop nest, one can define the lifetime of an array element as the difference (in loop iterations) between the time it is first assigned (written) and the time it is last used (read) [410]. For a given array index a (which might be multidimensional), the start of its lifetime is referred to as S(a), whereas the end of its lifetime is denoted using E(a)—both in terms of loop iterations. Using these definitions, the lifetime vector for this array element can be given as s = E(a) − S(a), where "−" denotes vector subtraction. Note that the lifetime of a is expressed as a vector, as in general there might be multiple loops in the next, and expressing lifetime as a vector allows the compiler to measure the impact of loop transformations on it. As an example, if an array element (that is accessed in a nest with two loops) is produced in iteration $(1\ 3)^{T}$ and consumed (i.e., last-read) in iteration $(4\ 5)^{T}$, its lifetime vector is s = $(4\ 5)^{T}$ − $(1\ 3)^{T}$ = $(3\ 2)^{T}$. It should be noted that before S(a) and after E(a), the memory location allocated to this array element could be used for storing another array element (which belongs to the same array or to a different array). Obviously, the shorter the difference between E(a) and S(a) the better, as it leaves more room for other array elements.

However, it is to be noted that reducing lifetimes of array elements does not guarantee memory location reuse. This is because in order to have memory location reuse, the lifetimes should be shortened such that the different elements can share the same memory location. As a result, it is important for the compiler to make sure that this happens (when lifetimes are reduced). That is, the compiler can check whether the lifetime reduction will really be beneficial; if not, it may choose not to apply it.

Along these lines, Lefebvre and Feautrier [411] propose a method for automatic storage management for loop-based parallel programs with affine references. They show that parallelization by total data expansion can be replaced by partial data expansion without distorting the degree of parallelism. Grun et al. [412] present memory size estimation techniques for multimedia applications. Strout et al. [413] introduce the universal occupancy vector that allows schedule-independent storage space optimization. Their objective is to reduce data space requirements of a given code without preventing subsequent locality optimizations from being performed. Wilde and Rajopadhye [414] investigate memory reuse by performing static lifetime computations of variables using a polyhedral model. Lim et al. [415] define an optimization method called the array contraction. The work done by Song et al. [416] and Catthoor and his colleagues [360, 417] shows how loop fusion can be used for minimizing data space requirements. Along similar lines, Fraboulet et al. [418] also present an algorithm to reduce the use of temporary arrays by loop fusion. They demonstrate that although the complexity of their algorithm is not polynomial, it is highly efficient in practice. Thies et al. [419] describe a unified general framework for scheduling and storage optimization. One of the problems that they address, namely, "choosing a store for a given schedule" tries to assign storage locations to array elements with storage size reduction in mind. It should also be observed that reducing memory space requirements alone might help improve data locality (cache behavior). This is because in array-intensive applications a significant portion of data cache misses (called capacity misses) occurs because large datasets are accessed in a short period of time (e.g., within a loop). By reducing the number of array elements, one can also improve cache behavior.

We also need to point out an important tradeoff that needs to be made for an MPSoC-based execution environment. In general, more memory space means more parallelism, as discussed in studies such as those of Tu and Padua [420] and Barthou et al. [421]. This is because a large percentage of data dependences in a given application are anti-dependences and output-dependences, which occur mainly due to reuse of the same memory location for different data items. By allocating a separate memory location for each data item, one can eliminate these dependences. However, as we have discussed above, reusing the same memory location for different data items is an important optimization as far as memory space utilization is concerned. Therefore, an optimizing compiler should be able to resolve this tradeoff considering the performance and memory space constraints at the same time.

Power/Energy Optimization

Power/energy-directed and performance-directed optimizations often conflict as far as an MPSoC-based environment is concerned. For example, the two graphs shown in Figure 9-6 give the execution cycles and energy consumption behavior when an array-based application (tomcatv from the Spec benchmark suite) is parallelized over a different number of processors (ranging from 1 to 8). Each bar in these graphs represents a result (for the corresponding loop nest), normalized with respect to the single processor case (that is, the energy consumption or execution

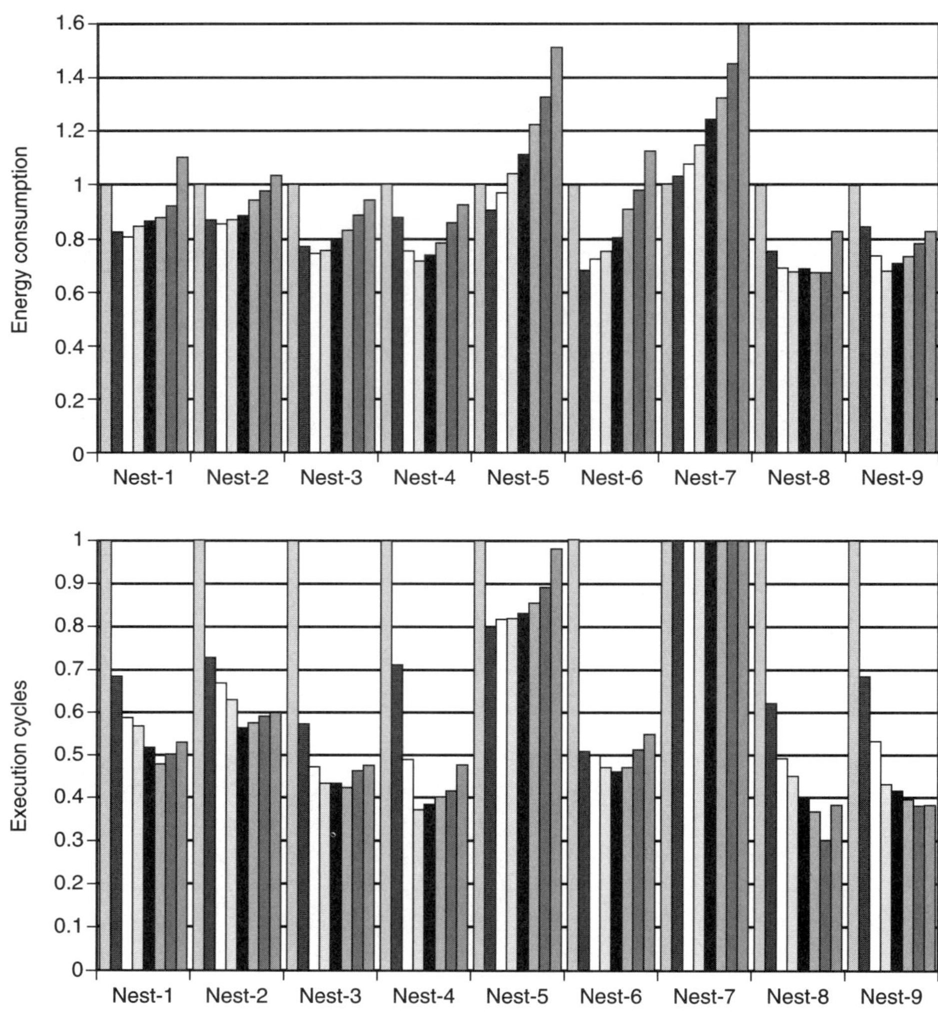

9-6 FIGURE

Energy consumption (top) and execution cycles (bottom) for different nests of tomcatv. For each nest, each bar corresponds to a specific number of processors (from one processor, the leftmost bar, to eight processors, the rightmost bar).

cycles when the nest is executed on a single processor). More details on the experimental setup and benchmark codes can be found in refs. 399 and 422.

From these graphs, one can make several observations. First, energy and performance trends are different from each other. That is, in a given nest, the best processor size from the energy perspective is (in general) different from the best processor size from the execution time perspective. This is because in many cases increasing the number of processors beyond a certain point may continue to improve performance (execution cycles), but the extra energy to power up the additional processor and its caches may offset any potential energy benefits coming from the reduced execution cycles (as a result of increased parallelism). Second, for a given nest, increasing the number of processors does not always reduce execution cycles. This occurs because there is an extra performance cost of parallelizing a loop, which includes spawning parallel threads, interprocessor synchronization during loop execution if there are data dependences, and synchronizing the threads after the loop execution. If the loop in question does not have a sufficient number of iterations to justify the use of a certain number of processors, this extra cost can dominate overall performance behavior. Third, the best number of processors from the energy or performance perspectives depends on the loop nest in question. That is, different nests may require different processor sizes for the best results, and it is not a good idea to use the same number of processors for all the nests in a given application.

Given these observations, one can see that it is not easy to select the ideal processor sizes for each nest in a given code to optimize an objective function, which might have both performance and energy elements. As an example, consider a compilation (parallelization) scenario in which we would like to optimize energy consumption of the application keeping the execution time under a predetermined limit. Important questions in this scenario are (1) whether there are processor sizes for each nest to satisfy this compilation constraint, and (2) if there are multiple solutions, how one can choose the best one.

Most published work on parallelism for high-end machines [398] is static; that is, the number of processors that execute the code is fixed for the entire execution. For example, if the number of processors that execute a code is fixed at 8, all parts of the code (for example, all loop nests) are executed using the same eight processors. However, this may not be a very good strategy from the viewpoint of energy consumption. Instead, one can consume much less energy by using the minimum number of processors for each loop nest (and shutting down the caches of the unused processors). In adaptive parallelization [423], the number of processors is tailored to the specific needs of each code section (e.g., a nested loop in array-intensive applications). In other words, the number of processors that are active at a given period of time changes dynamically as the program executes. For instance, an adaptive parallelization strategy can use 4, 6, 2, and 8 processors to execute the first four nests in a given code. There are two important issues that need to be addressed in designing an effective adaptive parallelization strategy for an on-chip multiprocessor:

- ✦ mechanism: How is the number of processors to execute each code section determined? There are at least two ways of determining the number of processors per nest: the dynamic approach and the static approach. The first option is adopting a fully dynamic strategy whereby the number of processors (for each nest) is decided in the course of execution (at run time). Kandemir et al. [400] present such a dynamic strategy. Although this approach is expected to generate better results once the number of processors has been decided (as it can take run-time code behavior and dynamic resource constraints into account), it may also incur some performance overhead during the process of determining the number of processors. This overhead can, in some cases, offset the potential benefits of adaptive parallelism. In the second option, the number of processors for each nest is decided at compile time. This approach has a compile-time overhead, but it does not lead to any run-time penalty. It should be emphasized, however, that although this approach determines the number of processors statically at compile time, the activation/deactivation of processors and their caches occurs dynamically at run time. Examples of this strategy can be found in refs. 399 and 422.

+ policy: What is the basis of the criterion on which we decide the number of processors? An optimizing compiler can target different objective functions such as minimizing execution time of the compiled code, reducing executable size, improving power or energy behavior of the generated code, and so on. In addition, it is also possible to adopt complex objective functions and compilation criteria such as "minimize the sum of the cache and main memory energy consumptions while keeping the total execution cycles bounded by M." For example, the framework described in ref. 400 accepts as objective function and constraints any linear combination of energy consumptions and execution cycles of individual hardware components.

Future Research

Perhaps the most important future research direction in compilation of MPSoC applications will be compiling applications under multiple constraints that involve power, energy, execution cycles, and memory usage. Unfortunately, current compilation techniques (even those developed in the context of high-end multiprocessor systems) do not directly extend to the MPSoC domain since they only focus on performance. Equally important is studying techniques that produce quick estimation of metrics of interest (e.g., power and memory usage). Optimizing compilers can use such estimates to rank different (and sometimes conflicting) optimizations and to select the best one given the constraints that need to be satisfied. That is, we envision future MPSoC compilers to have numerous estimation and optimization modules (each targeting at a different metric) and a solver, which selects the best optimization under multiple constraints.

9.4.   CONCLUSIONS

The significant interest in MPSoC architectures in recent times has caused researchers to study architectural optimizations from a different perspective. With the full advance knowledge of the applications being implemented by the system, many design parameters can be optimized and/or customized. This is especially true of the memory subsystem, in which a vast array of different organizations can be employed for application-specific systems, and the designer is not restricted to the traditional cache hierarchy. The optimal memory architecture for an application-specific system can be significantly different from the typical cache hierarchy of processors. In this chapter, we outlined different memory architectures relevant to embedded systems and strategies to customize them for a given application. Although some of the analytical techniques can be automated, much work remains to be performed before a completely "push-button" methodology evolves for application-specific customization of the memory organization in MPSoCs. We also discussed the current compiler support available (from parallelism, locality, memory usage, and power/energy consumption perspectives) and pointed out potential research directions.

ACKNOWLEDGMENTS

We thank Prof. Preeti Panda (IIT Delhi) for his assistance with the material on SPMs. This work is supported in part by NSF grants 0093082, 0103583, CCR0203813, and CCR0205712.

A SystemC-Based Abstract Real-Time Operating System Model for Multiprocessor Systems-on-Chips

Jan Madsen, Kashif Virk, and Mercury Jair Gonzalez

10.1 INTRODUCTION

With the increasing complexity of embedded systems and the capacity of modern silicon technology, there is a trend toward heterogeneous architectures consisting of several programmable as well as dedicated processors, implemented on a single chip, known as systems-on-chips (SoCs). As more applications are implemented in software that in turn is growing larger and more complex, dedicated operating systems will have to be introduced as an interface layer between the application software and the hardware platform [432]. Global analysis of such heterogeneous systems is a big challenge. Typically, two aspects are of interest when one is considering global analysis: the system functionality, in general, and the system timing and resource sharing, in particular.

As many embedded applications are *reactive* in nature and have real-time requirements, it is often not possible to analyze them statically at compile time. Furthermore, for single-chip solutions, we may need to use nonstandard *real-time operating systems* (RTOS) in order to limit the code size and hence the memory requirements, or to introduce special features interacting with the dedicated hardware, such as power management. When implementing an RTOS, we may wish to experiment with different scheduling strategies in order to tailor the RTOS to the application. For a multiprocessor platform, we may wish to study the system-level effects of selecting a particular RTOS implementation on one of the processors. To study these effects at the system level, before any implementation has been done, we need a system-level model that is able to capture the behavior of

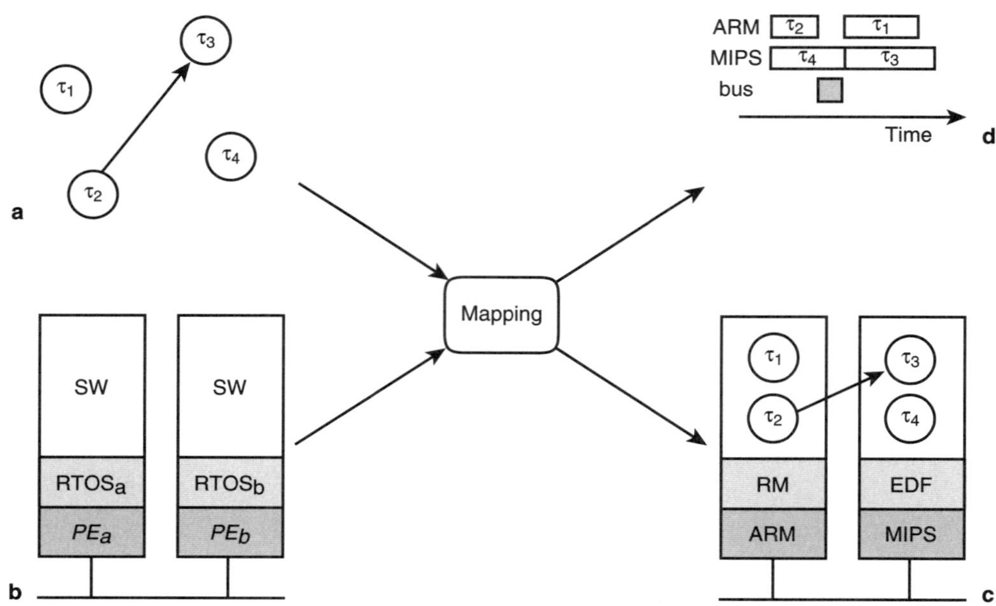

10-1

FIGURE

Mapping a set of tasks onto a multiprocessor platform. (a) Set of tasks. (b) Generic multiprocessor platform. (c) Task set mapped onto a specific multiprocessor platform. (d) Execution schedule.

running a number of RTOS on a multiprocessor platform. We refer to such a system-level model as an abstract model. In this chapter, we discuss an approach to model embedded applications represented as multithreaded applications executed on a multiprocessor platform running a number of, possibly different, RTOS.

Figure 10-1 illustrates a possible SoC design flow in which the application, modeled as a set of tasks, is mapped onto a *heterogeneous multiprocessor platform*. During the mapping, each task is assigned to an appropriate processor, and the scheduling policies of the RTOS are determined. By executing the application, properties of the system such as latency and resource utilization can be analyzed.

Validation of multiprocessor RTOS is typically done after implementation through an interactive process of experimentation on prototypes, measurement of performance, tuning of various parameters, and so on, until the performance meets the system requirements. In a more desirable development process, an extra step of modeling and simulation before (or during) implementation would shorten the time required for validation. However, as has been discussed in refs. 432 to 435, there is a lack of complete methodologies and tools that cover all the aspects pertaining to the modeling of modern heterogeneous systems in a satisfactory manner. Several approaches have been devised that tend to provide the designers with suitable modeling tools. For example, the approach followed by Sifakis [436] describes a methodology, based on the principles of *composition*, for modeling *real-time systems*, although the challenges implied by the modeling of real-time systems implemented on multiprocessor platforms are not mentioned. Another approach, followed by MetaH [437] and VEST [438], is based on the functional description of multiprocessor real-time systems giving modeling capabilities and automatic generation of different system components, including the operating system. However, the focus of both of these is at an abstraction level lower than the one we propose. In ref. 432, a high-level performance model for multithreaded *multiprocessor systems* is presented. This approach is based on modeling the layer of schedulers in an abstract manner, which resembles the aim of our approach.

Others have focused on providing RTOS modeling on top of existing system-level design languages (SLDLs), either for open languages such as SpecC [434] and the RTOS library of SystemC 3.0 [439] or for proprietary languages such as SOCOS [440] of OCAPI and TAXYS [441]. All these approaches offer functional models of the RTOS, allowing its emulation, on top of which functional models of software applications can be implemented.

In the early stages of design, even before a functional implementation exists, we can benefit from a methodology that can provide us with design strategy guidelines, in our specific case, helping us to select the right combination of scheduling, synchronization, and allocation algorithms for the RTOS running on multiprocessor platforms. A methodology, based on abstract modeling, has been presented in ref. 442. The method is based on SpecC and is limited to model preemptive schedulers for single-processor systems. Our approach is based on an abstraction of the functionality of the system. We logically partition a multiprocessor system into its components and devise a simple model for each component that will simulate its functionality using a set of parameters. We present a framework to model abstract application software executing on a multiprocessor platform under the supervision of abstract RTOS. The framework is based on SystemC [443], by the use of which we inherit all its properties, such as multiple abstraction levels, synchronous and asynchronous modeling facilities, predefined components, and progressive refinement of models.

The rest of the chapter is organized as follows. In Section 10.2, we introduce the basic concepts and terminology of real-time systems, with special emphasis on scheduling. Section 10.3 gives an overview of our abstract RTOS model. Section 10.4 gives a detailed description of the model based on the classical uniprocessor system. In Section 10.5, the uniprocessor model is extended to a more general multiprocessor model. Finally, Section 10.6 gives a summary and some concluding remarks.

10.2   BASIC CONCEPTS AND TERMINOLOGY

In this section, we present the basic principles of scheduling, as well as, in particular, a model for the tasks to be scheduled and a model of the platform architecture on which to schedule those tasks.

10.2.1   Platform Architecture

A *platform* architecture, A_k, can be modeled as a set of interconnected components. We distinguish among three types of components:

- processing elements (PE_k) are active components capable of executing one or more function(s). A processor may be a programmable general-purpose processor, a programmable dedicated processor such as a signal processor, or a coprocessor designed for a specific dedicated purpose.

- devices (D_k) are fixed operation components. A device may be a memory component or an input/output (I/O) component, for instance, a sensor, a display, or an actuator.

- communication (C_k) comprises communication channels between components, i.e., processing elements and devices. Communication channels may contain nodes (N) and a single link (L), as in a simple bus, or they may contain a complex structure of nodes and links, as in a network, such as a double torus network based on wormhole routing.

A component can be considered as a resource that may be in use when executing a task τ_i. In this chapter, we assume that a platform architecture is characterized by the resources used by the tasks.

10.2.2   Tasks

A task, τ_i, models a function (or parts of a function) to be executed and is the basic schedulable entity. A task may be implemented in software executing on a *PE* or directly as dedicated hardware, e.g., a coprocessor. In this chapter we assume that tasks are software tasks, although our model is not limited to software tasks. A task is characterized by its:

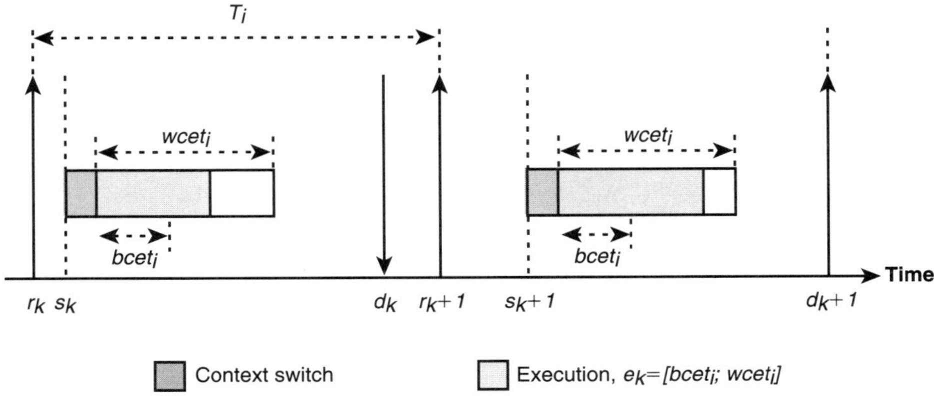

10-2

FIGURE

Task timing model showing the k'th and the $k + 1$'th instances of τ_i.

♦ *timing constraints* (Fig. 10-2).

  ♦ r_k, the release time at which the k'th instance of τ_i becomes active or available for execution.

  ♦ o_i, the offset from the time zero at which the first release of τ_i is initiated (this is also called the phase).

  ♦ s_k, the start time at which the k'th instance of τ_i actually starts execution.

  ♦ csw_i, the context switching time for τ_i.

  ♦ $wcet_i$, the worst-case execution time (WCET) for τ_i.

  ♦ $bcet_i$, the best-case execution time (BCET) for τ_i.

  ♦ $e_{i,k}$, the actual execution time of the k'th instance of τ_i. This may be calculated by a function returning a (random) value in the interval $[bcet_i; wcet_i]$.

  ♦ d_k, the deadline at which the k'th instance of τ_i should be completed. The relative deadline, D_i, is constant and is calculated as $d_k - r_k$.

  ♦ T_i, the period at which a periodic task is repeated (only valid for periodic tasks).

♦ *task execution frequency*, which may be *periodic*, in which case, the period defines the time between successive task executions; *aperiodic*, i.e., nonperiodic, typically corresponding to the tasks that handle the events from the environment that cannot happen at regular points in time; or *sporadic*, which is aperiodic but with a hard real-time constraint. From the point of view of modeling, an aperiodic task may be modeled as a periodic task whose period is a randomly generated value of the worst-case release or interarrival time (WCRT).

◆ *precedence constraints*, which impose a partial order on the execution of tasks. Precedence relations may be used to model data transmissions between tasks, i.e., when one task needs the result(s) from another task in order to execute. It may also be used to express explicit synchronization. A precedence relation from τ_i to τ_j is expressed as $\tau_i \prec \tau_j$.[1]

◆ *resource constraints*, which define the resources required by the task in order to execute. Hence, a task, τ_i, which reads data from a sensor, D_{sensor}, makes some computation, PE_{cpu}, and then stores the result in a memory device, D_{mem}, using the system bus, D_{bus}, will require all four resources to be available during its execution. We specify this resource requirement as $r(\tau_i) = \{PE_{\text{cpu}}, D_{\text{sensor}}, D_{\text{mem}}, D_{\text{bus}}\}$.

◆ *task assignment*, which defines the mapping of a task τ_i to a processing element PE_a. This mapping is expressed as $\{\tau_i\} \mapsto PE_a$[2] or $\tau_{i,a}$ in a short form.

10.2.3 Basics of Scheduling

From a system point of view, the major task of an RTOS is to determine the execution order of the tasks, i.e., to perform *task scheduling*. The process of mapping an application, represented as a set of tasks, onto a platform architecture involves three processes, which are highly interdependent:

◆ *allocation*, which determines the number and type of processors and resources. Typically, this is static and given by the platform architecture. However, in case of a reconfigurable platform, the process of allocation may be dynamic, i.e., performed during run time.

◆ *assignment/binding*, which binds tasks to processors, i.e., spatial binding. This process may be static, in which tasks are bound before running the application, or dynamic, in which tasks are bound during run time. The latter allows

---

[1] x $\prec$ y means that x precedes y.
[2] x $\mapsto$ y means that x is mapped onto y.

for task migration, whereby tasks may be moved between processors between or during executions.

◆ *scheduling*, which determines the execution order, i.e., temporal binding. This process may be static (off-line) or dynamic (on-line), as for the binding process.

In this chapter, we assume that allocation and binding have already been done as a part of the system design process, i.e., they are both static. Hence the problem we have to solve is that of scheduling.

Scheduling can be executed *off-line*, in which the schedule is obtained before the system is put into operation, i.e., at compile time, or *on-line*, in which the schedule is obtained during the execution of the system, i.e., at run time. On-line schedulers are often list-based in the sense that tasks are placed in a ready list as soon as they are released for execution. It is then the job of the scheduler to select the next task to execute among the ready tasks. The exact selection strategy is called the *scheduling policy*, and it is here that the schedulers differ.

An important classification of priority-based scheduling policies is:

◆ *competitive/preemptive scheduling*, i.e., the scheduler is allowed to stop the execution of a task, before it has finished execution, in order to execute another, possibly more important, task and then resume the execution of the preempted task.

◆ *cooperative/non-preemptive scheduling*, i.e., an executing task is allowed to run to completion regardless of its importance.

Another classification is based on *when* priorities are assigned to tasks:

◆ *static priority assignment*, whereby task priorities are determined and fixed off-line. This is used in rate monotonic scheduling (RMS) [435], in which priorities are determined based on the periods of the tasks, i.e., short periods are given highest priority.

◆ *dynamic priority assignment*, whereby task priorities are determined and updated online. This is used in earliest deadline first (EDF) scheduling [438], whereby the highest priority at any given time is given to the task with the closest deadline.

Whereas RMS is preemptive, EDF scheduling may be both preemptive and non-preemptive.

10.3   BASIC SYSTEM MODEL

At the system level, the application software may be modeled as a set of tasks, $\tau_i \in T$, which have to be executed on a number of programmable processors under the control of one or more RTOS.

Our system model is designed following the principle of composition, as described by Sifakis [436]. The principle of composition and a proof rule for composing modular specifications that consist of both safety and progress properties were defined by Abadi and Lamport [444,445]. They developed a semantic model for the principle of composition but did not tie it to any particular specification language or logic. It is a key tool for incremental system construction. A problem in specifying and verifying large systems is how to modularize a system and then prove that the components working together satisfy the overall system specification. This problem is compounded when a system is reactive and concurrent, as in a distributed operating system. An advantage of composing modular specifications is that each module can be specified and proven independently, instead of having to prove the entire system as a single large entity. Another advantage is that the system can be incrementally verified by composing a new module with a previously verified composition of modules. To realize these advantages, a system must be decomposable into well-defined, relatively simple modules. The interfaces between the modules must also be well defined so that the effect of the behavior of one module on the behavior of another can be completely and formally specified.

Our system model consists of three types of basic components: tasks, RTOS services, and links, with the links providing communication between other system components. The RTOS services are decomposed into independent modules that model different basic RTOS services, as illustrated in Figure 10-3:

◆ a *scheduler* models a real-time scheduling algorithm.

◆ a *synchronizer* models the dependencies among tasks and hence both intra- and interprocessor communications.

◆ an *allocator* models the mechanism of resource sharing among tasks.

In this chapter, we assume that each processor can run just a single scheduling algorithm. In a multiprocessor platform, we will have a number of schedulers representing the same number of processors, whereas synchronization and allocation may be represented by a single instance of each model.

One of the objectives of the framework is to provide the user with an easy way to describe an application as a set of tasks and the RTOS services, such as a

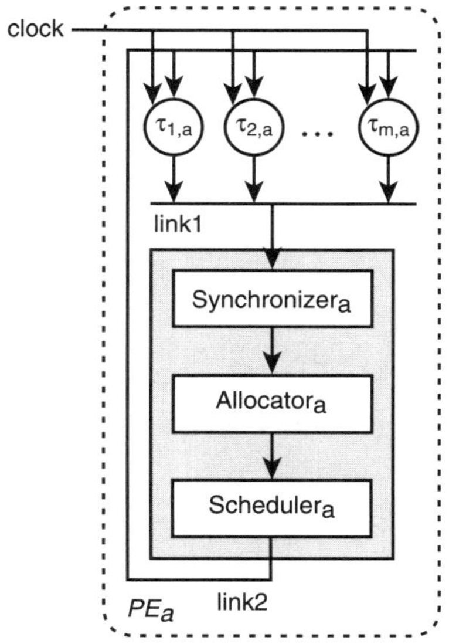

10-3

FIGURE

Architecture of the system model.

```
perTask t1=new perTask("task_1", id=1, period=50, bcet=5,
                         wcet=12, deadline=63, offset=8);
t1->clock(clock);
t1->out_message(link1);
t1->in_command(link2);
```

10-4

FIGURE

SystemC code for creating a periodic task.

scheduling algorithm. We model a task such that it carries all the pertinent information, i.e., computation time, period, deadline, and so on.

In Figure 10-4, **t1** is declared as a periodic task, **perTask**, with a period of 50 cycles, a BCET of 5 cycles, a WCET of 12 cycles, a deadline of 63 cycles, and an offset of 8 cycles. **t1** is then connected to the **clock** and the two links, **link1** and **link2**.

As a way to maintain composition, each component handles its relevant data independently of the others. For example, a task determines when it is ready to run and when it has finished. In this way, the RTOS scheduler behaves in a reactive manner, scheduling tasks according to the data received from them. Thus, we can add as many tasks and schedulers as we desire. The same is the case with the synchronizer and the allocator models. They hold the information regarding their services, i.e., which tasks depend on each other or, for the case of the allocator, what resources are needed by a given task.

As illustrated in Figure 10-3, we use a global clock to measure time in terms of clock cycles, i.e., using an abstract time unit. This allows us to identify the time at which a task is ready to be executed or when a task has finished its execution.

10.4   UNIPROCESSOR SYSTEMS

In this section, we will present the basics of uniprocessor scheduling, which are the prerequisites for understanding multiprocessor scheduling. We will describe how a uniprocessor system may be modeled using SystemC [446] and, in particular, compare two widely used on-line schedulers for uniprocessor scheduling; RMS and EDF.

Our basic system model consists of three types of components: task, RTOS, and link, with the link providing communication between the tasks and the RTOS. In this section, we present the scheduling service of the RTOS model, and in Sections 10.4.4 and 10.4.5, we extend the RTOS service model to include synchronization and resource allocation. Tasks can send the messages (**ready** and **finished**) to the scheduler, which, in turn, can send one of the three commands to the tasks (**run**, **preempt**, and **resume**). We start by describing how each of the three components and their interactions can be modeled in SystemC.

10.4.1   Link Model

The communication between the tasks and the scheduler is modeled as two channels: one to send messages from the tasks to the scheduler and one to send commands from the scheduler to the tasks, as illustrated in Figure 10-3. Our objective is to develop a convenient way to add tasks and schedulers to the system without having to create separate communication channels for each task/scheduler communication, i.e., new tasks and/or schedulers should be added by creating and connecting them, as expressed in Section 10.3. The Master-Slave library of SystemC [447] provides an elegant solution to this problem. The **sc_link_mp** component is a high-level implementation of a channel, providing a bus-like functionality, i.e., allowing the scheduler to communicate with n tasks through a single port, but without the complexity of a bus cycle-accurate implementation.

For each master port on one side of the link, there has to be at least one slave port on the other side. In our model, we have multiple master ports connected from the tasks to the slave port in the scheduler through **link1** and multiple master ports on the schedulers connected to the slave port in the tasks through **link2**.

From the Master-Slave library, we use the in-lined execution semantics (Remote Procedure Call) of the **sc_link_mp** channel. Here, a slave method executes in line with the caller process and returns to the caller after execution. The caller invokes the slave by writing a data value of type **message_type** to its master port.

In our model, each of the tasks and the scheduler have a master port and a slave port, whereby each task writes messages to its master port invoking the slave method of the scheduler, and the scheduler writes commands to its master port invoking the slave method of each task (Fig. 10-3).

If two tasks (or schedulers in a multiprocessor system) try to send a message at the same time, there will be two master ports writing to the same slave port of the scheduler. This is handled by the **sc_link_mp** as follows: two concurrent processes accessing the same link over the master ports access the link in a sequential but undetermined order. This allows the abstraction of any relationship of simultaneity between the two concurrent processes.

The **sc_link_mp** object does not provide built-in arbitration or privatization functions. It is a primitive (atomic) object that allows abstraction of such refinements. If any arbitration is needed, it has to be implemented in the channel interface. In our model, we do not use explicit arbitration; it is implicit in the behavior. This means that the scheduler just handles one request at a time (in real time), giving the feeling of arbitration at the time of simulation and making the implementation of a scheduler more intuitive, i.e., it is written as one would write the scheduling algorithm in pseudo-code, as explained in Section 10.4.3.

10.4.2 Task Model

At the system level, we are not interested in how a task is implemented, i.e., its exact functionality, but we need information regarding the execution of the task, such as the WCET ($wcet_i$), BCET ($bcet_i$), context switching overhead (csw_i), period (T_i), deadline (d_i), and offset (o_i), in order to characterize the execution of the task.

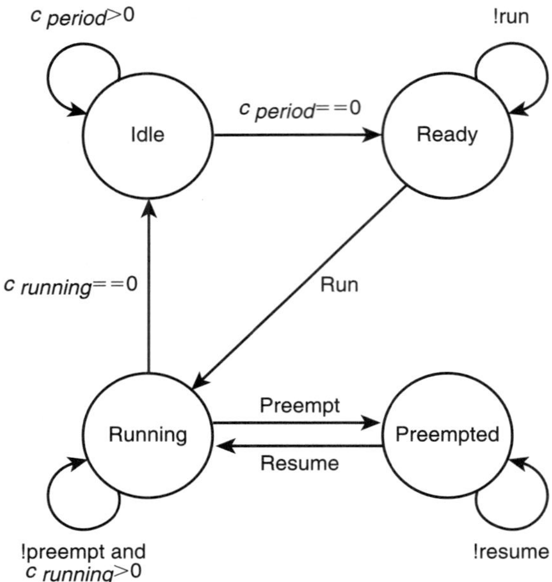

10-5   The task model.

FIGURE

Periodic Tasks

The behavior of a periodic task is modeled as a finite state machine (FSM) with four stages: **idle**, **ready**, **running**, and **preempted**, as illustrated in Figure 10-5.

We assume that all the tasks start in the **idle** state with a certain offset that can have any value including zero, in which case, the task goes immediately to the **ready** state, sending a **ready** message to the scheduler and waiting for the RTOS to issue a **run** command.

The task stays in the **ready** state until it receives a **run** command from the scheduler. It then goes to the **running** state, where it counts the number, c_{running}, of cycles. When entering the **running** state, c_{running} is initialized to the value of the task execution time e_i. When $c_{\text{running}} = = 0$, the task has finished its computation. It then issues a **finished** message to the scheduler and goes to the **idle** state. In all states, c_{period} is decremented each cycle. After reaching the **idle** state, the task stays there until $c_{\text{period}} = = 0$, indicating the start of a new period by making a transition to the **ready** state and setting $c_{\text{period}} = T_i$.

For preemptive scheduling policies, the task may be preempted by the scheduler at any time during the **running** state, i.e., the scheduler sends a **preempt**

```
SC_MODULE(perTask) {

    sc_outmaster<message_type>  out_indication;
    sc_inslave<message_type>    in_command;
    sc_in_clk                   clock;

    sc_event                    newStateEvent;

    void new_state();
    void update_state();
    void get_comm_from_scheduler();

    SC_HAS_PROCESS(perTask);
    perTask(sc_module_name name_, int id_, ...
      : sc_module(name_), id(id_), ...
    {
      SC_METHOD(new_state);
      sensitive << newStateEvent;

      SC_METHOD(update_state);
      sensitive << clock;

      SC_SLAVE(get_comm_from_scheduler, in_command);
    }
    private:
      t_state state, new_state;
};
```

10-6

SystemC code outline for the header file of the periodic task.

FIGURE

command to the task. When preempted, the task goes into the **preempted** state, where it waits for a **resume** command from the scheduler. During the **preempted** state, only c_{period} is updated.

The task model is implemented in SystemC using three processes. The basic code elements of the header file are outlined in Figure 10-6.

The **SC_SLAVE(get_comm_from_scheduler, in_command)** process takes care of commands from the scheduler and, based on these, determines what should be the new state of the task, i.e., **new_state**. The process **SC_METHOD(update_state)**, updates the actual state of the task (**state**) on the rising edge of the clock, i.e., when a cycle has been completed. It then notifies the process **SC_METHOD(new state)**, by issuing an event (**newStateEvent**) that immediately triggers the method **SC_METHOD(new state)**, i.e., it does not have to wait for the abstract time to advance. An interesting feature of this model is that, within a single cycle, the state of a task may change several times. The state changes during a cycle are handled by updating a local variable, **new_state**. At the next clock tick, when the system is stable, it is the last updated state value that determines the new state of the task. By this way of modeling, we avoid any dependency on the sequencing of concurrent messages sent to the scheduler.

Aperiodic and Sporadic Tasks

In contrast to the periodic tasks, an aperiodic task executes at irregular intervals and has only a soft deadline. In other words, the deadlines for aperiodic tasks are not rigid, but adequate response times are desirable. For example, an aperiodic task may process user input from a terminal. A sporadic task is an aperiodic task with a hard deadline and minimum interarrival time.[3]

The aperiodic and sporadic tasks are modeled in the same way as the periodic task, i.e., a four-state FSM implemented by three processes in SystemC, one for handling the FSM, a second receiving the commands from the schedulers and updating the **next_state** variable, and a third one sensitive to the clock edge, setting the new state of the FSM. The major difference is that the period counter, c_{period}, is changed to an interval counter, which models the time to the next execution of the task. The value of the interval counter, which is based on random numbers (or by an event sequence provided by the user), is calculated every time the task finishes its previous execution.

10.4.3   Scheduler Model

From a system point of view, the major job of the RTOS is to determine the execution order of the tasks, i.e., to perform task scheduling. The scheduler maintains a list of tasks ready to be executed. In our model, the list is a priority queue in which each task is given a priority according to the scheduling policy of the scheduler [435,448]. For example, for the RMS, the task priority is based on the period of the task, whereas for the deadline monotonic scheduling (DMS) and the EDF scheduling, the task priority is based on the task's deadline. In the following, we use $p(\tau_i)$ to denote the priority of task τ_i.

The scheduler is modeled as an event-based process that runs whenever a message (**ready** or **finished**) is received from a task. In the case of a **finished** message received from a task τ_i, the scheduler selects, among the ready tasks, the one with the highest priority, τ_j, and issues the command **run** to τ_j. In the case of an empty list and no running task, the scheduler just waits to receive a **ready** message. As soon as it receives this message, it issues a **run** command to the task, τ_i, which issued the **ready** message. If the list is empty, but a task, τ_k, is currently running, then:

---

[3] The time interval between the release times of two consecutive instances of a task.

1. If $p(\tau_k) \geq p(\tau_i)$ then τ_i enters the list of ready tasks.

2. If $p(\tau_k) < p(\tau_i)$ then τ_k is preempted by issuing a **preempt** command to τ_k and placing τ_k in the ready list. The scheduler then issues a run command to τ_i.

If the list is not empty, τ_i is only executed if it has the highest priority; otherwise the task, τ_j, with the highest priority is selected.

The scheduler is designed to service several messages in zero simulation time. When two or more different tasks send a ready message simultaneously to the scheduler in the same simulation cycle, the scheduler will choose the task with the highest priority to run and enqueue the others. This is handled by the Master-Slave library and the SystemC simulation kernel. Tasks are actually served sequentially during the simulation cycle.

Rate Monotonic Scheduling

In the case of RMS, the list is a priority queue, and each task is given a unique priority based solely on the period of the task, i.e., it is independent of e_i. The task with the shortest period gets the highest priority. The priority scheme is fixed, meaning that it is determined at compile time and is known to be optimal [435].

Let $p(\tau_i)$ be the priority of the task τ_i. RMS assumes that tasks are independent and that they can be preempted. Furthermore, deadlines are equal to the period, and the context switching overhead is ignored.

The implementation of the RM scheduler is rather straightforward. Figure 10-7 shows the SystemC code for the RMS algorithm, in which **ti** is the task sending the message received through **in_message**, **tj** is the task with the highest priority from the priority queue, **Q**, of tasks that are ready to be executed, and **tk** is the currently running task.

The scheduling algorithm is executed within a slave process, i.e., it is executed every time the scheduler receives a message on its input slave port. In line 2 the received message is set to task **ti** followed in line 3 by a check, which indicates whether the received message was actually meant for the scheduler.[4] If a **ready** message is received (line 4), **ti** is placed in the priority queue (line 5). If a **finished** message is received, the currently running task, **tk**, is marked as not running (line 6). The scheduler is now ready to decide which task to execute. In line 9, **tj** is set to the highest priority task (note that this may be **ti**). If the queue does contain tasks, which is checked in line 10, and there is a task currently running (line 11), it

---

[4] This check is needed, as the composition principle of our model allows for a scenario whereby all the tasks are connected to all the schedulers. Such a case is useful if we want to model aspects like task migration.

```
1   void RM scheduler: :doShedule() {
2     ti=in_message;
3     if(ti.snum==id){
4        if (ti.comm==ready)
             // if a ready command is received, put ti in the queue
5           Q.push(ti);
6        } else {
             // ti finished, select next task to run
7           tk. id=0;
8        }
9        tj=Q top();
10       if (tj.id !=0) {
             // if the queue is not empty
11          if (tk.id !=0) {
12             if (tk<tj) {
                  // if the priority of the running task,
                  // tk, is lower, preempt tk
13                out command=*(PreemptTask(&tk, &Q));
14                tk=tj;
                  // and run tj
15                out_command=*(runTask(&tj, &Q));
16             } else { /* leave tk running and tj in the queue
17          } else {
                  // no currently running task, so run the
                  // highest priority task from the queue
18             tk=tj;
19             out_command=*(runTask(&tj, &Q));
20          }
21       } else { //* queue is empty, so do nothing
22    } else { //* message is not intended for this scheduler,
             /* do nothing
23 }
```

10-7

SystemC code for the RMS algorithm.

FIGURE

is checked if the running task should be preempted (line 12). In the case of pre-emption, the running task is stopped (line 13) and the new task is started (line 15). In line 14, the new task is set to the currently running task. If the running task should not be preempted, it is just left running (line 16). In the case of no currently running task, the highest priority task from the queue is started (lines 18, 19). Finally, if the queue is empty, a situation that is only possible when a **finished** message is received, the scheduler does nothing.

Example 1. *Consider a task set $\{\tau_1, \tau_2, \tau_3\} \mapsto \mathrm{PE}_a$ with no dependencies. An important issue in real-time scheduling is that we have to guarantee that all tasks meet their deadlines, independent of the order of their release. The critical instance of a task occurs when the task and all the higher priority tasks are released simultaneously. Hence, the worst-case scenario is when all tasks are released at the same time, for*

Task	T_i	d_i	e_i	o_i
τ_1	50	50	12	0
τ_2	40	40	10	0
τ_3	30	30	10	0

10-1

TABLE

A simple tasks set to Illustrate RMS.

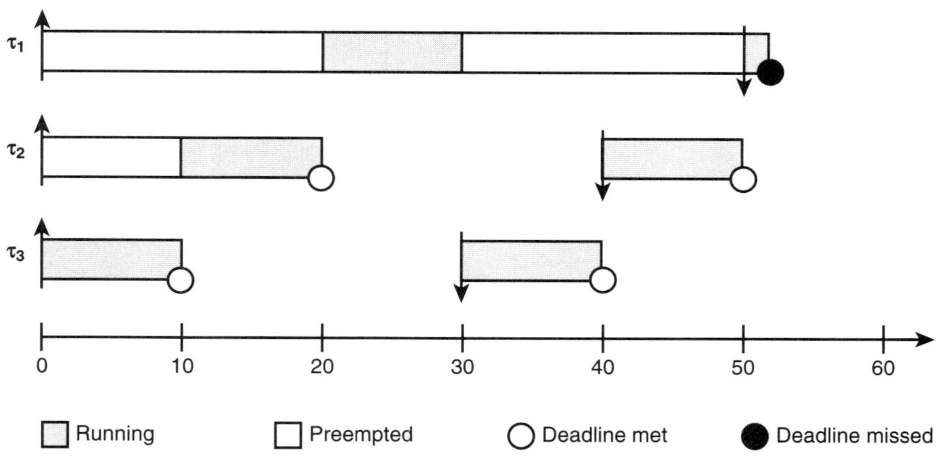

10-8

FIGURE

RMS of the tasks from Table 10-1. Note that τ_1 misses its deadline.

example, at time zero. Table 10-1 shows the periods, priorities, and execution times (WCET) for the three tasks.

Figure 10-8 shows the activities for each task. All three tasks are released at time 0. As τ_3 has the highest priority, it starts executing immediately, whereas τ_2 and τ_1 are preempted. After 10 cycles, τ_3 finishes execution, and τ_2 starts executing, whereas τ_1 is still preempted. After 30 cycles, τ_1 has been running for 10 cycles (its WCET is 12 cycles). At this point, τ_3 becomes ready again as its period of 30 cycles is

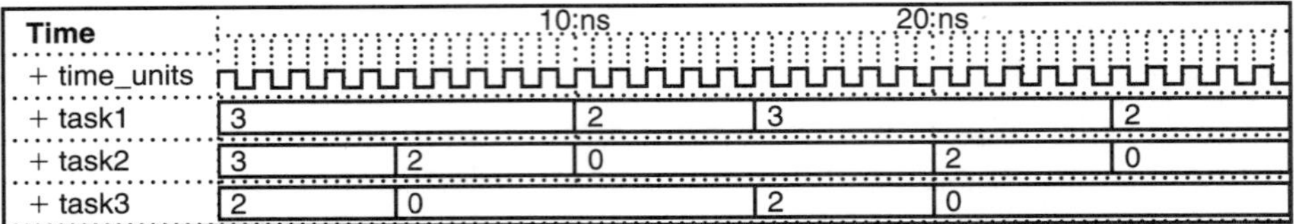

10-9

FIGURE

RMS of Figure 10-8 as viewed through a waveform viewer after simulation. Symbols: 0 = idle; 1 = ready; 2 = running; 3 = preempted.

repeating. As τ_3 has higher priority than τ_1, τ_1 is preempted and τ_3 is executed. After 50 cycles, both τ_3 and τ_2 have been executed twice, whereas τ_1 is still missing the last two cycles of its first run. Since the period (and hence deadline) of τ_2 is 50 cycles, it misses its deadline.

Figure 10-9 shows the schedule of Figure 10-8 as a waveform produced by simulating the model. The numbers in the waveform correspond to the various states of the task, i.e., 0 = idle, 1 = ready, 2 = running, and 3 = preempted.

Earliest Deadline First Scheduling

Another popular scheduling policy is the EDF scheduling. This is a dynamic priority assignment scheme whereby priorities are changed during execution, based on the release times. Tasks are prioritized according to their deadline such that the task with the closest deadline gets the highest priority.

Our model can handle both static and dynamic priority policies. Both approaches follow the above-mentioned algorithm, in which the model for handling dynamic priorities has an extra method that is sensitive to the global clock. At each clock cycle, the priority of the currently running task, $p(\tau_k)$, is updated as well as the priorities of the tasks in the ready queue. For the simplest case of EDF scheduling, the task priority is incremented by one at each clock cycle. For the static approaches, there is no need for this extra method as the priority scheme is fixed, i.e., it is determined at compile time.

Example 2. *Figure 10-10 shows the result of scheduling Example 1 using the EDF scheduling algorithm.*

10.4.4   Synchronization Model

Another of the basic services provided by an RTOS is synchronization among cooperative tasks that are running in parallel. This synchronization enables the

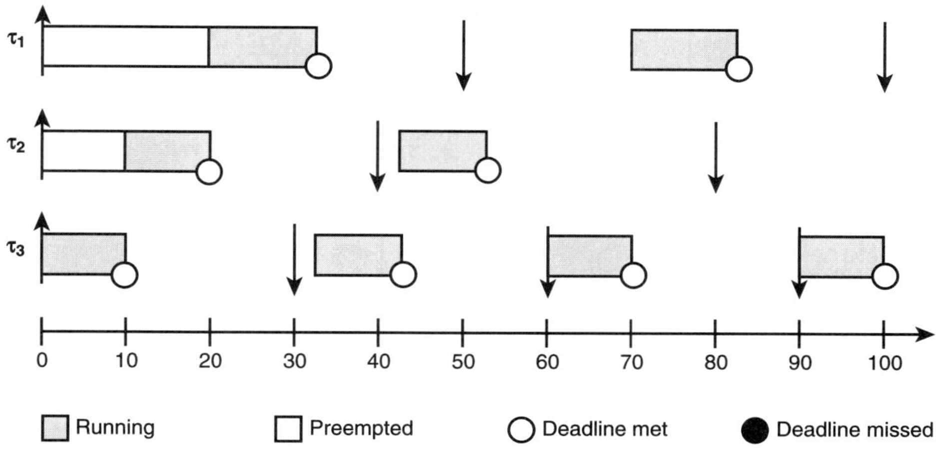

10-10   EDF scheduling of the tasks from Table 10-1. Note that τ_1 now meets its deadline.

FIGURE

simulator to model a real-time system with data dependencies among tasks. For example, if τ_i needs data computed by τ_j, then τ_i has to wait until the completion of τ_j in order to execute.

Task dependencies can be of various types, but, at the system level, we do not care about the nature of a dependency. We can formulate an abstraction and assert that the task τ_j is eligible to be released just after the task τ_i has finished its execution.

As we have designed our framework to support multiprocessor platforms, the synchronizer handles intra- and interprocessor dependencies as well as multirate systems.[5]

A synchronization *protocol* governs how the tasks are released for execution such that their *precedence constraints* are satisfied. One of the more commonly used synchronization protocols is the direct synchronization (DS) protocol, which was proposed by Sun and Liu [449]. According to the DS protocol, when an instance of a task completes, a synchronization signal is sent to the processor, where its immediate successor executes, and an instance of its immediate successor is released immediately.

In our implementation of the DS protocol, the synchronizer can be seen as a "message filter" between the tasks and the schedulers, letting other schedulers

[5] The period of the producer is different from that of one of the consumers.

know when a task is "really ready," i.e., when its dependency constraints have been resolved. The **finished** message will always pass, but the **ready** message will pass only when the dependency constraints have been resolved.

Each time a task issues a **ready** or a **finished** message, that message is passed to the synchronizer. If the synchronizer receives a **ready** message, it searches its **dependency_database** to check whether the task issuing that message is dependent on any other task. If so, the synchronizer checks the **finished_task_list** to see whether the dependent task has already finished its execution. If it has, the dependent task is erased from the **finished_task_list**, and the ready message is passed to the scheduler; otherwise, the information about the task issuing the ready message is stored in the **waiting_dep_list**.

If the synchronizer receives a **finished** message, it searches its **dependency_database** to check whether the task issuing that message has any other task dependent on it. If so, it scans the **waiting_dep_list** to see if that dependent task is ready to execute. If it is, the task waiting for the resolution of dependency is released to its scheduler for execution, but if it is not ready to execute, the information about the task issuing the **finished** message is stored in the **finished_task_list**.

Some synchronization protocols may require more complex modeling, whereas for others (e.g., if multirate is not needed) a simpler model would be enough. The use of the data structures is determined by the synchronization protocol; for this, the synchronization model defines standard interfaces to the data structures in order to separate the functionality of the model from the functionality of the data structures. This allows for change in the synchronization protocol without the need to modify the functionality of the data structures. Other synchronization protocols reported in the literature are the phase modification (PM) protocol and the release guard (RG) protocol, which have been proposed by Sun and Liu [449]. These protocols can also be modeled in SystemC and used in our framework.

10.4.5   Resource Allocation Model

In an embedded real-time system, we often find a number of non-preemptable resources used by a number of tasks. For example, we can have several concurrent tasks competing for the utilization of a shared memory unit, which, in principle, is a non-preemptable resource. Resource allocation is a basic service provided by the RTOS that coordinates access to the resources in a real-time system.

In a real-time system, the lack of a proper resource allocation mechanism can result in the phenomena of *priority inversion*[6] and *deadlocks*,[7] which lead to timing anomalies (unpredictable timing behavior). Therefore, a resource access control protocol is needed to keep the duration of each priority inversion and hence keep the blocking times of the tasks bounded. For a uniprocessor system, Sha et al. [450] have developed the priority inheritance protocol and the priority ceiling protocol (PCP). Baker [451] has developed the stack-based protocol (SBP), which gives the same upper bound on the blocking time of each task as with the PCP. In addition, the SBP is applicable to both static- and dynamic-priority scheduling policies.

In our model, the resource allocator implements the priority inheritance protocol for allocating the resources requested by the tasks. The priority inheritance protocol ensures that, in the absence of deadlocks, no task is ever blocked for an indefinitely long time because an uncontrolled priority inversion cannot occur. When a task requires a resource, it sends a **request** message to the allocator, which issues either a **grant** message to the scheduler if the requested resource is available or a **refuse** message if it is not. In both cases, the priority of the task requesting the resource is updated in accordance with the priority inheritance protocol and is notified to the scheduler by an **updatePriority** message. In a similar way, when a task has occupied a resource for its designated duration, it sends a **release** message to the allocator, which updates its resource database and issues an **updatePriority** message to the scheduler as demanded by the priority inheritance protocol.

10.5   MULTIPROCESSOR SYSTEMS

In this section, we will extend our model for uniprocessor scheduling to handle multiprocessor scheduling [452]. As for uniprocessor scheduling, the primary objective is to ensure that all deadlines are met. However, validating that the hard timing constraints are met is extremely difficult in the case of multiprocessor systems.

In the following discussion, we assume that each processor has its own scheduler, which uses a uniprocessor scheduling algorithm, and that the schedulers on

---

[6] As resources are allocated to tasks on a non-preemptive basis, a priority inversion may occur when a higher priority task is blocked by a lower priority task due to resource contention, even when the execution of both tasks is preemptable.

[7] Two tasks can create a deadlock when one of them holds a resource needed by the other and vice versa.

different processors may use different scheduling algorithms. Based on this assumption, we identify the following two important issues in multiprocessor scheduling:

+ Task assignment. Most real-time systems are static in the sense that tasks are partitioned and statically bound to processors.

+ Interprocessor synchronization. This ensures that the precedence relations between the tasks running on different processors are always satisfied.

Task assignment is often done off-line, i.e., at compile time, and is based on a detailed analysis of the application and the target architecture. The assignment may be based on one or more criteria, such as execution time, resource requirements, data dependencies, timing constraints, communication cost, and so on. In our model, all resource allocations are local, and remote resource requests are handled by the proper application of task assignment and interprocessor synchronization policies.

Example 3. *In this example, which is based on the example from Sun and Liu [449], we will illustrate the capabilities of our framework by analyzing a small multiprocessor example. Figure 10-11a shows a system with two processors, running two tasks each, i.e., $\{\tau_1, \tau_2\} \mapsto PE_a$ and $\{\tau_3, \tau_4\} \mapsto PE_b$. A single dependency exists between the tasks τ_2 and τ_3, i.e., $\tau_2 < \tau_3$. As seen in Figure 10-11, it is an interprocessor dependency. The figure also shows the characterization of each of the tasks.*

Figure 10-11b shows how we model the example using our abstract RTOS model. Note that, in this example, we have omitted the resource allocator.

In the first approach, we use RMS as the scheduling policy for the RTOS on both processors. Figure 10-12 shows the behavior of each task in the system, giving information about the release times and the deadlines. As τ_1 has a shorter period than τ_2, it has, according to the RMS policy, a higher priority and, hence, starts executing at time 0. After 2 time units, τ_1 has completed its execution well ahead of its deadline. τ_2 can then run until it completes after 2 time units at time 4. Due to the dependency between τ_2 and τ_3, τ_3 has to wait until time 4, which is the same time that task τ_4 is released. As τ_3 has higher priority than τ_4,[8] τ_3 starts executing at time 4. Having a period of 6 time units and a delay of 2 time units, τ_3 finishes just in time. At time 6, τ_3 finishes and τ_4 starts its execution, which should run for 3 time units and end before its deadline. However, at time 6, τ_2 starts its second iteration, which ends at time 8, immediately releasing τ_3. Since τ_3 has higher

---

[8] Although τ_3 and τ_4 have the same period, we assume that τ_3 is given a higher priority in this example.

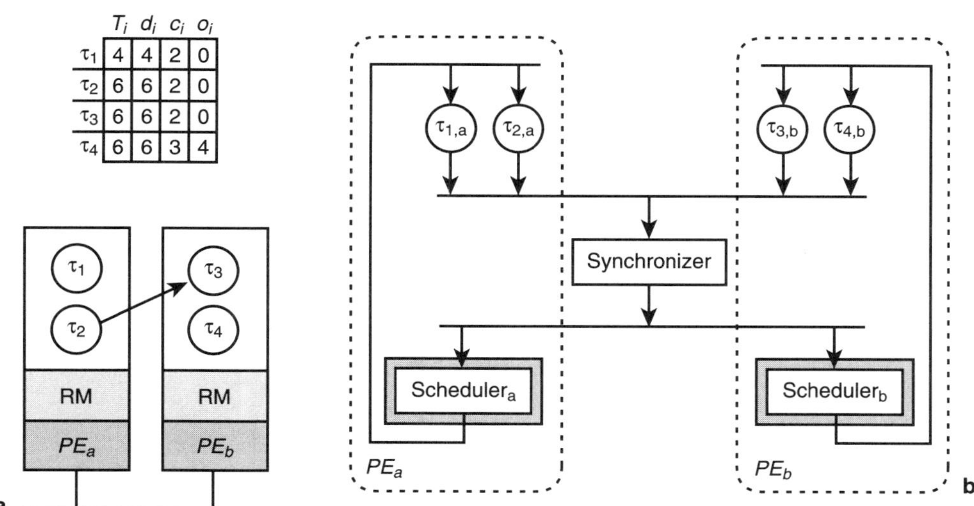

10-11
FIGURE

(a) Task characterization and allocation for the multiprocessor example; (b) Modeling the example.

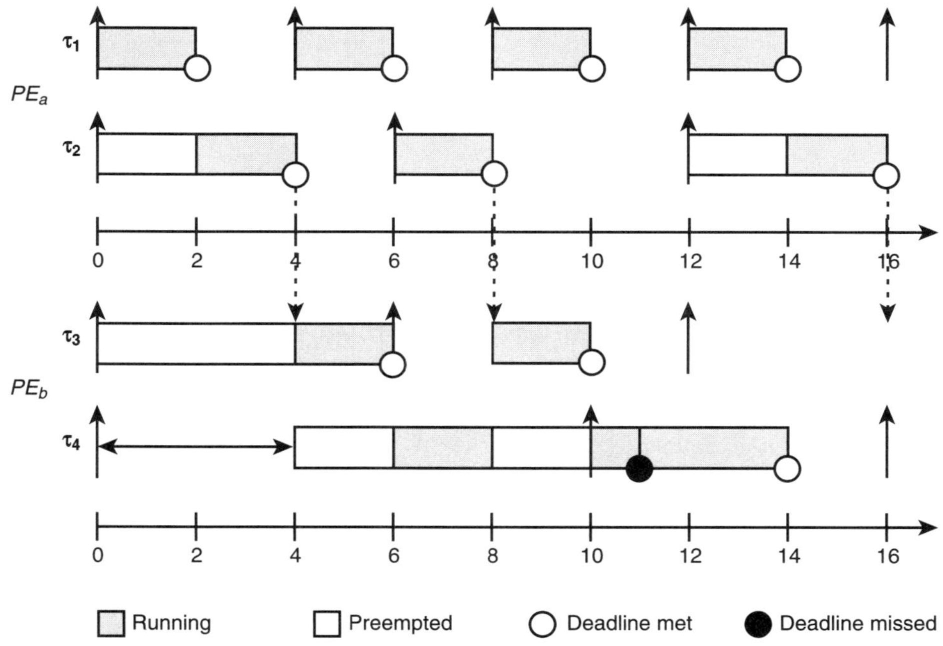

10-12
FIGURE

Schedule of the example from Figure 10-11 in which both processors are running the RMS.

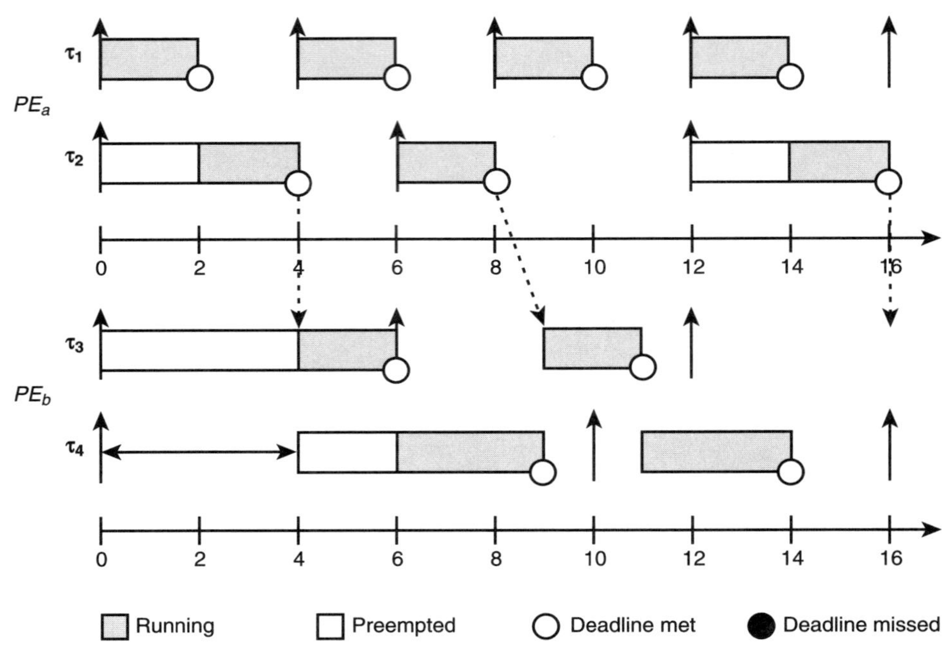

10-13

FIGURE

Schedule of the example from Figure 10-11 in which the scheduler on PE_b has been changed to EDF.

priority than τ_4, τ_4 is preempted at time 8 and cannot resume execution before τ_3 has finished at time 10, effectively missing its deadline by 1 time unit.

If the scheduling policy of the RTOS on processor PE_b is changed to a non-preemptive earliest deadline first (EDF) scheduling, τ_4 will complete its first execution at time 9, as shown in Figure 10-13, delaying the second execution of τ_3 by 1 time unit. In this scenario, all four tasks meet their deadlines. Although this is a very simple example, it demonstrates the modeling capabilities of our framework.

10.5.1   Multiprocessing Anomalies

An interesting aspect of multiprocessor scheduling that presents complications is that of anomalies. Let us consider a set of tasks that is optimally assigned and scheduled on a multiprocessor system that has a fixed number of processors, a fixed execution time for each task, and contains precedence constraints. Then, implementing any of the following intuitive "improvements"

- ✦ Changing the priority list of the scheduler
- ✦ Increasing the number of processors in the architecture
- ✦ Reducing the execution time of one or more tasks
- ✦ Weakening the precedence constraints

may *increase* the scheduling length! Hence, most on-line scheduling algorithms are subject to experience anomalies, as tasks may complete before their WCET.

Example 4. *Consider a task set of five tasks, $\{\tau_1, \tau_2, \tau_3, \tau_4, \tau_5\}$ assigned to two processing elements, $\{PE_a, PE_b\}$ such that $\{\tau_1, \tau_2\} \mapsto PE_a$ and $\{\tau_3, \tau_4, \tau_5\} \mapsto PE_b$. Let the dependency relations be as follows, $\tau_1 \prec \tau_2$ and $\tau_3 \prec \tau_4 \prec \tau_5$. Furthermore, let τ_2 and τ_4 share the same resource, D_x, i.e., $r(\tau_2) = r(\tau_4) = \{D_x\}$. Figure 10-14a shows an optimal schedule for the task set. Assume that we are able to find a better implementation of τ_1 such that its execution speed is doubled. Intuitively, we would assume better system performance; however, as illustrated in Figure 10-14b, the latency is actually extended from 12 to 15 due to the mutual exclusion of τ_2 and τ_4.*

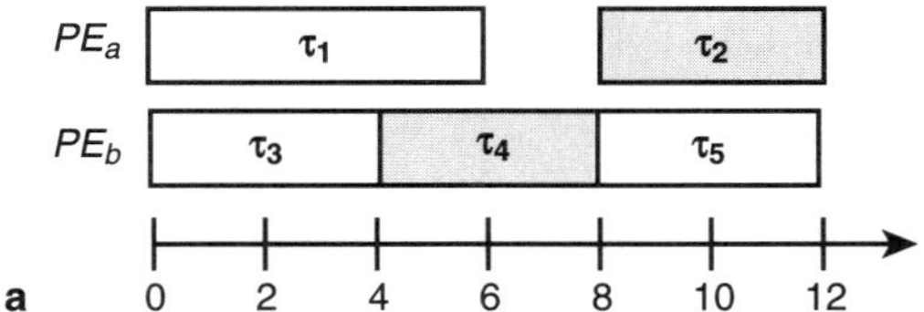

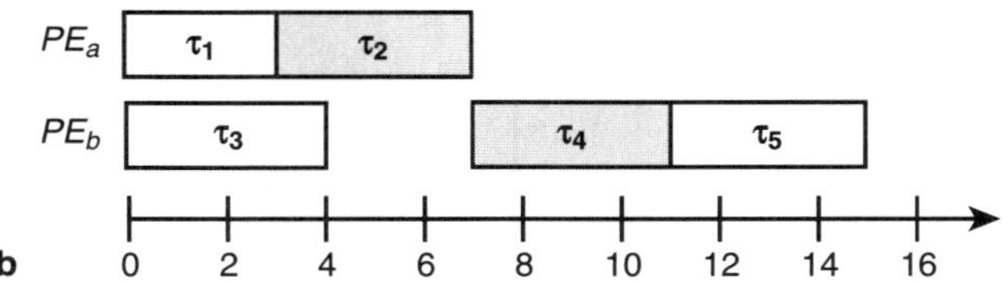

10-14 FIGURE   Example of anomalies; task τ_2 and τ_4 are sharing a resource. (a) Optimally scheduled task set. (b) Consequence of improving the execution time of τ_1.

10.5.2   Interprocessor Communication

Interprocessor communication overhead is a major factor that limits the performance of multiprocessor systems. All multiprocessor systems, no matter how powerful their interprocessor communication mechanism, suffer from this overhead. Communication overhead is exacerbated by the resource contention in the interprocessor communication network, which comprises the *node* and the *link* contentions. Node contention arises when a node attempts to transmit or receive several messages simultaneously. Link contention is caused by the sharing of a communication link by two or more messages. The network resource contention arises in all but the simplest communication requirements. It can be minimized or, in some cases, eliminated, by proper scheduling of messages. However, apart from incurring scheduling overhead, this requires synchronization among all the processors in a multiprocessor system, thereby incurring synchronization overhead. The mapping of an application on a multiprocessor system thus has to be optimized to satisfy the following conflicting requirements:

+ A completely contention-free schedule will incur substantial synchronization overhead.

+ A completely synchronization-free schedule will result in heavy contention overhead.

Clearly, there is a need to find a balance between the two types of overhead in order to minimize the overall execution time of the application running on a multiprocessor system [453].

In our multiprocessor *modeling framework*, instead of explicitly including the interprocessor communication costs, we model the interprocessor communication network as a communication processor on which message transmission tasks are scheduled non-preemptively on a fixed-priority basis (see p. 338 in ref. 454). In this way, the interprocessor communication costs are implicitly taken into account by the response times of the message transmission tasks scheduled on the communication processor.

We model a communication event within an interprocessor network as a message task (τ_m) that executes on the communication processor. When one *PE* wants to communicate with another *PE*, a τ_m is released for execution on the communication processor. Each τ_m represents communication only between a fixed set of two predetermined PEs. Since an interprocessor communication network supports concurrent communication, the τ_ms are synchronized, allocated network resources, and scheduled accordingly. Therefore, in our model, the network allocator reflects the topology, and the network scheduler reflects the interprocessor communication protocol. A resource database, which is unique to each network, contains information about all its resources (nodes, links, buffers, and so on). The network allocation and scheduling algorithms map a τ_m onto the available network resources.

From the implementation point of view, our interprocessor communication model has exactly the same structure as the abstract RTOS model described earlier but with some necessary modifications to its constituent module blocks.

10.5.3   Multiprocessor Example

In this section we demonstrate the capabilities of our system-level multiprocessor SoC modeling framework by presenting the simulation results of a practical example derived from a real-life application of an MP3 Decoder [455]. The preprocessing steps for abstracting the application code, like the extraction of *task graph* parameters through code profiling, and mapping the task graphs to the SoC architecture have been performed manually. The resulting task graph is shown in Figure 10-15, and the characterization of each task is listed in Table 10-2.

Initially, the task graph for the MP3 Decoder is mapped on a uniprocessor architecture and simulated using the modeling framework. The simulation results reveal that some tasks miss their deadlines. Therefore, the application task graph is *task-partitioned* and mapped on a multiprocessor SoC platform. The partitioned task graph is shown in Figure 10-15. This partitioned task graph is mapped on an architecture comprising two general-purpose processors (PE_a and PE_b) interconnected by a bus. The mapping results in the exchange of communication messages among the two PEs over the bus. The simulation results are shown in Figure 10-16.

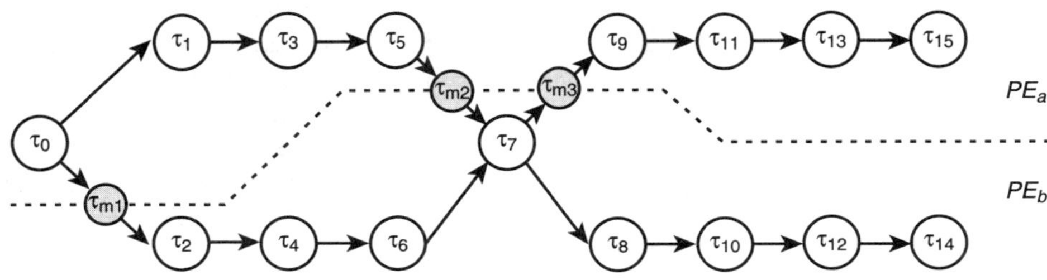

10-15 FIGURE   Task graph for the MP3 Decoder. The dotted line shows the partitioning of the tasks onto processors PE_a and PE_b.

Task	T_i	d_i	$bcet_i$	$wcet_i$	o_i
τ_0	30,000	2,500	43	45	0
τ_1	30,000	2,500	19	20	0
τ_2	30,000	2,500	19	20	0
τ_3	30,000	2,500	1,471	1,545	0
τ_4	30,000	2,500	1,471	1,545	0
τ_5	30,000	2,500	567	595	0
τ_6	30,000	2,500	567	595	0
τ_7	30,000	2,500	2,557	2,685	0
τ_8	30,000	2,500	103	108	0
τ_9	30,000	2,500	103	108	0
τ_{10}	30,000	2,500	852	895	0
τ_{11}	30,000	2,500	852	895	0
τ_{12}	30,000	2,500	5,797	6,087	0
τ_{13}	30,000	2,500	5,797	6,087	0
τ_{14}	30,000	2,500	10,667	11,200	0
τ_{15}	30,000	2,500	10,667	11,200	0

10-2

Characterization of the MP3 Task Set.

TABLE

10-16

The MP3 Decoder application of Figure 10-15 scheduled on a two-processor system. The RTOS of each processor is using the RMS policy. Symbols: 0 = idle; 1 = ready; 2 = running; and 3 = preempted.

FIGURE

10.6   SUMMARY

We have presented an abstract modeling framework based on SystemC that supports the modeling of multiprocessor-based RTOS. The aim of the framework is to provide the system designer with a user-friendly and efficient modeling and simulation environment in which one can experiment with different RTOS policies and study the consequences of local decisions on the global system behavior.

An RTOS is modeled as a combination of three interacting modules defining basic RTOS services: scheduling, synchronization, and resource allocation. Although the three services are interdependent, the separation of concerns allows the system designer to mix different design implementations easily in order to develop an efficient dedicated RTOS. The model presented supports periodic, aperiodic, and sporadic tasks, data dependencies among tasks, task offset, context switching, and dynamically determined execution times (based on a uniform distribution of execution times between the BCET and WCET limits).

The composition principle of our RTOS model allows for various multiprocessor models, including centralized RTOS, distributed RTOS, and task migration in the case of a pool of tasks supported by more than one RTOS. In addition, the effect of interprocessor communication can also be effectively modeled.

ACKNOWLEDGMENTS

Work on the SystemC-based modeling framework described in this chapter was supported by SoC-MOBINET (IST-2000-30094) and ARTIST (IST-2001-34820), both EC-funded projects. We wish to thank Marcus Schmitz from Linkoping University for providing us with details of the MP3 Decoder example.

11 Cost-Efficient Mapping of Dynamic Concurrent Tasks in Embedded Real-Time Multimedia Systems

Peng Yang, Paul Marchal, Chun Wong,
Stefaan Himpe, Francky Catthoor, Patrick David,
Johan Vounckx, and Rudy Lauwereins

11.1 INTRODUCTION

The merging of computers, consumer, and communication disciplines gives rise to fast growing markets for personal communication, multimedia, and broadband networks. Technology advances lead to platforms with enormous processing capacity that are, however, not matched with the required increase in system design productivity.

One of the most critical bottlenecks is the dynamic and concurrent behaviors of many current multimedia applications. Normally first specified in software-oriented languages (like Java, UML, SDL, and C++), these applications must be executed at real time in a cost/energy-sensitive way, probably on heterogeneous system-on-chip (SoC) platforms. A systematic way of mapping this software specification onto an embedded multiprocessor platform is required. The fully design-time-based solutions, as proposed earlier in the compiler and system synthesis communities, cannot handle the problem properly. They can only solve the problem by assuming the worst-case situations, which results in very costly designs, nor can the existing run-time solutions in present (real-time) operating systems solve the problem efficiently. They do not take cost optimization (especially energy consumption) into account.

The more recent quality-of-service (QoS) aspects make the multimedia and networking applications even more dynamic. Prominent examples of this can be found in the recent MPEG4/JPEG2000 standards and especially the new MPEG21 standard. In order to deal with these new dynamic applications in which tasks and complex data types are created and deleted at run time based on nondeterministic events (typically at the rate of tens of milliseconds), a novel system design paradigm is required. This chapter focuses on the new requirements and presents our work on the system-level synthesis of this kind of application. In particular, a task concurrency management (TCM) problem formulation is proposed, with special emphasis on power consumption reduction while meeting hard and/or soft real-time constraints. However, our methodology can be used in any situation in which run-time tradeoff exploration is possible and beneficial.

The promising results that can be obtained with such a methodology are illustrated with an MPEG21-based demonstrator mapped to a multi-V_{dd}, multiprocessor simulation platform with a hierarchical shared memory organization.

11.2   PLATFORM-BASED DESIGN

The future of embedded multimedia applications lies in low-power heterogeneous multiprocessor platforms. In the near future, the silicon market will be driven by low-cost, portable consumer devices that are integrated with multimedia and wireless technology. Most of these applications will be implemented as compact and portable devices, putting stringent constraints on the degree of integration (i.e., chip area) and on their power consumption (0.1 to 2 W). The applications will require a computational performance on the order of 1 to 40 GOPS [456]. Current PCs do offer this performance. However, their power consumption is extremely high (10 to 100 W). We must reduce the power consumption at least two to three orders of magnitude while keeping the computation performance. This can be achieved by exploring the parallelism provided by the SoC platform. Moreover, embedded systems are also subject to stringent real-time constraints, which complicates their implementations considerably. Finally, the challenge of embedding these applications on portable devices is increased even further because of user interaction: at any moment the user will be able to trigger new services, change the configuration of the running services, or stop existing services. Hence, the execution behavior changes dynamically at run time.

In summary, the future embedded system should have sufficient computing performance, consume an extremely low energy, and be flexible enough to cope with the dynamic behavior of future multimedia applications. Existing single-processor systems (e.g., those found in personal digital assistants [PDAs]) cannot deliver enough computation power at a sufficiently low energy level. Full custom hardware solutions can achieve a good performance/energy consumption ratio, but they are not flexible enough to cope with the dynamic behavior and have a relatively long time-to-market delay. Also, as the semiconductor technology scales down, the explosion of the cost of masks and the physical design (especially due to timing closure and test) implies that custom ASICs are only feasible for high-volume designs. A combination of the bests of both worlds is required for most of the new applications [457]. This explains the growing interest in platform-based design [458].

The benefits of platform-based design can, however, only be fully exploited when applications are efficiently mapped to the different processors of that heterogeneous multiprocessor platform. Unfortunately, current design technologies fall behind these advances in computer architecture and processing technology. When looking at contemporary design practices for systems implemented with these heterogeneous multiprocessor platforms, one can only conclude that these systems are nowadays designed in a very ad hoc manner. A systematic methodology and its corresponding tool set are definitely needed.

In this paper, we present the current status of our systematic design methodology and tool support to map multimedia applications dynamically to multiprocessor platforms. With our design methodology, we target especially the management of dynamic and concurrent tasks and their data on heterogeneous multiprocessor platforms. Given an application that has its own specific control and data structure, as well as a template for the platform architecture, it provides a way to decide the platform architecture instance and a means of mapping the application efficiently onto such an architecture (Fig. 11-1, top). In this chapter we focus only on the concurrency management issues, even though mapping and scheduling the data structures to the platform [459] are as crucial. Platform-based design also encompasses a physical implementation problem that can be called platform integration (Fig. 11-1, bottom), but this issue is not addressed here either. Compared with an earlier publication (460), this chapter introduces the scenario concept and presents the result of a more realistic multimedia application.

11.3 RELATED WORK

Task scheduling in a task concurrency management context has been investigated extensively in the last three decades. When a set of concurrent tasks, that is, tasks that can overlap in time, have to be executed on one or more processors, a predefined method called a scheduling algorithm must be applied to decide the order in which those tasks are executed. For a multiprocessor system, another procedure, assignment, is also needed to determine on which processor one task will

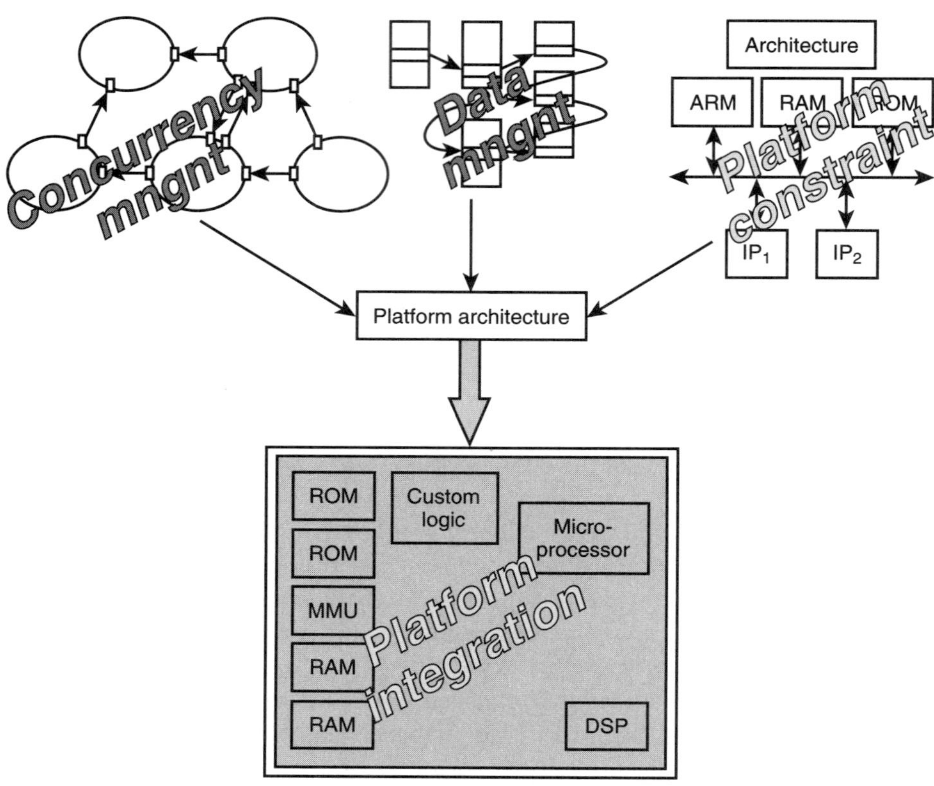

11-1   The platform integration.

FIGURE

be executed. In this chapter, the terminology task scheduling is used for both the ordering and the assignment. A good overview of task scheduling algorithms can be found in ref. 461.

Traditionally, in the real-time community, researchers tend to represent an application as task graphs, and look tasks as black boxes. The internal operations and data structures of tasks are not exposed to the designers [461–464]. This representation is at too high a level and it does not allow some important information (e.g., the number of iterations of a loop) to be exposed inside the tasks. In contrast, in the embedded system community, many papers focus on white-box task descriptions [465, 466], which are at the operation level. This level is too low and has too many unnecessary details, which are typically unavailable at the early design stage. Instead, we use a gray-box model, which is an abstraction level between the white- and black-box models mentioned above [467]. This already contributes one important distinction from related work, but the details of the gray-box model are not discussed here.

Scheduling algorithms can be roughly divided into dynamic and static scheduling. In a multiprocessor context, when the application has a large amount of nondeterministic behavior, dynamic scheduling has the flexibility to balance the computation load of processors at run-time and make use of the extra slack time coming from the variation from worst-case execution time (WCET). However, the run-time overhead may be excessive and a global optimal scheduling is difficult to find due to the difficulty of the problem. We have selected a combination of the design-time and run-time scheduling here to take advantage of both.

A large body of scheduling algorithms stems from Liu and Layland's classic paper [468], in which they gave five basic assumptions and studied the optimizability and schedulability of rate monotonic (RM) and earliest deadline first (EDF) algorithms under fixed and dynamic priority environments, respectively. Most of the later works can be classified as one of these two, namely, fixed priority or dynamic priority; they aim to relax the assumptions given in Liu and Layland's paper and try to deal with shared resources, aperiodic tasks, and tasks with different importance levels. Most importantly, Leung and Whitehead [469] suggested deadline monotonic to relax the deadline assumption, and Chetto et al. [470] introduced a transformation to the original task set to include the precedence constraints in EDF. To provide service for aperiodic or sporadic tasks, several bandwidth-preserving algorithms have been proposed [471, 472]. The problem of resource sharing is tackled with a priority inheritance protocol in refs. 463 and 473. All this work provides the basis of scheduling theory.

Since more and more embedded systems are targeted at multiprocessor architectures, the multiple processor scheduling plays an increasingly important role. El-Rewini et al. [474] give a clear introduction to the task scheduling in multiprocessing systems. Hoang et al. [475] suggest maximizing the throughput by balancing the computation load of the distributed processors. All the partition and scheduling decisions are made at compile time. This approach is limited to pure data flow applications. Yen et al. [476, 477] propose combining the processor allocation and process scheduling into a gradient-search cosynthesis algorithm. It is a heuristic method and can only handle periodic tasks statically.

In the above work, performance is the only concern. In other words, they only consider how to meet real-time constraints. For embedded systems, cost factors like energy must be taken into account as well. Gruian et al. [478] have used constraint programming to minimize the energy consumption at system level, but their method is purely static, and no dynamic policy is applied to exploit more energy reduction.

Recently, considerable interest has arisen in dynamic voltage scaling (DVS) (see ref. 479 for a good survey). Since the energy consumption of CMOS digital circuits is approximately proportional to the square of the supply voltage, decreasing the supply voltage is advantageous to low-power design, although it will also slow down the cycle speed.[1] Traditionally, the CPU works at a fixed supply voltage, even at light workloads. In fact, under such situations, the fast speed of the CPU is unnecessary and can be traded for a lower energy/power consumption by reducing the supply voltage. Based on this investigation, several real-time scheduling algorithms are provided, e.g., by Hong et al. [480] and Okuma et al. [481]. Chandrakasan et al. [482] have compared several energy saving techniques involving DVS. Recently Transmeta has released a commercial processor that allows the dynamic voltage scheduling technique [483]. Normally the scheduling techniques developed in the real-time community, e.g., fixed priority scheduling, EDF, slack stealing, and so on are still usable in DVS, only an extra voltage decision step is needed. A good survey of the recently proposed DVS algorithms can be found in refs. 479 and 484. Most related work (e.g., refs. 484, 485, and 486) concentrates on saving energy of independent tasks or on a single processor, whereas the tasks in real-world applications always exhibit control or data dependencies and are mapped to multiple processors, heterogeneous or homogeneous.

DVS in the multiple processor context is treated in ref. 487 by evenly distributing the workload. However, no manifest power and performance relation is used to steer the design-space exploration. In addition, they also assume a continuously scalable working voltage. In ref. 488, for a multiprocessor and multiple link platform with a given task and communication mapping, a two-phase scheduling method is proposed. The static scheduling is based on the slack time list scheduling, and a critical path analysis and task execution order refinement method is used to find the off-line voltage scheduling for a set of periodic real-time tasks. The run-time scheduling is similar to resource reclaiming and slack stealing, which can make use of the variation from the WCET and provide a best-effort service to aperiodic tasks. In ref. 489, an EDF-based multiprocessor scheduling and assignment heuristic is given, which is shown to be better than the normal EDF. After the scheduling, an integer linear programming (ILP) model is used to find the voltage scaling approximately by simply rounding the result from a linear programming (LP) solver; an accuracy within 97% is claimed. The method can be used for both continuous and discrete voltage.

---

[1] In the long term, the technologically available range for this scaling will decrease, as already mentioned.

Our task scheduling methodology (first introduced in ref. 467) is different from the methods described above in several ways. First, we also use a two-phase off-line and on-line scheduling, as do the more advanced existing approaches. However, the design-time off-line step is more a design-space exploration than a simple scheduling, because it provides a series of different scheduling tradeoff points, rather than only one, as is delivered by a conventional design-time scheduler (or the few existing combinations of a design- and run-time phase, as in, e.g., ref. 488). Second, we avoid having to use the WCET estimation, which is inaccurate and pessimistic due to the dynamic features of the applications. Thus, in the run-time phase, we can follow the real behavior and resource requirements much more closely. Third, we consider only discrete voltages, which is a more practical assumption in terms of future process technology and circuit design possibilities, compared with a continuous range, which is becoming more and more unrealistic to work with in the same circuit instance. It also involves less energy overhead in the V_{dd} generation, especially if we can put several processors in parallel, each with a fixed V_{dd} and a fully optimized circuit. In fact, as illustrated in ref. 489, the discrete V_{dd} case is even more difficult to solve because it corresponds to an ILP problem, not LP. Fourth, we apply voltage scaling at the intratask level (gray-box model) and make use of the run-time application information. The intratask voltage scaling has only recently been considered by other researchers [490].

11.4   TARGET PLATFORM ARCHITECTURE AND MODEL

In an up-to-date embedded processing platform like that shown in Figure 11-2 (in which we focus on the digital core only), one or more (reconfigurable)

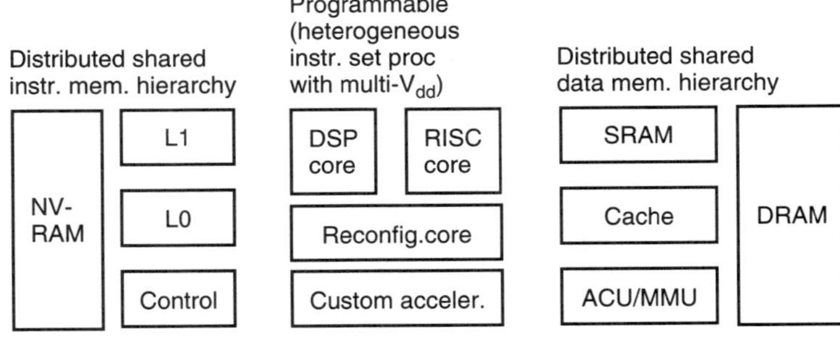

11-2   A typical heterogeneous processor platform.

FIGURE

programmable components, either general-purpose or DSP processor cores or ASIPs, the analog front end, on-chip memory, I/O, and other ASICs, are all integrated into the same chip (SoC) or in the same package (system-in-package). Furthermore, platforms typically also contain some software blocks (API, operating system, and some other middleware solutions), a methodology, and a tool set to support rapid architectural exploration. Examples of such platforms are the TI OMAP, the Philips Nexperia, the ARM PrimeXsys, and the Xilinx Virtex-II Pro [458].

Platforms obtain a high performance at a low energy cost through the exploitation of different types of parallelism [479, 491]. In the context of this chapter, we mainly focus on the task-level concurrency, but also the instruction-level and data-level parallelism should be exploited in the mapping environment. For the task-level concurrency, we can reduce the system power consumption by exploiting V_{dd} and frequency scaling in the processing cores. We will assume that only a limited set of V_{dd}s are supported with a not too large range because future process technologies will heavily limit that range. Large memories do not even allow any V_{dd} scaling, but in that case the power consumption can be controlled by an efficient mapping on a distributed memory architecture [492]. However, the heterogeneous multiprocessor architecture offers extra design tradeoff space (see ref. 493 for early results).

A multiprocessor platform simulator was used to test the effectiveness of our methodology and to answer "what-if" questions to explore the platform architecture design space. For the experiments described in this chapter, we focus on the processing modules. In order to demonstrate the impact of our MATADOR-TCM approach [494], we will assume that either several StrongARM cores are available, each with a different and fixed V_{dd}, or a single StrongARM core with a few discrete V_{dd}s is used. The ranges of the V_{dd}s will vary from 1.2 to 2.4 V, which is motivated by the data sheet information. The power models are also based on these data sheets too as well as instruction counts obtained from profiling [495]. The cycle counts of the thread nodes are also obtained by profiling, based on an ARMulator environment. A clock frequency of 266 MHz is assumed at 2.4 V to compute the execution times.

11.5   TASK CONCURRENCY MANAGEMENT

This approach addresses the dynamic and concurrent task scheduling problem on a multiprocessor platform for real-time embedded systems, in which energy consumption is a major concern. Here we propose a two-phase scheduling method, which is a part of our overall design methodology and can provide the required flexibility at a low run-time overhead. The global methodology is briefly introduced first, followed by the scheduling method. Then the scenario concept is explained. More information can be found in ref. 467 and especially ref. 460. Finally, we explain how a platform simulator is used to support and verify this methodology.

11.5.1   Global TCM Methodology

The TCM methodology comprises three stages (Fig. 11-3). The first is concurrency extraction. In this stage, we extract and explicitly model the potential parallelism and dynamic behavior of the application. An embedded system can be specified at a gray-box abstraction level in a combined MTG-CDFG model [464,496], in which MTG is the acronym for multitask graph and CDFG is the acronym for control/data flow graph. With MTG, the application can be represented as a set of concurrent thread frames[2] (TFs) that exhibit a single thread of control. Each of these TFs consists of many thread nodes (TNs) that can be looked at as a more or less independent section of the code. In the second stage, we apply concurrency improving transformations on the gray-box model. Neither of these two stages is the focus of this chapter. The third stage mainly consists of a two-phase scheduling approach. First, the design-time scheduling step is applied to each of the identified TFs in the system. Different from traditional design-time scheduling, it does not generate a single solution but rather a set of possible solutions, each of which is a combination of different mappings, orderings, and even voltage settings in the case of DVS. They represent different possible cost-performance tradeoff points. Finally, we integrate an application-specific run-time scheduler in the real-time operating system (RTOS) of the application. The run-time scheduler dynamically selects one of these tradeoff points for each running TF and then combines them to find a global energy-efficient solution.

11.5.2   Two-Phase Scheduling Stage

The design of concurrent real-time embedded systems, and embedded softwares in particular, is a difficult problem, which is hard to perform manually due to the complex consumer-producer relationships, the presence of various timing

---

[2] Please refer to ref. 460 for terminologies.

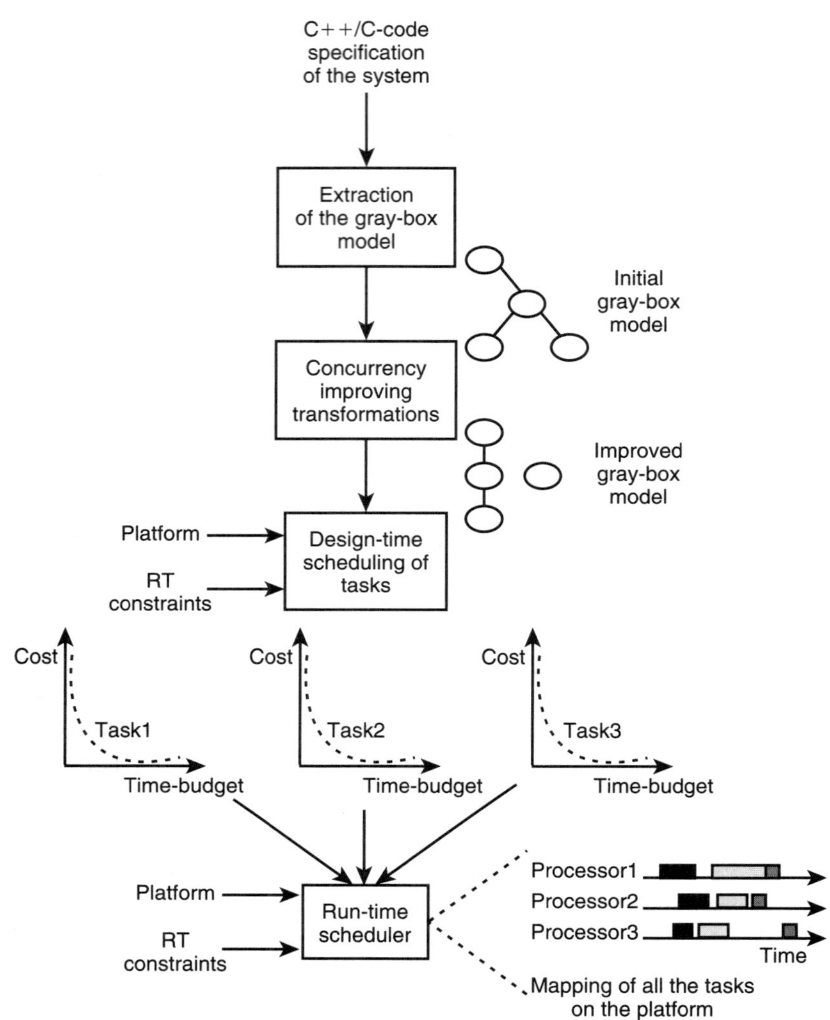

11-3   Task concurrency management.

FIGURE

constraints, the nondeterminism in the specification, and the sometimes tight interaction with the underlying hardware. Here we present a new cost-oriented approach to the problem of concurrent task scheduling on multiple processors.

The design-time scheduling is applied on the TNs inside each TF at compiling time, including a processor assignment decision of the TNs in the case of multiple processing elements. On different types of processors of a heterogeneous

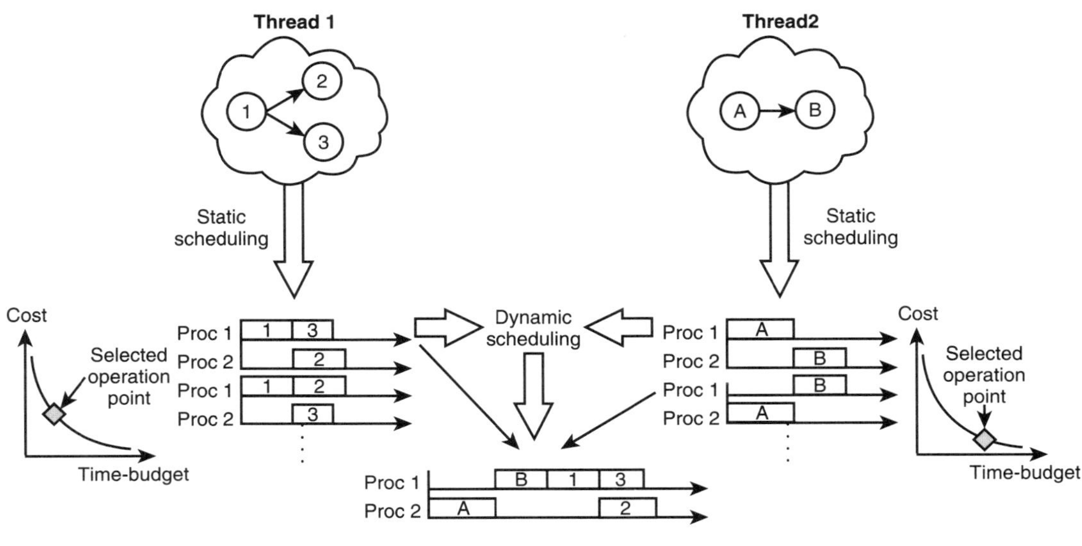

11-4

A two-phase scheduling method.

FIGURE

platform, the same TN will be executed at different speeds and with different costs, i.e., energy consumption in this chapter. These differences provide the possibility of exploring a cost-performance tradeoff at the system level. The idea of our two-phase scheduling is illustrated in Figure 11-4. Given a TF, our design-time scheduler will try to explore different assignment and ordering possibility and generate a Pareto-optimal set [497], in which every point is better than any other one in at least one way, i.e., either it consumes less energy or it executes faster. The Pareto-optimal set is usually represented by a Pareto curve. Since the design-time scheduling is done at compiling time, computation efforts can be paid as much as necessary, provided that better scheduling results and the computation efforts of run-time scheduling are reduced in the later stage. However, if highly data-dependent behavior is present inside the TF, the design-time exploration still has to assume worst-case conditions to guarantee hard real-time requirements. In order to improve the match with the real behavior even more, we have introduced the scenario concept, as discussed in Section 11.5.3.

At run time, the run-time scheduler will then work at the granularity of TFs. Whenever new TFs are initiated, the run-time scheduler will try to schedule them to satisfy their time constraints and minimize the system energy consumption as well. The details inside a TF, like the execution time or data dependency of each TN, can remain invisible to the run-time scheduler, and this reduces its complexity significantly. Only some essential features of the points on the Pareto curve will be passed to the run-time scheduler by the design-time scheduling results and be used to find a reasonable cycle budget distribution for all the running TFs.

In summary, we separate the task scheduling into two phases, namely, design-time and run-time scheduling, for three reasons. First, this gives more run-time flexibility to the whole system. We can indeed accommodate more unforeseen demands for more execution time by any TF, by "stealing" time from other TFs, based on their available Pareto sets. Second, we can minimize energy for a given timing constraint that usually spans several TFs by selecting the right combination of points. This will be illustrated in the realistic application of Section 11.7. Finally, it minimizes the run-time computation complexity. The design-time scheduler works at the gray-box level but still sees quite a lot of information from the global specification. The end result hides all the unnecessary details, and the run-time scheduler can operate mostly on the granularity of TFs, not single TNs. Only when a large amount of slack is available in between the TNs does a run-time local refinement on the TF schedule points result in a further improvement.

11.5.3   Scenarios to Characterize Data-Dependent TFs

In most cases, dynamic behaviors will occur inside a TF because of input data dependencies. For example, in the MPEG-2 video coder, based on the type of image frame, a different coding engine will be used. Three types of frames exist: I (intra) frame, P (predicted) frame, and B (bidirectional predicted) frame. I frames are simply coded as still images, not using any past history, and can be decoded independently; P frames are from the most recently reconstructed I or P frame; and B frames are predicted from the closest two I or P frames, one in the past and one in the future. The computation powers needed for decoding these frames are significantly different, and quite different TNs have to be executed correspondingly. Also, some data-dependent loop boundary (while loop) can exist inside a TF. A TN can be executed, for example, between 10 and 100 times depending on some input data. It is easy to see that one single Pareto curve is not enough to represent the behaviors of these widely different situations, or at least it is not cost-efficient to do that. To capture the data-dependent dynamic behavior inside a TF, we introduce the scenario concept.

In our case, a scenario represents a combination of TF behaviors (for different data-dependent parameter sets) that have closely resembling Pareto curves. It is also extracted based on the profiling data indicating which combinations occur very frequently because it does not make much sense to provide a separate scenario (and Pareto curve) for a set of situations that seldom occur, even if

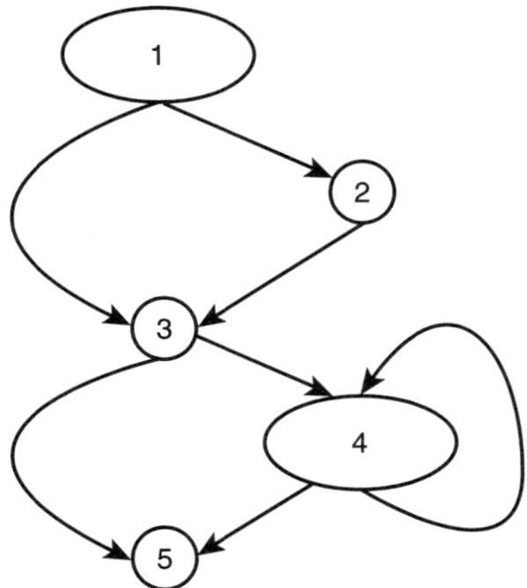

11-5   A simple example of the scenario.

FIGURE

they have a widely different Pareto curve location. It should be clear that by representing an entire set of behaviors with a single Pareto curve we lose some of our optimization potential. This single Pareto curve should indeed reflect the worst-case time behavior of that set to ensure real-time constraints. However, the clustering of the entire set into one scenario allows a nice (controlled) tradeoff between optimization potential and the overhead to store and scan the scenarios at run time.

An example is given in Figure 11-5, in which we have five TNs, 1 to 5, two conditional bypasses (1 to 3 and 3 to 5), and a loop over node 4. Only four execution sequences are possible at the TN level:

1. 1, 3, 5

2. 1, 2, 3, 5

3. 1, 3, 4, 5

4. 1, 2, 3, 4, 5.

If TN 2 needs much less computation power and consumes much less energy compared with TN 1, 3, and 5, the first two execution sequences can be merged and represented by a single Pareto curve. Actually, it is the Pareto curve of sequence 2. If the loop boundary over TN 4 has a large variation, for instance, from 1 to 30, and the execution time of TN 4 is comparable to that of the other nodes, it will be better to split the third execution sequence into two (or even more) scenarios according to its loop range (e.g., 1 to 10 and 11 to 30). Another possibility would be that since TN 2 is executed in execution sequence 4, the loop range over TN 4 would be affected and would be limited to, e.g., 2 to 6. Therefore no split is necessary in that sequence. At the end we obtain only four scenarios: one for the first two execution sequences combined, two for the third sequence split, and one for the last sequence. Note that the number of potentially different data-dependent parameter combinations would be $30 \times 2 + 2 = 62$. Clearly such an explosion is avoided by the use of these scenarios in our approach.

For each scenario, we apply our design-time scheduler to get a separate Pareto curve, which represents the tradeoff characteristics of that TF much more precisely. During the actual execution, the run-time scheduler will check the input data, select the active scenario based on the actual input data, and schedule it with all the other TFs. We have applied this scenario exploration on the QoS application, and the results are shown in Section 11-7.

11.5.4 Platform Simulation Environment

We have integrated the proposed run-time manager on our target platform with the help of a commercial RTOS [498]. In particular, the run-time manager uses the services of this RTOS to distribute the tasks/TFs in the application across the processors and to activate them in the correct order.

A multiprocessor simulator [494] is built on top of the standard ARM simulator, ARMulator. Besides extending it to the multiprocessor context, we also added an energy estimation model similar to that in ref. 495. The last extension is to the profiling and statistics collecting module, which allows us to monitor the execution of multiple code sections and get all the data such as energy consumption, average execution time, and WCET.

The simulation environment is also used to verify the functional correctness of the TCM approach. It also helps to quantify the effective energy reduction and performance gains that can be obtained with the TCM approach, including the overhead of the run-time scheduling phase.

We have applied our TCM approach to part of a real-life application, the QoS kernel of the 3D rendering algorithm that is developed in the context of an MPEG21 project, and we have simulated it on our platform. The application and experimental results are discussed in the following sections.

11.6   3D RENDERING QOS APPLICATION

To test the effectiveness of our approach, a real-life application, the QoS control part of a 3D rendering algorithm developed in the MPEG21 context, is used.

Figure 11-6 shows how 3D decoding/rendering is typically performed: a 2D texture and a 3D mesh are first decoded and then mapped together to give the illusion of a scene with 3D objects. This kind of 3D rendering requires that each frame of the rendering process be recalculated completely. The required computation power depends significantly on its number of triangles. When the available resources are not enough to render the object, instead of letting the system break down (totally stopping the decoding and rendering during a period of time), the corresponding mesh of the object can be gracefully degraded to decrease resource requirement, while maintaining the maximal possible quality.

The number of triangles that are used to describe a mesh can be scaled up or down. This can be achieved by performing edge collapses and vertex splits,

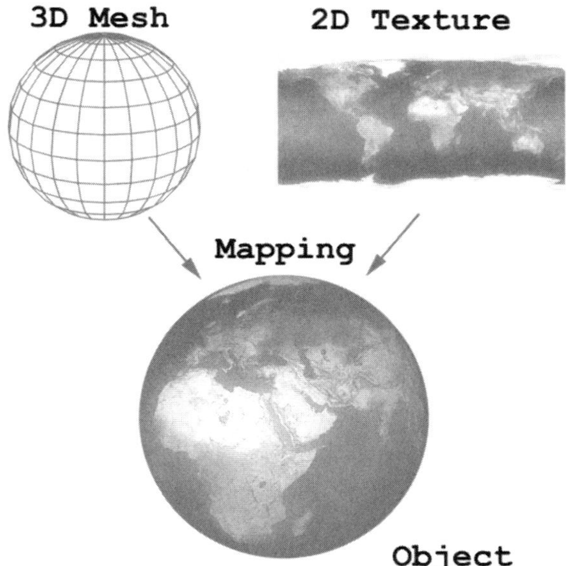

11-6   FIGURE   3D rendering consists of 2D texture and 3D mesh decoding. (© 2002 ACM, Inc. Reprinted by permission.)

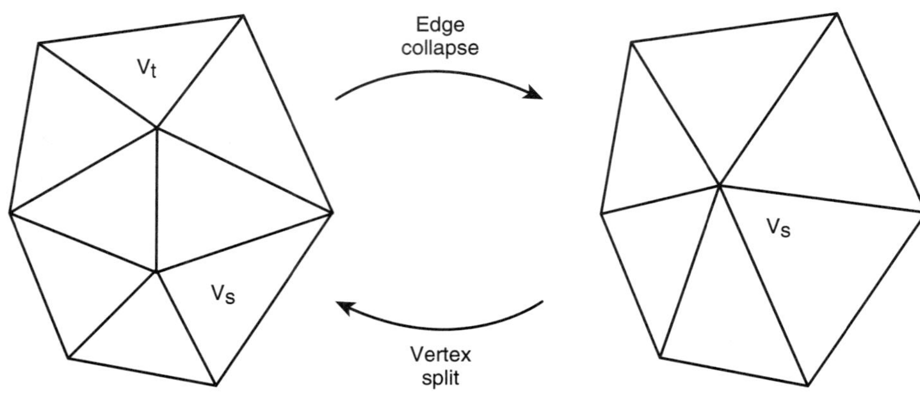

11-7   Edge collapse and vertex split.

FIGURE

respectively, as shown in Figure 11-7. To perform an edge collapse, and thus remove the edge (V_s, V_t), we first remove the triangles that V_s and V_t have in common and replace V_t with V_s in the triangles adjacent to V_t. We then recenter V_s, to keep the appearance of the new set of triangles as close as possible to the former one. The new set of triangles represents the same object with less detail but also with fewer triangles. The same principle but in a reversed direction is used to perform a vertex split. The edge collapse and vertex split approaches can be used repeatedly until the desired numbers of triangles are achieved.

For a 3D object, the more triangles that are used to represent its mesh, the more precise the description of the object. This increases the perceived quality. However, it slows down the geometry and rasterizing stages because more computation power is needed there. Consequently it decreases the number of frames that can be generated per second (FPS), whereas most videos or 3D game applications require a fixed FPS. Another issue that we have to consider here is that the same application can be run on different platforms, e.g., a desktop PC or a PDA, which provide completely different computation ability and power consumption features. Hence different qualities of the same service have to be supplied to achieve a similar FPS. For a given computation platform and a desired FPS, the number of triangles that can be handled in one frame is almost fixed. Based on the number of objects in the current frame and what these objects are, the QoS controller will assign the triangles to each object so that the user can get the best-effort visual quality at a fixed frame rate.

11.7   EXPERIMENTAL RESULTS

For the 3D rendering QoS application introduced in the previous section, we applied our complete design flow, from modeling to scenario selection to scheduling, and then compared the result with a few reference cases, by using our simulation environment.

11.7.1   Gray-Box Model

In the QoS kernel, for each visible object on the scene, a separate TF will be triggered, in which the number of triangles is adjusted to the number specified by the QoS algorithm. The gray-box model of that TF is shown in Figure 11-8, in which all the internal TNs are numbered as well. Table 11-1 gives the profiled execution time and energy consumption of each TN on a 2.4-V StrongARM processor.

11.7.2   Scenario Selection

From the gray-box model, we can see that, based on whether each object is the first time visible, which is true only once during the whole stream for each object,

Thread Node	Ex. Time (μs)	En. Cons. (mJ)
1	35.6	52.2
2	1120.2	1475.3
3	2819.5	4173.4
4	4963.9	8698
5	8037.8	14112.6
6	1619.7	2840.1
7	3.2	4.3
8	761.1	1080.9
9	794.9	1180
10	63.7	113.1
11	432.2	638.1
12	0.2	0.1
13	68.1	120.3
14	705.5	1045.9
15	0.2	0.2

11-1   Execution Time and Energy Consumption of TNs of the Quality Adjustment TF.

TABLE

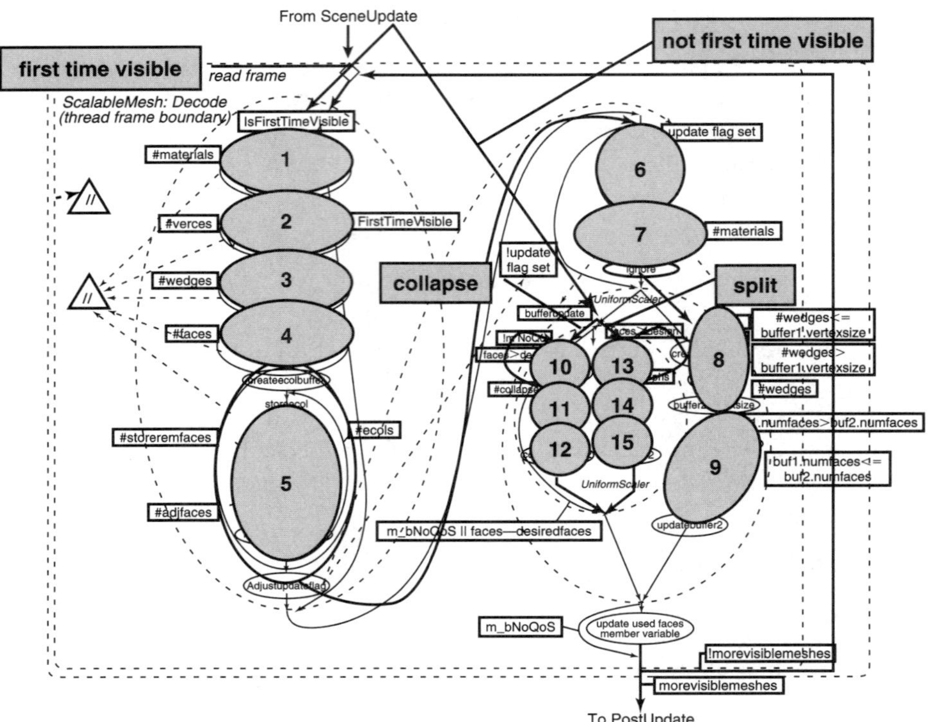

11-8

FIGURE

The gray-box model of quality adjust thread frame. (© 2002 ACM, Inc. Reprinted by permission.)

one of the branches will be taken. If it is the first time visible, the mesh and texture have to be parsed, generated, and bounded (TNs 1 to 9). If it is not, the current number of faces will be compared with the desired number of faces to decide whether to collapse edges (TNs 10, 11, and 12) or to split more vertices (TNs 13, 14, and 15). The edge collapse and vertex split are done in a progressive and iterative way to avoid abrupt changes of the object shape with a while loop over TN 10 or 13. The iteration number of this loop depends on the difference between the current and desired number of faces of that object, and it varies from 2 to 1000 based on the profiling data.

Clearly, one Pareto curve is not enough to represent these highly dynamic features. We have to distinguish first-time-visible or not-first-time-visible and the

	first time visible	collapse or split	range of iteration
scenario 0	yes		
scenario 1	no	collapse	2–12
scenario 2	no	collapse	13–30
scenario 3	no	collapse	31–180
scenario 4	no	collapse	181–1000
scenario 5	no	split	2–4
scenario 6	no	split	5–12
scenario 7	no	split	13–60
scenario 8	no	split	61–1000

11-2   Scenario Selection.

TABLE

while loop iteration numbers. For the latter, if we would not differentiate the iteration number over the loop body, we would have to consider the implementation for the worst case, which is 1000 iterations and much bigger than the average case. To avoid that, we have introduced the concept of "scenario selection" (see Section 11.5.3) in which different Pareto curves are assigned (in an analysis step at design time) to run-time cases that motivate a set of different implementations. Based on this analysis, we have decided to use nine different scenarios and hence also nine Pareto curves in the QoS application to represent the run-time behavior of one object: the first one is when the object is first time visible; the others are when it is not and has to be collapsed or split. For "collapse" and "split," each are assigned four curves with different implementations, corresponding to different iteration subranges. For example, the first curve of "collapse" will be selected if the actual iteration number falls between 2 and 12. Therefore, we only have to consider the worst case of that subrange, which is 12 in this example, not the worst case of the whole range, which is 1000. Extra code has been inserted to enable this. We have selected these ranges based on the profiling data from the application, and they are illustrated in Table 11-2.

11.7.3   Reference Cases for Comparison

In the QoS kernel, which is typical for future object-based multimedia applications, we know at the beginning of each frame the characteristics of its content. In this case we know how many objects we have to render and also (in our approach) the best matched scenario of each object. Each scenario is represented by a Pareto curve computed at design time. From these, the run-time scheduler uses a heuristic algorithm [499] to select an operating point from each curve and activates it with the help of the RTOS.

We have run the application for 1000 scenes and collected the data as a representative experiment.

For comparison reasons, we have also generated a reference case, REF2, to show how well a state-of-the-art DVS scheduler (like the one in ref. 500) can do. We assume it also has full access to the available application parameters at run time, but it does not exploit the scenario selection concept. In other words, it knows the number of objects it is going to schedule in that frame, but it does not exploit the scenarios. Therefore, it has to take an implementation that matches the worst case (TN 13 loops for 1000 times, with an execution time of 68 ms) for each object. However, when one object finishes, the slack time (the difference between the real execution time and the worst case) will be reclaimed and used by the scheduler for the subsequent tasks (i.e., slack stealing [5005] is exploited). Whenever the estimated remaining execution time is smaller than the desired deadline, a continuous DVS method is used to save energy.

Another reference case, REF1, is also generated, in which all code is executed on the highest voltage processor. This is also the outcome if no information of the application is passed to the run-time manager, i.e., neither the number nor the kinds of the objects are known. Since we have a very dynamic application (the number of objects varies from 2 to 20), to handle the worst case and still meet the stringent deadline, the code has to be run completely on the high-voltage processor. Most earlier techniques (see, e.g., ref. 479 for an overview) that use precharacterized task data in terms of WCET times and corresponding energy would lead to the REF1 result for applications with a very dynamic behavior. Only a few would come close to REF2, as currently none of them combines all the ingredients that were used to compose REF2.

11.7.4   Discussion of All Results

The energy consumptions for all cases are shown in Table 11-3, and the numbers of deadline misses are shown in Table 11-4. All results are collected after the application has been run for 1000 sequential scenes and for different FPS requirements. The voltages we have used here are V_{dd} = 1.2 and 2.4 V for the two-voltage case and V_{dd} = 1.2, 1.6, 2.0, and 2.4 V for the four-processor case. Normally, with a higher FPS, to satisfy a more stringent time constraint, parts of the code that are executed at lower voltages have to be moved to a higher voltage, resulting in an increase

	REF1 (1 CPU, 1 V_{dd})	REF2 (1 CPU, 2 V_{dd})	TCM (1 CPU, 2 V_{dd})	TCM (4 CPU, 4 V_{dd})
fps = 5	17.53	14.32	6.21	4.93
fps = 10	17.53	14.65	9.49	6.22

11-3 TABLE Energy Consumption.

	REF1 (1 CPU, 1 V_{dd})	REF2 (1 CPU, 2 V_{dd})	TCM (1 CPU, 2 V_{dd})	TCM (4 CPU, 4 V_{dd})
fps = 5	5	5	5	0
fps = 10	26	26	26	1

11-4 TABLE Deadline Miss.

in energy consumption. Also, the chance that a deadline is missed increases correspondingly. Compared with REF1, for the single processor situation, the TCM approach consumes much less energy (saving 65% when FPS is 5 and 46% when FPS is 10), whereas the deadline miss ratio remains the same. The latter is easy to understand because when the time constraint is really stringent, the TCM method will automatically schedule all TNs to the highest possible voltage processor, which is just what REF1 does. When the time constraint is less tight, a much cheaper solution will be found by the TCM method. Comparing REF2 and REF1, you will find that for this type of dynamic multimedia applications, state-of-the-art DVS cannot gain much because it does not exploit different combinations of TF realizations. This is especially so for a heterogeneous multiprocessor context in which all the prestored implementations of TN schedules and processor assignments can no longer be computed at run time without too much run-time overhead. From the result we can also see that by increasing the number of processors, we can reduce the energy consumption even more while now meeting nearly all the deadlines.

The distribution of the scenario selection is given in Figure 11-9, and Figure 11-10 gives the distribution of the selected Pareto points in scenario 6, both when FPS is 5. From the figures we can see that the TCM scheduler activates different scenarios dynamically and selects the optimal Pareto point from the activated

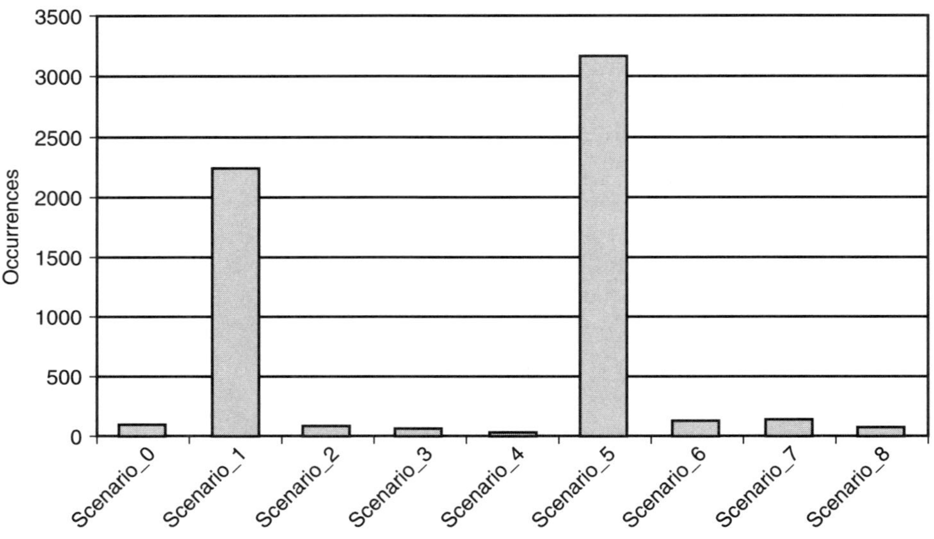

11-9 FIGURE   Distribution of the scenario activations.

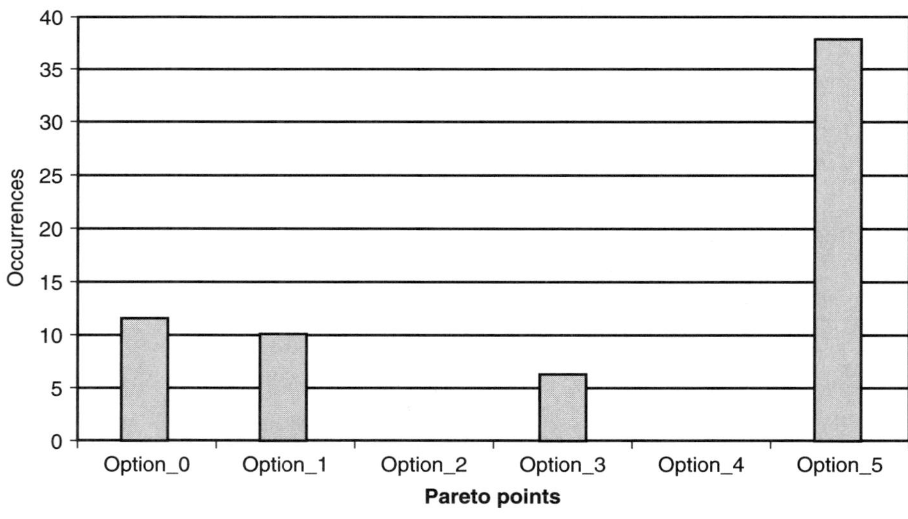

11-10 FIGURE   Distribution of the Pareto points activation in scenario 6.

scenario depending on the run-time situations, i.e., the resource available and the number of competitors. Most of the time, scenarios 1 and 5, which are the least time-consuming, will be selected. Therefore, we avoid the worst-case estimation and have more opportunities to scale down the voltage, compared with REF2. In scenario 6, the least energy-consuming solution, Pareto point 5, is selected as long as it is possible; otherwise a more expensive one is chosen to meet the time constraint. The combination of scenario and Pareto point selection gives us the advantage of heavily exploring the design space at design time and finding the most energy-efficient solution exactly for that situation at run time.

The same methodology has also been applied for several other realistic test applications at IMEC with similar promising results. They are, however, not included in this chapter.

11.8 CONCLUSIONS

In this chapter we have presented a novel approach toward the management of concurrent tasks in dynamic real-time applications. We have discussed a methodology to map the applications in a cost- (especially power-) efficient way onto a heterogeneous embedded multiprocessor platform. This approach is based on a design-time exploration, which results in a set of schedules and assignments for each task, represented by Pareto curves. At run time, a low-complexity scheduler selects an optimal combination of working points, exploiting the dynamic and nondeterministic behavior of the system.

This approach leads to significant power saving compared with state-of-the-art DVS techniques because of three major contributions. First, we effectively combine an intratask detailed design-time exploration and a low-overhead run-time scheduler. Second, our design-time scheduler provides a whole range of energy-time tradeoff points (Pareto curves) instead of a single fixed solution to use at run time. Third, by considering the information provided on run-time applications, we introduce a scenario approach to avoid the use of WCET estimation, while still meeting hard real-time constraints. In the future, we will extend this work to provide automatic tool support for code synthesis and RTOS integration. Also, research is under way on concurrency improving transformations to produce a better gray-box model as a starting point for the scheduler stage.

12 ILP-Based Resource-Aware Compilation

CHAPTER

Jens Palsberg and Mayur Naik

12.1 INTRODUCTION

Compilers are an important part of today's computational infrastructure because software is ever increasingly written in high-level programming languages like C, C++, and Java. Early on, compilers were successful for desktop computing, and now they are also used for mission-critical software, trusted mobile code, and real-time embedded systems. The field of compiler design is driven by the advent of new languages, new run-time systems, new computer architectures, and the need for compilers to be fast, to translate into efficient code, and to be correct. Even the smallest of bugs introduced by a compiler can lead to failures, security breaches, and crashes in an otherwise correct system.

This chapter focuses on compilers for embedded software. For embedded systems, predictability and resource awareness are of greater importance than execution efficiency. For example, if a compiler is not aware of the size of the instruction store for generated code, opportunities to squeeze code into the available space can be lost. A programmer can often hand-optimize generated code, but this is tedious and error-prone and does not scale to large systems. We present a snapshot of the state of the art and the challenges faced by resource-aware compilers. We focus on a technique commonly used in resource-aware compilation, namely, *static analysis* based on *integer linear programming* (ILP).

A static analysis determines facts about a program that are useful for optimizations. One can present a static analysis in a variety of ways, including *dataflow equations*, set constraints, type constraints, or integer constraints. The use of integer constraints is an emerging trend in approaches to register allocation, instruction scheduling, code-size minimization, energy efficiency, and so forth. In general, integer constraints seem to be well suited for building resource-aware compilers. The resources in question can be registers, execution time, code space, energy, or others that are in short supply.

Resource-aware compilation is particularly important for embedded systems, in which economic considerations often lead to the choice of a resource-impoverished processor. Integer constraints, in the form of ILPs, have been used to assist in the compilation of embedded software since the early 1990s. Satisfaction of an ILP is *NP-complete*, but there are many research results and tools concerned with solving ILPs efficiently. Off-the-shelf tools like CPLEX [501] make ILP well suited for use in compilers because they provide analysis engines that need to be written once and for all. In effect, a compiler writer sees a clean separation of "*what* the analysis is supposed to do" from "*how* the analysis is done." Once the analysis is specified, the ILP solving tool does the rest.

An ILP-based analysis is an instance of the more general notion of a constraint-based analysis. The idea of a constraint-based analysis is to do static program analysis in two steps: (1) generate constraints from the program text, and (2) solve the constraints. For ILPs, a constraint is of the form $C_1 V_1 + \ldots + C_n V_n \sim C$ where $C_1, \ldots, C_n, C$ are integers, $V_1, \ldots, V_n$ are integer variables, and $\sim$ is one of $<, >, \leq, \geq$, and $=$. A particularly popular subcase is the one in which all variables range over $\{0,1\}$. Satisfiability of general ILPs and ILPs with just 0-1 variables are both NP-complete [502].

ILPs are quite different from type constraints or set constraints. *Type constraints* use variables that range over types, that is, finite or infinite trees, whereas *set constraints* use variables that range over sets. Integers have less structure than trees and sets but remain highly expressive. For a fair comparison of ILPs and set constraints, consider set constraints of the form $e \subseteq e'$ where the set expressions e, e' can use union, intersection, and complement, and variables that range over a set of constants $\{c_1, \ldots, c_n\}$. Satisfiability of such set constraints is NP-complete, just like satisfiability of ILPs. If X, Y range over subsets of $\{c_1, \ldots, c_n\}$, then the set constraint $X \subseteq Y$ can be translated into n integer linear constraints $x_1 \leq y_1, \ldots, x_n \leq y_n$ where $x_1, \ldots, x_n, y_1, \ldots, y_n$ are 0-1 variables, $x_i = 1$ represents that $c_i \in X$, and $y_i = 1$ represents that $c_i \in Y$. So, are ILPs merely a verbose way of formulating a certain form of set constraints? No! Integers are more convenient for specifying resource bounds, simply because such things as code size, number of registers, voltage levels, deadlines, and so on can be naturally phrased or represented as integers. Resource constraints therefore become integer constraints, rather than type constraints or set constraints.

In the past decade, there has been widespread interest in using ILP for compiler optimizations such as instruction scheduling, software pipelining, data layout, and, particularly, register allocation. Some of the reasons are as follows:

(1) different optimizations like register allocation and instruction scheduling can be combined and solved within the same ILP framework, thereby avoiding the *phase-ordering problem* [503, 504]; (2) ILP guarantees optimality of the solution, as opposed to heuristic methods [505]; and (3) even though ILP is NP-complete, significant advances in the integer and combinatorial optimization field (namely, branch-and-bound heuristics and cutting-plane technology), in conjunction with improvements in computation speed, have permitted its effective application to compiler optimizations [504].

Goodwin and Wilken [506] pioneered the use of ILP for register allocation, in the setting of processors with uniform register architectures. Kong and Wilken [507] use ILP for irregular register architectures like the IA-32. Appel and George [508] partition the register allocation problem for the Pentium into two subproblems: optimal placement of spill code followed by optimal register coalescing; they use ILP for the former. Stoutchinin [509] and Ruttenberg et al. [505] present an ILP framework for integrated register allocation and software pipelining in the MIPS R8000 microprocessor. Liberatore et al. [510] perform local register allocation (LRA) using ILP. Their experiments indicate that their approach is superior to a dynamic programming algorithm that solves LRA exactly but takes exponential time and space, as well as to heuristics that are fast but suboptimal. Sjödin and von Platen [511] model multiple address spaces of a certain processor using ILP, and Avissar et al. [512] use ILP to allocate global and stack data optimally among different heterogeneous memory modules.

12.2   EXAMPLES

In this section we present the formulation of four recent ILP-based analyses in a uniform format for a simple example language. The four analyses are concerned with instruction scheduling, energy efficiency, code-size minimization, and register allocation. Casting the analyses in a common framework makes it easier to understand fundamental similarities and differences. For each of the four analyses, we specify the following items.

Processor: The key resource constraint of the intended processor.

Source program: Information about the source program that we assume is given to the ILP-based analysis, such as *liveness information*. Liveness information expresses, for each program point, which variables will be needed later in the computation.

Target program: The required format of the target program, including special instructions provided by the target language.

Problem: The optimization problem that the ILP-based analysis is intended to solve.

0-1 Variables: All four analyses use only 0-1 variables.

Constraints: We use ILPs. In this proposal, we will use constraints of just four forms:

$$V_1 + \ldots + V_n = 1$$
$$V_1 + \ldots + V_n \leq C$$
$$|V_1 - V_2| \leq V_3$$
$$V_1 + V_2 = V_3 + V_4$$

where $V_1, \ldots, V_n$ are variables that range over $\{0,1\}$, and C is an integer constant. It is easy to see that all such constraints can be encoded using the general format of ILPs.

Objective function: The analysis will try to minimize the value of the objective function.

Code generation: When a solution to the ILP has been found, code can be generated.

Example: We have a running example that is a little program in our example language. For each of the four optimizations, we show the code generated for the example program.

A program in our example language is an instruction list $s_1, \ldots, s_n$. Each instruction is an assignment of the form $x := c$ or $x \mathrel{+}= y$, where x,y range over integer variables, and c ranges over integer constants. Our example language can be viewed as a means for expressing basic blocks in a conventional language like C. In the future, we hope to extend the language.

A program in the example language is specified using sets. **P** is the set of program points, where a point lies between two successive instructions or precedes s_1 or follows s_n. **Pair** $\subseteq (\mathbf{P} \times \mathbf{P})$ is the set of pairs of adjacent points, and **Succ** $= \{(p_1,p_2,p_3) \mid (p_1,p_2) \in \mathbf{Pair} \land (p_2,p_3) \in \mathbf{Pair}\}$. **V** is the set of integer variables used in the program. **U** $\subseteq (\mathbf{P} \times \mathbf{P} \times \mathbf{V})$ is the set of triples (p_1,p_2,x) such that $(p_1,p_2) \in$ **Pair** and the instruction between p_1 and p_2 is $x := c$, and **B** $\subseteq (\mathbf{P} \times \mathbf{P} \times \mathbf{V} \times \mathbf{V})$ is the set of quadruples (p_1,p_2,x,y) such that $(p_1,p_2) \in$ **Pair** and the instruction between p_1 and p_2 is $x \mathrel{+}= y$. Additional sets specifying, for instance, data dependencies or liveness information, will be defined when required.

(a)	(b)	(c)	(d)	(e)	(f)
$u := c_u$	$u := c_u; v := c_v$	**svl 1**	**srp 1**	$u := c_u$	**srp 2**
$v := c_v$	$x := c_x; y := c_y$	$u := c_u$	$r1 := c_u$	$r := c_v$	**svl 1**
$v \mathrel{+}= u$	$v \mathrel{+}= u; y \mathrel{+}= x$	$v := c_v$	$r2 := c_v$	$r \mathrel{+}= u$	$R11 := c_u$
$x := c_x$		$v \mathrel{+}= u$	$r2 \mathrel{+}= r1$	$x := c_x$	$R12 := c_v$
$y := c_y$		**svl 2**	**srp 2**	$r := c_y$	$R12 \mathrel{+}= R11$
$y \mathrel{+}= x$		$x := c_x$	$r1 := c_x$	$r \mathrel{+}= x$	$r1 := c_x$
		$y := c_y$	$r2 := c_y$		**svl 2**
		$y \mathrel{+}= x$	$r2 \mathrel{+}= r1$		$r2 := cy$
					$r2 \mathrel{+}= r1$

TABLE 12-1   Example Program.

For the example program in Table 12-1, $\mathbf{P} = \{1, \ldots, 7\}$, $\mathbf{Pair} = \{(i, i + 1) \mid i \in [1 \ldots 6]\}$, $\mathbf{V} = \{u, v, x, y\}$, $\mathbf{U} = \{(1,2,u),(2,3,v),(4,5,x),(5,6,y)\}$, and $\mathbf{B} = \{(3,4,v,u),(6,7,y,x)\}$.

12.2.1   Instruction Scheduling

This section presents the ILP-based analysis for instruction scheduling by Wilken et al. [513], recast for our example language.

Processor: The processor can schedule R instructions simultaneously.

Source program: The source program is provided in the form of the standard sets and the set $\mathbf{Depn} \subseteq (\mathbf{P} \times \mathbf{P} \times \mathbf{P} \times \mathbf{P} \times \mathbb{N})$ consisting of 5-tuples (p_1, p_2, p_3, p_4, W) such that $(p_1, p_2) \in \mathbf{Pair}$, $(p_3, p_4) \in \mathbf{Pair}$, and the instruction between p_1 and p_2 can begin executing no earlier than W cycles after the instruction between p_3 and p_4 has begun executing.

Target program: The target program is a list of groups of instructions (as opposed to a list of instructions). Each group consists of at most R instructions, all of which are scheduled simultaneously. The schedule length of a target program is the length of the list. We are given (1) a lower bound L on the schedule length, and (2) a target program with schedule length U possibly computed using a heuristic.

Problem: Generate a target program that has the minimum schedule length. The ILP formulation (described below) is parameterized by the schedule length (M). If $U = L$, the schedule is optimal, and the formulation is not instantiated. If $U > L$, the formulation is instantiated with schedule length $U - 1$. If the resulting ILP is infeasible, the schedule of length U is optimal. Otherwise, a schedule of length $U - 1$ was found, and the formulation is instantiated with schedule length $U - 2$. This procedure is repeated until a schedule of minimum length is found.

0-1 Variables: Variable x is defined such that for each $(p_1, p_2) \in$ **Pair** and each cycle $i \in [1 \ldots M]$, $x_{p1,p2,i} = 1$ if the instruction between p_1 and p_2 is scheduled in cycle i.

Constraints: Each instruction must be scheduled in exactly one of the M cycles, and at most R instructions can be scheduled simultaneously.

$$\forall (p_1, p_2) \in \textbf{Pair}. \sum_{i=1}^{M} x_{p_1, p_2, i} = 1$$

$$\forall i \in [1 \ldots M]. \sum_{(p_1, p_2) \in \textbf{Pair}} x_{p_1, p_2, i} \le R$$

For each $(p_1, p_2) \in$ **Pair**, the cycle in which the instruction between p_1 and p_2 is scheduled is given by:

$$\sum_{i=1}^{M} i \, x_{p_1, p_2, i}$$

Using this formula, a constraint is expressed for each $(p_1, p_2, p_3, p_4, W) \in$ **Depn** to ensure that the instruction between p_1 and p_2 is scheduled no earlier than W cycles after the instruction between p_3 and p_4 has been scheduled.

$$\forall (p_1, p_2, p_3, p_4, W) \in \textbf{Depn}. \sum_{i=1}^{M} i \, x_{p_3, p_4, i} + W \le \sum_{i=1}^{M} i \, x_{p_1, p_2, i}$$

Code generation: The target program generated from a solution to the ILP obtained from the above ILP formulation and the source program is a list of groups of instructions, namely, if $(p_1, p_2) \in$ **Pair**, and i is the cycle such that $x_{p1,p2,i} = 1$, then the instruction between p_1 and p_2 in the source program is scheduled in the i'th group in the target program.

Example: Suppose the example program has four dependencies, namely, that $v \mathrel{+}= u$ must begin executing no earlier than one cycle after $u := c_u$ and $v := c_v$ have begun executing, and that $y \mathrel{+}= x$ must begin executing no earlier than one cycle after $x := c_x$ and $y := c_y$ have begun executing, i.e., **Depn** = $\{(3,4,1,2,1), (3,4,2,3,1), (6,7,4,5,1), (6,7,5,6,1)\}$. With $R = 2$, the optimal solution to the ILP obtained from the above ILP formulation and the example program yields a target program with schedule length 3. The target program generated using one such solution is shown in Table 12-1. Note that the program respects the stated dependencies: $v \mathrel{+}= u$ is scheduled two cycles after $u := c_u$ and $v := c_v$, and $y \mathrel{+}= x$ is scheduled one cycle after $x := c_x$ and $y := c_y$.

12.2.2   Energy Efficiency

This section presents the ILP-based analysis for energy efficiency by Saputra et al. [514], recast for our example language.

Processor: The processor has N voltage levels.

Source program: The source program is provided in the form of the standard sets and, for each $(p_1, p_2) \in$ **Pair** and $i \in [1 \ldots N]$, $E_{p_1,p_2,i}$ (resp. $T_{p_1,p_2,i}$) is the energy consumption (resp. execution time) of the instruction between p_1 and p_2 when voltage level i is used. Also, the overall execution time must not exceed a deadline D.

Target program: The target language differs from the source language in that it provides a special instruction **svl** i ("set voltage level to 'i'"), with execution time C_t and energy consumption C_e, to change the voltage level at any program point.

Problem: Generate a target program that consumes minimum energy.

0-1 Variables:

- variable VLVal is defined such that for each $(p_1, p_2) \in$ **Pair** and each $i \in [1 \ldots N]$,

- $\mathrm{VLVal}_{p_1,p_2,i} = 1$ if voltage level i is selected for the instruction between p_1 and p_2.

- variable SetVL is defined such that for each $(p_1, p_2, p_3) \in$ **Succ**, $\mathrm{SetVL}_{p2} = 1$ if the voltage level changes at p_2.

Constraints: Exactly one voltage level should be selected for each instruction.

$$\forall (p_1, p_2) \in \textbf{Pair}. \sum_{i=1}^{N} \mathrm{VLVal}_{p_1,p_2,i} = 1$$

For each $(p_1, p_2, p_3) \in \textbf{Succ}$, if the voltage level at the instruction between p_1 and p_2 differs from that at the instruction between p_2 and p_3, then $\text{SetVL}_{p2} = 1$.

$$\forall (p_1, p_2, p_3) \in \textbf{Succ}. \ \forall i \in [1 \ldots N]. \ |\text{VLVal}_{p_1, p_2, i} - \text{VLVal}_{p_2, p_3, i}| \leq \text{SetVL}_{p_2}$$

The overall execution time must not exceed D.

$$\sum_{(p_1, p_2) \in \textbf{Pair}} \sum_{i=1}^{N} T_{p_1, p_2, i} \text{VLVal}_{p_1, p_2, i} + \sum_{(p_1, p_2, p_3) \in \textbf{Succ}} C_t \, \text{SetVL}_{p_2} + C_t \leq D$$

The term C_t on the *l.h.s.* of the above constraint accounts for the execution time of an **svl** instruction introduced at the start of the target program (see code generation below).

Objective function: Minimize the energy consumption of the target program:

$$\sum_{(p_1, p_2) \in \textbf{Pair}} \sum_{i=1}^{N} E_{p_1, p_2, i} \text{VLVal}_{p_1, p_2, i} + \sum_{(p_1, p_2, p_3) \in \textbf{Succ}} C_e \, \text{SetVL}_{p_2}$$

Code generation: The target program generated from a solution to the ILP obtained from the above ILP formulation and the source program $s_1, \ldots, s_n$ is identical to the source program with the following modifications:

+ if $(p_1, p_2, p_3) \in \textbf{Succ}$, $\text{SetVL}_{p_2} = 1$, and i is the voltage level such that $\text{VLVal}_{p_2, p_3, i} = 1$, then we introduce instruction **svl** i at p_2.

+ if $(p_1, p_2) \in \textbf{Pair}$ where p_1 is the program point before s_1 and i is the voltage level such that $\text{VLVal}_{p_1, p_2, i} = 1$, then we introduce instruction **svl** i at p_1.

Example: With $N = 2$, $D = 38$, $C_e = 7$, $C_t = 2$, and the energy consumption and execution time of the unary and binary instructions at the two voltage levels given below:

Voltage Level:	1	2
instruction in **U**	10	20
instruction in **B**	15	30

12-2   Energy Consumption.

TABLE

Voltage Level:	1	2
instruction in U	6	3
instruction in B	10	5

12-3

TABLE

Execution Time.

The optimal solution to the ILP obtained from the above ILP formulation and the example program yields a target program whose energy consumption is 119 units. The target program generated using one such solution is shown in Table 12-1. Note that the program has an overall execution time of 37 cycles and thereby meets the specified deadline of 38 cycles.

12.2.3   Code-Size Minimization

This section presents the ILP-based analysis for code-size minimization by Naik and Palsberg [515], recast for our example language.

Processor: The processor has a banked register file of N banks and M registers per bank, as well as a register pointer (RP) indicating the "working bank."

Source program: The source program is provided in the form of the standard sets.

Target program: The target language differs from the source language in two respects:

1. It provides a special instruction **srp** b ("set RP to bank b"), occupying 2 bytes, to change the working bank at any program point.

2. It uses registers instead of variables. A variable allotted register d in bank b is denoted rd or Rbd, depending on whether b is the working bank or not, respectively. The target instruction corresponding to $x := c$ occupies 2 or 3 bytes depending on whether x is allotted a register in the working bank or not, respectively, and the target instruction corresponding to x $+= y$ occupies 2 or 3 bytes depending on whether x and y are allotted registers in the working bank or not, respectively.

Problem: Generate a target program that occupies minimum space.

0-1 Variables:

+ Variable r is defined such that for each $v \in \mathbf{V}$ and $b \in [1 \ldots N]$, $r_{v,b} = 1$ if v is allotted a register in bank b.

+ Variable RPVal is defined such that for each $(p_1, p_2) \in \mathbf{Pair}$ and $b \in [1 \ldots N]$, $\text{RPVal}_{p_1,p_2,b} = 1$ if the value of RP at the instruction between p_1 and p_2 is b.

+ Variable SetRP is defined such that for each $(p_1, p_2, p_3) \in \mathbf{Succ}$, $\text{SetRP}_{p_2} = 1$ if the value of RP changes at p_2.

+ Variable UCost_{p_1,p_2} is defined such that for each $(p_1, p_2, v) \in \mathbf{U}$, $\text{UCost}_{p_1,p_2} = 1$ if the value of RP at the instruction between p_1 and p_2 is not the bank in which the register for v is allotted.

+ Variable BCost_{p_1,p_2} is defined such that for each $(p_1, p_2, v_1, v_2) \in \mathbf{B}$, $\text{BCost}_{p_1,p_2} = 1$ if the value of RP at the instruction between p_1 and p_2 is not the bank in which the register for v_1 or v_2 (or both) is allotted.

Constraints: A variable must be stored in exactly one bank, the total number of variables must not exceed the total number of registers, and RP must be set to exactly one bank at every instruction.

$$\forall v \in \mathbf{V}. \sum_{b \in [1 \ldots N]} r_{v,b} = 1$$

$$\forall b \in [1 \ldots N]. \sum_{v \in \mathbf{V}} r_{v,b} \leq M$$

$$\forall (p_1, p_2) \in \mathbf{Pair}. \sum_{b=1}^{N} \text{RPVal}_{p_1,p_2,b} = 1$$

For each $(p_1, p_2, p_3) \in \mathbf{Succ}$, if the value of RP at the instruction between p_1 and p_2 differs from that at the instruction between p_2 and p_3, then $\text{SetRP}_{p_2} = 1$.

$$\forall (p_1, p_2, p_3) \in \mathbf{Succ}. \ \forall b \in [1 \ldots N]. \ \left| \text{RPVal}_{p_1,p_2,b} - \text{RPVal}_{p_2,p_3,b} \right| \leq \text{SetRP}_{p_2}$$

Any instruction in $\mathbf{U}$ occupies 2 or 3 bytes depending on whether the operand is stored in the working bank or not, respectively. The following constraint characterizes the space cost of each such instruction.

$$\forall (p_1, p_2, v) \in \mathbf{U}. \ \forall b \in [1 \dots N]. \quad \left| \mathbf{r}_{v,b} - \mathrm{RPVal}_{p_1,p_2,b} \right| \leq \mathrm{UCost}_{p_1,p_2}$$

Any instruction in $\mathbf{B}$ occupies 2 or 3 bytes depending upon whether both operands are stored in the working bank or not, respectively. The following constraints characterize the space cost of each such instruction.

$$\forall (p_1, p_2, v_1, v_2) \in \mathbf{B}. \ \forall b \in [1 \dots N]. \quad \left| \mathbf{r}_{v_1,b} - \mathrm{RPVal}_{p_1,p_2,b} \right| \leq \mathrm{BCost}_{p_1,p_2}$$

$$\forall (p_1, p_2, v_1, v_2) \in \mathbf{B}. \ \forall b \in [1 \dots N]. \quad \left| \mathbf{r}_{v_2,b} - \mathrm{RPVal}_{p_1,p_2,b} \right| \leq \mathrm{BCost}_{p_1,p_2}$$

Objective function: Minimize the space occupied by the target program:

$$\sum_{(p_1,p_2,v)\in\mathbf{U}} \mathrm{UCost}_{p_1,p_2} + \sum_{(p_1,p_2,v_1,v_2)\in\mathbf{B}} \mathrm{BCost}_{p_1,p_2} + \sum_{(p_1,p_2,p_3)\in\mathbf{Succ}} 2\,\mathrm{SetRP}_{p_2}$$

Code generation: The target program generated from a solution to the ILP obtained from the above ILP formulation and the source program $s_1, \dots, s_n$ is identical to the source program with the following modifications:

+ If $v \in \mathbf{V}$ and b is the bank such that $\mathbf{r}_{v,b} = 1$, then v is allotted a unique register in bank b.

+ If $(p_1, p_2, v) \in \mathbf{U}$ and v is allotted register d in bank b, then the instruction between p_1 and p_2 is denoted $rd := c$ or $Rbd := c$, depending on whether $\mathrm{UCost}_{p1,p2}$ is 0 or 1, respectively.

+ If $(p_1, p_2, v_1, v_2) \in \mathbf{B}$ and v_1, v_2 are allotted registers d_1, d_2 in banks b_1, b_2, respectively, then the instruction between p_1 and p_2 is denoted $rd_1 \mathrel{+}= rd_2$ or $Rb_1 d_1 \mathrel{+}= Rb_2 d_2$, depending on whether BCost_{p_1,p_2} is 0 or 1, respectively.

+ If $(p_1, p_2, p_3) \in \mathbf{Succ}$, $\mathrm{SetRP}_{p_2} = 1$, and b is the bank such that $\mathrm{RPVal}_{p_2,p_3,b} = 1$, then we introduce instruction $\mathbf{srp}\ b$ at p_2.

+ If $(p_1, p_2) \in \mathbf{Pair}$ where p_1 is the program point before s_1 and b is the bank such that $\mathrm{RPVal}_{p_1,p_2,b} = 1$, then we introduce instruction $\mathbf{srp}\ b$ at p_1.

Example: With $N = M = 2$, the optimal solution to the ILP obtained from the above ILP formulation and the example program yields a target program that occupies 16 bytes. The target program generated using one such solution is shown in Table 12-1 (variables u and v are stored in the first and second registers of bank 1, whereas variables x and y are stored in the first and second registers of bank 2).

12.2.4   Register Allocation

This section presents the ILP-based analysis for register allocation of Appel and George [503], recast for our example language.

Processor: The processor consists of K registers.

Source program: The source program is provided in the form of the standard sets and **Live** $\subseteq (\mathbf{P} \times \mathbf{V})$, the set of pairs (p,v) such that variable v is live at program point p, and **Copy** $= \{(p_1,p_2,v) \mid (p_1,p_2) \in \mathbf{Pair} \wedge (p_1,v) \in \mathbf{Live} \wedge (p_2,v) \in \mathbf{Live}\}$.

Target program: The target language differs from the source language in two respects: (1) it uses registers or memory addresses instead of variables: a variable in register d is denoted rd, and a variable in memory is denoted by the variable name itself, and (2) it provides special instructions **ld** rd, x ("load variable from memory address x to register d") and **st** x, rd ("store register d into memory address x"), each occupying 3 bytes. The cost of loading (resp. storing) a variable is C_l (resp. C_s) cycles. The cost of fetching and decoding one instruction byte is C_i cycles. The target instruction corresponding to $x := c$ occupies 1 or 2 bytes depending on whether x is a register or memory address, respectively. The target instruction corresponding to $x \mathrel{+}= y$ occupies 2, 3, or 4 bytes depending on whether both x and y are registers, one is a register and the other is a memory address, or both are memory addresses, respectively.

Problem: Generate a target program that executes in minimum time.

0-1 Variables: Variables r,m,l,s are defined such that for each $(p,v) \in \mathbf{Live}$ we have:

$r_{p,v} = 1$ iff v arrives at p in a register and departs p in a register.

$m_{p,v} = 1$ iff v arrives at p in memory and departs p in memory.

$l_{p,v} = 1$ iff v arrives at p in memory and departs p in a register.

$s_{p,v} = 1$ iff v arrives at p in a register and departs p in memory.

Constraints: For each $(p,v) \in \mathbf{Live}$, v must arrive at p either in a register or in memory and, likewise, depart p either in a register or in memory.

$$\forall (p,v) \in \textbf{Live}.\ r_{p,v} + m_{p,v} + l_{p,v} + s_{p,v} = 1$$

Since registers are scarcer than memory, all the stores will be performed before all the loads at each program point. Therefore, for each $p \in \textbf{P}$, the sum of (1) the variables stored at p (i.e., variables arriving at p in registers and departing p in memory), and (2) the variables arriving at p in registers and departing p in registers must not exceed K.

$$\forall p \in \textbf{P}.\ \sum_{(p,v) \in \textbf{Live}} s_{p,v} + r_{p,v} \le K$$

If variable v is live at successive program points p_1 and p_2, then it either departs p_1 in a register (resp. memory) and arrives at p_2 in a register (resp. memory). If it departs p_1 in a register, then either it must have already been in a register ($r_{p1,v} = 1$), or it must have been loaded at p_1 ($l_{p1,v} = 1$). Likewise, if it arrives at p_2 in a register, then either it must continue in a register ($r_{p2,v} = 1$), or it must be stored at p_1 ($s_{p2,v} = 1$).

$$\forall (p_1, p_2, v) \in \textbf{Copy}.\ l_{p_1,v} + r_{p_1,v} = s_{p_2,v} + r_{p_2,v}$$

Objective function: Minimize the execution time of the target program:

$$\sum_{(p,v) \in \textbf{Live}} ((C_l + 3C_i)l_{p,v} + (C_s + 3C_i)s_{p,v}) + \sum_{(p_1,p_2,v) \in \textbf{U}} (C_s + C_i)(m_{p_2,v} + l_{p_2,v}) +$$
$$\sum_{(p_1,p_2,v_1,v_2) \in \textbf{B}} ((C_l + C_i)(m_{p_1,v_1} + s_{p_1,v_1}) + (C_l + C_s + C_i)(m_{p_1,v_2} + l_{p_1,v_2}))$$

The first component accounts for the cost of each 3-byte load and store instruction that is introduced. The second component accounts for the "overhead" cost of each unary instruction $x := c$, namely, if x is in memory, it accounts for the cost to store x and the cost to fetch and decode an extra byte of the instruction. The third component accounts for the "overhead" cost of each binary instruction $x \mathrel{+}= y$, namely, (1) if y is in memory, it accounts for the cost to load y and the cost to fetch and decode an extra byte of the instruction, and (2) if x is in memory, it accounts for the cost to load and store x and the cost to fetch and decode another extra byte of the instruction.

Code generation: The target program generated from a solution to the ILP obtained from the above ILP formulation and the source program is identical to the source program with the following modifications (performed in that order):

- ✦ if $(p,v) \in$ **Live** and $l_{p,v} = 1$, then v is allotted a unique register (say d) and instruction **ld** rd, v is introduced at p.

- ✦ if $(p,v) \in$ **Live** and $s_{p,v} = 1$, then v is allotted some register (say d) and instruction **st** v, rd is introduced at p.

- ✦ if $(p_1,p_2,v) \in$ **U**, then $(p_1,v) \notin$ **Live** and $(p_2,v) \in$ **Live**, and there are two cases dictating how the instruction between p_1 and p_2 must be denoted:

  if $r_{p_2,v} + s_{p_2,v} = 1$, then v is allotted a unique register (say d) and the instruction is denoted $rd := c$.

  otherwise, $m_{p_2,v} + l_{p_2,v} = 1$, and the instruction is denoted $v := c$.

- ✦ if $(p_1,p_2,v_1,v_2) \in$ **B**, then $(p_1,v_1) \in$ **Live** and $(p_1,v_2) \in$ **Live**, and there are four cases dictating how the instruction between p_1 and p_2 is denoted:

  if $r_{p_1,v_1} + l_{p_1,v_1} = 1$, then v_1 was allotted some register (say d_1) before the instruction.

  if $r_{p_1,v_2} + l_{p_1,v_2} = 1$, then v_2 was allotted some register (say d_2) before the instruction.

  then the instruction is denoted one of $rd_1 \mathrel{+}= rd_2$, $rd_1 \mathrel{+}= v_2$, $v_1 \mathrel{+}= rd_2$, and $v_1 \mathrel{+}= v_2$, depending on whether v_1 and v_2 were both allotted registers, only v_1 was allotted a register, only v_2 was allotted a register, or neither v_1 nor v_2 was allotted a register, respectively.

Example: With $K = 1$, $C_i = C_l = C_s = 2$, **Live** $= \{(2,u),(3,u),(3,v),(5,x),(6,x),(6,y)\}$, and **Copy** $= \{\}$, the optimal solution to the ILP obtained from the above ILP formulation and the example program yields a target program with execution time 16 cycles. The target program generated using one such solution is shown in Table 12-1.

12.3   OPEN PROBLEMS

In this section we discuss three possible research directions. First, how can we design combinations of ILP-based analyses? This would, for instance, allow us to combine the phases of a compiler, thereby solving the *phase-ordering problem.* Second, how can we prove the correctness of ILP-based analyses, including combinations of such analyses? Third, how can we prove equivalence results that clarify the relationships among ILP-based analysis, type-based analysis, and set-based analysis? This would be a step toward a unified framework and a better understanding of when each of the approaches is the best fit for a given problem.

12.3.1 Combination

Performing different ILP-based transformations sequentially on a source program might result in the violation of certain constraints on the resource usage of the target program. For instance, consider the ILP formulations for energy efficiency (Section 12.2.2) and code-size minimization (Section 12.2.3), denoted I_E and I_S, respectively. The I_E-based transformation, when performed in isolation, ensures that the target program will consume the least energy. However, performing the I_S-based transformation following the I_E-based transformation does not ensure this. Likewise, the I_S-based transformation, when performed in isolation, ensures that the target program will occupy the least space. However, performing the I_E-based transformation following the I_S-based transformation does not ensure this.

Precise control over both the energy usage and the space usage of the target program can be achieved by a single transformation based on an ILP formulation that is a combination of I_E and I_S, denoted I_{ES}. The constraints in I_{ES} are those in I_E and I_S, except that constraint (1), which constrains the overall execution time of the target program, must also account for the execution time (say C_t) of each **srp** instruction that is introduced:

$$\sum_{(p_1,p_2)\in\text{Pair}}\sum_{i=1}^{N} T_{p_1,p_2,i}\text{VLVal}_{p_1,p_2,i} + \sum_{(p_1,p_2,p_3)\in\text{Succ}} C_t(\text{SetVL}_{p_2} + \text{SetRP}_{p_2}) + 2C_t \leq D$$

However, there is no one obvious way of dealing with multiple objective functions. Williams [516] suggests that one approach is to take a suitable linear combination of the objective functions. We will illustrate this approach by choosing the objective function of I_{ES} as the sum of the objective functions of I_E and I_S. Let the energy consumption of the **srp** instruction be C_e and the space occupied by the **svl** instruction be 2 bytes. The energy usage E and the space usage S of the target program are:

$$E = \sum_{(p_1,p_2)\in\text{Pair}}\sum_{i=1}^{N} E_{p_1,p_2,i}\text{VLVal}_{p_1,p_2,i} + \sum_{(p_1,p_2,p_3)\in\text{Succ}} C_e(\text{SetVL}_{p_2} + \text{SetRP}_{p_2}) + 2C_e$$

$$S = \sum_{(p_1,p_2,v)\in\text{U}} \text{UCost}_{p_1,p_2} + \sum_{(p_1,p_2,v_1,v_2)\in\text{B}} \text{BCost}_{p_1,p_2} + \sum_{(p_1,p_2,p_3)\in\text{Succ}} 2(\text{SetRP}_{p_2} + \text{SetVL}_{p_2}) + 4$$

Then the objective function of I_{ES} is $E + S$.

The optimal solution to the ILP obtained from I_{ES} and the example program in Table 12-1 yields the target program in Table 12-1, which has space usage 21 bytes and energy usage 116 units.

12.3.2   Correctness

Until now, there have been only a few results on the correctness properties of ILP-based analyses. This is in marked contrast to type-based analyses and set-based analyses, for which many theoretical results are available. Key correctness properties include *soundness*, preservation, and combination, that is: (1) is the analysis sound with respect to formal semantics? (2) is the analysis preserved after program transformations and optimizations? and (3) can analyses be combined in ways that preserve basic properties of the program? Foundational results about the correctness of ILP-based analyses can lead to improved compiler technology, quicker debugging of compilers, fewer bugs and increased confidence in generated code, principles for developing new analyses, increased understanding of how to combine analyses, and ILP-based code certification, in the spirit of proof-carrying code, typed assembly language, and Java bytecode verification.

An approach to correctness can take its starting point in the well-understood foundations of type-based analysis and set-based analysis. First, there are results on soundness, that is, the analysis is sound with respect to formal semantics. Soundness is the key lemma for proving correctness of the optimization that is enabled by the analysis. For type systems, a typical soundness theorem reads:

A well-typed program cannot go wrong.

This is a safety result: Bad things cannot happen. For set-based analyses, a typical soundness theorem reads, informally:

If an expression e *can produce a value* v, *then the set-based analysis of* e *gives a set that approximates* v.

Again, this is a safety result: unexpected things cannot happen. Milner [517] proved the first type soundness theorem, Sestoft [518,519] proved that a set-based control-flow analysis is sound with respect to call-by-name and call-by-value semantics, and Palsberg [520] proved that the same analysis is sound for arbitrary beta-reduction, a proof that was later improved by Wand and Williamson [521]. A well-understood proof technique for type soundness [522,523] uses a lemma that is usually known as type preservation or subject reduction. Intuitively, such lemmas state that if a program type checks and it takes one step of computation using a rewriting rule, then the resulting program type checks. A similar proof technique works for proving soundness for set-based analysis: Palsberg [520] demonstrated that a set-based analysis is preserved by one step of computation. Some soundness results have been automatically checked, e.g., Nipkow and Oheimb's type soundness proof for a substantial subset of Java [524]. Among the correctness results for optimizations enabled by a program analysis is that of Wand and Steckler [525] for set-based closure conversion.

The *preservation* of typability and set-based analysis may go beyond merely steps of computation. In particular, the field of property-preserving compilation is concerned with designing and implementing compilers whereby each phase preserves properties established for the source program. Intuitively, the goal is to achieve results of the form:

If a program has a property, then the compiled program also has that property. Meyer and Wand [526] have shown that typability is preserved by CPS transformation, and Damian and Danvy [527] and Palsberg and Wand [528] have shown that a set-based flow analysis is preserved by CPS transformation. Examples of end-to-end type-preserving compilers with proofs include the TIL compiler of Tarditi et al. [529] and the compiler from System F to typed assembly language of Morrisett et al. [530].

Types and set-based analysis can be *combined* into so-called flow types. The idea is to use types that are annotated with sets. Flow types have been studied by Tang and Jouvelot [531], Heintze [532], Wells et al. [533], and others, who show how to prove soundness for various kinds of flow types. Correctness results for other kinds of annotated-type systems have been presented by Nielson et al. [534]. One can go further and establish the equivalence of type systems and set-based flow analysis [532,535–537]. In these cases, a correctness result for one side can often be transferred easily to the other side. Lerner et al. [538] combined five dataflow analyses into one analysis, thereby overcoming the phase-ordering problem. Their compiler produces results that are at least as precise as iterating the individual analyses while compiling at least five times faster.

It remains to be seen whether the proof techniques used for type-based and set-based analysis can also be applied to ILP-based analysis.

12.3.3 Relationships with Other Approaches

For object-oriented and functional languages, most static analyses are based on a set-based flow analysis. This leads to the suggestion that resource-aware compilation for, say, Java can best be done using a combination of set-based analysis and ILP-based analysis. This is particularly interesting if the two kinds of constraints interact in nontrivial ways. Currently, there is no practical method for combining these; set-based analysis is usually formulated using set constraints and solved using one kind of constraint solver, whereas ILPs have an entirely different kind of constraint solver. Should the constraints be solved using a joint constraint solver or should one kind of constraints be translated to the other? How can correctness be proved for such a combined analysis?

For a typed language, perhaps ILP-based analyses can be phrased as type-based analyses [539]. Focusing on typable programs tends to make correctness proofs easier in the setting of set-based analysis for object-oriented and functional languages; perhaps also for ILP-based analyses?

12.4   CONCLUSIONS

Devices such as medical implants, smart cards, and so on lead to a demand for smaller and smaller computers, so the issues of limited resources will continue to be with us. Writing the software can be done faster and more reliably in high-level languages, particularly with access to a resource-aware compiler. One of the main approaches to building resource awareness into a compiler is to use ILP-based static analysis. We have presented the formulation of four recent ILP-based static analyses in a uniform format for a simple example language. We have also outlined several open problems in ILP-based resource-aware compilation.

ACKNOWLEDGMENTS

This work was supported by Intel and by a National Science Foundation ITR Award number 0112628.

METHODOLOGY AND APPLICATIONS

*T*his section concerns both methodology and applications. These two topics are fairly closely related—the most effective design methodologies are often tied to applications in order to leverage knowledge about the characteristics of the application space. Because MPSoCs make use of many different component technologies—processors, hardwired logic, embedded software, and so on—methodologies are critical to the success of an MPSoC design project. A methodology tells us what to measure, when to measure it, and what to do with the results.

Chapter 13, by Wander O. Cesário and Ahmed A. Jerraya, describes a component-based design methodology. These components may be either hardware or software. The hardware components include both processors and communication. Interfaces are used to abstract the components. Communication refinement is used as the basis for the design methodology. The chapter uses a VDSL design as an example to illustrate the methodology.

Chapter 14, by Santanu Dutta, Jens Rennert, Tiehan Lv, Jiang Xu, Shengqi Yang, and Wayne Wolf, looks at video as an application domain. It introduces some real-time video algorithms to illustrate the types of capabilities required from a video MPSoC. It uses the Philips Nexperia™ processor as a case study in MPSoC design. It also briefly considers the design of MPSoCs for smart cameras that perform real-time analysis on video data.

The book closes with two final chapters on methodology and models of computation. Chapter 15, by JoAnn M. Paul and Donald E. Thomas, describes system modeling and its relationship to models of computation. The chapter compares several different models of computation and evaluates their usefulness at various stages in system design. It then describes the MESH environment for hardware and software modeling.

Chapter 16, by Felice Balarin, Harry Hsieh, Luciano Lavagno, Claudio Passerone, Alessandro Pinto, Alberto Sangiovanni-Vincentelli, Yosinori Watanabe, and Guang Yang, describes the Metropolis design environment. Metropolis captures heterogeneous system designs in a unified semantic model. The design can be successively refined for both functional and nonfunctional properties. The chapter uses a multimedia design as an example of the use of Metropolis.

Component-Based Design for Multiprocessor Systems-on-Chips

Wander O. Cesário and Ahmed Amine Jerraya

13.1 FROM ASIC TO SYSTEM AND NETWORK ON CHIP

The ITRS roadmap [540] predicted that by 2000 most application-specific integrated circuits (ASICs) would include at least one embedded instruction-set processor, which meant that most ASICs would be systems-on-chips (SoCs). Some existing designs confirm and strengthen this prediction; many applications include several processors with different instruction sets: mobile terminals (e.g., GSM [541]), set-top boxes (e.g., PNX8500 from Philips [542]), game processors (e.g., PlayStation2 from Sony [543]), and network processors [544]. Most system and semiconductor houses are developing platforms involving several cores (microcontroller unit [MCU], digital signal processor [DSP], Internet protocol [IP], and so on) and sophisticated communication networks (hierarchical bus, time-division multiple access [TDMA]-based bus, point-to-point connections, and packet-routing switches) on a single chip.

The trend is now toward interconnecting standard components, as was performed for boards a few years ago. This evolution is creating several breaking points in the design process and new challenges for the electronic design automation (EDA) industry:

1. Systems are becoming more and more complex: communication cannot be specified manually at the register-transfer level (RTL), as before, without introducing many errors. At this level, bus structures (data, address, and control) and clock cycles need to be detailed to verify logical and electrical constraints. However, this level is not well adapted for checking communication protocols because this would be too time-consuming and not necessary

(verify at the most appropriate level). For these reasons, higher abstraction levels, to model and verify the interconnection between components, are required to master the complexity, i.e., reduce design time and errors.

2. SoCs will include many processors with different instruction sets to execute dedicated functions in order to increase the flexibility of the whole system. The software will become more complex than the hardware and in the near future will require several hundreds or thousands person-years [540]. This software will not be programmed at the assembler level as it is today. SoC can have different types of processors and any arbitrary topology of interconnection of processors. Thus, programming for SoC architectures is more complicated than conventional parallel programming. To cope with the difficulty, it is necessary to use efficient models and methods for programming SoCs.

3. Complex, on-chip, hardware/software interfaces are required to implement an application-specific communication network. The hardware (microprocessor interfaces, bank of registers, memories) and software (drivers, operating systems [OS]) elements required to perform the communication protocols need to be adapted to the communication network according to the type of core.

This chapter presents a systematic approach toward assembling components to build application-specific SoCs, as well as the definition of a composition model to abstract the interconnection between hardware (cores) and software (tasks) components. Section 13.2 introduces the basic concepts for complex SoC design and the component-based approach. Section 13.3 details a new component-based specification model, and Section 13.4 presents the design flow that allows multiprocessor SoC (MPSoC) design at a higher level than that of RTL. Section 13.5 presents the application of this flow onto the design of a very high data rate digital subscriber line (VDSL) circuit and the analysis of the results. Finally, Section 13.6 presents our conclusions.

13.1.1   Applications for MPSoC

The expression SoC is used to designate an ASIC that combines heterogeneous components (central processing units [CPUs], DSPs, IPs, memories, busses, and so on) on the same chip. This is similar to what we used to see on system boards. An SoC is application-specific and imposes tight design and performance constraints. The designed architecture should be tuned, specialized, and optimized to a specific application or a highly restricted set of related applications. An SoC gen-

Components/application	Data processing	Control	On-chip communication	Typical designs
Telecom terminals and GSM	1 DSP	1 MCU	Bridge (hierarchical bus)	STEP 1 VDSL [577]
Multimedia	Few DSPs	1 MCU	Network switch crossbar	TRIMEDIA [578]
Network processors	Many DSPs	Few MCUs	On-chip network	IXPIZDE [544]
Game processors	Few DSPs	Few MCUs	On-chip hierarchical network	PlayStation2 [543]

13-1 State-of-the-Art Multiprocessor SoC Products.

TABLE

erally also imposes tight time-to-market constraints. Thus, the architecture design cycle should be limited.

Table 13-1 shows examples of state-of-the-art SoC products. These SoCs may include several embedded processors. As shown in the table, application-specific operations such as data processing or control are performed by DSPs, MCUs, or CPUs. To meet the required communication bandwidth, several types of communication networks such as hierarchical busses, crossbar networks, and others are used on the same chip. All these designs require more than a megabyte of on-chip memory, and, in addition to the use of standard components (memory, processors), they all include application-specific hardware that may amount to several million gates.

13.2   BASICS FOR MPSoC DESIGN

The classic literature on multiprocessor systems provides all the basic concepts used in this chapter [545, 546], but it is too general to be efficient for SoCs. Figure 13-1 shows a typical multiprocessor SoC architecture with heterogeneous processors and the on-chip communication network. A key difference from classic computer architecture is that, based on their utilization, this model distinguishes two kinds of processors (CPUs): those used to run the end application and those dedicated to the execution of specific functions that could have been designed in hardware. The programming and interfacing of these two kinds of processors are quite different, as we will explain later.

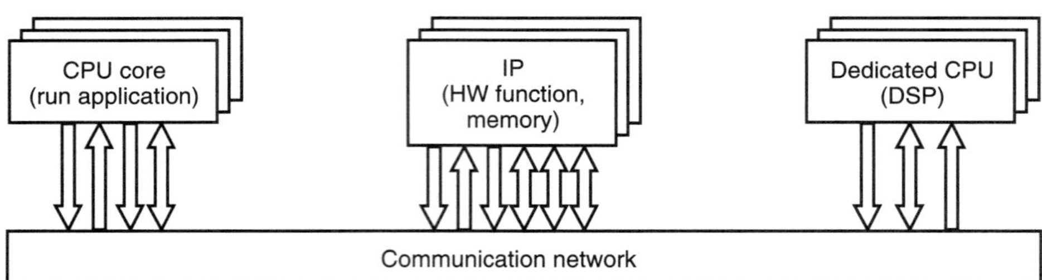

13-1   A typical multiprocessor SoC architecture.

FIGURE

This architecture is quite different from that of early SoCs, in which a single processor played the role of master in communication over a shared bus, and the other components (e.g., ASICs, memory, peripherals, and so on) played the role of slaves. In early SoCs, most system design efforts focused on the design of the software running on the master processor and on communication optimization over the shared bus. In future SoC design, compared with conventional embedded systems design, the implementation of system communication becomes much more complicated, since (1) heterogeneous processors are involved in communication, and (2) complex communication protocols and networks are used:

1. Single-processor architectures are made of a single CPU generally acting as a master and may include several slave peripherals. Multiprocessor architectures allow integration of different types of CPUs. A key difference is that when we have several masters we generally need synchronization that is more sophisticated.

2. In single-processor architectures, the communication structure is generally based on master/slave shared-bus interconnections. In multiprocessor architectures, it often happens that several complex system busses are used.

3. In single-processor architectures, generally the processor controls the communication protocols. This represents a heavy load for the CPU. In multiprocessor architectures, specialized hardware is often used to free processors from executing communication code [547]. This also allows us to separate computation and communication.

4. With single-processor architectures, communication is generally performed through memory-mapped input/output (I/O). In multiprocessor architectures, we need a communication controller to allow the use of high-level

primitives (e.g., first-in, first-out [FIFO], broadcasting, and so on) that may be executed by hardware in parallel with the computation executed on the processor.

Additionally, dedicated processors require application-specific communications and memory schemes due to performance optimization reasons. These specific architecture optimizations are generally related to the application domain. For instance, application-specific interfaces are required to respect tight resource/performance constraints (e.g., area, power, run time, and so on).

13.2.1   MPSoC Software Architectures

Processors used to run the end-user application and those dedicated to execute specific functions employ a different software organization. The application software is generally organized as a stack of layers on top of the hardware, as shown in Figure 13-2a.

The lowest software layer provides the drivers and the low-level architecture controllers. This is also called the software support package and is generally provided by the SoC design team. The upper layers, OS services and the SW application layer, are provided by the application designer. The OS layer generally reuses an existing embedded OS, which may be customized to the application. The top layer is application-specific software that may be designed using high-level languages and code generation techniques.

Presently, the software running on the dedicated processors is generally written in Assembly or the C programming language. When multitasking and

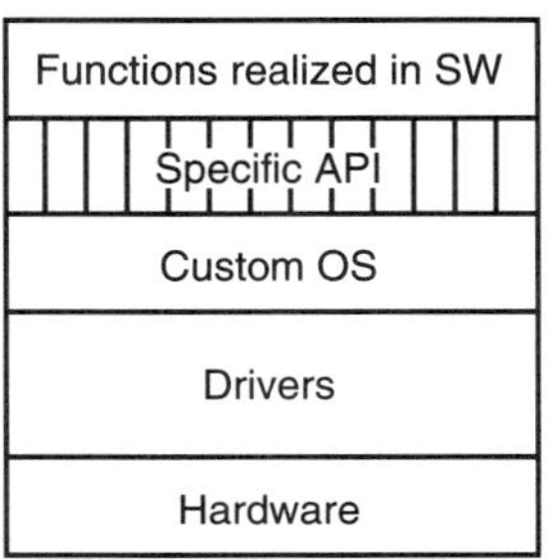

13-2   Software organization.

FIGURE

complex I/O are required, specific code is provided to perform scheduling and to manage I/O. This code is generally mixed with the functional code, and there is no separated OS layer. This low-level programming is one of the main bottlenecks in SoC design. Additionally, this scheme lacks flexibility/portability and makes it difficult to change the hardware part without a complete redesign.

In this chapter, we consider a higher level programming approach for dedicated software. Each dedicated CPU will use a software stack (Fig. 13-2b). Of course, for code size and performance reasons, it will not be realistic to use any existing generic OS as an isolation layer. In this case, a *custom OS* supporting an application-specific program interface (API) is required. Custom OS design automation is a new research area brought on by SoC design [548]. With this scheme, both dedicated and application software can be written independently from the hardware implementation. Another benefit of this approach is that dedicated software can be independent from hardware/software interfaces on architecture components and OS choices. Finally, the main advantage is to allow this level of flexibility/portability without any loss of performance.

13.2.2   MPSoC Design Methods

Currently, most SoC designs start at the RTL for the hardware interfaces and at the instruction-set architecture (ISA) level for the dedicated software components. Validation of the overall design uses co-simulation techniques that generally combine one hardware simulator and one or several instruction-set simulators (ISS). As explained by Keutzer [549], two design automation approaches are competing to improve productivity: top-down and bottom-up approaches. The most popular top-down approaches are synthesis from system level models [550–552] and platform-based design [553–555]. These approaches start with an architectural solution, target architecture, or architectural platform. In contrast, the bottom-up approach, also called *component-based SoC design*, starts with a set of components and provides a set of primitives to build application-specific architectures and communication APIs [556]. The key idea of component-based design is to increase the abstraction level when designing component interconnections. Even if this approach does not provide much help on automating the architecture exploration phase, it may provide a considerable reduction of design time for hardware/ software communication refinement and component integration and may facilitate IP reuse.

Figure 13-3 shows a full design flow from a system-level specification to the RTL architecture. The design starts with an informal model of the application, which is the contract between the end-customer and the system designer

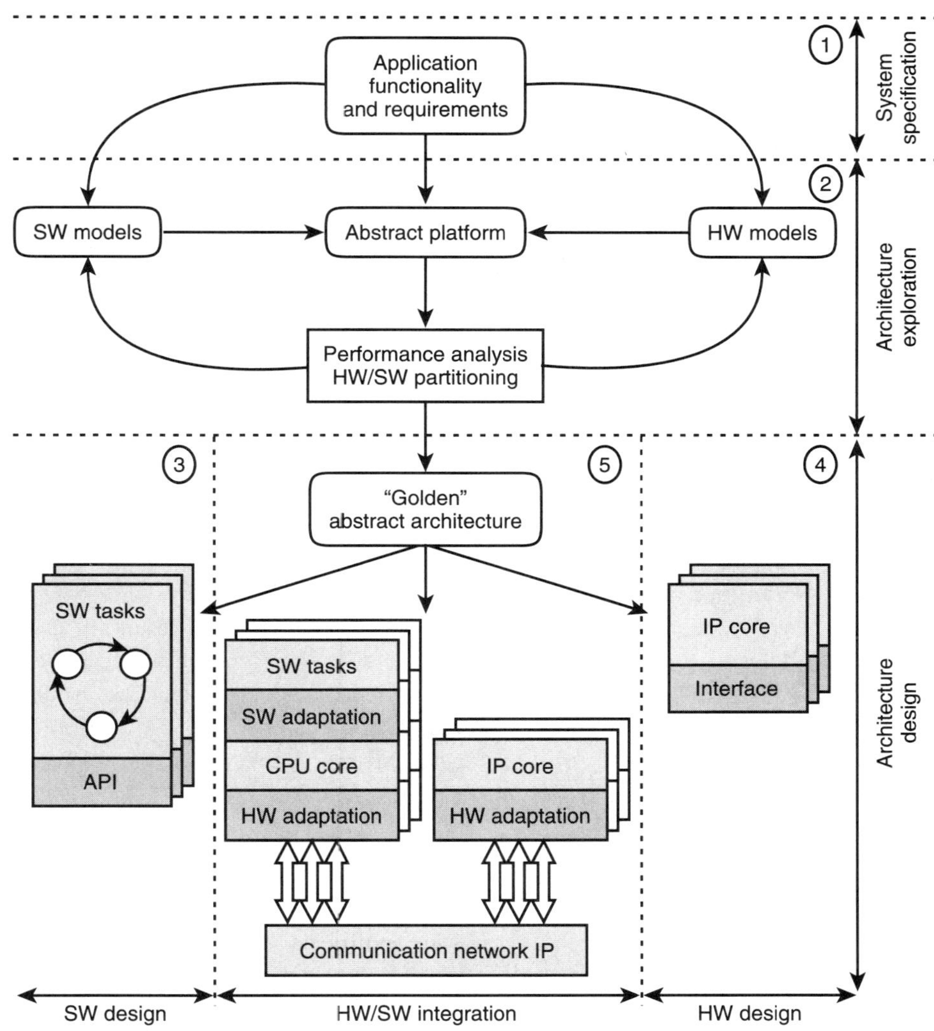

13-3   System-level design flow for SoC.

FIGURE

(Fig. 13-3, 1). System designers then build a more formal SoC specification that
the end-customer can validate. Then system designers fix the system architecture
and decide about the hardware/software partitioning for the application functionality (Fig. 13-3, 2). This step generally uses an executable specification model
that allows designers to go through a performance analysis loop. In order to
achieve an executable specification model at this stage, system designers build a

simulation model using software profiles for parts of the application and abstract models for the hardware components.

System architecture exploration fixes the specification of the hardware components (e.g., selection of existing processors or specific hardware) and the global structure of the on-chip communication network. This step produces a golden architecture model that guides the next step: SoC architecture design. SoC designers need to interconnect the hardware and software components while respecting the performance constraints described in the golden architecture model (Fig. 13-3, 5). In parallel, hardware (Fig. 13-3, 4) and software (Fig. 13-3, 3) designers implement those components in conformance with the golden architecture model. This overall flow is quite complex, and to our knowledge very few efforts cover both system-architecture exploration and system-architecture design.

In this chapter, we present a design methodology aimed mainly at automation of the work of the SoC designer for multiprocessor platforms. The key point in such a flow is the use of an abstract architecture whereby communication is separated from the component on the hardware side and from the functions on the software side. This abstract architecture may be used by the software programmer as an API. On the other hand, the abstract architecture may use hardware components through an abstract API. This is generally called separation between communication and computation for component-based design [554]. With this scheme, SoC design consists of adapting each of the hardware and software components to the rest of the system. The key technology will be the generation of customized hardware/software (HW/SW) interfaces to adapt components to the communication network.

13.2.3   Component Interface Abstraction

In order to separate component and SoC design, we need software and hardware adaptation layers to isolate both software and hardware components from the communication network. Figure 13-4 presents a decomposition of these adaptation layers that can be used to implement HW/SW interfaces. With this scheme, the software design team uses APIs for both application and dedicated software development. The hardware design team uses abstract interfaces provided by CPU wrappers. The HW/SW integration design team can concentrate on implementing hardware and software interface layers for the selected communication network. Designing these HW/SW interface layers represents a major effort, and design tools are lacking. Established EDA tools are not well adapted to this new MPSoC design scenario, and consequently the challenges listed in Section 13.2.1 need to be addressed.

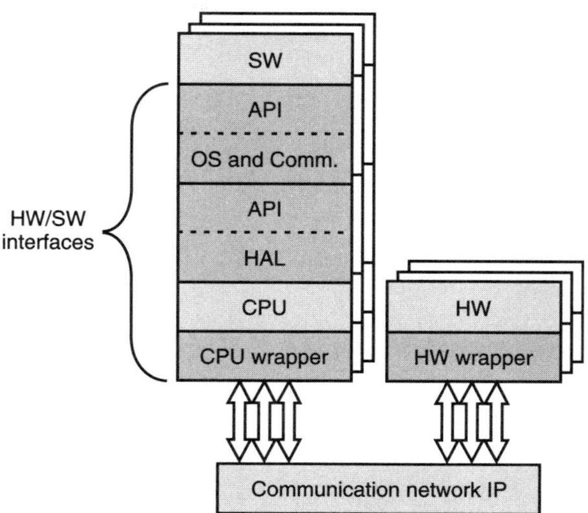

13-4   HW/SW interface layering for MPSoC.

FIGURE

The key issue in success with component-based SoC design is the automatic generation of customized HW/SW interfaces from abstract models. The next subsection introduces the component-based design approach, and Section 13.3 details different abstraction levels for HW/SW interfaces.

13.2.4   Component-Based Approach

In component-based SoC design methodologies [557], the goal is to allow the integration of heterogeneous processors and communication protocols by using abstract interconnections. Behavior and communication must be separated in the system specification. Thus, system communication can be described at a higher level and refined independently of the behavioral system.

There are two component-based design approaches: use of a standard bus protocol and use of a standard component protocol. IBM defined a bus architecture called *CoreConnect* [558] that adopts the first approach. To connect heterogeneous components to the bus architecture, a wrapper is designed to adapt the protocol of each component to *CoreConnect* protocols. The Virtual Socket Interface Alliance (VSIA) promotes the second approach, with virtual component interface (VCI) and functional interface (FI) as standard component protocols [559]. In this case, the designer can choose a bus protocol and then design wrappers to interconnect using this protocol. Leijten et al. [560] also present methods for on-chip network communication design in the case of multiprocessor SoCs.

Recently, some commercial tools began to use component-based design concepts: N2C from CoWare [561], Silicon Backplane μNetwork from Sonics [562], and virtual component co-design (VCC) from Cadence [563]. Still, designers are not able to obtain significant reduction in system design cycle and optimized multiprocessor architectures. This is mainly due to the nonseparation of functional and application software and to the nonavailability of custom *software wrapper* generation tools. Other limitations, given below, appear when the target platform is composed of heterogeneous processors and communication protocols.

Hardware-Oriented IP Integration Assuming On-Chip Bus Usage

In most existing methods for component-based design (standard bus/component protocol), wrapper design is required to connect each component to an on-chip bus. In current methods, the wrapper is implemented to achieve one-to-one protocol conversion between the protocol of the component and that of the on-chip bus considered. However, in practical systems, components can have multiple communication protocols and can be connected with multiple components via multiple communication networks. Thus, the wrapper should play many-to-many protocol conversions (between the protocol[s] of the component and that (or those) of the communication network[s]).

Support of Mixed Hardware/Software Wrappers Is Not Handled

Another problem in current wrapper design is that the wrapper is considered only as a hardware interface. In the case of a processor connected to a communication network, the functionality of the wrapper can be implemented in the form of software, hardware, or mixed HW/SW to achieve more optimal wrapper design in terms of system run time, power consumption, area cost, and so on. More generally, such a HW/SW tradeoff can include the implementation of a specific partitioning of the protocol.

Automatic Architecture Generation Is Not Considered

In current component-based approaches, system integration, i.e., wrapper design and connection of components and communication networks with wrappers, is performed manually using hardware description language (HDL) at the RTL. Thus, the system integration process is time-consuming and error-prone. Such a manual process prevents the designer from trying other choices of system design, i.e., extensive design space exploration cannot be done. To allow design space exploration in SoC design, automatic generation of wrappers and automatic interconnection of components, wrappers, and communication networks starting from high-level models are required. In the industry and in the literature, some automated component interface synthesis methodologies are proposed. In Synopsys Protocol Compiler [564], PIG [565], POLARIS [566], and ProGram [567], interfaces are created from an interface specification using some kind of formal model. Usually, the specification is converted into finite state machines (FSMs) that implement the formal model. The UIC [568] tool generates a customized bus for the data transfer. Coral [569] connects *CoreConnect* compliant IPs through predefined busses. However, with these approaches, the interface is generally limited to a single communication protocol. In library-based approaches, wrappers are created by composing predefined library cells. However, it is hard to cover all possible protocols without exploding the size of the library. In Vercauteren et al. [570], components are connected using a handshake protocol. The COSY [571] system uses a layered communication model for HW/SW communication refinement.

System Validation Method Unable to Achieve True Modular Design

Component-based design allows modular design whereby a subsystem can be refined from a high abstraction level to a lower level, independent of the other parts of the system. After subsystem refinement, the whole system consists of models at different abstraction levels (e.g., models of the refined subsystem at a low level and models of the other parts of the system at a high level). To validate the intermediate implementation of the whole system, mixed-abstraction-level (in short, mixed-level) co-simulation is required. However, current component-based design methods have limitations in supporting mixed-level co-simulation in the case of heterogeneous multiprocessor SoCs. Several HDLs such as SystemC [572] and SpecC [552] have emerged as system-level design languages. However, they allow the software application to be described only at a functional level. There is no native support to model HW/SW interfaces including features of the OS (e.g., scheduling policy) and the *hardware abstraction layer* (HAL, e.g., interrupt service routines, device drivers).

13.3   DESIGN MODELS FOR COMPONENT ABSTRACTION

This section introduces ROSES, a system-level methodology for multiprocessor SoC design starting from a *virtual architecture* and using a communication refinement approach. The system is described as a virtual architecture made of a set of *virtual components* interconnected via communication channels. A virtual component consists of a *wrapper* and a component (or module). The component corresponds to software tasks or a hardware function. The wrapper adapts accesses from the component to the channels connected to the virtual component. The component and the channel(s) can be different in terms of (1) communication protocol, (2) abstraction level, and (3) specification language. Depending on the difference, the functionality/structure of the wrapper is determined and automatically generated.

13.3.1 Conceptual Design Flow

The ROSES design flow starts with a virtual architecture model that corresponds to the golden architecture in Figure 13-3 and allows automatic generation of CPU wrappers, device drivers, OS, and APIs.

The design flow starts with a virtual architecture model that captures the global organization of the system into modules and architecture configuration parameters (as indicated in Fig. 13-5a). This input model may be manually coded or may be generated automatically by specification analysis tools (e.g., refs. 553 and 573).

The goal is to produce a synthesizable RTL model of the MPSoC that is composed of processor cores, IP cores, the communication network IP, and HW/SW wrappers. The wrappers are automatically generated from the interfaces of virtual components (as indicated by the arrows and dark gray boxes in Fig. 13-5). Software written for the virtual architecture specification runs without modification on the implementation because the same APIs are provided by the generated custom OS.

13.3.2 Virtual Architecture Model

The virtual architecture [574], referred to as a macroarchitecture, represents a system as an abstract netlist of virtual components (Fig. 13-5a). Each virtual component has a wrapper that is composed of a set of virtual ports. A virtual port has internal and external ports. There could be an n to m (n and m are natural numbers) correspondence between the internal and the external ports. Internal and external ports are not directly connected because they can use different communication protocols and/or transmit signals at different abstraction levels. For this reason, in some cases the wrapper will do a conversion between different internal/external communication protocols and/or abstraction levels.

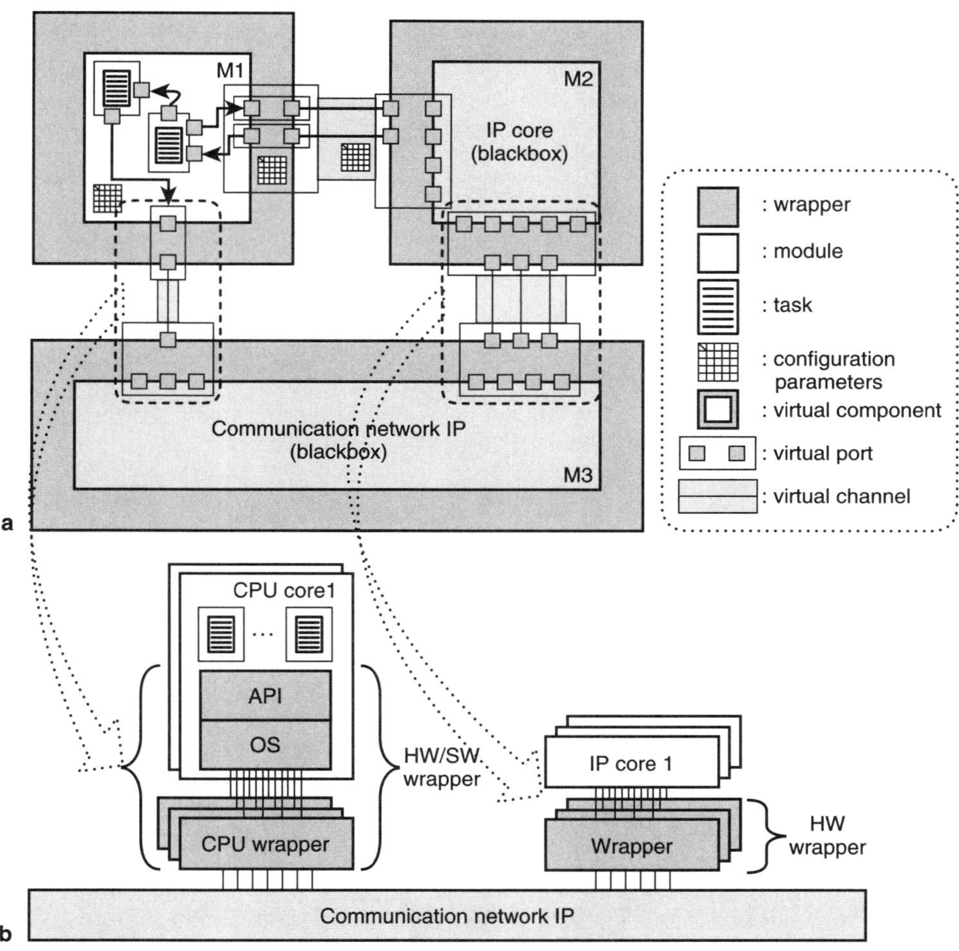

13-5   ROSES design models for MPSoC. (a) Virtual architecture, (b) Target platform.

FIGURE

Channels hide many details of communication protocols, for instance, FIFO communication is realized using high-level communication primitives (e.g., *put* and *get*). In our design flow, the virtual architecture is described using extensions layered on top of SystemC. SystemC [572] is a C++ library that provides hardware modeling concepts for simulation: time, events, netlist, and concurrency. Three new concepts are used:

1. Virtual module: consists of a module and its wrapper.

2. Virtual port: groups the corresponding internal and external ports having a conversion relationship. A wrapper may be composed of a hierarchy of several virtual ports.

3. Virtual channel: grouping several channels having a logical relationship (e.g., multiple nets belonging to the same communication protocol).

For system refinement, this model needs to be annotated with architecture configuration parameters (e.g., protocol and physical addresses of ports). This model is not directly synthesizable/executable because behaviors of wrappers are not described. The main goal of this design methodology is to automate the generation of these wrappers, in order to produce a detailed architecture that can be both synthesized and simulated.

13.3.3   Target Architecture Model

We use a generic MPSoC architecture whereby processors are connected to a communication network IP via wrappers (Fig. 13-5b). In fact, processors are separated from the physical communication IP by wrappers that act as communication controllers or bridges. Such a separation is necessary to free the processors from communication management, and it allows parallel execution of computation tasks and communication protocols. Software tasks need also to be isolated from hardware through an OS that plays the role of SW wrapper. When defining this model, our goal was to have a generic model in which both computation and communication may be customized to fit the specific needs of the application. For computation, we may change the number and kind of components, and, for communication, we can select a specific communication network IP and protocols. This architecture model is suitable to a wide domain of applications; more details can be found in Lyonnard et al. [575].

13.3.4   The Hardware/Software Wrapper Concept

The wrapper is made of a software part and a hardware part, as shown in Figure 13-6. On the hardware side (Fig. 13-6b), the internal architecture of the wrapper consists of a processor adapter (PA), channel adapters (CAs), and an internal bus. The number of channel adapters depends on the number of channels that are connected to the corresponding virtual module. On the software side (Fig. 13-6a),

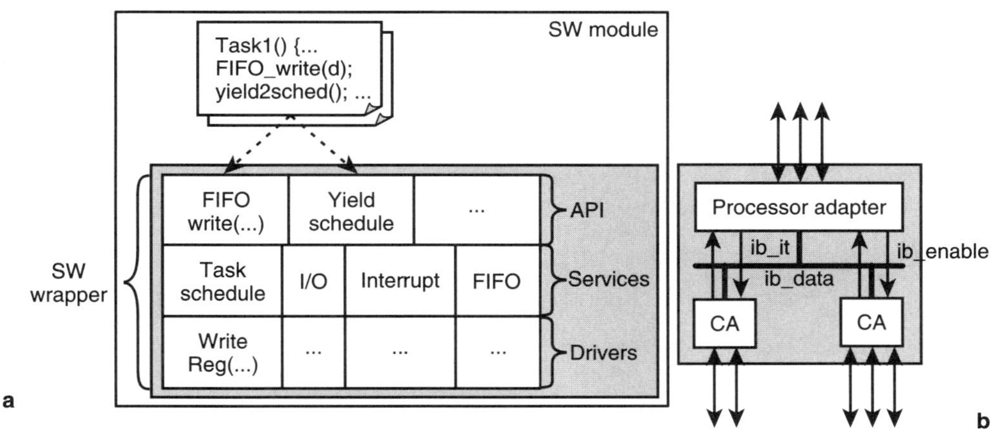

13-6   Software (a) and hardware (b) wrapper architecture.

FIGURE

wrappers provide the implementation of high-level communication primitives (API) used in the software module and the drivers to control the hardware. If required, the software wrapper will also provide sophisticated OS services such as task scheduling and interrupt management.

13.4   COMPONENT-BASED DESIGN ENVIRONMENT

Figure 13-7 shows an overall view of the ROSES component-based design environment. The initial SoC model is a virtual architecture annotated with configuration parameters. There are three tools for automatic wrapper generation: the co-simulation wrapper generator, the *hardware wrapper* generator, and the custom OS generator. A thorough explanation of these tools can be found in refs. 548, 575, and 576, respectively; some details are presented in the next subsections.

13.4.1   Hardware Generation

Figure 13-8 shows the detailed hardware generation flow for generation of the RTL architecture from a virtual architecture specification. There are two steps: refinement of processor modules and refinement of the communication network. These steps are explained in the next subsections.

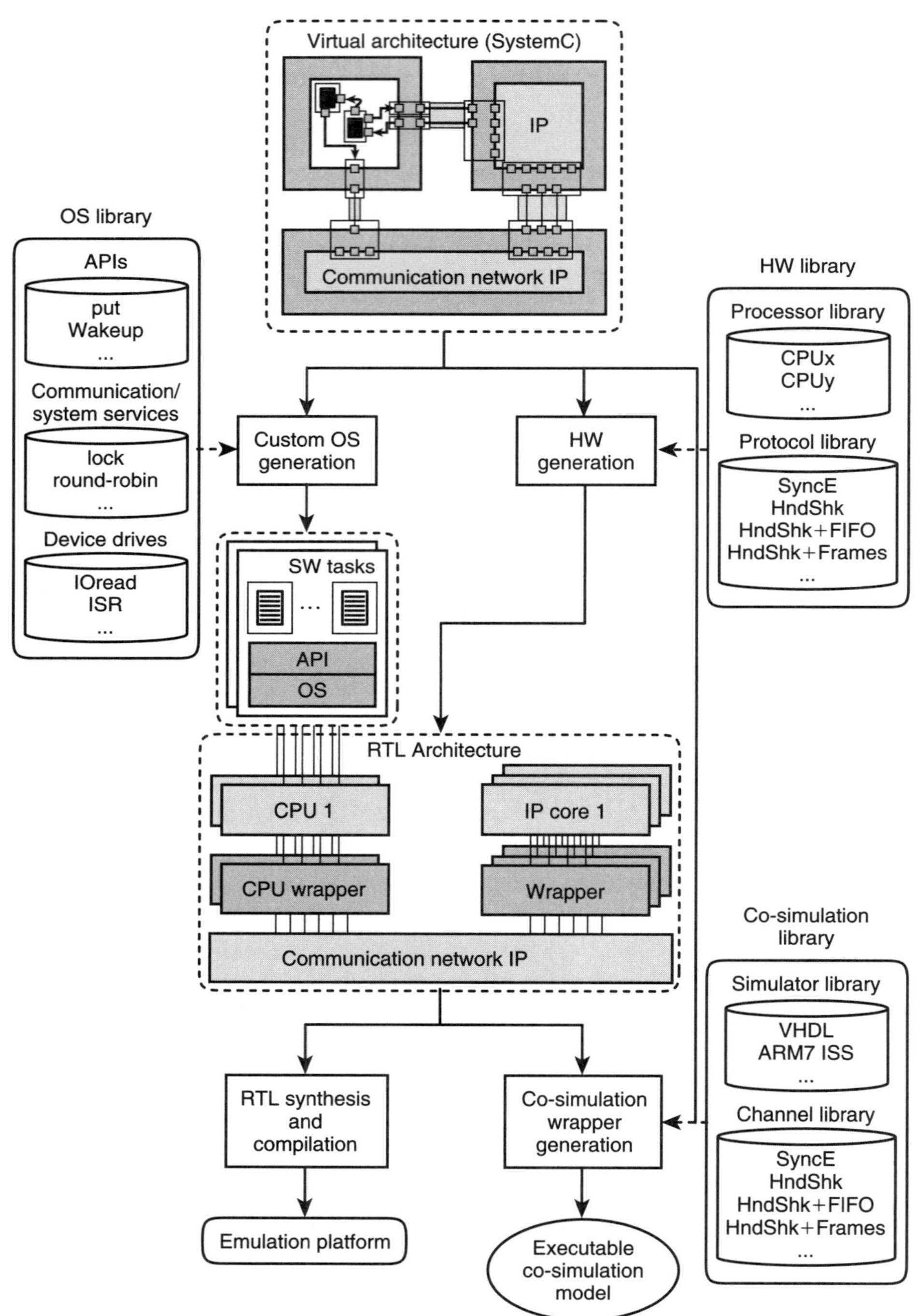

13-7 Component-based design environment for MPSoC.

FIGURE

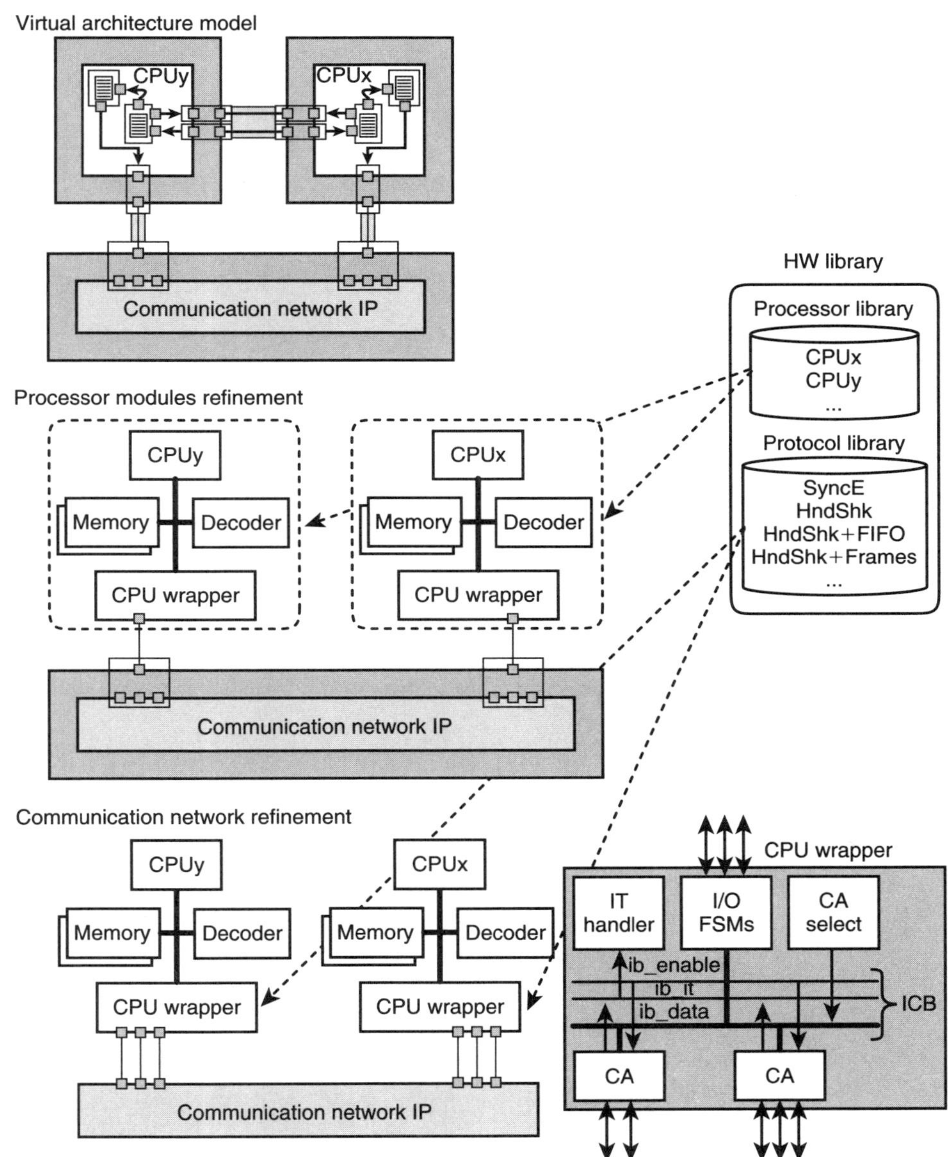

13-8   Hardware generation flow.

FIGURE

The hardware library has two parts: the processor library and the protocol library. All models stored in this library are synthesizable; they are instantiated and configured using architecture configuration parameters. The processor library contains processor adapters, template architectures for processors, processor cores, local memories, and peripherals. For instance, the following configuration parameters are used for processor adapter instantiation: port addresses, number of interrupts, and their priorities. The protocol library contains channel adapters and communication network models. Channel adapter configuration uses the following configuration parameters: I/O type, master/slave operation, type of data transmitted, buffer size, and interrupt parameters.

Instantiation of the Processor Local Architecture

In Figure 13-8, the macroarchitecture specification has two processors, CPUx and CPUy. A microarchitecture is generated from an architecture template by setting the parameters for the four types of elements: processor local architectures, CPU wrappers, IP components, and communication network(s). To instantiate a processor local architecture from the processor library, parameters such as size of local memory and channel-access addresses are used. The CPU wrapper adapts the processor to the communication network, and it implements the protocols of communication channels. Thus, to instantiate CPU wrappers, both processor and protocol libraries are used.

CPU Wrapper Generation

The processor adapter in the CPU wrapper, shown in Figure 13-8, is selected from the processor library and configured with the corresponding architecture parameters (allocated addresses, number of interrupts, and their levels). It performs (1) channel access detection by address decoding (using an I/O FSM), (2) channel selection, and (3) interrupt management. The address decoder in the processor adapter is configured by a lookup table generated from the parameters for allocated addresses. The processor adapter enables a channel adapter by sending to it an enable signal when the channel adapter, i.e., its channel, is accessed by the processor. Since enable signals are used, the internal communication bus (ICB) of the CPU wrapper does not have address lines, which provides an easy and area-efficient implementation of CPU wrappers. The interrupt controller is configured from the architecture parameters related with interrupts (e.g., the number of interrupts, their levels).

The generic channel adapter structure. The CA implements (1) the communication protocol of the macroarchitecture level channel and (2) the protocol of the connected communication network at the microarchitecture level. There are two types of CAs, depending on the direction of channel communication, input and output. For each channel in the macroarchitecture level specification, a pair of CAs is selected from the protocol library with the parameter of the communication protocol (type of communication protocol), and they are configured with the architecture parameters (e.g., I/O, master/slave, data type, buffer size, interrupt usage).

Although CAs can vary greatly, there are some similarities between them. We can decompose each CA into two interfaces, connected by a storage block (Fig. 13-9a). One interface connects the CA to the internal communication bus (ICB–read-write interface [RWI]), and the other connects it with the network (network-RWI). Each interface consists of read-write synchronization and processing parts. Protocols, available for the internal communication bus side of the CA, are a subset of ones used for external communication. Thus, we have to describe only three types of blocks in the library: *RWI*, *Processing*, and *Storage*. Figure 13-9b describes in a simplified unified modeling language (UML) class diagram the set of interfaces that must be implemented by each of these blocks.

The RWI is responsible for the synchronization, activation, and processing task calling. It is a control block and does not process data itself. The processing block (*Processing*) is responsible for incoming data preparation for storage or for transforming stored data to output format. It can split or join long bit vectors, or, for example, parity check. This block does not produce any new data itself. The execution of this block can take from zero to several clock cycles, so the block should inform other RWIs about its state after each operation. The storage block (*Storage*) is responsible for storing and retrieving the data. It manages the storage buffer itself. The block provides some information to other blocks about the status of the buffer. It acts as a joining part between internal and external interfaces.

13.4.2   Memory Wrapper Generation

Hardware generation is also used to adapt global memories (IPs) to the communication network (Fig. 13-10). The memory-wrapper library consists of (1) CAs and (2) memory port adapters (MPAs). CA configuration uses the following allocation parameters: read/write type, master/slave operation, type of transmitted data, and access mode. Parameters used for the MPA configuration correspond to the address/data bus size, the memory interface signals, and static or dynamic. Every time we add a new component to the library of memory IP, we have to write all the specific parts related to this new memory, as well as some other MPA functionality (refresh, bank selection, and so on).

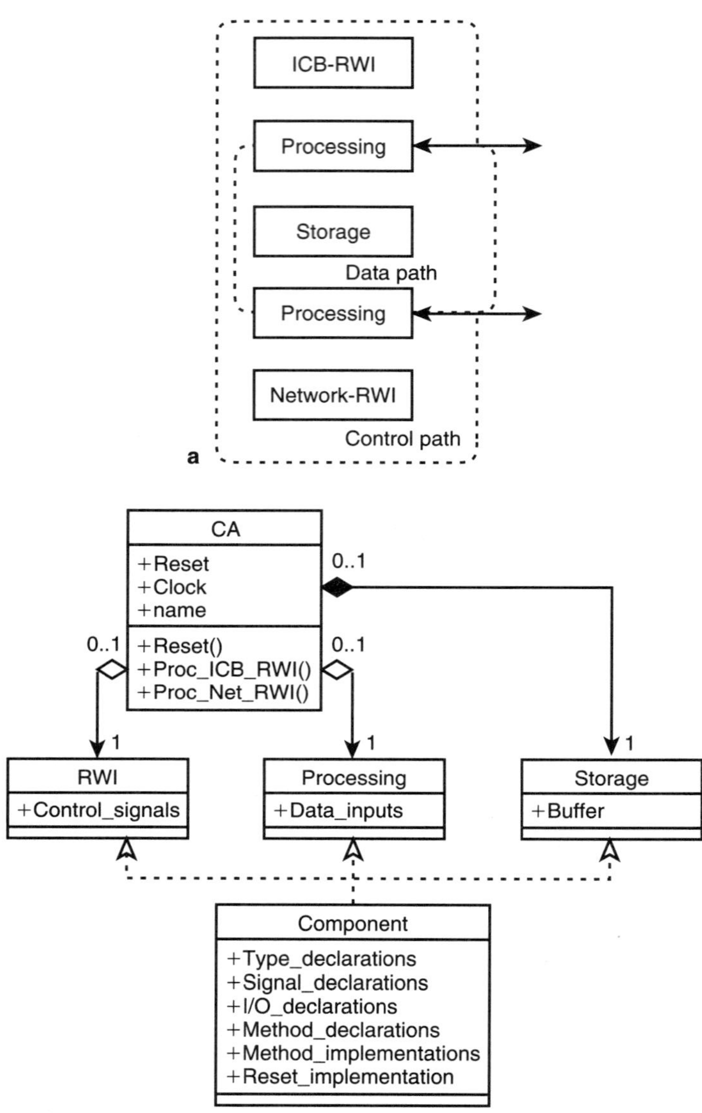

13-9

FIGURE

Decomposition of channel adapter. (a) Data/control flow. (b) Simplified UML model.

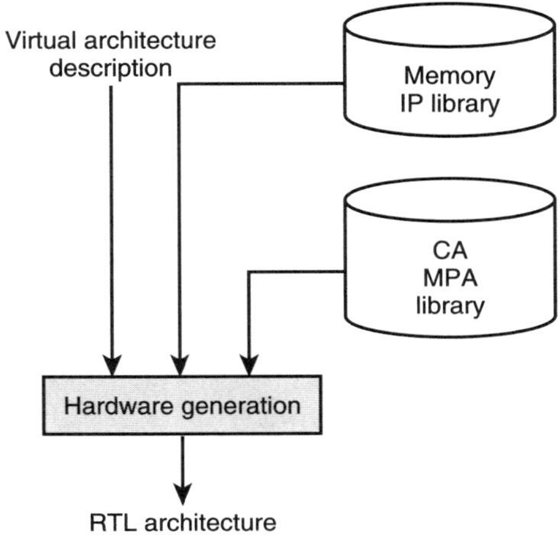

13-10   Memory wrapper generation flow.

FIGURE

Generic Memory Wrapper Architecture

Figure 13-11 shows a generic model of a *memory wrapper* that connects a global memory shared by several processors through the communication network. It is composed of two main parts. The first part, the MPA, is specific to the memory. The second part depends on the communication protocol used by the communication network, and it contains several modules (CAs). These two components communicate through an ICB.

Memory port adapter (MPA): includes a controller and several memory-specific functions such as control logic, address decoder, bank controller, and other functions that depend on the type of memory (refresh management in DRAM). In addition, the MPA performs type conversion and data transfer between an internal communication bus and a memory bus.

Channel adapter (CA): implements the communication protocol. Its implementation depends on many parameters such as communication protocol (FIFO, burst, and so on), channel size (int, short, and so on), and port type (in, out, master, slave, and so on). The CA manages read and write operations requested by the other modules accessing the memory. The number of channels in the virtual architecture equals the number of CAs.

Internal communication bus (ICB): interconnects the two parts of the memory wrapper (MPA part and the CA part). This internal bus is usually

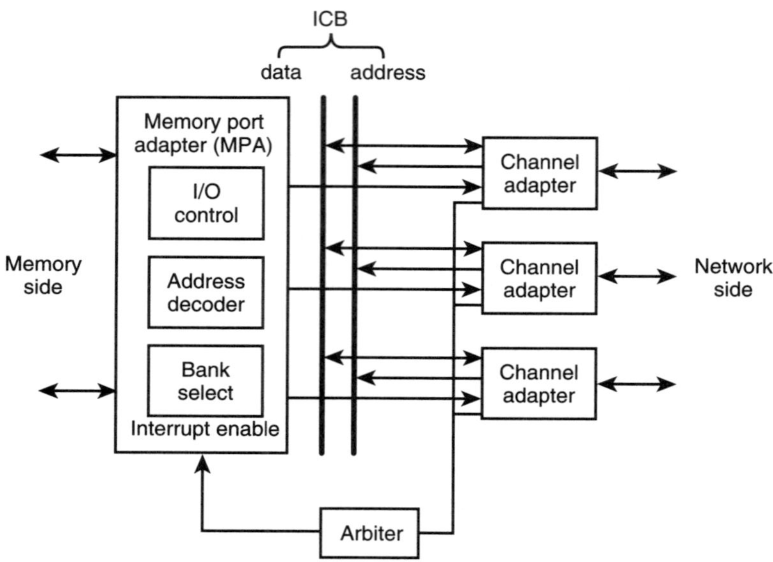

13-11   Generic memory wrapper architecture.

FIGURE

composed of address bus, data bus, and control signals. The size of these internal busses depends on the memory bus size and the channel size. This size is determined for each instance of memory wrapper at the wrapper generation step.

Arbiter: the arbiter manages the conflicts generated by multiple connection of CAs to the internal bus of memory wrapper. In the case of multiaccess to a shared memory port, the arbiter must give the access permission to only one channel. In our model, this is managed using a priority access list.

13.4.3   Software Wrapper Generation

The software wrapper generator [548] produces OS streamlined and preconfigured for the software task(s) that run(s) on each target processor. It uses an OS library that is organized in three parts: APIs, communication/system services, and device drivers. Each part contains elements that will be used in a given software layer in the generated OS. This library contains a dependency graph between services/elements that is used to determine the minimal set of elements necessary to implement a given OS service. This mechanism is used to keep the size of the generated OS at a minimum, by avoiding inclusion of unnecessary elements from the library.

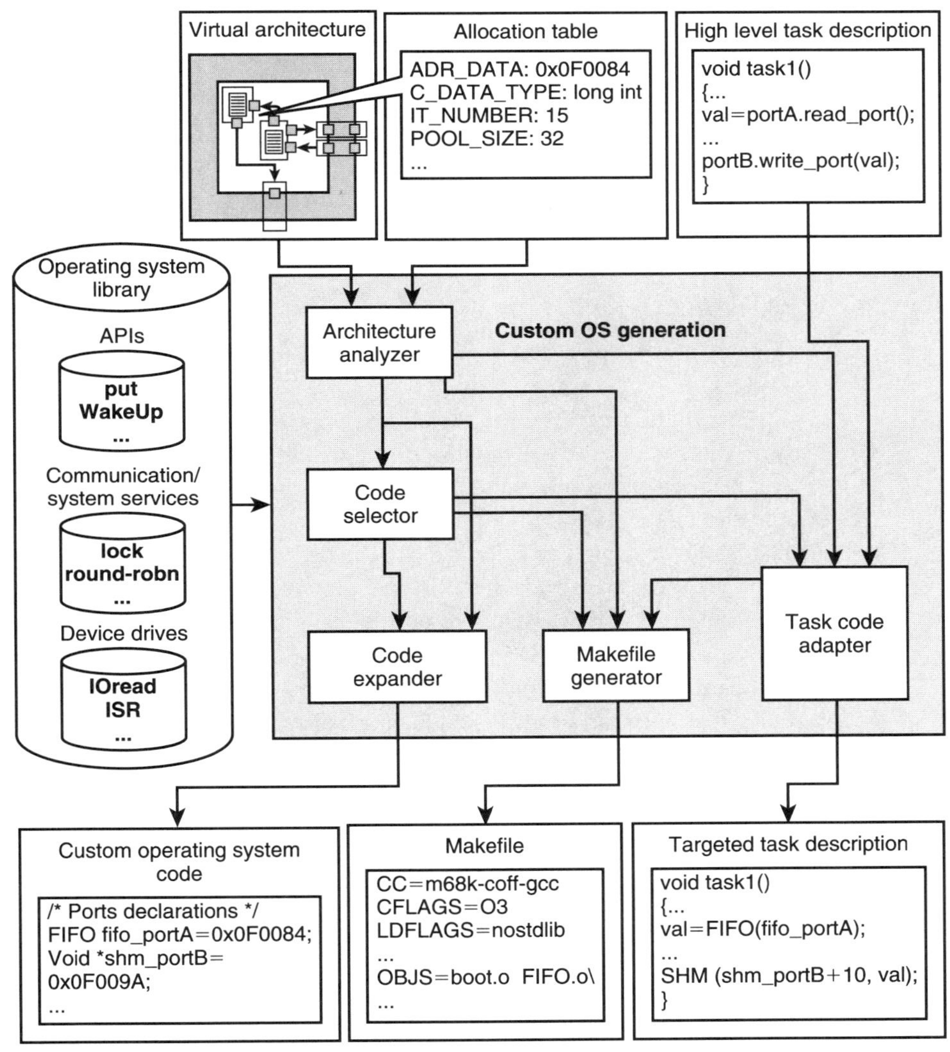

13-12 The custom OS automatic generation flow.

FIGURE

Figure 13-12 shows our design flow of automatic OS generation. The input to this flow is the virtual architecture description. As shown in the figure, *Architecture Analyzer* reads the structural information and the memory allocation information. *Code Selector* receives a list of services required by the application and finds the full list of (original and deduced) services. *Code Expander* generates the source code of the OS. *Task Code Adapter* performs software targeting, and *Makefile Generator* produces makefiles. Thus, the outputs (for each processor) are the source code of generated OS, targeted application code, and a makefile. To obtain the binary code to be downloaded onto the target processor memory, the designer runs a compilation of both generated OS and targeted application using the generated makefile.

OS Library

The OS library provides (1) a very small and flexible OS kernel and (2) OS services.

The main function of the OS kernel is scheduling multiple tasks. Several preemptive schedulers are available in the OS library such as the round-robin scheduler, priority-based scheduler, and so on. In the case of the round-robin scheduler, time-slicing (i.e., assigning different CPU load to tasks) is supported. To make the OS kernel very small and flexible, (1) the task scheduler is customized to the application requirements and (2) a minimal amount (less than 10% of kernel code size) of processor-specific assembly code is used (for context switching and interrupt service routines). The OS library contains all basic functions for the OS kernel. For instance, function *ContextSwitch* performs context switching between the currently running task and the next task to be executed. Since context-switching operation differs from processor to processor, the function consists of two kinds of code: C code and assembly code. The C code part is called by other schedulers in C code (e.g., by a priority-based scheduler or a round-robin scheduler). The assembly code part performs processor-specific context switching operations.

The OS library also provides services that are specific to embedded systems: communication services (e.g., FIFO communication), I/O services (e.g., AMBA bus drivers), memory services (e.g., cache or virtual memory usage), and so on. Services can have dependency between them. For instance, communication services are dependent on I/O services. There are two types of service code: reusable (or existing) and expandable. As an example of existing code, a FIFO service code can exist in the OS library in the form of C language. As an example of expandable code, OS kernel functions can exist in the OS library in the form of macrocode.

Architecture Analysis

Architecture Analyzer finds the following information from the input system description:

- ✦ application-specific services and their parameters

- ✦ module-specific parameters.

Application-specific OS services are found in the attributes of modules, channels, and ports in the input virtual architecture. For instance, if a channel has an attribute for FIFO implementation, FIFO service is selected to be included into the OS to be generated. Parameters of required services are also found on the allocation table (virtual architecture parameters). For instance, the address range of FIFO communication and the interrupt priority of interrupt-driven port can be found on the allocation table. The information of required services is sent to the Code Selector. Module-specific parameters (e.g., task priority, CPU load, type of mapped processor, and so on) are also found from module attributes. The type of processor is sent to Makefile Generator to choose the right compiler and to Code Expander to target the OS code to the processor.

Code Selection

Code Selector takes as input a list of required services from the Architecture Analyzer. It looks up the OS library to check service dependency and finds all the services that have a dependency relation with the required services. For instance, a required service, FIFO communication, may need interrupt handling service. In this case, the interrupt handling service should also be chosen to be included in the OS to be generated. Since there are two types of code (existing and expandable), after the whole list of required services is determined in this way, Code Selector first looks up the OS library to find the existing codes for required services and sends the list of existing codes to Makefile Generator. Then it sends the list of services with expandable codes to Code Expander.

Macrocode Expansion

Code Expander takes as input a list of required services from the Code Selector and processor type information from the Architecture Analyzer. It generates the final OS code (in C and assembly code) as follows. For each service, it finds a corresponding macrocode(s) in the OS library. Then it expands the macrocode to a source code(s) (in C or assembly code). Processor-specific code expansion is limited to functions such as context switching, synchronization primitives (e.g., semaphore functions), and interrupt service routines.

API Bindings

To target the application SW code to the OS, Task Code Adapter replaces function calls of communication and synchronization in the original application software by OS service calls (i.e., system calls). For instance, the function call of communication called *portA.read_port( )* is replaced by an OS service function call *FIFO(fifo_portA)*. Note that original function arguments are also replaced by arguments specific to the OS service function. In the example, the original function argument is replaced by a FIFO service-function argument *fifo_portA* that is a pointer to the FIFO address in this case.

Integration of Existing OS

Existing OS can be integrated into the proposed flow for automatic generation (to be more specific, in this case, automatic configuration) of application-specific OS. For automatic configuration of existing OS using the proposed flow, there is no change on execution of service extraction (by Architecture Analyzer), makefile generation, and adaptation of application code to the automatically configured OS. In terms of code quality (size, execution time, and so on) of automatically configured OS, it depends on the granularity of OS services in the existing OS. Thus, if the existing OS supports as fine a granularity in OS services as our OS kernel and services, the quality of automatically configured OS can be comparable to that of automatically generated OS.

13.4.4   Simulation Model Generation

Figure 13-13 shows the flow used to generate (mixed-level) co-simulation models alongside with refinement from the initial virtual architecture, passing by intermediate architectures, to the final RTL architecture [576]. In this flow, we use two simulation libraries: one (called co-simulation library) for high-level simulation models of modules, channels, and the other (called synthesizable code library) for simulation models of synthesizable components (at RTL). Our co-simulation tool selects, for each module/channel in abstract/intermediate/RTL architectures, an appropriate simulation model from the libraries according to the abstraction level of module/channel. Then it generates co-simulation codes with the selected simulation models in possibly different languages (e.g., SystemC, VHDL, and so on).

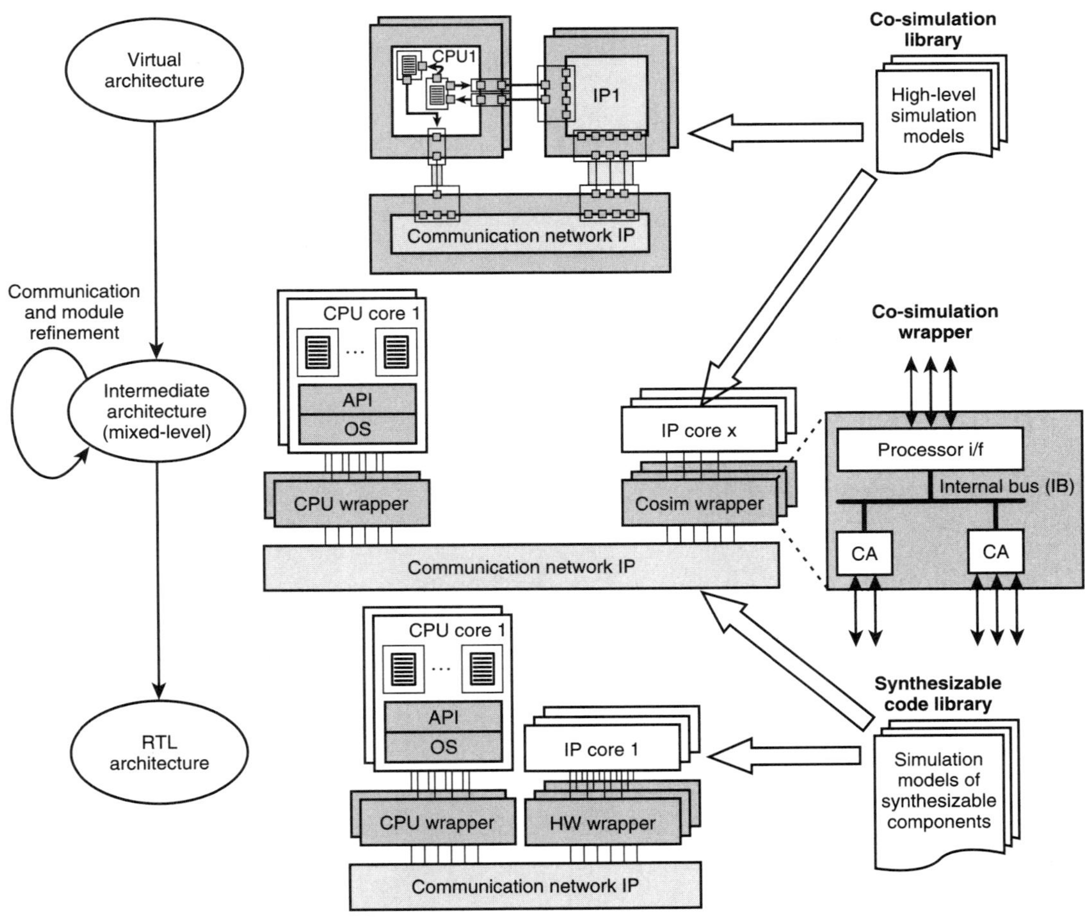

13-13   Flow for generation of co-simulation models.

FIGURE

With the initial virtual architecture, high-level simulation models are used in co-simulation. During incremental communication and module refinement, we obtain intermediate architectures in which synthesizable components and high-level simulation models coexist; simulation models of both libraries are used in mixed-level co-simulation. In Figure 13-13, shaded regions represent simulation models of synthesizable components. To generate wrappers for mixed-level co-simulation, we use the *co-simulation wrapper* generator for a module that requires a simulation wrapper, as exemplified in Figure 13-13. After communication refinement, only simulation models of synthesizable components are used to simulate the physical architecture at RTL.

Generic Co-simulation Wapper Architecture

Figure 13-14 shows the generic co-simulation wrapper architecture. It consists of internal and external ports on each side, module adapter (MA), a CA, and an IB. For each case of adapting different protocols or different abstraction levels, an instance of the generic wrapper architecture is constructed. In this instance, ports can be given any protocols/abstraction levels (in the figure, protocols/abstraction levels are exemplified by Pi/Lj). IB is used to transfer data between MAs and CAs. To do that, it has three kinds of signal: data, enable, and ready. Data signal is bidirectional and has a generic data type. A specific data type (e.g., int, short, logic vector, and so on) of data signal is determined for each instance of the generic wrapper architecture. Enable signals are set/reset by module adapter. They determine one CA and enable it to write/read data to/from data signal. Each CA sets its ready signal when it receives data from its connected port or when it is ready to receive data from the MA (to send the data to the channel via its connected port). In communication between MAs and CAs, the MA acts as a master to control CAs. This model allows one to isolate the module from the communication network and permits the interconnection of any number of modules.

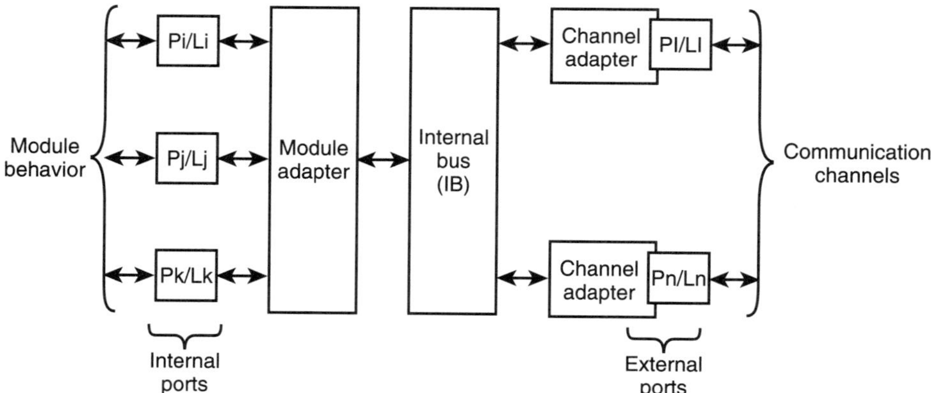

13-14   Generic co-simulation wrapper structure.

FIGURE

13.5   COMPONENT-BASED DESIGN OF A VDSL APPLICATION

This section demonstrates the application of the ROSES high-level component-based methodology using a VDSL application as a design example.

13.5.1   The VDSL Modem Architecture Specification

The design we present in this section was taken from the implementation of a VDSL modem using discrete components [577]; the block diagram for this prototype implementation is shown in Figure 13-15a. The subset we will use in the rest

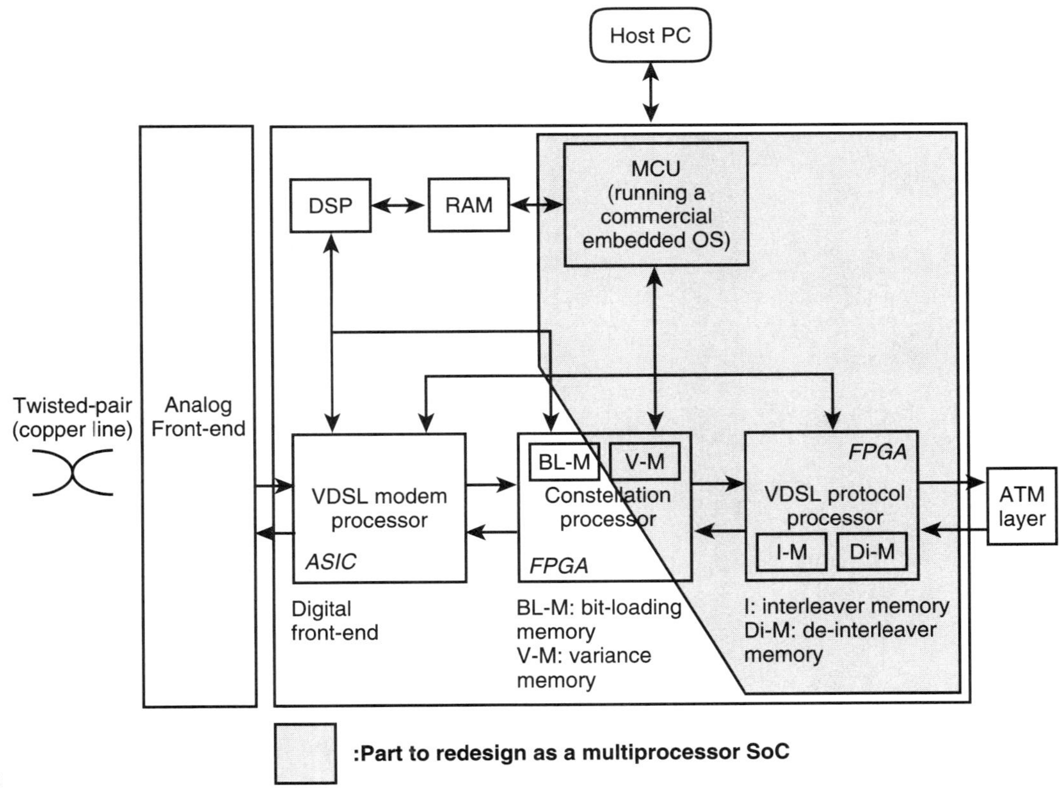

13-15   (a) VDSL modem prototype (*Continued*).

FIGURE

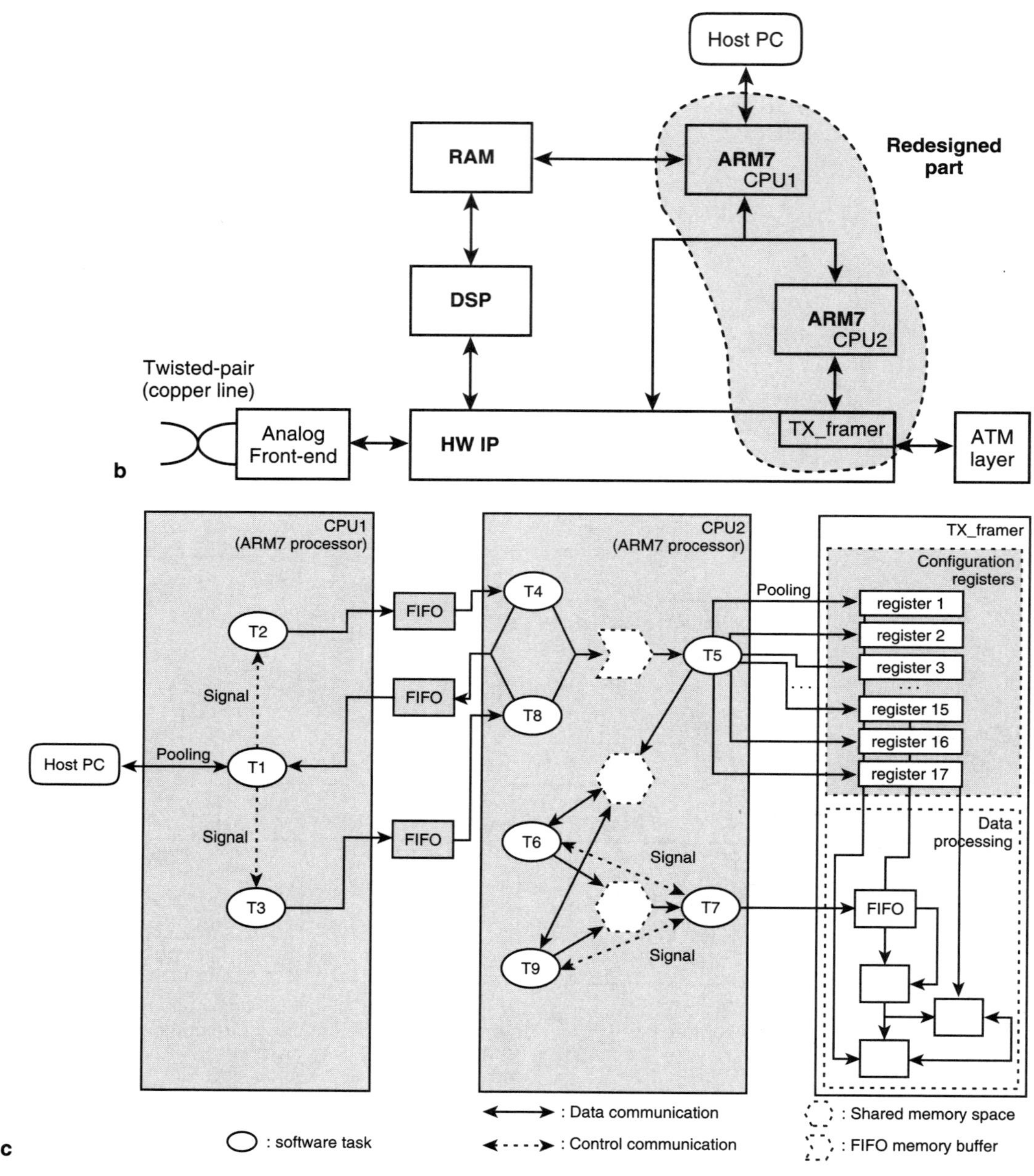

13-15   Continued. (b, c) MPSoC architecture.

FIGURE

of this section is shaded in Figure 13-15a. It is composed of an MCU and part of the original datapath, the TX_Framer, which is described at the RTL and used as a blackbox in this experiment.

To configure, monitor, and synchronize the DSP and the TX_Framer block, we decided to split the host interface task, the control tasks, and the high-level VDSL code over two ARM7 processors (**CPU1** and **CPU2** in Fig. 13-15b). This partition is a very important design decision and is not presently part of our design flow, so we adopted a partition of processors/tasks suggested by the design team of the VDSL-modem prototype. Some details of the application are given in Figure 13-15c: CPU1 runs three concurrent tasks (**T1**, **T2**, and **T3**), and CPU2 runs six tasks (**T4** to **T9**). CPU1 and CPU2 exchange data using three asynchronous FIFO buffers. The TX_Framer block can be configured through 17 configuration registers, and CPU2 can input a data stream to this block via a synchronous FIFO buffer.

Tasks running in each embedded processor communicate and/or control each other by making system calls to the embedded OS. In order to reduce embedded memory requirements, the two OS provide only the set of services required by the cluster of tasks running in each processor. Tasks use a variety of control and data-transmission protocols to communicate. For instance, task **T1** can use control communication to change the execution status of tasks **T2** and **T3**, and it can block/unblock the execution of these tasks by sending them an OS *signal*. For data transmission, tasks use FIFO memory buffer, shared memory (with or without semaphores), and direct register access (pooling).

Despite all simplifications, the design of the selected subset remains quite challenging. In fact, this application uses two processors executing parallel tasks. The control over the three modules of the specification is fully distributed. All three modules act as masters when interacting with their environment. Additionally, the application includes some multipoint communication channels requiring sophisticated OS services.

13.5.2   Virtual Architecture Specification

Figure 13-16 shows the virtual architecture model that captures the VDSL specification. Modules **VM1** and **VM2** correspond to the ARM7s on Figure 13-15c, and module **VM3** represents the TX_Framer block. (Only the interface is known so it is represented as a blackbox.) The virtual architecture can be mapped onto different architectures depending on the configuration parameters annotated in modules, ports, and nets. For instance, the three point-to-point connections (**VC1**, **VC2**, and **VC3**) between **VM1** and **VM2** can be mapped onto a bus or onto a shared

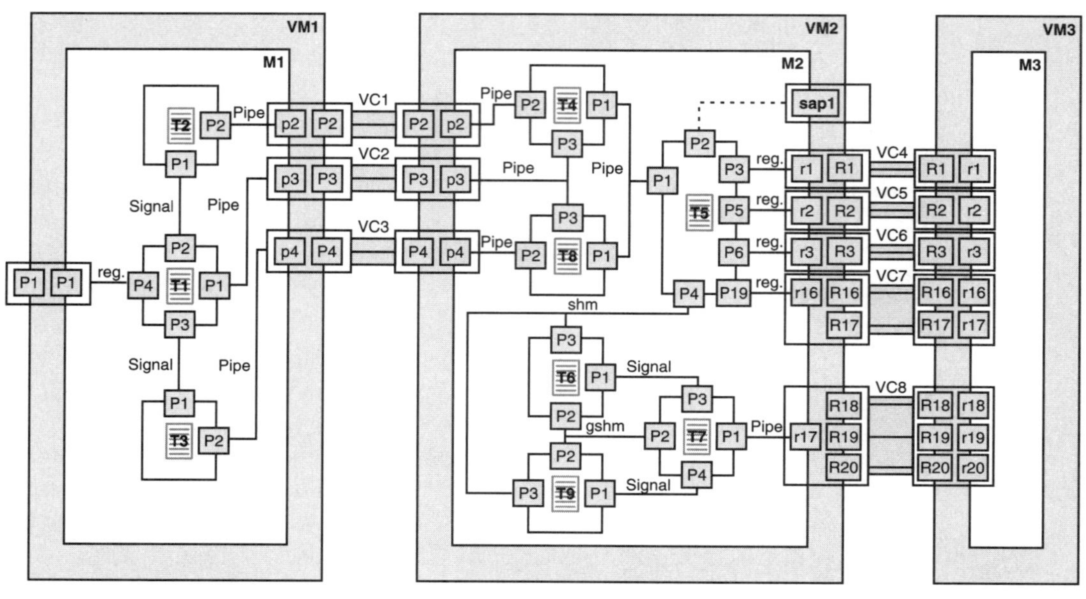

13-16

VDSL virtual architecture specification.

FIGURE

memory if the designer changes the configuration parameters placed on these virtual channels and virtual ports.

13.5.3 Resulting MPSoC Architecture

The manual design of a full VDSL modem requires several person-years; the presented subset was estimated as a more than five person-years effort. When using this high-level component-based approach, the overall experiment took only one person during 4 months (not counting the effort to develop library elements and debug design tools). This corresponds to a 15-fold reduction in design effort. Running all wrapper generation tools takes only a few minutes on a Linux PC 500 MHz; most of the time was spent in writing the virtual architecture model with all the necessary configuration parameters. The behavior of each task was described using the SystemC C++ library [572].

Figure 13-17 shows the RTL architecture model obtained after HW/SW wrapper generation: two ARM7 cores with their local architectures, the

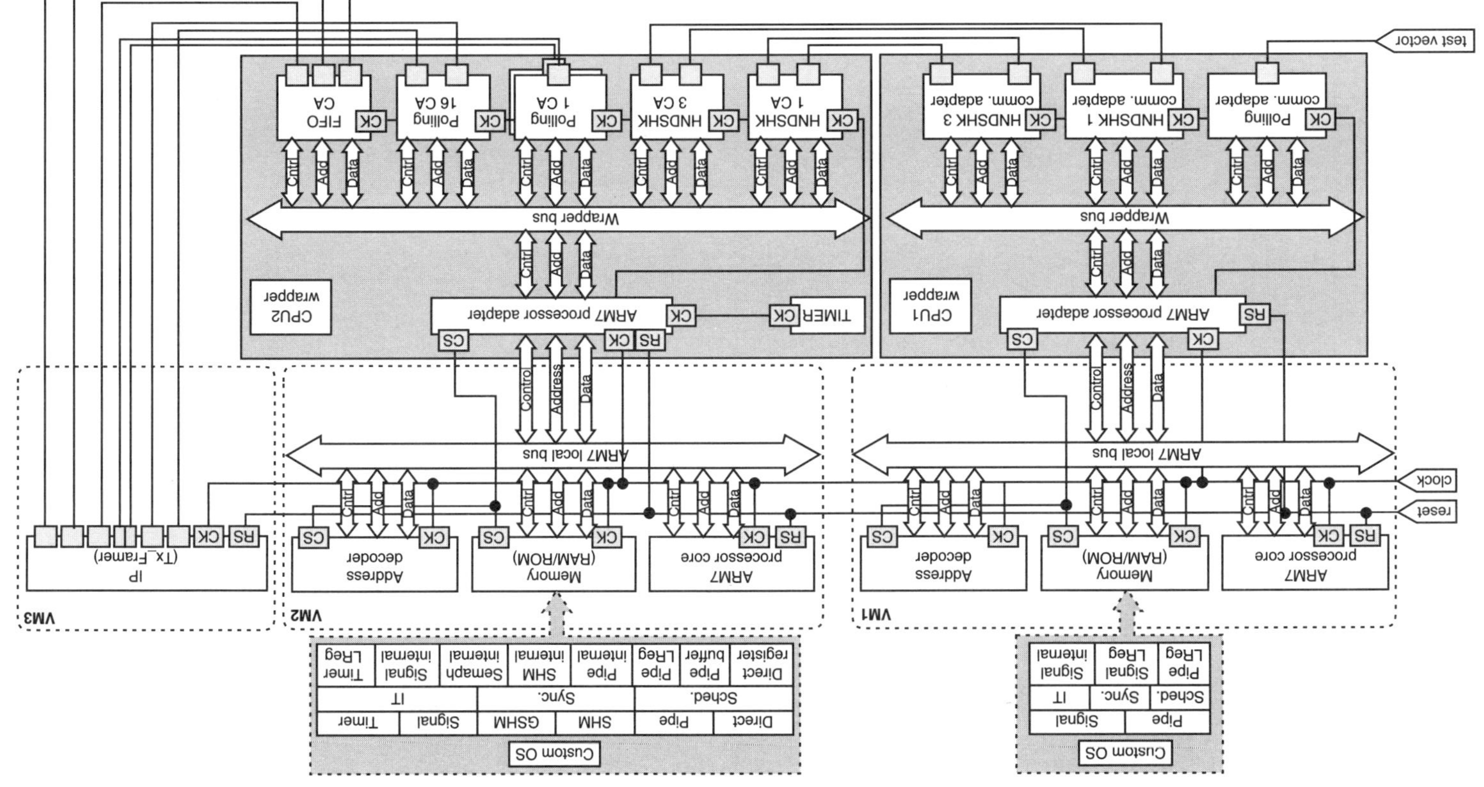

13-17 FIGURE

Generated MPSoC architecture.

TX_Framer block. Each CPU wrapper acts as a communication controller for the ARM7, it contains an ARM7 processor adapter that bridges the ARM7 local bus to the CAs. There is a CA for each virtual channel in the virtual architecture specification. For instance, module **VM1** reads test vectors from the environment through a simple register using the *Pooling* CA and communicates with **VM2** through asynchronous FIFOs using three *HNDSHK* CAs (corresponding to virtual channels **VC1**, **VC2**, and **VC3**).

Each software wrapper (custom OS) is customized to the set of tasks executed by the processor core. For example, software tasks running on **VM1** access the custom OS using an API composed of two functions: *Pipe* for communication with **VM2** and *Signal* to modify the task scheduling on run time. The custom OS contains a round-robin scheduler (*Sched*) and resource management services (*Sync, IT*). The driver layer contains low-level code to access the CAs (e.g., *Pipe LReg* for the *HNDSHK* CA) and some low-level kernel routines.

The CPU wrapper for processor **VM2** includes a timer module because task **T5** (see Fig. 13-16) must wait 10 ms before starting its execution. A hardware interrupt is generated by the TIMER block; the task can configure this block using the *Timer* API provided by the service access port **sap1** (see Fig. 13-16). The custom OS for **VM2** provides a more complex API: the *Direct* API is used to write/read to/from the configuration/status registers inside the TX_Framer block; *SHM* and *GSHM* are used to manage shared-memory communication between tasks.

Application code and generated OS are compiled and linked to execute on each ARM7 processor. CPU wrappers can be synthesized using RTL synthesis. Table 13-2 presents the results regarding the generated OSs. Part of the OS is written in Assembly; it includes some low-level routines (e.g., context switch and processor boot) that are specific to each processor.

OS results	# of lines C	# of lines assembly	Code size (bytes)	Data size (bytes)
VM1	968	281	3829	500
VM2	1872	281	6684	1020

Context switch (cycles)	36
Latency for interrupt treatment (cycles)	59 (OS) + 28 (ARM7)
System call latency (cycles)	50
Resume of task execution (cycles)	26

13-2 Results for OS Generation.

TABLE

HW interfaces	# of gates	Critical path delay (ns)	Max. freq. (MHz)
VM1	3284	5.95	168
VM2	3795	6.16	162
Latency for read operation (clock cycles)			6
Latency for write operation (clock cycles)			2
Number of code lines (RTL VHDL)			2168

13-3

TABLE

Results for Hardware Generation.

If we compare the numbers presented in Table 13-2 with configurable commercial embedded OS, the results are still very good. Generally, the minimum size for commercial OSs is around 4 KB, but with this size, few of them could provide the required functionality. Performance was also very good: context-switch takes 36 cycles, latency for hardware interrupts is 59 cycles (plus the 4 to 28 cycles needed by the ARM7 to react), latency for system calls is 50 cycles, and task reactivation takes 26 cycles.

Table 13-3 shows the numbers obtained after RTL synthesis of CPU wrappers using a CMOS 0.35-μm technology. These results are good because wrappers account for less than 5% of the ARM7s core surface and have a critical path that corresponds to less than 15% of the clock cycle for the 25-MHz ARM7 processors used in this case study.

13.5.4   Evaluation

The results extracted from the RTL model show that our approach can generate HW/SW interfaces and OS that are as efficient as manually coded/configured ones. Wrappers are built up from library components so the HW/SW interface in wrapper implementation can be displaced. This choice is transparent to the final user since everything that implements the communication API (hardware interfaces and OS) is generated automatically. Designers do not need to rewrite the application code because the API does not change (only its implementation does). Furthermore, correctness and coherence can be verified inside tools and libraries against the API semantics without having to impose fixed boundaries to the HW/SW interface (in contrast to standardized interfaces and busses).

The generation of layered wrappers by assembling modular library components provides considerable flexibility for this design methodology. One of the most important consequences of having this flexibility is that the design environment can be easily adapted to accommodate different languages to describe system behavior, different task scheduling and resource management policies, different communication network topologies and protocols, a diversity of processor cores and third-party blocks, and different memory architectures. Modular library components are at the roots of the methodology; the principle followed by these components is to contain a unique functionality and to respect well-defined interfaces that allow easy composition. In most cases, inserting a new design element in this environment only requires the writing of the appropriate library components. The modular wrapper structure prevents library size explosion since components whose behavior depends on the processor are separated from components that implement communication protocols.

13.6   CONCLUSIONS

This chapter presented a summary of a high-level component-based methodology and design environment for application-specific multiprocessor SoC architectures. The system specification is a virtual architecture annotated with configuration parameters. The ROSES component-based design environment has automatic wrapper-generation tools able to synthesize hardware interfaces, device drivers, and OS that implement a high-level communication API. Results show that wrappers generated automatically by these tools have performances close to the commercial/handcrafted equivalents. The key issues are:

1. Automatic wrapper generation eliminates most errors caused by manually managing all details of a multiprocessor SoC architecture and associated OS.

2. The use of a virtual architecture model is a step in the right direction by raising the abstraction level when one is dealing with complex designs like the VDSL modem.

3. Flexibility is very important in such design environments; fixed communication topologies or restricted sets of protocols will hardly be accepted by designers because they want to reuse the internal designs that frequently are tied to in-house (nonpublic) standards.

ACKNOWLEDGMENTS

This research was sponsored in part by the ToolIP-A511 and SpeAC-A508 projects of the Europe MEDEA+ (Microelectronics for European Applications) program. Many individuals have contributed to the work described in this chapter. The authors would like to acknowledge the contributions of Paul Amblard, Iuliana Bacivarov, Amer Baghdadi, Aimen Bouchhima, Anouar Dziri, Ferid Gharsalli, Arnaud Grasset, Frédéric Hunsigner, Lobna Kriaa, Damien Lyonnard, Gabriela Nicolescu, Yanick Paviot, Frédéric Rousseau, Adriano Sarmento, Arif Sasongko, Ludovic Tambour, Sungjoo Yoo, Mohamed-Wassim Youssef, and Nacer-Eddine Zergainoh.

MPSoCs for Video

Santanu Dutta, Jens Rennert, Tiehan Lv,
Jiang Xu, Shengqi Yang, and Wayne Wolf

ABSTRACT

This chapter looks at the design of multiprocessor systems-on-chips (MPSoCs) for video applications. Video and multimedia require huge amounts of computational power and real-time performance. Furthermore, video algorithms are still evolving; hence, multiprocessor architectures must be designed for flexibility as well as performance. One of our case studies is the Philips Nexperia™ architecture, a leading commercial MPSoC framework for set-top box, digital television and connected home applications. We also consider other applications, including real-time gesture recognition.

14.1   INTRODUCTION

The growing demand in multimedia video processing and its applications owes its origin to and, at the same time, is responsible for the further development of both hardware design and software techniques. Aided by advancements in very large-scale integrated circuit (VLSI) manufacturing technology that has made possible the integration of increased functionality in smaller circuits, it is primarily the development of novel signal-processing architectures and design techniques that has brought audio, video, graphics, image, speech, and text processing together and prompted advanced multimedia video applications such as high-definition digital television, digital set-top boxes with time-shift functionality, *3D* games, *H.26x* video conferencing, *MPEG-4* interactivity, and so forth. The computational requirements of multimedia video processing being dominated by signal-processing tasks that require complex and real-time processing on high volumes of data, this chapter attempts to take a closer look at some of the recent trends in designing integrated circuits (ICs) for such systems.

This chapter considers MPSoC architectures for advanced video applications. Video applications are rapidly evolving along with the increases in computational power supplied by Moore's Law. Although MPSoC must be tailored to their primary application in order to squeeze the maximum amount of performance from the available silicon, the architecture should also be designed for flexibility in order to maximize the utility and longevity of the design. The computational requirements of multimedia video processing being dominated by signal-processing tasks that require complex and real-time processing on high volumes of data, we attempt a closer look at some of the recent trends in designing ICs for such systems.

We first look at several video applications in order to understand the requirements better on video MPSoCs. We of course consider video compression, the dominant application today of digital video. We also look at one of our own applications, the real-time gesture recognition system designed as part of the Princeton Smart Camera Project, as an example of a new generation of video applications. We then spend a great deal of time identifying some of the recent trends in the design of multimedia SoCs and use the Philips Nexperia™ Home Entertainment Engine as a case study. The specific topics touched on are: processor architectures, central processing unit (CPU) configurations, system and chip integration, intellectual property (IP) reuse, platform-based designs, communication bus structures, and design-for-testability (DFT) issues. We close with a brief discussion of trace-driven analysis of applications and architectures as part of the design of video MPSoC architectures.

14.2   MULTIMEDIA ALGORITHMS

In this section, we study the basic characteristics of some multimedia applications. We start with video compression, the dominant digital video application today. We then look at a next-generation smart camera application.

14.2.1   Compression

Video compression has been developed over the last 30 years and is now a mass-market item. Satellite television, terrestrial digital television, digital video cameras, and personal video recorders all make use of video compression methods. The original video compression systems occupied racks of equipment. Today, a great deal is known about how to reduce video compression algorithms to VLSI.

Most image and video compression methods are lossy. These methods take advantage of the fact that the human visual system is not equally sensitive to all features and changes to an image or image sequence.

Video and image compression methods evolved in parallel in their early days, but modern standards make use of image compression techniques. The best-known family of image compression standards is JPEG. The JPEG-2000 standard incorporates wavelet-based compression, but a basic image compression technique that is also used in video compression is the discrete cosine transform (DCT). The DCT is a frequency transform that is applied to blocks of images, typically 8×8 blocks. The DCT yields frequency decompositions of the block in two dimensions: x and y. Some of the frequency components can be discarded—a process known as quantization—to reduce the amount of information transmitted. Perceptual coding strives to throw away coefficients such that the perceptual difference between the original and compressed images will be minimal; this generally means that high-frequency components, which correspond to fine details, are discarded.

The discrete cosine transform has the form:

$$F(u,v) = \sqrt{\frac{2}{N}} \sqrt{\frac{2}{M}} \sum_{0 \le i \le N-1} \sum_{0 \le j \le M-1} \lambda(i)\lambda(j) \cos\left[\frac{\pi u}{2N}(2i+1)\right] \cos\left[\frac{\pi v}{2M}(2j+1)\right] x(i,j)$$

where $\lambda(a) = \frac{1}{\sqrt{2}}$ for $a = 0$, 1 otherwise.

Quantization is the term for the elimination (zeroing out) of some transform coefficients. After quantization, a lossless compression method (usually some combination of a Huffman coding and a run-length coding) is applied to the quantized coefficients in order to reduce their representation for transmission or storage. Because high-frequency components are often the first to be eliminated in a DCT coefficient set, the DCT coefficients are generally read in a zigzag pattern as shown in Figure 14-1. This pattern reads the coefficients starting at the DC value (the 0,0 coefficient) to the highest frequency component (the 7,7 value). If high-frequency coefficients are zeroed out, this pattern produces longer strings of zeroes than would be true from row- or column-oriented patterns; these longer strings of zeroes can be compressed by lossless compression methods.

Several video compression families exist: the H.x26x standards for teleconferencing and the MPEG standards for video broadcast and distribution. Each of these families includes several standards, developed at different times for varying levels of hardware support and bandwidth. In this section, we will concentrate on MPEG-style compression—the basic techniques used in MPEG-1 and

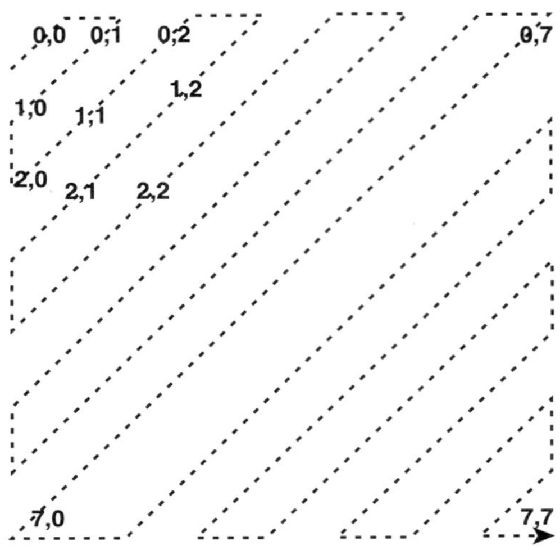

     A zigzag pattern in DCT coefficients.

-2. The MPEG standard defines a bit stream standard but does not determine the exact algorithms used to generate those bit streams. This allows developers to improve their implementations of the standard—for picture quality, compression rate, power consumption, and so on—while maintaining compatibility with other manufacturers' devices. MPEG-1 and -2 are also designed for asymmetric applications, in which the transmitter is assumed to have more computational power than the receiver. This is typical in broadcast, in which the transmitter is less cost-sensitive than the consumers' receivers; videoconferencing, in contrast, typically uses terminals of equal computational power at each end.

Figure 14-2 shows the block diagram for an MPEG-1/2 style encoder. MPEG takes advantage of DCT-based compression. The other major compression operation is motion estimation/compensation. Whereas DCT works entirely within a frame, motion estimation compares data between frames. Two important data structures in MPEG video coding are the *block* (an 8 × 8 set of pixels) and the *macroblock* (a 16 × 16 set of pixels). DCT is performed on blocks; motion estimation is generally performed on macroblocks.

Motion estimation reduces a macroblock to a motion vector that describes how the macroblock is displaced from another frame. (Block motion estimation handles only translational motion.) The receiver can then read the macroblock from its position in the reference frame in the receiver's frame store and apply the motion vector to reconstruct the *motion-compensated* frame. As shown in Figure 14-3, the

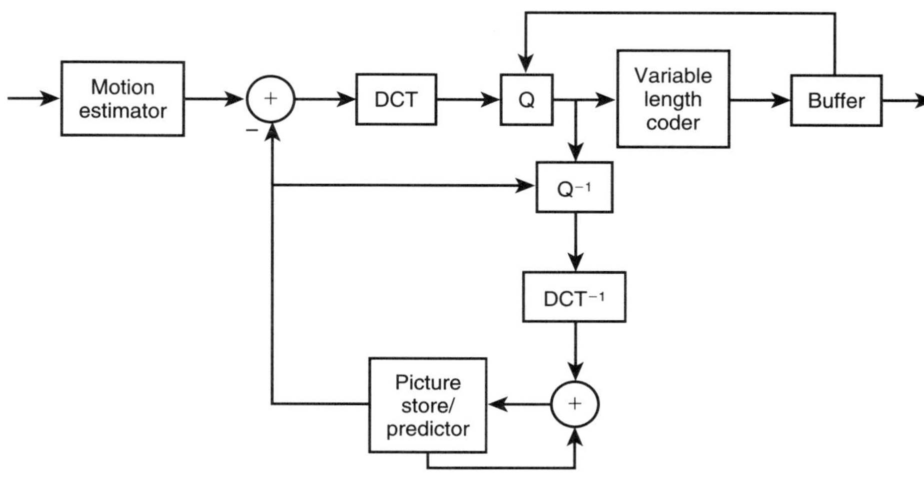

14-2
FIGURE

Block diagram of an MPEG-1/2 style encoder.

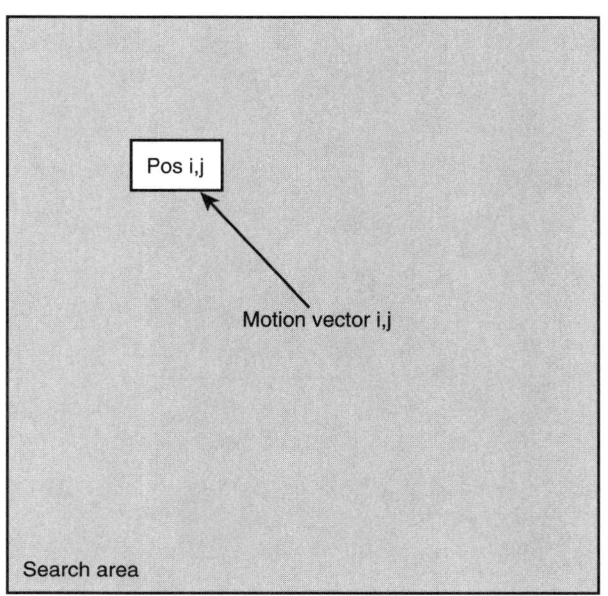

14-3
FIGURE

Motion estimation and motion vectors.

encoder compares the reference block with a search field at a number of different i, j offsets. At each point it computes a two-dimensional correlation:

$$D(i,j) = \sum \sum \left| r(i,j) - s(i+m, j+n) \right|$$

where r and s are the reference and search macroblocks, respectively. This computation is known as a sum-of-absolute-difference (SAD) computation. The smallest SAD value is selected to provide the motion vector. This formula assumes full search of all possible locations in the search area. A number of other search schemes have been proposed to reduce the number of correlations performed during motion estimation.

An astonishing number of algorithms have been developed for motion estimation. A comprehensive review of motion estimation algorithms is beyond the scope of this chapter, but representative algorithms include three-step search [579], four-step search [580], diamond search [581], one-dimensional full search [582], and modified log search [583]. Motion estimation developers measure both the image quality produced by their estimates of motion and the number of search steps. Key to reducing the time required for motion estimation is to reduce the number of points tested. Fast search algorithms tend to have less uniform memory access patterns and be less pipelineable than simple full search. The most sophisticated algorithms test a small number of candidate positions; not only does the exact pattern of points vary depending on the visual content of the frame, but some algorithms also vary the number of candidate positions tested.

Macroblocks often do not appear totally unchanged from one frame to another; to handle this problem, the encoder decodes the frame, compares it with the original, and then produces an error stream to describe the differences between the motion-compensated image and the original. The motion estimator may not find a sufficiently good match for a macroblock, in which case it transmits the block directly.

MPEG-1/2 define three types of frames. The I (for inter) frame does not use motion estimation, it only uses DCT. P (for predictive) frames use motion estimation in forward time—earlier frames are used to predict later frames. B (for bidirectional) frames use motion estimation in both forward and backward time. Relatively few encoders available today produce B frames because of the large amount of memory and high computation rates required to perform bidirectional motion estimation.

The MPEG bit stream has a rich syntax. The system layer describes the relationship among audio, video, and additional material. The video layer is organized into *groups of pictures* (GOPs). Each GOP may consist of various combinations of I, P, and B frames.

Compressed video is generally accompanied by synchronized audio, so a word about MPEG audio encoding is appropriate. MPEG-1 defines three layers or levels of audio encoding. Layer 1 is the simplest layer; it applies subband coding followed by Huffman coding. A subband coder uses a filter bank to decompose the input into multiple frequency bands. Each of the subbands tends to have higher correlation than the entire stream, allowing more efficient entropy coding. Layer 2 adds quantization to layer 1. Layer 3 is the most complex; it applies perceptual coding to achieve high-quality encoding at relatively low bit rates. (The term MP3 is derived from MPEG audio layer 3.)

MPEG-4 includes object-based encoding. Various coding operations can be performed on arbitrarily shaped blocks, not just square regions. Although some of the promise of MPEG-4 remains unfulfilled, object-based coding has become popular in multimedia application VD menus.

MPEG-7 is designed to describe multimedia libraries. MPEG-21 concentrates on rights management. The other numbers are unused.

14.2.2   Recognition

Video recognition builds on image recognition techniques. Image recognition relies on a hierarchy of operations, such as color segmentation, edge detection, and contour generation. Video adds motion information that can often be very useful in identifying important subjects.

A number of recognition problems have been defined. Human recognition is clearly an important category. Separate techniques have been developed for other types of subjects, ranging from animals to mechanical components. Face detection and face recognition are two important human recognition problems. A face detector determines when a face is visible. A face recognizer, in contrast, identifies a person based on facial features. In practice, most face recognizers will need face detectors because they will operate in relatively unstructured environments with people in a variety of positions, and so on. Gesture recognition is another important category of human recognition that identifies positions or movements of the body and classifies them into known types of gestures.

As systems-on-chips become more powerful, we expect to see new applications of digital video that will place their own demands on MPSoC architectures. Figure 14-4 shows the basic flow of the gesture recognition system developed at Princeton University [584]. Each frame is processed through five major phases. The first phase subtracts out the background and classifies pixels by color. This phase identifies flesh-tone and non-flesh-tone pixels. The *contour following* phase extracts out a continuous contour for each major region. *Ellipse fitting* creates a

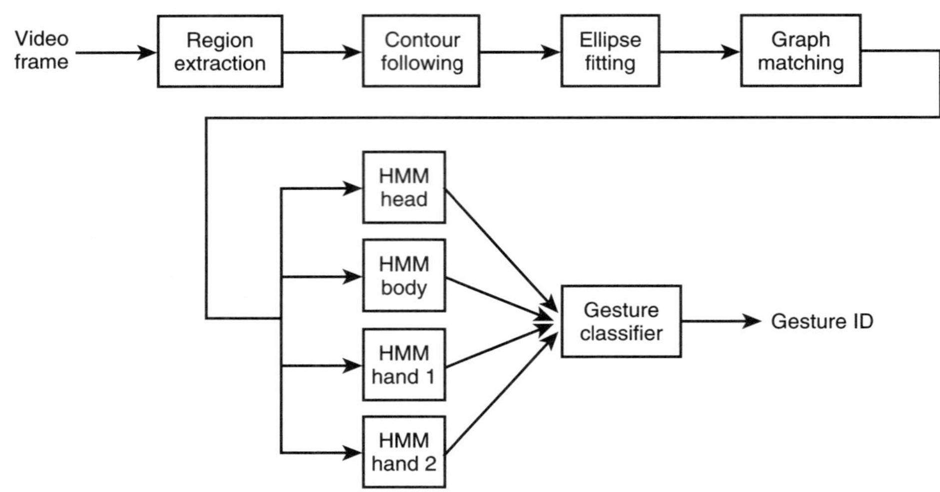

14-4

Block diagram of our gesture recognition process.

FIGURE

closed curve to represent each contour. Graph matching creates an annotated graph that describes the relationships of the curves generated by ellipse fitting, and then compares the graph against a library of known graphs in order to determine the identity of the regions: head, torso, and so on. We then apply hidden Markov models (HMMs) to the major body parts in order to analyze their behavior over time. The results of the HMMs are combined in a classifier to determine finally the gesture being performed at that stage.

This application requires an even greater diversity of processing than does video compression. Although some of the early operations are performed on 23-bit full-color pixels, later stages can be performed on 1-bit or 2-bit pixels that represent various color classifications. At even later stages we operate on graph models. The final stage uses a significant amount of floating-point arithmetic.

14.3 ARCHITECTURAL APPROACHES TO VIDEO PROCESSING

Early architectures for video concentrated on optimizing a single operation. To this end, different types of array processors and single-instruction multiple data (SIMD) machines were proposed for video operations, and SIMD has been widely

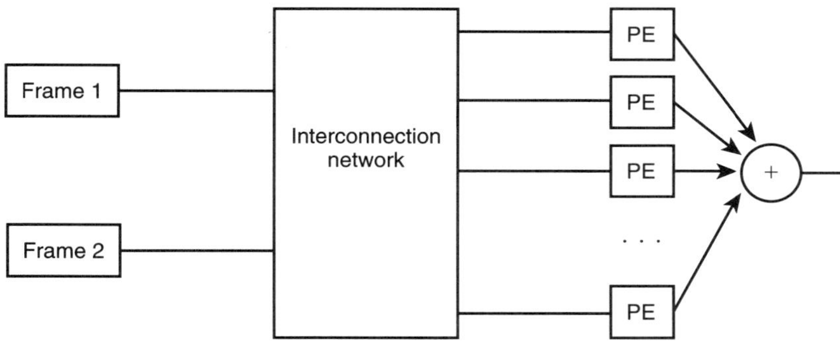

14-5

FIGURE

The motion estimation architecture of Yang et al (ref. 585).

used for motion estimation. Figure 14-5 shows the SIMD motion estimation architecture proposed by Yang et al. [585]. This machine uses an interconnection network to shuffle the data values as required between processing elements (PEs) in order to minimize the number of frame-memory accesses to pixels.

SIMD machines can implement regular video operations very efficiently. However, they generally offer only limited programmability. Furthermore, they generally restrict the maximum sizes of certain data objects. Although size restrictions may be acceptable for systems built to implement established standards, they may not accommodate new standards or applications.

As video algorithms became more sophisticated and chip sizes increased, attention moved to chips that implemented more complete video applications. In general, these early VLSI video systems were built as heterogeneous multiprocessors. In some of the earliest systems, every major operation was built as a separate processing element. Since different blocks received very different utilizations, this caused some hardware to become idle. As chips became larger, lower rate operations tended to be swept into *programmable multimedia processors*.

Multimedia processors are defined by Bhaskaran et al. [586] as a class of programmable processors that provide "multimedia on a chip" and are meant to accelerate the simultaneous processing of several different multimedia data types. Early implementation of such multimedia-specific computing engines saw an effort toward supporting the multimedia applications (e.g., MPEG video playback, JPEG still image display, and so on) on general-purpose processors. However, owing to the inefficiency inherent in trying to map multimedia operations (e.g., multiply-accumulate, saturation arithmetic, and so on) and data types (e.g., audio and video data requiring 8, 10, 16, 24, or 32 bits of precision) onto general-purpose computers, this approach soon gave way to multimedia enhanced processors, whereby the instruction-set architecture of a general-purpose processor was augmented via multimedia-specific instructions. Instruction-set extensions to CPUs also take advantage of SIMD structures to offer improved implementation of regular video operations on traditional CPUs. The CPU's datapath is split into subwords. The major change to the datapath is to allow for the carry chain to be cut during subword operations. The CPU instruction is applied to the split datapath: for example, an ADD instruction applied to operands that are split into four subwords will provide four results, all packed into the destination. Two of the earliest adopters of the philosophy were the PA-RISC [587] from HP and the UltraSPARC [588] from Sun. Other quick inclusions in this exalted company were the MMX^{TM} [589, 590] enriched Pentium processors [591] from Intel, the Media-GX processor [592] from Cyrix (now a part of National Semiconductors), and the K6 processor [593] from Advanced Micro Devices.

In spite of offering a significant performance boost to general-purpose processors, however, the multimedia-enhanced processors had difficulty keeping up with the constant evolution of multimedia standards, sophisticated algorithms, and new applications. This led to the shift toward developing programmable single-chip multimedia processors such as the Mpact [594] from Chromatic Research (now a part of ATI Technologies), the Trimedia [595, 596] from Philips, the Alpha 21164 [597] from Digital Equipment Corporation (that later became a part of Compaq and is now a part of Hewlett Packard), and the advanced products from Texas Instruments (TI) [598].

Programmable architectures comprise both functional and memory (both on- and off-chip) units that allow processing of different tasks under software control, thereby trading area for flexibility. Programmability thus incurs additional hardware cost not only for extra control units and program storage, but also software development. The big advantage, however, is that not only can many different algorithms now run on the same programmable hardware, the flexible control mechanism can also support execution of algorithms with irregular and unpredictable data and operation flows. Such programmable media-processing architectures, according to Pirsch and Stolberg [599], are typically designed to utilize the data-, instruction-, or task-level parallelism inherent in the application and algorithms; special instructions and/or hardware are also designed at times to improve the processing efficiency. Based on the design and operational principles involved in making effective use of the available parallelism, the following architectural concepts have found widespread use in the media-processing world:

* SIMD: SIMD stream architectures are based on data parallelism. They are characterized by multiple datapaths executing the same operation in parallel

on different data entities. An example of the SIMD concept can be found in the Multimedia Video Processor (MVP) [600] from Texas Instruments.

+ Split-ALU: The split-arithmetic and logic unit (ALU) concept makes use of subword parallelism, whereby a number (e.g., four) of lower precision (e.g., 8-bit) data items are processed in parallel on a higher precision (e.g., 32-bit) ALU. Of course, the ALU needs hardware extensions, for example, to prevent carry signals in addition operations to propagate across data boundaries. An example of the split-ALU concept can be found in Sun's Visual Instruction Set (VIS) [601] design for its UltraSPARC processors.

+ VLIW: A very long instruction word (VLIW) machine provides a means to exploit instruction-level parallelism of multimedia algorithms by specifying, in a long instruction word, the concurrent execution of multiple operations on multiple functional units. In contrast to superscaler machines that also try to extract instruction-level parallelism (dynamically in hardware), VLIW processors employ static instruction scheduling performed by software at compile time. An example of a VLIW machine is the Philips TriMedia processor [595].

+ MIMD: The multiple instruction and multiple data (MIMD) stream architectures try to exploit parallelism at both the task and the data level. An MIMD machine features multiple datapaths, each having its own control unit; different datapaths, therefore, can be programmed at the same time to perform different processing on different data streams. MIMD machines can be further classified as *tightly coupled* with a shared memory or *loosely coupled* with a distributed memory. The SGI Power Challenge [602], from Silicon Graphics Inc., is an example of a shared-memory MIMD machine.

+ Specialized instructions: The idea here is to study specific multimedia algorithms, identify common operations, and introduce special hardware functional units in order to replace a longer sequence of standard, frequently occurring instructions by a specialized hardware-supported instruction in an effort to reduce instruction count and speed up program execution. An example is the MMX™ technology [589] designed to accelerate multimedia applications for the Intel Pentium processors.

+ Co-processors: By incorporating one or more separate dedicated hardware modules adapted to specific tasks, co-processors allow execution of regular, compute-intensive tasks on dedicated hardware, whereas the less compute-intensive but irregular control and processing tasks are executed on one or more programmable processor cores. The PNX-8500 media processor [603] from Philips features a number of co-processors for audio and video processing.

We conclude from the above data that a variety of different processor architectures have been conceived and designed to date to support the emerging demands of multimedia.[1] However, as new algorithms and applications continue to put increasing demands on multimedia processors, in terms of dealing with data dependence, multiple media streams, and irregular data and control flow, exploration of even more innovative architectural concepts have become necessary. Simultaneous multithreading [599] and reconfigurable computing [604] are some of the current research and development efforts.

14.4   OPTIMAL CPU CONFIGURATIONS AND INTERCONNECTIONS

For the ever increasing set of media-processing applications, improving the performance of the execution of a single instruction stream often results in only a limited overall gain in the system performance. Intuition and experiments suggest that for these applications, much better performance can be achieved by employing multiple processors that share the burden of controlling the necessary real-time and non-real-time tasks.

14.4.1   Monolithic CPUs

In addition to integrating various audio, video, and peripheral interface units, an emerging trend for multimedia SoCs is to muster enough processing power by utilizing multiple CPU cores. In fact, quite a few of the currently available ICs already feature two CPUs. For a typical digital television, digital video set-top box, or DVD recorder system, Philips provides silicon that integrates a MIPS core and a TriMedia processor [603]; for a mobile handset baseband processor, Philips offers a combination of an ARM core or multiple ARM cores and an Adelante digital signal processor (DSP).[2] The OMAP5910 SoC from TI also incorporates two CPU cores—a TMS320C55x digital signal processing core and a TI-enhanced ARM (925) core [605].

---

[1] An excellent survey and classification of the prevalent multimedia architectures have been compiled by Furht [606]. In his survey, Furht also considers hybrid architectures that combine several of the different architectural principles mentioned above.

[2] The DSP is based on a R.E.A.L. core that Philips Semiconductors developed internally and then spun off to Adelante, a startup initially founded via the merger of Philips' DSP division and Frontier Design.

For the PNX-8500, an architectural decision was made quite early in the design cycle to use the TM32 TriMedia core together with the standard MIPS32 reduced instruction set computing (RISC) core. The choice was guided by the pre-existing (external) software stacks for the target application and Philips' own portfolio of processor cores with existing applications and standard compilers. The requirements on the RISC processor were high performance, capability to run popular embedded operating systems, and efficient control of infrastructure peripherals. The VLIW processor, on the other hand, was required to have a very high performance and a multimedia-enhanced instruction set suitable for audio and video processing. This allows balancing of the system and distribution of tasks among the two CPUs. The scheduling task in PNX-8500 is usually assigned to TM32—the CPU with the faster response time. Besides running all the audio decoding and processing functions, the TriMedia core also implements other nontrivial multimedia algorithms that are not supported directly by the hardware functional units. The MIPS core, on the other hand, runs the operating system and, on top of it, the software application provided by the service provider. The application software deals with the accessibility of the service (conditional access) and the general control functions. All graphics-related functions are also handled by the MIPS processor.

14.4.2   Reconfigurable CPUs

Today's SoC architectures, as mentioned above, frequently exhibit only a few more or less powerful embedded processors that, depending on the application, might be RISC processors, VLIW cores, or DSPs. Recent trends in the embedded CPU IP market, however, show an increasing preference toward reconfigurable computing [604] and, therefore, toward compile-time-configurable CPUs like Tensilica [607] and ARC [608]. Extensible and configurable (tailored) processors offer many of the benefits of hardware accelerators (adding hardware for specific processing problems), while solving some of the problems associated with the design of hardware accelerators [609]. These reconfigurable CPUs are generated with software tools and customized with regard to not only the cache and memory size but also the number and kind of peripherals that need to be supported. Another very helpful feature of these CPUs is the capability to add custom operations associated with hardware extensions such as lookup tables, x-y memories, add-compare-select units, and multiply-accumulate (MAC) units. These extensions make the CPUs very "DSP-like" and therefore well suited for baseband communication or media-processing tasks.

Reconfigurable hardware is believed by many [610] to be an ideal candidate for use in SoCs because it offers a level of flexibility not available with more traditional circuitry. *Hardware reuse,* for example, allows a single configurable architecture to implement many potential applications. Another flexibility offered is easy *postfabrication modification* that allows alterations in the target applications, bug fixes, and reuse of the SoC across multiple similar deployments to amortize design costs. Reconfigurable logic, Gries [611] stresses, should, however, be used for the execution of tasks with moderate timing constraints, in order to benefit from the saving of silicon area owing to a more efficient utilization. Gries argues that general CPUs are equipped with many functional units, thereby making it very difficult to obtain an optimal exploitation of the hardware resources; with reconfigurable hardware, however, it is possible to "synthesize" the required units at the right time and to occupy only the chip area that is needed for the execution of the current task(s). Besides, dynamically reconfigurable systems also permit adaptive adjustments during run time [611].

14.4.3   Networked CPUs

The choice of the CPU architecture is determined not only by the raw processing power required but also by the chosen software architecture and the set of use cases that need to be fulfilled. A few big monolithic CPUs may be very suitable for traditional, high-performance, general-purpose processing tasks, but they are less capable of efficiently performing many small (and special) tasks, as required, for instance, in the realm of mobile multimedia communications; the task-switch overhead and the special functions needed to perform the multitude of tasks (e.g., in baseband processing) can impact and reduce the performance of a single monolithic general-purpose CPU.

In such a situation, it might be a better choice to disassemble the required algorithm into its base components and then map them individually onto more specialized hardware. A network of small DSPs or DSP-like CPUs with special extensions targeted toward the specifics of the application provides a more flexible platform for the mapping process, but it comes with its own set of challenges. Problems to be solved in the new scenario comprise sharing the available storage and automating the dataflow between the CPUs, without having to introduce a new bottleneck in the form of a *shared-memory* resource. The usage of a shared memory among multiple CPUs, although often welcomed by software programmers, is probably best avoided wherever possible, because it is generally associated with high-throughput demands on the off-chip memory subsystem. *Data streaming* is a much more desirable alternative from a hardware point of view and is acceptable as long as such streaming is assisted by the hardware components and rendered semitransparent to the software drivers. In fact, Philips is advocating [596], via the Nexperia™ platform, the use of multiple smaller streaming embedded processors controlled by another embedded processor to ensure the speed and reliability of a given design while saving silicon costs.

14.4.4   Smart Interconnects

Data streaming usually assumes use-case-dependent flexible interconnections between the software and the hardware components. The software view of these connections should be abstracted to the concept of pipelines (blocking reads and writes). Tree interconnect structures, which loosely connect highly concentrated clusters of CPUs, are an efficient way to keep the wiring within reason. However, to offload the CPUs from the interprocessor communication tasks, s*mart interconnects* are desirable. These interconnects can automatically manage the buffers and dataflows associated with two partners in the network without any CPU intervention. Data-triggered software tasks, via interrupt mechanisms, result in very efficient data-driven communication and processing within such a system. Note that the network processing elements, or processing *nodes* as they are called, are usually of a heterogeneous nature; depending on the use cases, however, a mix of general-purpose CPUs to perform control tasks, DSP-like compute elements with specific extensions for signal processing, and highly specialized hardware functions for high-performance computations (e.g., large fast Fourier transforms [FFTs] or filters for high data rates) is usually desirable. The mix can be further augmented by field programmable gate arrays (FPGAs) and/or highly programmable array-configurable hardware in order to perform functions such as interfacing to off-chip peripherals. Size and power considerations become extremely important for such heterogeneous multiprocessors, and so does a detailed use-case analysis to identify the algorithmic requirements.

14.4.5   Software Support

The processing networks mentioned earlier are very demanding with respect to the choice and use of software tools. Scheduling of resources and programming of the processing chain is a particularly nontrivial task. To alleviate the problem, a static mapping of a given algorithm to the hardware can be initially performed. However, as the tools (software) get more sophisticated and as the hardware becomes capable of automatic resource scheduling and buffer management, new functions such as dynamic hardware-resource allocation at run time, real-time task switching, and time sharing of the hardware resources between different applications become possible. Advanced software tools can potentially ease the task of mapping a certain algorithm onto a computing (processor) array. A configurable custom library of common functions (e.g., FFT, finite impulse response filtering, Viterbi decoding, Reed Solomon decoding, modulation, and so on) can also assist the mapping process.

Flexibility—programmability, adaptability, and upgradability—in multimedia systems mandate that the system-level functionality be implemented more in software than hardware. Market data indicate that this is already the case today, with more than 80% of the system development efforts being in software [612]. Software productivity, however, is rapidly becoming a bottleneck in multimedia SoC designs because even though the amount of software is increasing exponentially, the efficiency of software design is unable to keep pace [613]. Efficient software design environments, effective software reuse standards, easy software portability, and widespread software compatibility among products and product generations are, therefore, absolutely essential for the successful design of future multimedia systems.

14.5   THE CHALLENGE OF SOC INTEGRATION AND IP REUSE

As the electronics industry demands ever more powerful and cheaper products, the latest chip design trend, thanks to the fabulous growth in the semiconductor industry, is to build all the circuitry needed for the complete electronic system on a single chip. Remarkable advances in manufacturing technology have blurred the traditional separation between component design and system design by allowing the merger of various components—pre-designed modules, hard or soft IP blocks, and so on—on the same silicon substrate. The buzzword used to describe such designs is SoC design. As depicted in Figure 14-6, these chips are no longer stand-alone components but complete silicon boards comprising nonhomogenous circuit components and encapsulating complex system knowledge. SoC designs require all functions from the front-end to the back-end design to be integrated into a seamless flow. Figure 14-7 shows a high-level abstraction of such a design flow.

Successful SoC designs today are assembled at the IP block level, and they demand concurrent hardware-software design (co-design) and careful IP integration. To this end, reusability or recycling of an IP core, with an easy-to-use and/or an easy-to-modify interface, becomes extremely important. A reusable IP core,

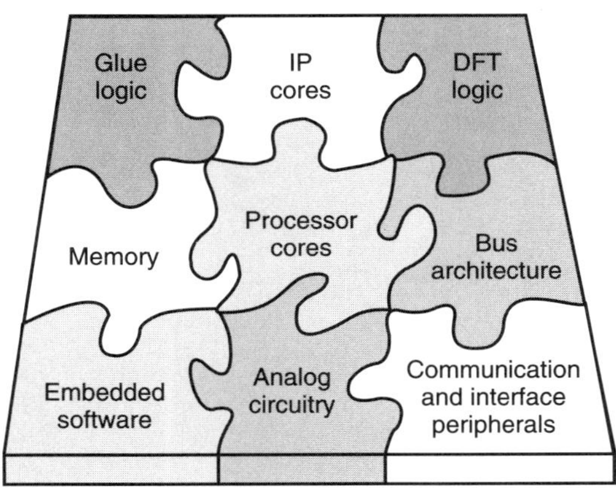

14-6   SoC design example.

FIGURE

often called a *virtual component*, typically refers to a preimplemented, optimized, and reusable module with a standardized interface that can be quickly adapted and reliably integrated, with little or no modification, to build single-chip systems. Reusable hard (placed, routed, and verified logic), firm (a synthesized netlist with floorplanning or physical-placement guidance), or soft (synthesizable RTL description) IP cores are usually obtained from independent vendors, or they get exchanged and/or sold between various departments and cost centers of a company. The latest design trend is to implement true systems on chips that depend on realizing the system functionality by integrating and composing pre-existing, plug-and-play IP cores (each of which may potentially be designed in an ASIC flow) from different vendors and/or design groups, and possibly porting them to advanced technologies with smaller geometries. A digital television or a digital set-top box SoC can, for example, make use of IP cores for video process-ing (e.g., video capture, picture enhancement and scaling, picture artifact and noise reduction, MPEG encoding and decoding, display processing, and so on), IP cores for audio processing, memory controllers, on-chip processors, and bus interfaces (e.g., USB, 1394, UART, and so on). The design of such a chip is pri-marily a design-integration or a design-composition process, where a co-design methodology is followed to identify various virtual components, the components are stitched together by designing the necessary glue logic between them, and the assembled logic is pushed through a tightly coupled set of electronic design automation (EDA) tools in an integrated synthesis environment.

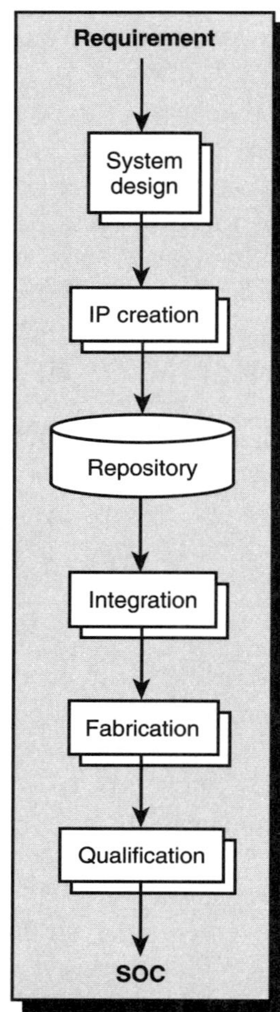

14-7   SoC design challenges.

FIGURE

With the ability to integrate millions of gates on a single chip, the SoC design bottleneck is no longer in obtaining higher density on the chip—the bottleneck is more in IP selection, design planning, design optimization, hardware-software co-design, design verification, and the low-level electrical problems which one needs to cope with. Some of the electrical problems, which did not pose that serious a threat in traditional designs but now need to be carefully considered, are signal integrity, signal noise, harmonic frequencies, transmission-line effects, thermal and voltage gradients (across the chip), clock speeds, self and mutual inductances, and delay uncertainties. Besides, an SoC implemented using IP cores can potentially lead to a design with multiple clock domains that pose difficulties for accurate timing analysis and DFT implementation. Therefore, some of the main (and new) requirements of a successful SoC design are:

+ a well-defined on-chip interconnect architecture and communication subsystem.

+ adequate supply of high-quality, reusable (i.e., should have standardized interfaces and should allow easy creation of derivatives), retargetable (i.e., should easily be mapped to different processes and architectures), and reconfigurable (i.e., should be possible to tailor different parameters to meet functionality and performance requirements) IP cores.

+ IP-core compliance to certain rules (for architecture, design, verification, packaging, and testing) that enable integration with minimal effort.

+ availability of IP-evaluation frameworks to evaluate a third-party IP before committing to use it.

+ efficient use of design plan synthesis methods for early exploration of alternative design topologies in order to determine an optimal design flow.

+ hardware-software partitioning and co-design.

+ logic synthesis linked with physical design.

+ performance-driven place-and-route integrated with timing/power analysis.

+ a well-crafted verification and emulation environment that facilitates hardware-software coverification and reuse of drivers and tests across simulation, emulation, and validation.

+ an in-depth system-level design and analysis.

14.6   THE PANACEA/PROMISE OF PLATFORM-BASED DESIGN

In order to emphasize systematic reuse, lower development cost, minimize development risks, and reduce time-to-market, by spinning off quick derivatives, another recent trend is to exploit the benefits of a *platform-based design*. The basic idea behind the platform-based approach is to avoid designing a chip from scratch; some portion of the chip's architecture, as Richard Goering [605] points out, is predefined for a specific type of application. Usually, there is the hardware architectural platform comprising one or more processors, programmable IP cores, a memory subsystem, a communication network, and an input output (I/O) subsystem; an application programming interface (API) then provides the required software abstraction by wrapping the essential parts (of the architectural platform) via device drivers and a real-time operating system (RTOS) [612]. Depending on the platform type, users might customize by adding hardware IP, programming FPGA logic, or writing embedded software.

As Claasen [613] points out, when a company designs a variety of similar systems, it is really advantageous to incorporate the commonalities between different designs into a template from which the designers can derive individual designs; a *platform*, for Claasen, is, therefore, a restrictive set of rules and guidelines for the hardware and the software architecture, together with a suite of building blocks which fit into that architecture. The Virtual Socket Interface Alliance's (VSIA) Platform-Based Design Development Working Group (PBD DWG) defines an SoC platform as "a library of virtual components and an architectural framework, consisting of a set of integrated and pre-qualified software and hardware virtual components (VCs), models, EDA and software tools, libraries and methodology, to support rapid product development through architectural exploration, integration and verification." Another common definition [612] is the "creation of a stable microprocessor-based architecture that can be rapidly extended, customized for a range of applications, and delivered to customers for quick deployment."

Whatever may be the definition, the primary goal is to hide the details of the design through a layered system of abstractions and allow maximum reuse of architecture, hardware components, software drivers, and subsystem functionality, whereby different hardware and/or software components can be added easily to the base platform, and several derivatives can be created quickly. In this manner, a family of products can be delivered instead of a single product, and future generations can build on the original hardware and software investments.

There are essentially four basic steps to using a platform-based design methodology [614]: defining the platform design methodology, creating the platform, defining the derivative design methodology, and creating the derivative SoC product(s). Since it is usually easier to de-configure than configure, the starting point in a platform-based approach is often a de-configurable and extensible prototype reference design built from reusable IP components; some of the existing IP cores are then modified (extended), some are removed (de-configured), and some new ones added during the integration process.

One of the best known examples of a full SoC applications platform is the Nexperia™ Digital Video Platform (DVP) [603] from Philips Semiconductors. It is

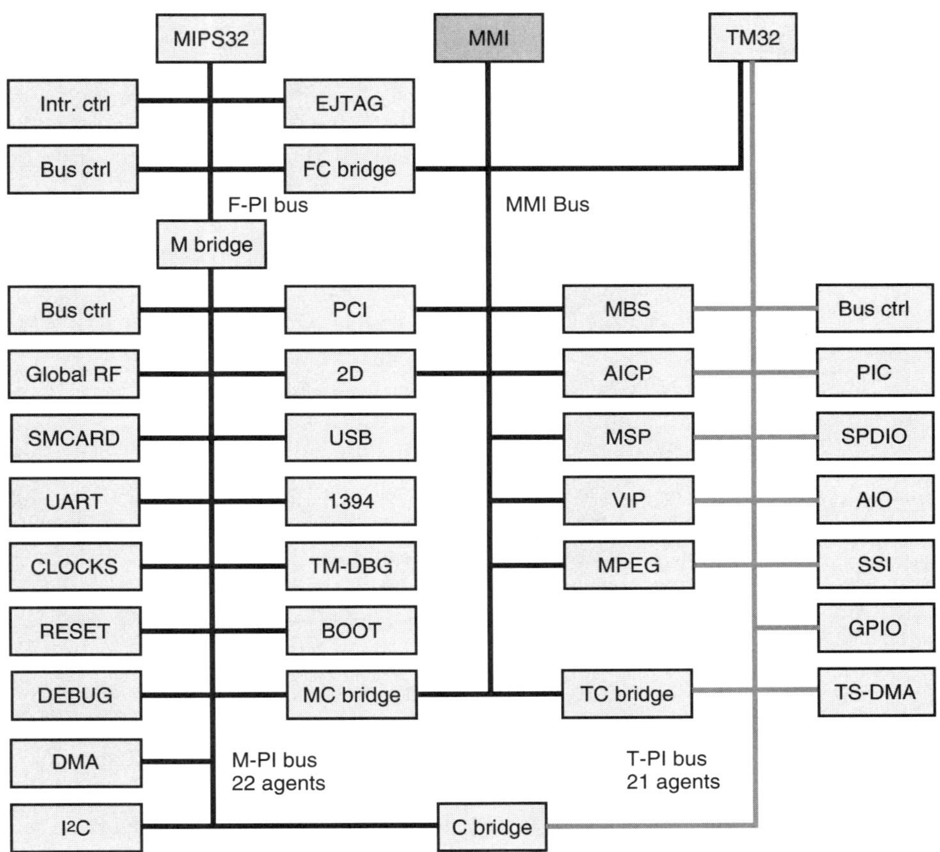

14-8   Architecture block diagram of the PNX-8500 digital television processor.

FIGURE

a scalable platform architecture for a wide range of digital video applications. Time-to-market savings was a strong motivation for the design team that created both the platform and the first iteration of that architecture, the PNX-8500 media-processing chip [603]. The chip was introduced in 2001, and it provides a highly integrated multimedia SoC solution for the integrated digital TV (iDTV), the home gateway, the set-top box market, and the emerging connected-home market.

As illustrated in Figure 14-8, the PNX-8500 media-processing chip, in the true spirit of a platform IC, incorporates various IP blocks and entails a high level of IP and design reuse. Besides featuring multiple processors—a 32-bit VLIW

TriMedia core (TM32) together with a standard 32-bit MIPS RISC core (MIPS32)—the PNX( )-8500 SoC integrates separate audio, video, peripheral, and infrastructure subsystems that were assembled from IP blocks designed or available in-house or obtained from outside. The *peripheral subsystem* comprises general-purpose on-chip peripheral IPs such as a universal serial bus (USB) host controller, three universal asynchronous receive and transmit (UART) interfaces, two multi-master inter-integrated circuit (I^2C) interfaces, one synchronous serial interface (SSI) for implementing soft modems, a general purpose I/O (GPIO) module with infrared remote-receive capability, a 1394 link-layer controller that provides a PHY-LINK interface, and a PCI/XIO expansion-bus interface unit to connect to a variety of board-level memory components.

The *audio subsystem* is centered around an audio I/O (AIO) block featuring three audio input and three audio output IP modules and an SPDIO IP block that provides the Sony Philips Digital Interface (SPDIF) input and output capabilities.

The *video subsystem* is quite pronounced, with well-defined IP blocks for high-quality video processing and display, and boasts of two instances of a video input processor (VIP) to capture standard definition video (NTSC, PAL), a memory based scaler (MBS) that not only deinterlaces, scales, and converts video but also filters graphics, two instances of an advanced image composition processor (AICP) that is responsible for combining images from the main memory and composing the final picture that is displayed, a slice-level MPEG2 video decoder (MPEG) that is suitable for HD decoding, an MPEG system processor (MSP) that, besides filtering, demultiplexing, and processing transport streams (packets), also provides conditional access, and a 2D rendering engine (2D) that is used to accelerate graphics functions. The *infrastructure subsystem* comprises a memory management interface (MMI) that provides and controls access to the external memory and a hierarchical on-chip bus system whose segments are connected by local bridges.

14.7   THE EVER CRITICAL COMMUNICATION BUS STRUCTURES

Multimedia SoCs frequently house multiple processor cores that share the task of running the operating system and controlling the critical and noncritical on-chip functional-unit resources. In this context, an efficient bus architecture and arbitration (for reducing contention) play important roles in maximizing system performance. Besides, for many applications, the performance of multiprocessor systems relies heavily on an efficient communication between the processors and a balanced load distribution (of computing tasks) among them.

With multiple CPUs and a plethora of functional units, on-chip communication poses a critical design problem that is often solved using multilevel hierarchical busses connected via local bridges, the bridges primarily serving as protocol converters between different bus systems and/or connectors between busses with different speeds (e.g., high-speed processor bus and low-speed peripheral bus). CoreConnect, for example, has three levels of hierarchy: processor local bus (PLB), on-chip peripheral bus (OPB), and device control register (DCR) [615, 616]. PLB provides a high-performance and low-latency processor bus with separate read and write transactions, whereas OPB provides low speed with separate read and write data busses to reduce bottlenecks caused by slow I/O devices; the daisy-chained DCR offers a relatively low-speed datapath for communicating status and configuration information [617].

The advanced microcontroller bus architecture (AMBA) from ARM has two levels of hierarchy: the advanced high performance bus (AHB), similar to PLB, and the advanced peripheral bus (APB), similar to OPB. CoreConnect and AMBA are both pipelined busses with bridges to increase the communication efficiency between the high- and low-speed busses (for data transfer between them). Core-Frame from Palmchip Company, on the other hand, is a nonpipelined bus that also has two independent bus types: Mbus for memory transfer and Palmbus for I/O devices [616, 617]. The PNX-8500 SoC from Philips is no exception; it also features, as shown in Figure 14-8, an on-chip hierarchical bus structure supported by local bridges. An interesting analysis of different bus structures has been provided by Keol in one of his papers [617] that compares five different high-speed busses for single-chip multiprocessor systems.

Even though the bus structures across various SoCs (mentioned above) resemble one another at a coarse level of comparison, they do differ in their characteristics and implementation details (e.g., function, width, delay, throughput, pipelining, utilization, number of segments, number of busses, number and type of bridges, number of bus agents, and so on); an in-depth analysis of the target application, the desired timing, and the on-chip communication pattern determine the exact implementation. To this end, the next few sections offer insights into the analysis that guided the final bus implementation in the PNX-8500 SoC.

14.7.1   PNX-8500 Structure

In PNX-8500, the multimedia processing and the control processing functions are split between two CPUs—the TriMedia CPU (TM32) and the MIPS RISC CPU (MIPS32). Thus, each CPU is responsible for the peripherals that belong to its task domain. This leads to the concept of separate processor busses, whereby each CPU controls all the devices on its local bus. Not all the peripheral devices, however, can be owned by one CPU in all user cases (applications), and so provisions have been made so that every peripheral is still accessible from both the CPUs, but with a preference. If a peripheral is indeed shared at run time, the CPUs must negotiate its availability through the use of *Semaphores*. Combining both CPUs in one system (from an SoC point of view) lowers the overall cost of computing by sharing system resources such as main memory, disk, and network interfaces.

All the on-chip functional units (peripherals) are programmable via CPU writes to their control registers. These control registers being memory-mapped, programmable reads from and/or writes to the registers are commonly referred to as memory mapped input output (MMIO) or programmable input output (PIO) transactions. Even though each peripheral can be addressed by both of the CPUs, it is "usually" read or written by the CPU to whose local bus it is connected.

Bus System Requirements

A variety of bus-architecture options were explored for PNX-8500 before deciding on the exact bus structure. The implementation was guided by the following architectural requirements:

◆ the cache traffic of a CPU must be separated from its register-access traffic.

◆ the register-access traffic from the two CPUs should be separated.

◆ the CPUs must have a high-performance and low-latency path to memory (relative to the peripherals).

◆ each CPU must have a low-latency access to the peripherals on its local bus.

◆ all the registers in the various peripheral units must be accessible from the two CPUs, the PCI block, the BOOT block, and the EJTAG block.

With the bus requirements nailed down, the next step was to decide on the bus topology. However, before exploring different topologies, a study of both *tristate* and *point-to-point* bus implementations was conducted in order to find out which of the two better suited the needs.

Tristate Versus Point-to-Point Bus Design

A comparative study of tristate versus point-to-point bus implementations, as outlined in Table 14-1, shows that a point-to-point bus architecture is desirable for

Item	Tristate bus	Point-to-point bus
Wiring	Number small	Number large
	Long	Short
Testability	Complicated	Simple
Layout	Complicated	Simpler
Performance	Low	Higher
Speed	Low	Higher
Modularity	Good	Poor
Timing	Difficult	Simple
Observability	Easy	Difficult

14-1

TABLE

Comparison of Tristate and Point-to-Point Busses.

designs requiring high performance, simple testability, and reduced layout. However, one main problem with the point-to-point bus architecture is that it does not easily allow for multiple-master access to peripherals. For example, if a peripheral requires access by four masters, it is required to have four slave interfaces; adding an additional master will require changes to the peripheral to support five masters. Thus, the point-to-point bus is not very modular or scalable.

For PNX-8500, it was decided to use a high-performance point-to-point memory bus—the MMI bus—for bandwidth- and latency-critical access to the external SDRAM. For access to modules' slave control registers and for lower bandwidth direct memory access (DMA) peripherals, a tristated bus—the PI bus— seemed more appropriate. Another advantage of using the PI tristate bus was that the PI bus had been used extensively throughout Philips Semiconductors and, therefore, a large portfolio of IPs with this interface was already available.

Comparison of Bus Topologies

As mentioned before, high-bandwidth peripherals and peripherals requiring low-latency access to the external memory clearly call for a direct interface to the point-to-point memory bus. For the other peripherals that do not require very high DMA bandwidth, the following three architectural options, as illustrated in Figure 14-9, were evaluated:

+ shared PIO and DMA on a common PI bus.

+ split PIO and DMA on separate PI busses.

+ split PIO and DMA on the PI and the MMI bus, respectively.

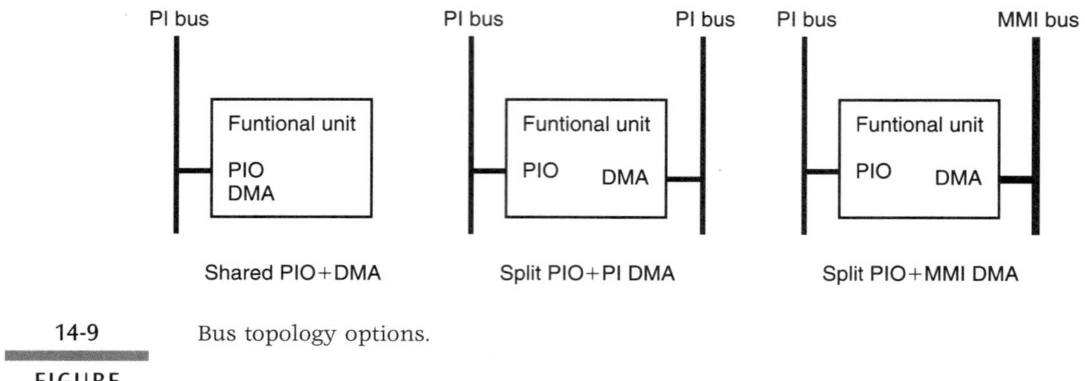

14-9

FIGURE

Bus topology options.

Parameter	Shared PIO + DMA	Split PIO + PI DMA	Split PIO + MMI DMA
Expandability	Very good	Good	Average
Low-latency PIO	Average	Very good	Very good
Security	Good	Very good	Very good
DMA latency	Good	Good	Very good
DMA throughput	Good	Good	Very good
Number of busses	Very good	Average	Bad
Layout complexity	Very good	Average	Bad
Portfolio (compatibility)	Good	Bad	Bad

14-2

TABLE

Comparison of Bus Topologies.

A comparison of the different options is shown in Table 14-2.

For PNX-8500, it was decided to go for Option 1, with *shared PIO and DMA on a common PI bus,* for the modules that are not bandwidth-hungry. The main reason is that most of the existing portfolio of IPs (inside Philips) already supported this topology and, therefore, this option had the least risk.

The Final Structure

The backbone of the on-chip communication infrastructure in PNX-8500 was finally provided by two separate bus systems: a 64-bit point-to-point high-performance MMI bus, also called the memory bus or the DMA bus, and a 32-bit tristated PI bus (Peripheral Interconnect Open Processor Initiative Standard 324).

The MMI bus provides high-speed memory access to those on-chip units that require high bandwidth and low latency. There is no PIO traffic on this bus. The bus connects to the external memory (SDRAM) via a 64-bit, 143-MHz memory management interface that generates the required SDRAM protocol but isolates the on-chip resources by using the proprietary DVP protocol; the MMI also controls access (by the on-chip components) to the memory highway via a round-robin arbitration algorithm with programmable bandwidths.

Unlike the memory bus, the PI bus is not only used for MMIO reads and writes to memory-mapped control registers of the various peripherals, but it also provides a medium-bandwidth DMA path via a bridged gateway connection to the MMI bus. The PI bus itself is divided into three different segments:

+ F-PI (Fast PI) bus: used for low-latency access to the memory and selected peripherals by the MIPS CPU.

+ M-PI (MIPS-PI) bus: used for access to the peripherals typically controlled by the MIPS CPU.

+ T-PI (TriMedia-PI) bus: used for access to the peripherals typically controlled by the TriMedia TM32 CPU.

The MMI bus and the various PI bus segments are connected via a number of bridges, as shown in Figure 14-8. The FC-Bridge, the MC-Bridge, and the TC-Bridge act as gateways providing memory access to the corresponding PI segments, whereas the C-Bridge acts as a crossover PI-to-PI MMIO bridge that allows memory-mapped I/O access from each processor to control and/or observe the status of all peripheral modules. The M-Bridge basically bridges transactions between the fast PI bus and the not-so-fast peripheral PI bus segments on the MIPS side.

14.8   DESIGN FOR TESTABILITY

With the push toward increased product performance and higher design complexity, the current "giant" SoCs not only incorporate multiple IP cores, they also mix diverse circuits such as digital random logic, functional units, processor cores, static memories, embedded dynamic random access memories (DRAMs), and analog circuits on a single chip. Also, new circuit types such as FPGAs, flash memories, radio frequency (RF) devices, and microwave devices are also almost at the point of becoming a regular feature of the on-chip circuitry. Zorian [618] believes that it is only a matter of time before we move beyond the realm of conventional circuits and "electronics-only" ICs and start integrating optical devices and micro-electro-mechanical (MEM) elements in a regular SoC.

Today's core-based SoC designs, unfortunately, pose multiple testability problems. For one, the nonhomogenous circuit types and cores exhibit different defect behaviors and require different test solutions. Second, the cores may originate from widely different sources and therefore have varying degrees of "test friendliness." An easy, cost-effective, and widely accepted method to alleviate these problems is to embed in the design, besides the traditional scan and built-in-self-test (BIST) circuitry, special hardware, *Sources* and *Sinks* [617] (that provide easy accessibility—controllability and observability—of internal points) corresponding to each circuit type and/or each IP core.

To guarantee adequate testability of cores in an SoC, as well as their test interoperability and test reusability, a new test standard—the IEEE P1500 [619, 620] embedded core test standard—has emerged in recent years. The P1500 standard suggests using module-level boundary-scan structures, called *wrappers*, that allow *intercore* and *intracore* test functions to be carried out via a test access mechanism (TAM). The wrapper isolates an IP core from its environment and ensures that:

◆ the IP core itself can be tested after it has been instantiated in the SoC.

◆ the interconnect structures between the cores can also be tested.

In case of the PNX-8500, it was not only a very large suite of functional tests, a full-scan design methodology, and a BIST of the larger memories and the caches (in the CPUs), but also a *test-shell isolation* of each IP core that finally led to a very high test coverage. The test-shell isolation guarantees that every IP core is completely testable. Test isolation is obtained by ensuring that each IP core input and output is both controllable and observable. Controllable inputs and observable outputs facilitate stand-alone as well as parallel testing of a core. Observable inputs and controllable outputs, on the other hand, allow development of interconnect tests that verify bus connections between different IP cores. Busses and all interconnectivity between the on-chip peripherals and the CPUs in PNX-8500 were tested using *interconnect tests*; these tests are relatively straightforward to implement when every peripheral contains a specially designed test shell.

14.9 APPLICATION-DRIVEN ARCHITECTURE DESIGN

In this section, we consider characterization of applications and architectures. Trace-driven simulation is widely used to evaluate computer architectures and are useful in MPSoC design. Because we know more about the application code to be executed on an application-specific MPSoC design, we can use execution traces to refine the design, starting with capturing fairly general characteristics of the application and moving toward a more detailed study of the application running on a refined architectural model.

Because video applications are computationally intensive, we expect that more than one platform SoC will be necessary to build video systems for quite some time. For some relatively simple applications, it is possible to build a single platform that can support a wide variety of software. However, fitting a video application on a chip often requires specializing the architecture not just to generic video algorithms but more specifically to the types of operations required for that application. As we design video SoCs, we need to evaluate both the application and the architecture. At the early stages of the design process, we want to evaluate the application in order to understand its characteristics. As we add detail to the design, we want to evaluate the application's properties while running on the candidate architecture. As we design SoCs for leading-edge applications, we will use all the available computing resources. We can squeeze extra capability out of the system with some specializations, but we need to know where to specialize the architecture. Of course, we also want the architecture to be as flexible as possible since video algorithms are still evolving. Overspecialization may prevent an SoC from being adapted to a new algorithm, limiting its lifetime. So we need to characterize the application carefully to understand the right points for architectural assists and where more general-purpose solutions can be used.

14.9.1 Application Characterization

Fritts [621] conducted a number of studies on programmable media processors. These studies used the Impact compiler and simulation system and the Media-Bench benchmark set to study both the properties of multimedia applications and processor architectures for multimedia. A few results are particularly interesting for this discussion.

First, Fritts found that nearly 20% of operations are branches and that the average size of a basic block was 5.5 operations. The average size of basic blocks varied widely across the MediaBench set. The mean basic block size was consistent with general-purpose applications, but MediaBench showed a much wider variance. Static branch prediction was found to work well; when the pgpcode benchmark was removed, the average static hit rate was 88%. A great deal of total execution time was spent on the two innermost loop levels; loops averaged about 10 iterations per loop. The path ratio, the average number of instructions executed per iteration divided by the total number of loop instructions, was found to be lower than expected. The average path ratio was 78%, indicating that multimedia applications have a significant amount of control in their loops.

Fritts studied various instruction issue mechanisms including VLIW, in-order superscalar, and out-of-order superscalar. These experiments showed that out-of-order processors performed significantly better than either in-order or VLIW processors. A study of instructions per cycle (IPC) as a function of instruction width showed that issue width flattened out relatively quickly, around 5, and confirmed that out-of-order was significantly better than either in-order or VLIW.

14.9.2   Architectural Characterization

As we refine the SoC architecture, we want to evaluate our target applications on the architecture under consideration. An MPSoC for an advanced video application such as video analysis will require multiple processors; these designs are natural candidates for networks-on-chips. Network-on-chip design requires careful evaluation of the utilization of several candidate network architectures. Because activity in advanced applications is data-dependent, trace-based evaluation is an important methodology. We can perform some evaluation early by simulating each stage relatively independently or by executing it on a general-purpose processor. Even if this evaluation is not done with detailed timing information, it is still important in two senses. First, it gives communication traffic patterns, which are the base of partitioning a whole system into a collection of subsystems. Second, if a netlist has a tight performance without communication delays, it won't satisfy performance requirements with them.

Let us first use traces to evaluate the performance of several motion estimation algorithms, which would constitute one subsystem in a video compression system. Figure 14-10 compares the performance of seven popular motion estimation algorithms, in terms of clock cycles per block matching. The seven motion estimation algorithms are modified log search (MLS), diamond search (DS), four-

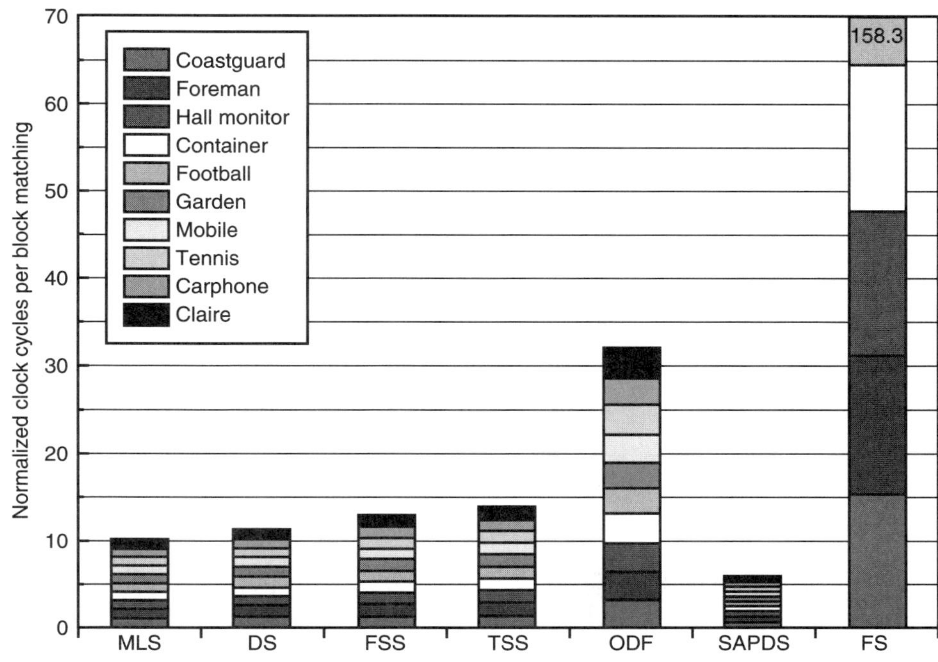

14-10

FIGURE

Performance of several motion estimation algorithms on a number of benchmark videos.

step search (FSS), three-step search (TSS), one-dimensional full search (ODF), sub-sampled motion field search with alternating pixel decimation (SAPDS), and full search (FS). Ten standard test sequences are used in this step, and they are Coastguard, Foreman, Hall Monitor, Container, Football, Garden, Mobile, Tennis, Carphone, and Claire. *Sim-outorder* is used to collect the timing information. It is a simulation tool that belongs to SimpleScalar, a suite of public domain simulation tools, and can generate timing statistics for a detailed out-of-order issue processor core with a two-level cache memory hierarchy and main memory. In Figure 14-10, all the data are normalized by the simulated results of an MLS algorithm. For example, the clock cycles per block matching for FS algorithm running the Coastguard sequence is normalized by that of the MLS algorithm running the same sequence. The figure for FS is truncated, and the value is 158.3.

The traditional way to compare search speed is just to compare the search points per block matching of different algorithms. However, this method is not accurate. For example, when compared in terms of search points per block matching, the speed ratio between FS and MLS is 19:1, on the other hand, when compared in terms of clock cycles per block matching, which exactly means running speed, the speed ratio is just 15:1. Moreover, the SAPDS algorithm, which uses 4.8 times more search points than the MLS algorithm, is the fastest algorithm when executed in the SimpleScalar simulator, as shown in Figure 14-10. This is because SAPDS uses the pixel decimation pattern when doing the criterion calculation. The above result strongly suggests that number of search points is not a good measurement to compare the speed performance, as is commonly used by most reported works. For MLS, DS, FSS, TSS, and ODF algorithms, the central search point is not known until the last search point is checked in the previous step. Then the data for the next step cannot be preloaded into the cache before the beginning of the search. What's more, there are many branches in these algorithms to determine which location around the central search point should be used in the search. These two aspects waste many clock cycles when doing the block motion estimation for MLS, DS, FSS, TSS, and ODF algorithms. Instead, FS and SAPDS algorithms avoid these problems and spend less clock cycles per search point. SAPDS uses pixel decimation and spends the fewest clocks to finish one block matching.

Because global communications is the performance bottleneck in SoCs, designers must pay attention to global communication traffic and localize communication within a small part of the network as much as possible. By grouping highly cooperative IP cores into subsystems, most communication traffics are localized. Localization is based on the traffic information gathered in the last step. As the communications are localized, the whole system is also partitioned into a collection of subsystems. These subsystems must be small enough that data can cross it in one clock cycle. This requirement is compatible with the communication localization requirements, and more importantly it helps to avoid asynchronous designs, which are usually very hard, in subsystems. Actually, asynchronous design is limited in global network designs. After the whole system is partitioned into subsystems, each subsystem can be separately designed, and a global network, which connects all the subsystems, is also designed. Each subsystem has its own clock and forms a local network. Busses could be main actors in local networks because they are synchronous and work efficiently in small areas. The communication between one subsystem with another is handled by communication interfaces. The interface is an agent that translates protocols and buffers data between a subsystem and the global network. Global network is the networks-on-a-chip we discussed before. It connects all the subsystems and transfer data between them fast and efficiently. Its nodes are synchronous subsystems, but it works asynchronously. This is a globally asynchronous and locally synchronous scheme. A further verification and performance estimation can test the results of the subsystem and global network design. A problem found in this step may mean that the design must be revised, and it may need several iterations before all the problems are solved.

One could simulate a multiprocessor system by building a simulator that models the whole system. In practice, the method is usually not an effective solution. Few multiprocessor systems can be modeled by existing simulators. As a consequence, developers have to develop a simulator, which results in high costs. In addition, when modeling details of the whole system, a simulator can be very slow to execute and results in high time cost in development. In our design process, we take a different approach. The whole system is evaluated in a hierarchical method. An architectural design is segmented into several components. Each component is evaluated by a detailed simulator. At the system level, the evaluation results and communication data between different components feed into a high-level simulator, which gives an overall evaluation of the system.

We often start the design process with a reference implementation that we can execute on a uniprocessor. We can use this uniprocessor model to generate *abstract traces* of system behavior. The abstract trace will not reflect detailed architectural behavior such as timing, but it does capture the inputs and outputs of the subsystems as they interact. We can add detailed information about timing, cache behavior, and so on, a subsystem at a time. By running a subsystem's abstract trace through a more accurate architectural simulator, we can generate additional detail as we need it without running the entire trace through a relatively slow detailed simulator.

Figure 14-11 shows the traffic as a function of time in the network-on-chip for a multiprocessor designed to support smart camera algorithms such as the Princeton gesture recognition application. The trace used to create this plot was generated in two steps: we first collected frame-by-frame results of each stage of the smart camera application; we then used SimpleScalar to simulate each stage in order to provide detailed timing information. After this, the trace files and the collected performance data are used in the Opnet™ simulation environment to analyze the communication costs and overall system performance.

Looking at the network activity allows us to evaluate how utilization varies as a function of time; workloads vary as each stage finishes the work on a frame at different times, depending in part on the data in that frame. We are also interested in overall performance, which we can evaluate by aggregating the trace data. Figure 14-12 shows overall performance for several candidate architectures for the smart camera MPSoC.

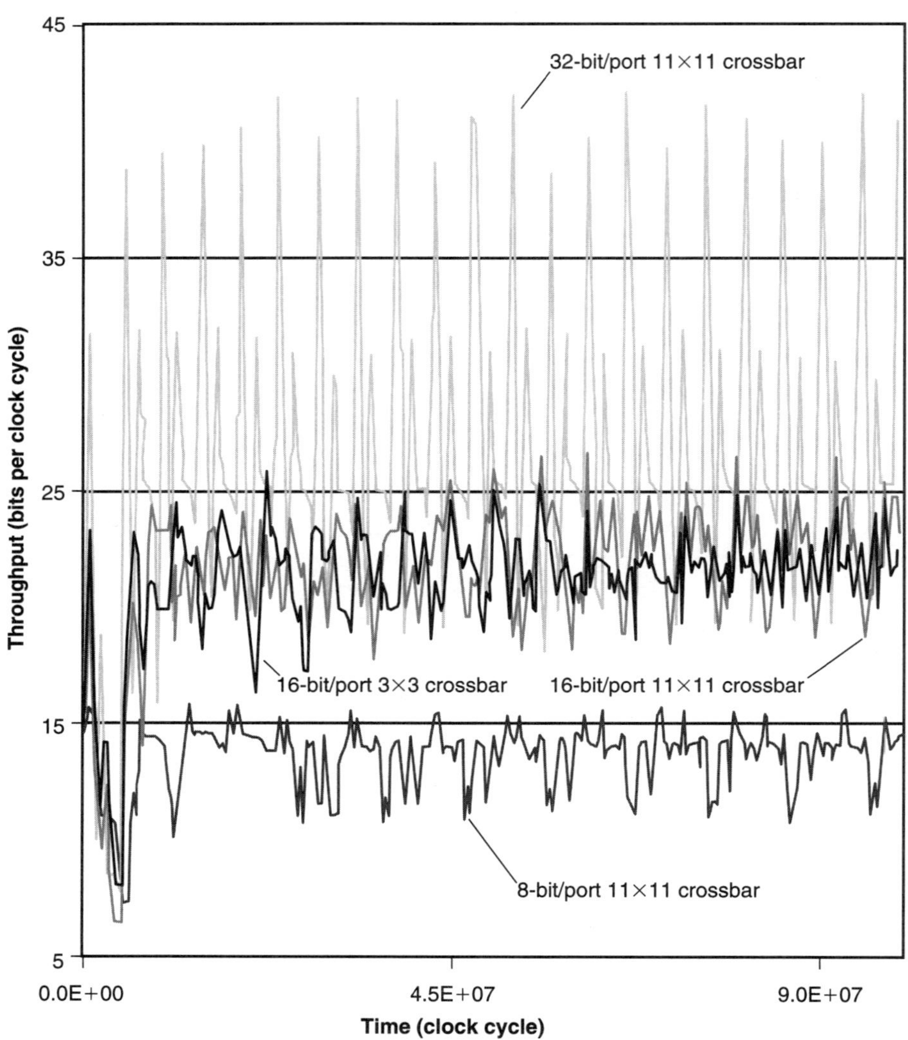

14-11

Network throughput as a function of time for a smart camera MPSoC.

FIGURE

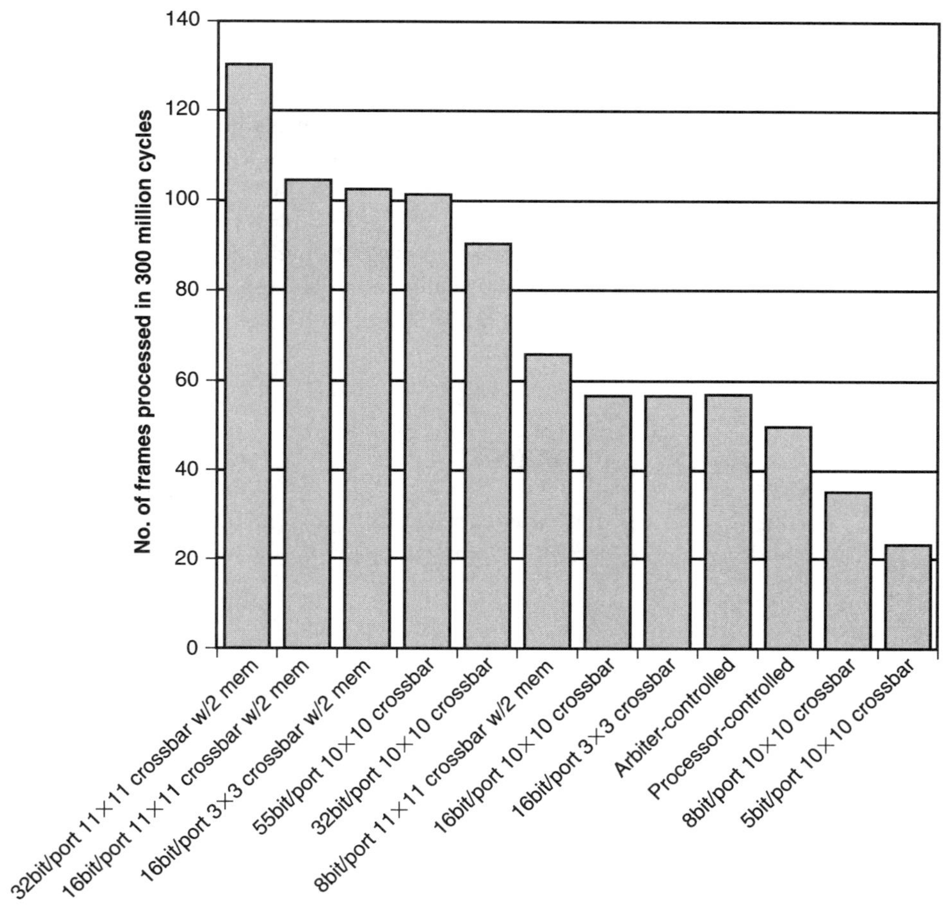

14-12 FIGURE   Measured performance for several candidate smart camera MPSoC architectures.

14.10   CONCLUSIONS

Although imagination and innovation have constantly pushed the envelope, leading to revolutionary ideas, new standards, novel algorithms, fancy displays, personalized services, and cooperative systems with blurring product boundaries, function and design complexity have also grown exponentially, leading to increased development costs and time to market. It is thus becoming increasingly clear that business success, in the new era of networked multimedia systems, will be determined by the ability to offer complete system solutions based on a combination of hardware and software, the agility to react quickly to changing market conditions, the capacity to serve more markets with fewer products, and the acuity to offer the right solution for the right market at the right price. To meet these challenges, an emerging trend is to resort to a platform-based SoC design approach that achieves functionality through co-design, shortens development time through reuse, provides flexibility through software, maintains continuity through architecture, and amortizes cost through derivatives.

ACKNOWLEDGMENTS

Shenggi Yang's work was supported in part by the National Science Foundation under grant CCR-0324869. Jiang Xu's work was supported in part by NEC.

Models of Computation for Systems-on-Chips

JoAnn M. Paul and Donald E. Thomas

15.1 INTRODUCTION

Intuitively, models of computation (MoCs) are abstract representations of computing systems. The term is almost as basic as computer science itself, going back to the earliest research in computer representations, including the work of Turing, Kleene, and Church in the 1930s. The earliest origins of the term are unclear, but it was not used by the 1930s researchers to describe their work. Although most textbooks and articles on MoCs seem to take for granted that its meaning is well understood, here we advance a definition, albeit mostly intuitive, discuss how MoCs have developed since the 1930s, and propose a possible future evolution.

An MoC captures, in an abstract manner, the relevant features of a computation. It defines how a new state evolves from a previous state, what operations produce the new state, and how the state is propagated to and from the operations. The operations can imply arbitrary amounts of computation complexity. The term "relevant features" suggests that there may be many uses for an MoC, and thus some features may not be equally detailed in all aspects; indeed, the area of MoCs has supported several distinct categories of research over the years. We also observe that the notions of state, operations, and propagation suggest a set of design elements for an MoC, i.e., a lingua franca for the user of a model. The design elements can have varying degrees of formality, as well as varying degrees of relationship to the design elements of real computation devices.

Many chapters, and even many books would be required to cover in full detail all the MoCs that have been investigated and utilized over the years. In many cases, detailed comparisons are required to compare different implementations of a given MoC fairly, let alone the distinct differences between describing a computer system in one MoC versus another. Several good texts exist that are devoted to that purpose [626–628].

Our purpose is to broaden the discussion of what it means to model a computer at all. Our thesis is that a more general category of computer modeling be motivated in lieu of the limited set of models more commonly associated with MoCs.

We feel it is more important to promote new ways of thinking about how models are used in the process of designing computers, even to the point that others may disagree with us, than to relate more conventional wisdom. We feel justified in so doing, since sys-tems-on-chips (SoCs) will soon be characterized by tens to hundreds of interacting, heterogeneous processors, their communications, their scheduling, and their programming—all of which contribute to the overall system performance, cost, and design time. This creates the foundation for an exciting next generation of single-chip computing, well beyond single processors and application-specific integrated circuits (ASICs). The next generation will potentially change the perspective of system designers, chip designers, tool designers, and computer architects. A discussion of the evolution of MoCs through what it means to model a computer seems well suited to prompt a new, fresh perspective.

15.1.1   Evolution of Models of Computation

The earliest reference to MoCs in the ACM and IEEE literature is in the 1960s to classify one of three basic areas of a computer science education [629]. (The other areas were data structures and programming languages.) Even though the term was not defined there, its meaning was motivated by topics that belong in that area, including the properties and classification of automata, behavioral analysis of switching circuits and sequential machines, and formal languages and grammars. These topics have roots back in the early organization of computing systems with such fundamental models as Turing machines, finite state machines (FSMs), pushdown automata, PRAMs, and logic circuits. (More detailed discussions of these models can be found in ref. 627). The goal of this early research was to answer the question "what is computable?" There were no physical restrictions in these models, as evidenced by the supposition of ideal compute times and physically unbounded tapes and stacks in many of the models.

Later, it became apparent that the issues behind computer design went beyond what was computable to "what is reasonable to compute on a real computer?" Thus, research in MoCs started to focus on computation complexity during the 1960s and 1970s, resulting in "big O" notation to represent the order of complexity of a compute problem and the classification of problems as P and NP-complete to capture whether a problem is computable in polynomial time or not.

In the late 1970s and early 1980s the term "models of computation for VLSI" began to appear in the literature [626] and the question became "what is the minimal area for a circuit to do a computation?" Gridded computational models of logic circuits, which had multiple levels to represent complex interconnection, were proposed to capture the layout aspects of circuits. Design elements such as arithmetic and logic units (ALUs) and registers appeared as the emphasis shifted to the architecting of computer systems. The clear trend over this period was toward supporting the representation and calculation of area and time, two important physical aspects for designing real, buildable computer systems.

In the 1980s and 1990s, MoCs evolved along a different tack. Concurrency became an issue at the application level as processors became capable of executing more functionality. At the same time, new forms of inter-processor communications made it possible to view collections of processors as systems. The emphasis shifted from MoCs for very large-scale integration (VLSI) to concurrent MoCs. Some of the more prominent concurrent MoCs include Petri nets [630], communicating sequential processes (CSP) [631], synchronous dataflow (SDF) [632], Kahn process networks (KPNs) [633], Statecharts [634], shared variables [635], and others, as well as meta-models for combining them such as Actors [636], Ptolemy [637] and the unified modeling language (UML) [638]. Some of these models target abstract representations of the multiprocessing nature of concurrent computation (Petri nets, KPN, shared variables), whereas others target abstract representations for classes of computing problems, such as digital signal processing (SDF), or reactive systems (Statecharts). These models represented a trend toward formal representations of computation but with less of a tie to the physical aspects of the system. The goal was to model the system functionally well in advance of building the actual computing system while also permitting the manipulation of an abstract set of design elements that were, themselves, functional representations. Thus, these models benefit the de-composition and analysis of concurrent system functionality only when the computation power of the underlying computation engine is so abundant that physical limitations are not a factor. However, the presumption of an ideal machine when many key decisions are being made in the design process of a single-chip computing device seems unsatisfactory. MoCs for single-chip programmable devices must allow designers to consider how the complexity and form of new levels and types of functionality interact with the physical capabilities of the underlying computation engine in real designs. Thus, the combination of new levels of programmable concurrency with the need to design the physical portion of single-chip systems has raised anew the question of what makes a good MoC.

15.1.2   What Makes a Good Model of Computation?

The answer to this question can be likened to the answer to the question of what makes a good programming language. The difficulties lie in separating what is objective from what is subjective. Most programmers, let alone language designers, have a preferred language, or even class of languages—they simply express themselves more easily in one style rather than other, mostly taken from the experience of designing what they need to design. Some languages, such as C and C++, are clearly more widely used than others. However, it is interesting that language purists can often be most critical of even these widely accepted languages for their many flaws. However, even with this criticism, designers adhere to what works, which may be in stark contrast to what many feel is "good for them." When it comes to MoCs, the problem of objective evaluation is no less complex—and perhaps is even more complex since computer programming might be considered a part of the modeling and design of a programmable computer system. As with computer languages, perfectly even-handed discussions of MoCs are difficult to find. Those who really understand the nuances are those most likely to espouse one view over another. In such a circumstance, the fairest thing to do might be to put out one's reasoning as a basis for provoking thought. This is our goal.

In the evolution from MoCs for VLSI toward concurrent MoCs the model elements became more abstract as the emphasis shifted toward specification of the desired behavior of a concurrent computation rather than the models that facilitate the architecting of the computer system. A benefit of viewing a computer system as a specification is that sound, formal modeling principles based on mathematics can be used. The limitation is that less formal aspects contributing to the computer system's physical behavior are not well captured.

Physical models are best captured as programmatic models that abstract the key features of the way model elements behave both individually and as an integrated collection in the formation of a system. The latter is more like any physical model (not just that of computer systems), since it is based on observable and measurable properties of design elements that do not always distill to a mathematical description. Physical models are often descriptive, relying on words, analogies, and simple diagrams to describe key features of what is observed. From the perspective of physical computer systems that are synthesized or designed, computers can automate or assist the design process either by mathematical transformation or programmatic translation from one model to another. However, the synthesis process even for pure, fully synchronous hardware designs relies on heuristics. A heuristic is a descriptive observation of a scenario that often works, but is not guaranteed to work, in making a decision toward achieving some desired objective. Like a programmatic description of how model elements interact, a heuristic is a programmatic description of model manipulation that cannot be described mathematically, and without which design would be severely constrained.

Programmatic models allow the design of highly complex systems because they permit a set of essential features of model elements, and the way they interact in the formation of a system, to be captured as executable descriptions. Non-computer physical models can often be ahead of the mathematics to describe them; descriptive models are often well ahead of mathematical ones when new concepts are discovered or architected. Similarly, the advancement in computation complexity has resulted in the need to capture rapidly new concepts in design tools that are more descriptive than mathematical. As new model elements are formed when new architectural solutions are discovered, and as complexity of design evolves, models must adapt to new feature sets. When models are not bound to adhere to an idealized, mathematical foundation, only the relevant characteristics for design need be included. This is also the basis of multi-level modeling, in which mixed levels of detail can co-execute, an important aspect of executable design languages for highly complex systems in which some portions of the system design are more critical than others. Simply, the formal model must not get in the way of the need to capture an ever changing design landscape in executable computer models that assist the design process.

The current state of design of SoCs largely involves programmatic models. Design languages such as SystemC are de -facto standards because they are really the C++ programming language with libraries. Because the simulator is written in C++, designers can programmatically extend their models to be customized for their own design situation. The use of SystemC with a lack of any real basis in either formal modeling or identifiable design elements at the system level emphasizes the need for programmatic models that capture salient features of the system design problem. The current state of the art in MoCs for SoCs is thus split between ad hoc solutions that get designers closer to workable designs and a host of formal models that seem far removed from real designs. Clearly something between these two extremes is required.

The problem of computer system modeling is even more difficult to characterize than the modeling of other physical systems because a computer can be viewed in part as an engine for execution of mathematical models that are not constrained by any physics other than the time required to carry out the calculation. The boundary between the computation the computer carries out, which may be highly mathematical, and the model of the physical computer is of central importance in MoCs for SoCs, since the computation engine is part of the design problem. The problem is still further complicated by constraints on the physical implementation, such as design complexity, design size, overall computation speed, and power consumed.

Thus, SoC design really contains two models, a logical, mathematical model with little or no physical restrictions, and a descriptive model of the physical elements imposed by the requirement that computer systems must be buildable and achieve logical computation in some optimal or constrained physical time. The logical model may be considered perfect in the achievability of a correct execution sequence on an abstract machine in proper order and imperfect in that its logical order as well as the time required to carry out its computation is affected by the computation engine used to carry out the computation. However, ideal assumptions about the capabilities of the computation engine severely limit the designer's ability to produce an efficient computing device.

Still, the question of whether a computer should be modeled by almost pure mathematics, as in a software program such as Matlab, or by more physical representations of observable and even measurable phenomena that arise when computing elements interact is at the heart of the question of what makes a good MoC. New forms of integration for SoCs, adding software programming, concurrency, and scheduling to what was once considered the physical design domain of a chip has provided an additional sense of urgency to getting the answer right so that a next generation of design is facilitated.

15.1.3   Toward an MoC for SoCs

The trend away from the idealized, highly abstract MoCs of the 1930s toward MoCs for VLSI designs in the 1970s permitted efficient design of performance-oriented computing systems. In much the same way, we envision a trend away from idealized, highly abstracted concurrent MoCs in the 1980s and 1990s toward MoCs for SoC designs. Ironically, the abundance of computation power on single-chip designs has brought back a need to model the physical characteristics of the computer that carries out the computation. Soon it will be possible to integrate the equivalent of hundreds of ARM processors or their equivalent on a single chip. At this point, programmable processors will be like registers were in the early days of VLSI design, except that the processors will, themselves, be programmable.

The possibility of integrating so much processing power on a single chip not only creates new possibilities for what single-chip designs can do, but also creates new barriers that make the design of the multiprocessing architecture on the chip more important than for a multi-chip collection of processors. MoCs that focus only on correctness for concurrent systems can rightfully consider processing power relatively cheap when the addition of one more processor bears the penalty only of the cost of the processor itself. However, SoC designs must optimize the numbers and types of processing elements, memory, and manner of communication in order to maximize the processing that can occur on a single chip. The penalty of adding one more processor can be prohibitive if that processor will not fit on the chip or if it will consume too much power—or if its presence will slow the overall system down because of a poor partitioning decision. Thus, design elements that capture the physical characteristics of real designs must be emphasized in MoCs for SoCs. In these multiprocessor, single-chip systems of the future it will be at least as important to model the numbers and types of heterogeneous processors, scheduling decisions, and methods of communications as it was to model the functionality of more traditional multiprocessing systems. Clearly mathematical, logical and descriptive, physical MoCs must be integrated for the next generation of SoCs—but how?

In this chapter, we discuss some of the MoCs mentioned above and classify them according to a new scheme. We introduce this scheme as a means of permitting critical evaluation of how MoCs may facilitate SoC design. We close by presenting a unique MoC for next-generation SoC computing systems. It attempts to answer this question by breaking free of long-standing assumptions on what makes a good MoC for SoCs.

15.2   MOC CLASSIFICATIONS

In this section, we outline several MoCs, classify them according to conventional views of timed versus untimed and synchronous versus asynchronous, and then classify them according to our view of how they impact on computer systems as well as computer system design. This will lead to a proposed new view of MoCs for SoCs.

15.2.1   Conventional Classifications

Major issues in classifying MoCs are how a state is updated and how a new state is propagated. In Figure 15-1 we classify several widely known MoCs based on whether their state update is synchronous or asynchronous and whether the completion of a state update function is related to an interval that captures the physical capacity of the machine being designed, such as the delays on registers, gates, or other design elements.

	Timed	Untimed
Synchronous	Finite State Machine (FSM)	Sync. Reactive (Statecharts, Sync. DF (Lustre) Esterel)
Asynchronous	Discrete Event	KPN TM CSP Petri Shared variable

15-1

Classification of models of computation.

FIGURE

In a timed model, the references to time are based on physical properties of the design elements, resulting in time that implies magnitude as well as order. Thus, time is typically ordered by the counting numbers. Time becomes a reference against which all computation can be defined and measured. When the complexity of state calculation and state propagation between any two time instants implies a relationship back to what is physically implementable in the time specified by the interval, the model can be used to support the physical design of the system. The design process of specifying design elements in a timed system implies real resources with time delays (e.g., transistors, gates, processors, and so on) to implement exclusively the functionality. Indeed, two computations overlapping in time require two resources for implementation. The important aspect here is that the design elements in these MoCs are physically based, requiring area and time to execute.

In an untimed model, there is only a notion of logical sequence—certain actions must occur before others to ensure dataflow precedences and causality, whereas others can, but do not need to, occur in parallel. State update is triggered by either a global event acting on all design elements or local events acting on individual elements. In either case, new state is immediately available. Executable, untimed models may have a global simulation cycle (or tick), yet the tick only acts to group sets of functional events so that causality may be established between them. The system is viewed as having ideal calculation and propagation times, in contrast to timed models, in which time is derived from the propagation of state through the delays of non-ideal model elements. Without the relationship back to model elements, there is no implication on the physical resources needed to implement the functionality in untimed systems.

An independent axis of the table captures whether a system is synchronous or not. A synchronous model has a tick indicating when new state may be propagated through all model elements. In an asynchronous model, new state appears and can be immediately available to model elements—there is no need to wait for propagation through the global barrier that the clock reference provides. In an untimed synchronous model, the tick has no relation to physical resources within the system and may not even relate to physical properties external to the system. In these models state update is considered to occur in an ideal, "zero time," in contrast to timed models, in which state update is considered to occur in intervals that relate back to propagation times of one or more design elements in the machine, such as registers, instruction cycles, or bus cycles.

Representative Models

Although many more models have been presented in the literature, we have selected the subset of models shown in Figure 15-1 because of the interest they hold in the SoC community. We will discuss these models with respect to how they are placed within the classification scheme, rather than providing a detailed overview of the models. For further details, the reader is referred to the literature.

In the synchronous, timed category, we consider an FSM. An FSM models a system as a set of states, transitions between the states, and outputs. The vast majority of FSMs are designed to the presumption that the transition to a new state is caused by a global sequencing event—a clock. Thus, we will assume that FSMs are synchronous. Although asynchronous FSMs exist, they are not one of the models we will consider in this chapter. In an FSM, the new state is determined by the current state and inputs. The outputs are determined by the state and possibly the inputs. Due to its wide use in integrated system design, it is closely tied to physical implementations and thus is often considered to be timed. Thus we consider an FSM to be a synchronous, timed MoC. For example, when an FSM is used to model a processor or ASIC, the propagation times of gates and registers are used to establish the clock cycle. The minimum clock tick is related to elements from within the system being designed, as opposed to arising from an external specification of desired behavior. FSMs can also be used to model the overall behavior of programmable processors. In this case, multiple instruction cycles are grouped by coarse-grained system-level states. However, the execution of an FSM -model on a programmable processor is more limited than that of a general programmable processor. A processor can be considered a physical representation of computation that draws on a mathematical foundation and yet compromises it. Where the tape is infinite in the Turing machine, the memory is finite in a processor. However, it is conceptually unbounded in the programmer's view of the memory space, since memory can be allocated and de-allocated during system execution. The result is that a powerful class of computing machines can be realized.

When there are multiple FSMs acting concurrently and sharing the same sequencing event (i.e., the clock), we say that they are synchronous FSMs. Because synchronous FSMs require a relationship back to a system clock event in which propagation times of design elements must be guaranteed, FSMs do not scale well to systems with large numbers of programmable processing elements. The cost of coordinating the design elements in such systems with global events becomes prohibitive, since the propagation time of the event becomes costly and does not scale well.

In the synchronous, untimed category are models and languages for synchronous reactive systems such as Statecharts [634] and Esterel [639]. Special cases of synchronous reactive systems include real-time systems which can be thought of as synchronous reactive systems subject to external real-time constraints [640], and synchronous dataflow systems, which are captured in languages such as Lustre [632]. A synchronous dataflow system includes the assumption that the system will process streams of values in bounded time, which characterizes many signal processing and control systems [641]. A dataflow MoC consists of a set of processes connected by arcs [642]. Events are generated by the processes and arrive at the next process, "firing it" (triggering its execution) when the requisite number of events has arrived. The events are consumed in strict order. In an SDF MoC, the amount of data received, processed, and produced per unit time is constant, making it highly applicable to signal processing applications [628].

The tick of a synchronous reactive model need not relate to any physical time basis, from either external or internal physical constraints, rather, synchronous reactive models are concerned with a strong mathematical basis that permits formal reasoning about models. These models target functional correctness and composition. Statecharts [634] provide a visualization scheme and MoC for concurrent, hierarchical FSMs, but the FSMs need not relate to actual system elements. They add to FSMs the notion that multiple concurrent machines can be interacting as defined by an extended set of state transition semantics. Furthermore, sets of states can be hierarchically grouped, capturing the notion of a subroutine in hardware. The concurrent machines are untimed but synchronous.

When the tick of synchronous, untimed MoCs in languages such as Statemate [643] and Esterel implies some physical external requirement, which fixes the processing time of the specification, a burden is placed on the design process in constructing a machine to meet the specifications implied by the model. It is possible for synchronous reactive system descriptions to be compiled to source code executing on processors so long as execution times between system cycles can be guaranteed. Since inter-process communication is essentially free on uni-processor systems, a single processor can almost be considered "ideal" for implementing such languages. However, the assumption of perfect synchrony and ideal communication times has proved difficult to extend to multiprocessor systems for which MoCs that accommodate asynchronous communications can permit greatly enhanced overall system performance.

The most prominent of the untimed, asynchronous MoCs are Petri nets [630], CSPs [631], and KPNs [633]. They permit formal reasoning about the system behavior and are flexible enough to accommodate non-deterministic execution times. However, they do not include any relationship back to either external system time constraints or the execution times of model elements. The oldest of these abstract models is Petri nets, which date back to 1962. They have similarities to the FSM and dataflow MoCs [644]. A computation is modeled as a set of places with transitions between places. Tokens move from one place to another along a transition when the transition "fires." Transition firing occurs when there is a sufficient number of tokens at its "input place" and corresponds to state change in the model. Since there can be many tokens in the system, concurrency is directly modeled. Petri nets typically only model the interactions between the concurrent systems; functionality is modeled only if it impacts on the means of concurrent interaction.

CSP served as the model of concurrency in the Ada programming language. To communicate, a process must arrive at a particular point in its execution and determine that another process has reached a corresponding point. When this occurs, a "rendezvous" has occurred, and state information may be shared through a channel implied by the rendezvous. As a practical matter, CSP requires a monitor function to co-ordinate the global state of tasks so that rendezvous can be established. The need for centralized co-ordination has made CSP challenging for multiprocessing systems. Milner [645] also developed a formal model to co-ordinate multiple communicating processes. A KPN [633] is a static interconnection of processes communicating through channels that are unbounded, first-in first-out queues with single readers and writers. KPN provides formal guarantees such as deadlock-free execution; however, the model requires that queue sizes be unbounded.

The shared variables MoC allows multiple processes to share memory locations directly for reading and writing [635, 636]. Although synchronization methods must be established to ensure data consistency and freedom from deadlock, this model allows for a wider range of shared memory systems to be described than do the other methods above. These untimed models are not considered synchronous in the traditional sense of a global clock signal, rather they become synchronized due to the actions of semaphore variables.

Finally, we include a Turing machine in the category of untimed asynchronous systems. The program and data interleaving are asynchronous. The Turing machine is a standard MoC since no other model performs tasks that it cannot. It is an abstract model of a computer that consists of a control unit and a tape (memory) with an infinite number of cells (locations). The cells hold symbols that the control unit can interpret. The control unit performs computations by moving the tape head to read and write these cells.

Alone in the category of timed, asynchronous models is discrete event (DE). Many do not consider DE to be a valid MoC because it does not have the formal properties of other models. For example, when two events occur at the same time the actual order of execution of events is arbitrary. However, DE must be included, since it is the workhorse of simulation, capturing the less than perfect effects of interactions of design elements such as deadlines missed when state does not propagate through designs constructed of elements modeled with delays. It does this in part by including mixed forms of timed state advancement and multi-level modeling. For instance, clock edges can capture state update between global system events, but this can be executed in the same model in which delays on gate-level models capture the asynchronous propagation, through combinational logic, that must occur within a given clock cycle. If the asynchronous portion of the model is not designed correctly, the simulation will correctly capture that the design elements come together to form an inadequate solution. In these systems, all state updates are tagged to occur at a time that is based on a reference to a propagation delay. Execution of these updates occurs in the order of the time tags. Languages such as Verilog, VHDL, and SystemC are based on this MoC.

Observations, Other Classifications, and Meta-Models

At this point we have summarized the more common MoCs that have been associated with SoCs. It is important to make a few observations.

First, in all cases, we represented only the original view of each MoC. Thus, we did not discuss variants such as timed Petri nets or KPNs that have bounded queue sizes. However, a substantial body of work exists in forming variants on the most basic models that is beyond the scope of this chapter. Interestingly, most of these variants exist because using the pure form of the model has proved to be too limiting in building real designs—the variants tend to include some physical constraint that takes the model one step closer to a real computer design.

Second, all the MoCs listed in Table 15-1 are based on a single process type and a single means of inter-process communications. The processes can be roughly separated into those that capture a logical, software process and those that capture a physical, hardware process. (In Section 15.4 we will motivate a means of uniting and resolving these heterogeneous process types throughout the design process.)

Third, many other classifications of MoCs exist. For example, Jantsch [628] classifies MoCs as untimed, synchronous, and timed, further subdividing synchronous models into perfectly synchronous and clocked synchronous, which correspond roughly to our untimed and timed categories of synchronous, respectively. In perfectly synchronous systems, neither computation nor communication takes time, whereas his clocked synchronous models take one clock cycle to produce and propagate new state. Our view of timed systems places added emphasis on time being a physical quantity that captures computation delay, allowing the implication of resources and a basis for performance analysis.

Most work in MoCs beyond that of the early work in the 1930s has applied to concurrent systems, and increasingly to programmable concurrent systems. In 1986, Agha [636] introduced an Actor model as a "concurrent model of computation" intended to provide a "general framework in which computation in distributed parallel systems can be exploited." Agha's Actors describe tasks as 3-tuples that include a tag, a target, and a communication. Agha considered the strength of Actors to be in their generality. He considered it possible to describe arbitrary sequential processes using an Actor system, but not the other way around. Actors were an important foundation for meta-models of concurrent computation. However, Agha admits that in switching from talking about "processors" to talking about "processes" not only is there a shift from talking about the physical machine to talking about an abstract computation, but also the utilization of the processors is "an issue for the underlying network architecture supporting the language."

The focus of most recent work in system-level design is toward providing a basis for uniting various MoCs. Most continue to adopt the premise of a perfect machine that carries out the behavior specified by the MoCs. This separation of models of the physical machine from MoCs requires that a mapping or a synthesis step eventually take place, as a physical machine must eventually be realized.

Lee proposes a framework for uniting concurrent MoCs [637] that adopts the synthesis approach. In his framework, he brings together the definition of an event model as a pair of data and time values and the notions that digital systems include both logical and physical sequencing [646] as well as partially and totally ordered sequences [647] by referring to the time value in an event tuple as a tag. He then considers specifications to have time tags that are totally ordered and that the tags correspond to real, physical time. By contrast, he considers a model to have partially ordered tags that correspond to a sequence of events. He notes that finding a physical time assignment for the partially ordered time tags corresponds to a synthesis process and thus specifications and models can co-execute; however, this step is not pursued, and so the practical aspects of implementation of the abstract models is left as an open question. The goal of composition of MoCs in Ptolemy continues to place the emphasis on the specification of the net physical behavior of a system in which a synthesis step to a real system is presumed [648]. The presumption of a synthesis step raises the question of how the design of computer system infrastructure might be included in a design flow that focuses on net system behavior as well as the overall efficiency of synthesis for complex designs.

The UML [638] can be considered another meta-model in that it unites familiar models on a collection of diagrams. However, UML was originally intended to be used to develop diagrams in the absence of any underlying executable semantic basis for the language. A collection of commercial and non-commercial tools has been developed that permits subsets of the UML to be executable by generating software for real systems for some subsets of the language such as Statecharts [649], thus resulting in a synthesis approach for some subsets of the language. Additionally, the lack of a common or well-accepted semantic basis for the language has proved to be limiting. The level of abstraction of UML is even farther away from capturing properties of the underlying machine that carries out the behavior it describes than other work done in meta-MoCs. Thus, the applicability of UML to SoC design is limited.

The separation of models of the physical machine from MoCs in the function/architecture (F/A) system design methodology of Keutzer et al. [650] results in a "meet in the middle" mapping to resolve the two separate design domains. In F/A, functional specifications of desired system behavior equate to MoCs. The premise is that system functionality can be specified using a variety of formal, mathematical models—a premise that holds only if the underlying machine can be considered perfect. Keutzer et al. [650] admit that when function and architecture "meet in the middle," as they eventually must in real designs, the overall system behavior may not be as ideal as the functional specification. The presumption of a perfect machine at the very beginning of the design process can force designers to produce a physically ove-designed machine or one that is limited to respond to a finite number of situations so that physical limitations of the machine will not get in the way.

By contrast to these approaches, computer system designers implicitly optimize the physical machine to the behavior it carries out using a rich set of design elements that encompass rather than avoid the impact of the physical machine on the net behavior that the system carries out. This results in a class of computer artifacts that neither model the physical machine nor presume an ideal physical machine. Schedulers are an example of this type of computer artifact. Schedulers have long been important design elements in concurrent systems. A key property of schedulers is that they are capable of dynamically optimizing functionality to available system resources because they reside at the boundary—

they provide a layer that permits some properties of the machine to appear nearly ideal to higher level software layers, while they also support run-time management of the real, physical elements of which the machine is comprised in that same software layer. The most effective schedulers can be those that take both system functionality and machine models into account [651]. The observation that schedulers are important design elements that do not fall into either category of function or architecture was first discussed by Paul et al. [652], who also introduceds a simulation basis for uniting rather than separating software (which does not necessarily presume an idealized machine), underlying hardware, and schedulers, starting from the very beginning of the design process.

Although most work in MoCs isolates ideal behavior from less than ideal machine models, the effectiveness of schedulers as an example of an important class of computer artifacts leads to the need to acknowledge them in the design process. We build on this observation to propose a new classification scheme for MoCs, discussed in the next section, and ultimately an entirely new MoC, as described in Section 15.-4.

15.2.2 Classification Based on Applicability to Computer Design

Table 15-1 outlines the differing objectives by listing MoCs from the previous section followed by the design elements that are manipulated in the MoC. These are followed by a classification showing whether the MoC leads to a well-recognized computer system artifact or a design artifact. Through this grouping, we propose four categories by which MoCs might be classified. We propose separate classes of MoCs that lead to hardware artifacts, software artifacts, design artifacts, or analytical models. Computer artifacts are models around which classes of computer systems or computer system elements are organized. Computer artifacts can include hardware artifacts and software artifacts, or artifacts such as schedulers that can encompass portions from each domain. Design artifacts are models around which computer system design tools are organized. Analytical models are used to facilitate reasoning about computer system behavior but do not directly result in widely accepted artifacts. Note that here we use the word "artifact" when we might have used the more familiar "object." Both describe a "material entity" or "material thing," but we avoid the obvious overloading of the term "object."

The class of MoCs that leads to hardware artifacts includes Turing machines and FSMs. A Turing machine is an idealized MoC in that it has an infinite tape.

MoC	Model elements	Hardware artifact	Software artifact	Design artifact
Turing machine	FSM, infinite tape, instructions, program	Processors		
Finite state machine (FSM)	States (registers), state update and output functions, global clock	FSM		
Communicating sequential processes (CSPs)	Processes, monitor function, dynamic channels		Ada Rendezvous	
Shared variables	Processes (threads), bus, Com mon memory		pthreads	
Discrete event (DE)	Processes, physical network, global time			Verilog, VHDL, SystemC
Synchronous reactive	Processes, global tick			Esterel
Synchronous dataflow (SDF)	Processes, logical network, glo Bal tick			Lustre
Statecharts	Processes, events			Statemate
Petri nets	Processes, places, transitions, tokens			
Kahn process networks (KPNs)	Processes, static single reader/writer channels, unbounded queues			

15-1   MoC, Model Elements, and Classification.

TABLE

However, the conceptually unbounded tape along with the interleaving of program and data by an underlying machine leads to the model of the most common of all computer hardware artifacts, a programmable processor. The design elements of a Turing machine, the parts of the system that designers think of in terms of manipulating (if they were actually to use a Turing machine!), are the FSM, the infinite tape, the instructions, and its contents (the program). An FSM with synchronous state update can be thought of as both an MoC and a computer artifact. The design elements for the FSM are the set of states, their update and output functions, and the clocking scheme. These are the design elements that designers think in terms of when designing to these MoCs.

The class of MoCs that leads to software artifacts includes CSP and shared variable. Both CSP and shared variable are methods of allowing multiple software processes (and threads) to co-ordinate in real systems according to a common scheduling framework. CSP and shared variable have resulted in languages or libraries that have been widely recognized. Whereas CSP has resulted in the

Ada Language Runtimes, the shared variable model is the basis for lightweight processes, or threads. There are many implementations of shared memory models with varying degrees of support for executing threads on multiple processors ranging from hardware-level to kernel-level to source-code-level. Many operating systems (OS) that support processes at the user level actually have a multi-threading model underneath so that information may be exchanged in a more efficient manner in a common memory space. Probably the most widely recognized standard for multithreading at the user level is pthreads. Since there are many ways processes can utilize communications channels to exchange information, but threads have only a few synchronization primitives, threads can be viewed as a more fundamental model of concurrency in computer software than process-based models.

In the category of design artifacts are DE, Statemate, Esterel, and Lustre. These are all design tools or the semantic foundation for design tools. Since DE is the foundation for simulation languages such as Verilog, VHDL, and SystemC, this category of MoCs has resulted in foundations for simulation languages used to support the design process. DE is a timed MoC that permits analysis and design of computer hardware artifacts. It does this by utilizing a foundation for physical design more fundamental than FSMs and processors—that of gates. The most basic design element of a DE model is a multi-input, single-output process triggered in a dataflow manner. The trigger is a time-stamped change in value on a net, thus capturing hardware gates. As a foundation for multiple levels of modeling computer hardware, DE provides the basis for analyzing and designing computer artifacts. As such, DE may be considered a meta-model of computer hardware design, with the ability to integrate and develop more complex models in its fundamental framework. In contrast to DE, Statemate, Esterel, and Lustre model the desired net behavior of the system. For example, Statecharts utilizes design elements that are abstract representations of behavior but that do not have a direct translation to computer artifacts such as gates, registers, or even processors. Thus, whereas Statecharts permits designers to analyze computer system functionality given design elements that represent abstract and ideally composable representations of functionality, DE permits designers to explore how elements might come together to form novel computer hardware artifacts using design elements that themselves relate to computer hardware artifacts.

We include Petri nets and KPN in the category of analytical models. Petri nets are probably the oldest analytical tool for concurrent computation. Its formal basis permits designers to analyze computation; however, it has not resulted in a widely accepted artifact. KPN can be likened to a kind of concurrent Turing machine. Like a Turing machine, a KPN is an ideal machine with unrealizable properties. In the case of the Turing machine, the tape is unbounded. In the case of the KPN, queue sizes are unbounded, but although it has been possible to approximate a Turing machine in building a processor, by preserving the conceptually unbounded tape as a device with a dynamically re-allocatable memory space, KPN has not resulted in a widely accepted artifact.

The rich history of computer engineering has benefitted by the flexibility to design novel computer artifacts by organizing large numbers of computer elements constrained by relatively simple models.

Students of computer engineering are consistently taken aback by the notion that all it takes to build the hardware of a computer is NAND gates and a graph for how they are interconnected by simple wires. Students of computer science are similarly impressed when first told that in order to build a programmable computer, all that is required is a few instructions to read and write from an addressable memory—(capturing the tape of a Turing machine)—and to load the program counter (—the equivalent of moving the tape in the Turing machine). The earliest work in MoCs answered the question of what is computable. All else in a computer design is relative, either permitting physical execution performance or for more complex systems to be designed more efficiently. Thus, in order to understand the value of different MoCs in computer system design, it is important to understand further the role played by MoCs in computer design, as discussed in the next section.

15.3   MODELS OF COMPUTATION AND COMPUTER MODELS

Here, we take a step toward a broader definition of what it means to model a computer system than the conventional views of MoCs. For this, we introduce the broader class of computer modeling, which we propose can be roughly separated into three areas: formal MoCs, computer artifacts, and computer design tools. All can be thought of as computer modeling, even though the formal models are traditionally the only ones considered to fall into the category of MoC.

A formal MoC is generally considered to be one with a mathematical basis. Simulations of formal models may be more efficient for large systems; however, the properties of formal models permit the representation of the system to be manipulated purely mathematically. For this reason, many of the recent formal models proposed for SoCs such as synchronous dataflow have actually been modeled in Matlab. Since many rely on a presumption of perfect synchrony, perfect execution times, or perfect communication times, they are actually very close to mathematical equations that represent the desired behavior of systems.

Since there are no physical resources in formal MoCs, they do not address how processes are scheduled to execute on and share the computation resources. Put another way, the limits of the computation engine are presupposed not to be factors. In contrast, many models of computer artifacts have benefitted by permitting resource scheduling decisions to be determined at system run-time. For this reason, computer artifacts tend to capture decisions made at execution time more effectively than formal MoCs.

Computer artifacts are the objects of computer architects. They can include software, hardware, or both. Like conventional architects, computer architects work to create what might be likened to a set of blueprints; they are concerned with making a model or even a class of models that captures the essence of buildable designs while falling short of a specification of an instance of a design. Computer artifacts can include processor styles (e.g., reduced instruction set computing [RISC], complex instruction set computing [CISC], very long instruction word [VLIW]), design elements such as caches and shared memory busses, organization concepts such as distributed memory networks, and the like. These artifacts are actually computer models because they represent a class of computers instead of any particular design instance. They are not classically considered MoCs. Many of these can actually be difficult to capture formally and mathematically, and they can also pose challenges when it comes to the construction of formal models that permit analysis. Thus, computer architects do not develop new computer artifacts through the use of ideal formal MoCs, rather, these artifacts are developed to provide new concepts that address some of the less ideal aspects of computer behavior, i.e., physical limitations. And yet, models of computer artifacts can be highly elegant representations of what we observe about the nature of computation as well as more effective means of organizing it. Thus, we refer to models of computer artifacts as computer models instead of MoCs. Many novel computer artifacts have arisen because of the intuition of computer architects in seeing how computation can be organized more effectively, if not more rigorously.

Consider, for instance, the computer network around which the internet is formed as an example of a computer artifact. The network is a blueprint for both software and hardware, around which a broad class of computer systems can be designed. Although it is a powerful model, a network is not a formal or rigorous model, founded in mathematics. Packets will sometimes be dropped from network nodes due to insufficient system resources. Dropped packets are re-sent by another portion of the network model in higher layers of the network stack. The queue sizes are one of many design trade-offs that affect network performance. Although the network would probably not be thought of as a formal MoC, it results in an elegant way to capture resource sharing and dynamic, run-time robustness; it can tolerate the complete elimination of large portions of the physical network and still allow for packet routing between two nodes in the remaining portion. Importantly, a network is a computer artifact in which hardware and software can be traded-off and in which the software is considered to be layered on the hardware. This is an important point that we will revisit in Section 15.4.

Design tools are computer programs that are used to assist the construction of instances of computers as well as the conceptualization of computer artifacts. Design tools may be considered synonymous with design artifacts, since they are objects, or entities, used to facilitate the design process. They may or may not have a formal mathematical basis. DE, for instance, is the foundation of the simulation semantic of the VHDL and Verilog languages. However, few would argue that DE or the hardware description languages (HDLs) built around them are formal MoCs. Similarly, hardware synthesis tools are used to translate hardware descriptions from one representation to another, often with a built-in set of highly informal assumptions about how designers will represent the input to the tool as well as a collection of heuristics used to make the tool more efficient. As computer programs that represent the behavior of real systems, design tools can also be considered models. However, they are not formal MoCs. Still, design tools have been essential to computer design because they permit manipulation (simulation, analysis, and translation) of physical elements in executable models rather than manipulating the actual system.

This leaves us with the observation that models arise in computer systems on a variety of levels, and these levels are not always associated with the term MoCs. We also observe that the less formal categories of computer artifacts and design tools have had enormous impact on computer systems designed during the past 30 years, whereas formal MoCs have had relatively less impact. With the new opportunity to integrate previously separate application domains onto single chips in the future, there is no reason to believe that the design of computer artifacts and design tools in this conventional sense will be any less important.

But what should the design elements be in a design tool that permits the effective design of next-generation computer systems?

Unlike VLSI-level designs whereby processes and their interconnect have a $1:-1$ correspondence to hardware components and their manner of interconnect, the SoC problem is one of considering the mapping of m tasks to n processor resources with the processor resources interconnected more often by shared busses and networks rather than dedicated point-to-point channels. In these systems, not only are the processing resources shared among processes, but the communications channels and memories are shared. The actual state trajectory (concurrent execution sequence) of such systems is affected not only by the physical performance of these shared resources, but also by logical scheduling decisions such as which task will execute next, which packet will be forwarded next, or which processor will have access to the bus or network next. Such highly complex interactions are often difficult to capture formally, unless the formal models are far removed from models of the integration of the design elements that actually create the need for the models.

Because software processes share and consume rather than supply a computational power in the form of SoC resources, pure process models are insufficient to capture the actual system state trajectory of most SoCs, rather, design tools that permit the design of novel SoC artifacts must utilize processors, schedulers, and threads as the most fundamental design elements—these are to programmable, highly concurrent computer system design what gates were to computer hardware design. These will be the basis for capturing the effects of resource sharing, data-dependent execution and performance modeling, as described in the next section.

15.4 MODELING ENVIRONMENT FOR SOFTWARE AND HARDWARE

Again, any article written that compares and contrasts MoCs is most likely to have been written by those that either have their favorite, or who have developed one. We are no exception. In this section, we discuss the semantic and modeling basis of our modeling environment for software and hardware (MESH) [653–655]. Currently, MESH is the foundation for a new form of simulation. Thus we consider it the basis for a new design tool that will permit designers to architect more efficiently the next generation of computing devices made possible by SoCs. Eventually, MESH will be developed to include a new language and associated design tools and methodologies.

MESH acknowledges that all programmable computer systems really contain at least two models—that of the logical system sequencing and that of the physical computation engine. Although the current trend in MoCs is to presume a perfect machine at the highest levels of abstraction, MESH unites logical sequencing (which may or may not presume a perfect machine) with models of the physical capabilities of the underlying machine—starting from the highest levels of the model of the programmable computer system and continuing in more detailed views of the design. MESH's origins are in a simulation foundation called frequency interleaving, which provides the basis for uniting rather than separating the impact of the physical machine on overall system performance and correctness at the highest levels of design, user-defined schedulers as design elements, and the resolution of logical and physical sequencing through layers [652]. A co-design virtual machine concept was also used to illustrate the importance of modeling the machine with the functionality it would carry out in the design of a programmable system [656].

MESH is unique in forming a basis for two process types, united in a layered relationship through schedulers. This captures the impacts of software and hardware artifacts on the net performance of programmable, concurrent systems. In MESH, logical sequencing is not specified as a physical computation, but rather it has properties that directly impact the net physical behavior of the system when it executes on an underlying layer of processors.

MESH is unique because it captures concurrent design elements that lead to the overall physical performance of the programmable, concurrent chip. It is less abstract than functional MoCs that target the net system behavior with many assumptions about the underlying hardware, but it is more abstract than instruction set simulator (ISS) models of processors and fully detailed software models that require advance specification of software. Thus it relates the contributions of the computation (software), co-ordination (scheduling and communications) and computation engine (processing resources) in an SoC, but at a new level of design element definition and simulation semantic—above that of gates and discrete event.

A significant aspect of MESH is the way it permits resources to be logically grouped and shared at run time through global contexts.

15.4.1 Programmable Heterogeneous Multiprocessors

In VLSI system design, two design abstractions have dominated the past several decades. The first is register-transfer-level (RTL), used to synthesize current FSM-style systems. The second is instruction set architectures (ISAs), which provide a basis for dynamic, data-dependent software programming that is far richer than FSMs. However, design for the widest range of behaviors for complex future devices that simultaneously interact with humans, the physical world, and other computers will require new views beyond these traditional abstractions. Many agree that future SoC designs will be programmable heterogeneous multiprocessors [657], and network-on-a-chip communication suggests relatively coarse-grained design elements communicating asynchronously in a data-dependent manner [658]. The central challenge for these systems is understanding how the programmatic co-operation of multiple elements to advance state affects overall system performance.

A new model and methodology beyond RTL-, FSM-, and ISA-based design is required. We refer to this abstraction as programmable heterogeneous multi-processing (PHM), emphasizing that not only must individual processing resources on a chip be considered programmable, but also the chip must be considered programmable as a whole. Other domains of scheduling and programming of groups of resources may also exist on the chip. A key is in providing a basis for understanding how programming and scheduling impact on both inter- and intra-processor resource sharing.

Performance of a system over time can be thought of as the system trajectory—progression of logical state advancement as measured against physical time advancement. Factors affecting the trajectory of a PHM include the number of processor resources, their unique (heterogeneous) features, the way they communicate, the way the software is mapped to the resources, the manner in which the software is written (data-dependent execution), and inter- and intra-resource scheduling decisions (e.g., which packet is forwarded next or which thread executes next). For instance, just adding a software thread or processor in a concurrent system may create a very different trajectory.

Our layered view of system modeling is illustrated in Figure 15-2. Here heterogeneous, hardware resource models are shown on the left as processors (P_i),

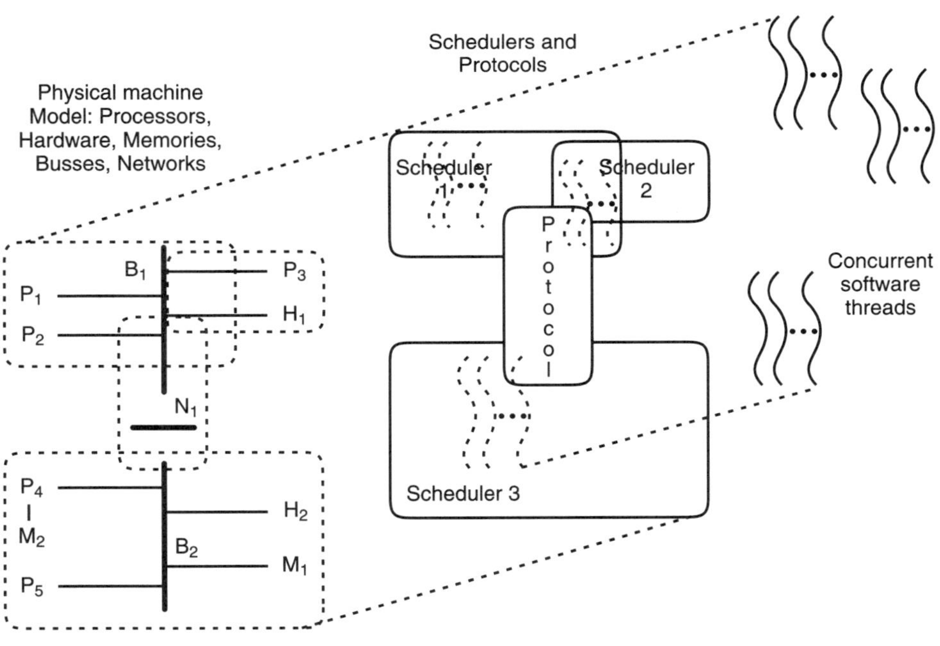

15-2   A layered view of system modeling.

FIGURE

memories (M_i), busses (B_i), hardware devices (H_i), and networks (N_i). In the middle is a layer of scheduler and protocol models that groups resources in an overlapping manner, capturing both inter- and intra-resource sharing decisions. At the right is the concurrent software layer. (Many more layers may be in a programmable system.) In permitting the most flexible foundation for the design of novel SoC artifacts, software and hardware should not be reduced to a common, graph-based wire model, nor should a single global clock reference be presumed.

15.4.2 A New Simulation Foundation

Although layers seem to capture software executing on hardware intuitively, their power as a modeling abstraction is also what poses the greatest challenge in capturing performance at high levels of design—that lower layer boundaries hide implementation details from higher layers. Thus, it is difficult to optimize net system performance through design layers unless high-level design elements provide layered timing relationships.

Most simulations can be classified into three types, based on the way they advance state according to the representation of the physical time that distinguishes a simulation from a program: continuous time (CT), discrete time (DT) and discrete event (DE).

In CT, all state is advanced each time tick. This tick approximates continuous change in models of the physical world with discrete intervals so that a digital computer can calculate integrals and differentials. For CT, the larger the interval, the greater the error, in general. The presumption is that as the interval sizes become smaller, the model error goes toward zero. With acceptably small interval sizes (and adequate numerical precision), the CT world can be successfully simulated on digital computers.

In DT, the models include a notion of a time period between which nothing of interest is presumed to happen in the system being modeled. Time is therefore discrete and not continuous. For instance, when all state advancement is calculated in each time period (interval) of a certain digital signal processing (DSP) system, the system is exact. Thus, if the complexity of the computation that occurs each cycle in a DT system does not exceed the physical capacity of a digital computer to compute it each and every cycle (accounting for numerical representation as well), the error is zero. The same is true for cycle-accurate descriptions of processors, so long as the functional state advancement that occurs between successive clock cycles is achievable. For these reasons, high-level models of processors and DSP systems are based on DT models. DT is the basis for the executable synchronous MoCs because they all presume the existence of an ideal synchronization system tick. The presumption of this global system tick carries over as a required property of the hardware (and software) platform of the system being designed.

DE differs from both CT and DT in that not all system state is re-calculated each simulation time. Instead, only the "interesting" state advancement functionality is advanced each unit simulation time. This is achieved by associating a time advance function with the state advance function. DE models can be thought of as a faster means of executing CT or DT models—if there are many elements in a system that do not update their local state each cycle, then it does not make sense to execute them each time unit. For example, in a DE model of a gate-level simulation, only a fraction of the many gates in a model typically actually update their outputs each cycle. So it makes more sense in terms of computation efficiency to maintain a list—a discrete event list—of events that are eligible to execute each time step. More than a faster way of executing DT, however, DE has evolved into an execution semantic in its own right. The time advance function is captured in HDLs as a delay (#delay in Verilog), which designers use to annotate gate and functional level blocks. Between this delay time, which now can be considered to be much larger than unit simulation time, presumably nothing of interest happens.

The difficulty with DE models is that there is little ability to relate the timing of elements to the way software design elements execute on hardware design elements. A time-annotated block of software added to a DE model has the undesirable property of adding performance to the system. This is appropriate for hardware design, in which physical system elements are specified by the model and so presumably added to the system with additional functionality, but inappropriate for software design. The addition of software can, and often does, adversely affect overall system performance by loading hardware resources—although—additional software functionality adds complexity to the system, it does not at the same time specify the physical machine to carry out the functionality.

Software timing is dependent on hardware timing. A new basis for simulation at this level is required—something that is not simply CT, DT, or DE. The software execution must be tied into the physical time-based models for performance analysis. A means of tying together timed and untimed models is needed.

15.4.3   Description of Layered Simulation

We describe, in more detail, our layered model foundation in MESH by developing relationships between logical and physical sequencing of events [653, 654]. The definition of an event model as a pair of data and time values is not new, and the notions that digital systems include both logical and physical sequencing [646] as well as partially and totally ordered sequences [647] are both well established. We adopt the nomenclature in Lee [637] referring to a time value in an event tuple as a tag. Our layered timing resolution distinguishes us from Lee [637]. That is, we do not merely presume that a synthesis step exists, but rather model the performance impacts of interacting design elements in a programmable multiprocessor system. High-level performance modeling requires an understanding of how to model logical events as they are interleaved by and resolved to resources (processors) in a data-depen-dent manner when the resources are, themselves, interleaved in simulation time.

An event has a tag and a value $e = (t, v)$. The value $v \in V$, the set of all values in the system, is the result of a calculation. The tag indicates a point in a sequence of events at which the value is calculated.

Threads are an ordered set of N events,

$$Th = \{e_1, \ldots, e_N\}$$

where the ordering is specified by the tags of the events, and N may be considered infinite. Event $e_i < e_j$ iff $T(e_i) < T(e_j)$, where $T(e_x)$ represents the tag of event e_x.

Computer systems contain two kinds of event ordering, —logical and physical [646]. The tags used in logical ordering specify a sequence that is not physically based; the ordering does not relate back to physical intervals or global time. These correspond to the untimed MoCs. Logical ordering often arises from functional language specifications at a high level of design. A basic assumption is that reordering the logical events of a thread (i.e., reordering the time tags) is allowable as long as data precedences are not violated. The tags used in physical ordering represent a physical time basis; there is a real, physical interval of time between tags i and j when i .neq. j. These correspond to the timed MoCs.

Logical-to-Physical Event Resolution

The means of resolving logical events to physical events impacts the design. The thread

$$Th = \{e_1, \ldots, e_N\}$$

is ordered based on its tags. Clearly, a physically ordered system is totally ordered. By contrast, a partially ordered system has at least two logical tags t and t' for which we do not know whether $t < t'$ or $t' < t$. Thus, assuming that events e_a and e_b are partially ordered, one resolution to a physical (total) order is the sequence

$$Th = \{\ldots, e_a, e_b, \ldots\},$$

whereas another correct resolution is

$$Th = \{\ldots, e_b, e_a, \ldots\}.$$

Another possible resolution is that they will have the same physical tag; they will be concurrent. Describing a system with a partially ordered sequence allows greater flexibility in the design of the system; partially ordered events give rise to alternate implementations of the system, where actual concurrency and ordering can be determined at run time.

Logical and physical thread sequences are denoted as Th_L and Th_P, respectively.

Assume that the thread sequence,

$$Th = \{e_1, e_2, \ldots, e_i, \ldots\}$$

is a high-level model. Because the events represent a relatively large amount of functional advancement, we term them macro states or macro events. These macro events imply several states or events that have relatively less functional advancement—we term these micro states or events. If the macro states are totally ordered, such as in a system in which time tags are ordered by physical intervals or are otherwise totally ordered as in a purely sequential software program, they allow for substitution on micro event sequences, allowing the sequence to be re-written as

$$Th = \{(e_{11}, e_{12}, \ldots, e_{1j}, \ldots), (e_{21}, e_{22}, \ldots, e_{2j}, \ldots), \ldots\}$$

Thus, each macro event, e_i, is seen to contain a sequence of micro events, $e_{i1}, e_{i2}, \ldots e_{ij}$. Each macro event triggers a sequence of micro events, which is presumed to be functional and atomic by the macro events. If the macro events are totally ordered, then the micro events must complete before the next macro event. Each micro event sequence in turn may contain another micro event sequence. This is physical decomposition, which ultimately results in simple functions, such as gates, separated by fine-grained events for computer hardware. When a system is decomposed to adequate detail, the logical events specified by Th_l may be substituted by a physical sequence Th_p with physical time tags—the system is "synthesized."

The hardware design process resolves logical events to physical events by binding them at design time, e.g., a logic synthesis tool binds Boolean algebraic functions (logical events) to gates (physical events). By contrast, in software design, the resolution of logical events to physical events does not occur until run time.

Optimization of concurrent software executing on concurrent hardware requires the software to be considered with respect to its underlying machine; the nuances of how the logical events resolve to physical events. Two observations support this. The first is that adding or subtracting a thread (software process) or processor from a concurrent system does not tell the designer whether the system will be faster or slower. The second is that relationships between containment and detail in hardware-like components are not preserved when modeling the logical on physical layering of software on hardware execution. To see this, consider the de-composed thread sequence, Th_1, as above. The ordering of events in the de-composed, more detailed micro -event sequence can be implied by the substitution on design elements for a physical design in which, for instance, registers can be de-composed into gates, which may in turn be decomposed into transistors. The execution sequence preserved as detail is added to a design—the micro event sequence is contained, compo-nent-style, by the macro event. However, hardware is not a more detailed model for software, rather, software executes on hardware. In a layered system the event ordering—the execution sequence—of higher design layers is affected by the execution sequencing of lower layers. One does not imply the other in the same way software does not imply hardware. For instance, a network may make a packet available to a waiting software task, permitting the interleaving of events in the more detailed model as

$$Th = \{e_{11}, e_{22}, e_{21}, e_{12} \ldots\}$$

In this case, the parentheses that imply the containment are removed since the order of events is not preserved by containment. Instead, the execution of software on hardware can be viewed as layered relationship on schedulers which permits a broader grouping of events to be made available to higher design layers. This is possible because the software logically groups shared resources in the underlying physical layer without containing the timing details of the underlying layer. This emphasizes the importance of capturing layered relationships in MoCs for SoCs as well as schedulers as design elements.

The Role of Scheduling

MESH has two dimensions of scheduling: that of the physical events to integer-valued global simulation time and that of self-timed scheduling of the logical events as they interact with each other in a layer resolved to physical time by schedulers. This permits performance modeling of concurrent software executing on concurrent hardware.

$$Th_{P1} = \{e_{11}, e_{12}, \ldots, e_{P1t}, \ldots\}$$
$$Th_{P2} = \{e_{P21}, e_{P22}, \ldots, e_{P2t}, \ldots\}$$
$$\vdots$$
$$Th_{PR} = \{e_{PR1}, e_{PR2}, \ldots, e_{PRt}, \ldots\}$$

Concurrent, multithreaded software system design requires scheduling the logical threads on concurrent hardware resources. Consider a system with M logical threads of execution executing on R resources, where M > R. We define the physical event sequences in the system as a sequence for each resource as shown above. Here we extend the thread notation of the previous section so that Th_{Pr} denotes the physical events of resource r, for r = 1, . . . , R. The events are a totally ordered sequence with time tag $T(e_{Prt})$.

We also extend the logical thread notation to let Th_{Lrm} denote the logical thread m that is mapped to resource r. Each physical resource, r, can in general support M_r logical threads (M_r indicates r as a subscript of M). Thus, M = M_1 + . . . + M_R. These threads are mapped by the resource's scheduler U_{Pr} to a physical thread Th_{Pr} . Each U_{Pr} (below) is a scheduling function that logically interleaves the M_r threads on resource r. M_r is typically unbounded.

$$U_{P1}(Th_{L11}, TH_{L12}, \ldots, TH_{L1M\_1}) \rightarrow Th_{P1}$$
$$U_{P2}(Th_{L21}, TH_{L22}, \ldots, TH_{L2M\_2}) \rightarrow Th_{P2}$$
$$\vdots$$
$$U_{PR}(Th_{LR1}, TH_{LR2}, \ldots, TH_{LRM\_R}) \rightarrow Th_{PR}$$

Logical events have no implication for physical interval sizes; logical event ordering in and of itself does not model performance. However, the schedulers map these logical events to physical events, thus capturing performance. In general, the events of the threads Th_{Lrm} are grouped by the schedule and assigned to execute on a resource. The current state and computational complexity of the functionality executed between logical events determines order and extent of computation that executes in a given physical time slice on a given resource. In so doing, each scheduler on a resource r has access to the logical event sequences of M_r threads, as shown below.

$$Th_{Lr1} = \{e_{Lr11}, e_{Lr12}, \ldots, e_{Lr1i}, \ldots\}$$
$$Th_{Lr2} = \{e_{Lr21}, e_{Lr22}, \ldots, e_{Lr2i}, \ldots\}$$
$$\vdots$$
$$Th_{LrM\_r} = \{e_{LRM\_r1}, e_{LRM\_r2}, \ldots, e_{LRM\_ri}, \ldots\}$$

Alternately, a logical collection of schedulers mapped to individual resources may be formed as

$$U_{Li} = \{U_{P1}, U_{P2}, \ldots, U_{PR}\}$$

allowing scheduling to be considered as a single, common logical scheduling context across multiple processing resources with cooperative scheduling.

Design Elements for SoCs

Figure 15-3 illustrates the above multithreading relationship with a single slice of the layering—a single resource in a multi-resource system. The threads in the figure represent how heterogeneous design elements inter-relate; MESH includes more than one process type, unlike the other models of Figure 15-2. At the top of the slice is a dynamic number of unrestricted software threads (labeled as logical threads $Th_{Li1} \ldots Th_{Lin}$). These are interleaved for execution on the resource labeled as a physical thread, Th_{Pi}, by a scheduler U_{Pi}, which resides on Th_{Pi}. U_{Pi} threads determine intra-resource timing of software executing on hardware by selecting which thread to execute and determining how long it is permitted to execute before the next resource is interleaved in the simulation. Thus, software is executed in both a data- and resource-dependent manner. U_{Li} threads can also logically group resources for inter-resource scheduling, as shown by the dashed oval going off of the figure, permitting M threads to be dynamically mapped to N

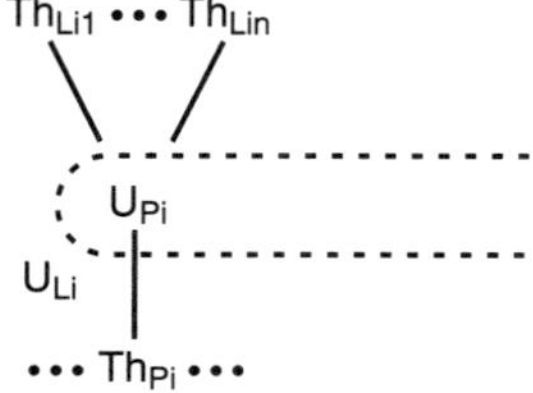

15-3   A thread slice.

FIGURE

resources, e.g., a pthread scheduler. This layer between the resource and software models brings together physically timed resource models and self-timed, (logically) interacting dynamic software threads [653, 654].

Performance from Layered Interleaving

Now consider R concurrent resources, three of which (1, 2, and R) are shown in Figure 15-4. As physical entities, the resources can be interleaved by the relative interval sizes implied between the physical events—this is the time-based scheduling. Consider the one sub-sequence of physical events, e_{P12}, e_{P23}, e_{PR1}, as shown from left to right. (Note that the second subscript of the events (time t) represents time within the time base of the physical thread; these are not global time tags. Global time is calculated from the relative rates of interleaving of the physical resources.) The scheduler of each resource selects logical threads to be executed in the physical time period implied by each event. Here, scheduler U_{P1} might select the events shown in the box, e_{L113}, e_{L122}, e_{L123}, implying a logical interleaving—resource sharing among threads—on resource 1. (Each of these logical events could represent a large amount of software functionality.) This is the self-timed scheduling in our approach since decisions about logical event sequences are made in concert with data dependencies from data generated elsewhere in the system.

Considering all three resources, the actual event sequence of logical threads in the example is $\{\ldots, e_{L113}, e_{L122}, e_{L123}, e_{L222}, e_{LR11}, e_{LR21}, e_{LR22}, \ldots\}$ as implied by the boxed events. The actual sequence of logical events is the system trajectory.

In Figure 15-4, the atomic groupings shown are executed left to right, as the sequence R_1, R_2, R_R is implied by the rate-based interleaving of the resources. The trajectory can be affected by many high-level factors such as resource rates, scheduling policies, and data dependencies. For instance, an increase in the

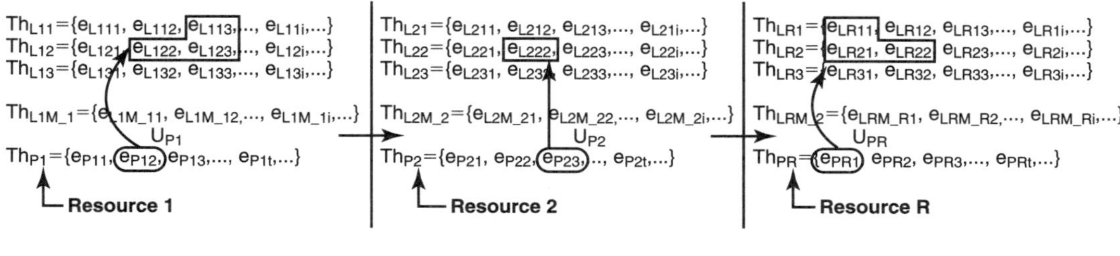

15-4   System trajectory.

FIGURE

computation power of a resource, say resource 2, could allow it to execute an extra logical event as part of e_{P23}. If this extra event, say e_{L233}, was being waited for by thread Th_{LR3} on resource R, then scheduler U_{PR} might make a different scheduling decision, executing e_{LR31} along with other logical events on R instead of those shown.

Significantly, the actual trajectory of the system over time (its performance) is calculated by the simulation through the interaction of heterogeneous design elements with the separate (logical and physical) time bases of software and hardware. The computation complexity of software as it executes on hardware can be physically descriptive while the software can be a mathematical representation.

15.5   CONCLUSIONS

No single chapter can summarize in detail the many MoCs that have been associated with SoCs. Instead, we attempted to provide a means of assessing specific MoCs according to a new classification scheme drawn from a broad historical perspective and projections about the nature of future SoC designs. We introduced familiar MoCs classified according to typical categories of synchronous versus asynchronous and timed versus untimed. More importantly, beyond that we proposed that there are classes of MoCs that have very different purposes; these can be as broad as to include models of types of computer artifacts as well as design tools. Most computer artifacts and design tools are conceived, developed, and used broadly with little or no formal, mathematical basis, yet they have had enormous impact on computer design over the years.

Currently, there is a gap between those doing real SoC designs, largely by intuition, educated guesses and ad -hoc, home-grown simulation programs, and those who suggest that real SoC designers must accept and adopt one or more formal models before they can build anything of which to be proud. We believe the real answer lies somewhere in between. What is required is a common language based on the design elements that will be used in real designs. This language of design is an MoC that will be more universal than synchronous versus asynchronous or timed versus untimed, but it will be less formal. It will be a design tool that will permit the design of computer artifacts, capturing the nuances of buildable designs by uniting the different time bases of software and the physical computation engine on which it executes.

We overviewed our tool, MESH, which we believe addresses some of the fundamental challenges in providing a simulation basis for these next-generation design elements. We believe it takes this step because it relates different types of processes, represented as threads in a model that resolves software and hardware design elements through scheduling decisions that capture the performance effects of resource sharing. We believe its design elements will have an enormous impact on SoC design.

Thus, with a new chapter in computer design unfolding, our objective is to broaden the discussion of what a computer model really is so that the arena of SoC—design may be more accessible to designers and thus result in a novel collection of software, hardware, and mixed computer artifacts that would not otherwise be possible.

<table><tr><td>**16**
</td><td>

Metropolis: A Design Environment for Heterogeneous Systems

</td></tr></table>

Felice Balarin, Harry Hsieh, Luciano Lavagno,
Claudio Passerone, Alessandro Pinto,
Alberto Sangiovanni-Vincentelli, Yosinori Watanabe,
and Guang Yang

16.1   INTRODUCTION

The ability to integrate an exponentially increasing number of transistors within a chip, the ever expanding use of electronic embedded systems to control more and more aspects of the "real world," and the trend to interconnect more and more such systems (often from different manufacturers) into a global network are all creating a nightmarish scenario for embedded system designers. Complexity and scope are exploding into three inter-related but independently growing directions at the same time that teams are shrinking in size to reduce costs. This problem can only be tackled by a revolutionary approach to system design, which dramatically raises the level of abstraction, while keeping close control of cost, performance, and power issues.

This chapter focuses on embedded systems, which are defined as electronic systems that use computers and electronics to perform some task, usually to control some physical system or to communicate information, without being explicitly perceived as a computer. Thanks to the phenomenal growth exhibited by the scale of integration, they are a preferred means to provide ever-improving services to a multitude of drivers, callers, photographers, watchers, and so on. Basically, embedded systems offer the revolutionary advantages of Moore's law to ordinary people, even those who are not skilled in the use of computers.

A key characteristic of such systems is the heterogeneity both of specification models, because of the habits and needs of designers from very different application fields, and of implementation mechanisms, because of the variety of tasks and environmental conditions. For example, block-based design environments, multi-CPU implementation platforms, critical safety needs, and hard deadlines are typical of the automotive world. On the other hand, a cellular telephone user interface is designed using software-oriented tools, such as compilers and debuggers, for implementation over a single micro-controller shared with protocol stack management tasks.

This heterogeneity has so far caused a huge fragmentation of the field, resulting in highly diverse design methodologies with often scarce tool support. De facto standards in terms of specification models and tools have emerged in only a few areas, namely, telecommunications (SDL), and automotive electronics (Simulink diagrams and Statecharts), essentially because of standardization and integration needs. This is of course problematic for a number of reasons, and a more unified approach to embedded system design, with possible ramifications to general electronic system design, would obviously be a benefit.

A well-structured design flow must start by capturing the design specifications at the highest level of abstraction and proceed toward an efficient implementation. The critical decisions are about the architecture of the system (processors, busses, hardware accelerators, memories, and so on) that will carry on the computation and communication tasks associated with the overall specification of the design. There is great value in understanding the application domain since the peculiarities of the design have to be served well, otherwise system designers will not be able to make efficient use of the flow.

Platform-based design [659] came about to allow the design process to be segmented into a series of steps that are similar in nature. The basic principles are to hide unnecessary details of an implementation, to summarize the important parameters of the implementation in an abstract model, and to limit the design space exploration to a set of potential platform instances from which to choose. The platform instances are chosen from a set of components in a library that includes as essential elements interconnections and communication mechanisms. The design process is then seen as a meet-in-the-middle approach whereby the successive refinement from specification toward implementation (the top-down part) is matched with a library of components whose models are an abstraction of possible implementations (the bottom-up part). Another important general principle (also known as component- or communication-based design) is how to handle the communication mechanism among the library elements. If indeed interfaces and communication mechanisms are carefully identified and specified, design reuse is greatly simplified and enhanced.

Today, the design chain is not supported adequately. Most system-level designers use a collection of tools that are not linked. The implementation is then carried out with informal techniques that involve a great deal of human language-based interactions, thus creating unnecessary and unwanted iterations among groups of designers in different companies or different divisions that share little understanding of their respective domains of knowledge. The tools are linked by manual or empirical translation of intermediate formats with no guarantee that the semantics of the design are preserved. This is of course the source of errors that are difficult to identify and debug.

Finally, the move toward programmable platforms shifts the design implementation task toward embedded software design. When embedded software reaches a certain degree of complexity typical of today's designs, the risk that the software will not be correct increases exponentially. This is due mainly to poor design methodologies and to fragile software system architectures, the result of growing functionality over an existing implementation that may be quite old and undocumented.

The goal of the Metropolis project is to cope with these problems in a unified framework, shown in Figure 16-1. We are not advocating a single, overarching

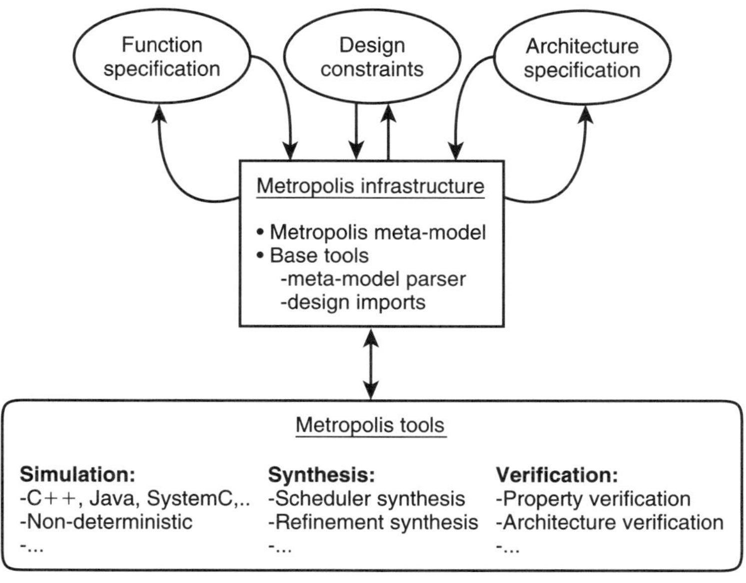

16-1   The Metropolis framework.

FIGURE

super-tool that could solve all design problems, since this is obviously not feasible. The idea is to provide an infrastructure based on a model with precise semantics but general enough to support the models of computation proposed so far and, at the same time, to allow the invention of new ones [660]. The model, called a meta-model of computation for its characteristics, is capable of supporting not only functionality capture and analysis, but also architecture description and mapping of functionality to architectural elements. Since the model has precise semantics, it can be used to support a number of synthesis and formal analysis tools in addition to simulation. Non-functional and declarative constraints are captured using a logic language.

The Metropolis approach has the following key characteristics. First of all, it leaves the designer relatively free to use the specification mechanism (graphical or textual language) of choice, as long as it has a sound semantical foundation (model of computation).

Second, it uses a single meta-model to represent both the embedded system and some abstract relevant characteristics of its environment.

Finally, it separates orthogonal aspects, such as:

1. Computation and communication. This separation is important because:

   a. Refinement of computation is generally done by hand, or by compilation, or by scheduling, and other complex techniques.

   b. Refinement of communication is generally done by use of patterns (such as circular buffers for FIFOs, polling, or interrupt for hardware-to-software data transfers, and so on).

2. Functionality and architecture, or "functional specification" and "implementation platform," because they are often defined independently, by different groups (e.g., video encoding and decoding experts versus hardware/software designers in multi-media applications). Functionality (both computation and communication) is "mapped" to architecture in order to specify a given refinement for automated or manual implementation.

3. Behavior and performance indices, such as latency, throughput, power, energy, and so on. These are kept separate because:

   a. When performance indices represent constraints they are often specified independent of the functionality, by different groups (e.g., control engineers versus system architects for automotive engine control applications).

   b. When performance indices represent the result of implementation choices, they derive from a specific architectural mapping of the behavior.

All these separations result in better reuse, because they decouple independent aspects, that would otherwise tie, e.g., a given functional specification to low-level implementation details, or to a specific communication paradigm, or to a scheduling algorithm. It is very important to define only as many aspects as needed at every level of abstraction, in the interest of flexibility and rapid design space exploration.

The first design activity supported by Metropolis is transfer of information both between people working at different abstraction levels (implementation and integration) and between people working concurrently at the same abstraction level (concurrent design). The meta-model includes constraints representing in abstract form requirements that are not yet implemented, or that are assumed to be satisfied by the rest of the system and its environment.

The second design activity is analysis, by using both simulation and formal verification, of how well an implementation satisfies the requirements. Proper use of abstraction can speed up verification dramatically, whereas always using very detailed representations can introduce excessive dependencies between developers, reduce understandability of interfacing requirements, and reduce efficiency of analysis mechanisms.

The third design activity is synthesis of an implementation (refinement) from a mixed declarative/executable specification.

One can argue that application domains and their constraints on cost, energy, performance, design time, safety, and so on are so different that there is not enough economy of scale to justify the development of tools to automate the design activities discussed above. The goal of the Metropolis project is to show that, at least for a broad class of domains and implementation choices, this is not true. Which particular technique or algorithm will be used for analysis and synthesis of a particular design depends both on the application domain and on the design phase. For example, safety-critical applications may need formal verification techniques that require significant human skills for use on realistic designs. On the other hand, simple low-level equivalence checks between various abstraction levels in hardware design (e.g., logic versus transistor levels) are routinely executed with (different) formal verification tools.

Metropolis thus is not aimed at providing algorithms and tools for all possible design activities. It offers syntactic and semantics mechanisms to compactly store and communicate all relevant design information, and it can be used to "plug in" the required algorithms for a given application domain or design flow. A meta-model parser reads meta-model designs, and a standard application programming interface (API) allows one to browse, analyze, modify, and augment those designs by additional information. For each tool integrated in Metropolis, a back-end generates whatever is required as input by the tool from the relevant portion of the design. This unified mechanism makes it easy to incorporate tools developed elsewhere, as demonstrated by integrating into Metropolis the software verification tool Spin [661].

A fundamental aspect that we considered throughout this work is the ability to specify rather than implement, execute reasonably detailed but still fairly abstract specifications, and finally use the best synthesis algorithms for a given application domain and implementation architecture. For these reasons we represented explicitly the concurrency available at the specification level, in the form of multiple communicating processes. We used an executable representation for the computation processes and the communication media, in order to allow both simulation and formal and semi-formal verification techniques to be used. Finally, we restricted that representation, with respect to a full-fledged programming language such as C, C++, or Java, in order to improve both analyzability and synthesizability.

Unfortunately, we were not able to remain fully within the decidable domain (nor, perhaps even worse, within the efficiently analyzable domain), because this goal is achievable only with extreme abstraction (i.e., only at the very beginning of a design), or for very simple applications, or in highly specialized domains. However, we defined at least a meta-model for which the task of identifying sufficient conditions that allow analysis or synthesis algorithms to be applied or speeded up could be simple enough. Hence our meta-model can be translated relatively easily, under checkable conditions or using formally defined abstractions, into a variety of abstract synthesis- or analysis-oriented models, such as Petri nets, or static dataflow networks, or synchronous extended finite state machines (FSMs).

The meta-model uses different classes of objects to represent the basic notions discussed above:

1. Processes, communications media, and netlists describe the functional requirements in terms of input-output relations first and then of more detailed algorithms later.

2. A mechanism of refinement allows one to replace processes, media, and sub-netlists with other subnetlists representing more detailed behavior.

3. Temporal and propositional logics can be used to describe constraints over quantities such as time or energy. They:

    a. Initially describe constraints over yet undefined quantities (e.g., latency constraints when describing processes at the untimed functional level using Kahn process networks [KPNs]).

b. Then specify properties that must be satisfied by more detailed implementations (e.g., while a KPN is being scheduled on a processor).

c. Finally define assumptions that must be checked for validity (e.g., using a cycle-accurate instruction set simulator [ISS] to run the scheduled processes in software).

In the rest of the paper we discuss all these notions in detail, and we apply them to a complex, realistic embedded system specification from the multimedia consumer electronics domain.

16.1.2   Related Work

VCC is a recent commercial example of a system-level design environment from Cadence, based on the idea of separation between functionality and architecture pioneered in POLIS [662]. The VCC model of computation is fixed, whereas Metropolis allows one to define communication primitives and execution rules most suitable for the application at hand. An architecture in VCC can be constructed only from a small, predefined set of components, whereas in Metropolis it can be any network of communicating processes, permitting a recursive layering of platform models. Finally, VCC uses C and C++ to define the behavior of processes, ruling out formal process analysis or optimization techniques beyond those used by standard compilers. On the other hand, Metropolis has a flexible and powerful formal underlying semantics (the meta-model) that allows the representation of the widest range of system design problems, from FSM-based models, to dataflow, to Petri nets, to discrete event. It also allows a precise formulation of optimization problems and a precise definition of classes of systems to which the problems apply.

Tools from Arexsys, Foresight, Artisan, and CardTools are very close in spirit and implementation to VCC. They all use a separation between functionality and architectural resources, and they all use a mapping to derive performance information. Of those, only ArchiMate from Arexsys has the capability to model system functionality using a formal language, SDL; the others use C or C++. SDL is a good modeling language for a specific class of systems like telecommunication protocols, but it lacks expressive power when it comes to modeling other aspects, such as digital signal processing and multi-media.

Ptolemy [663], SystemC [664], and SpecC [665] are all system-level modeling languages or frameworks, bearing several superficial similarities to the meta-model, since all share the notion of concurrent processes communicating through channels. Another similarity to SystemC and SpecC is the use of communication refinement as the basic mechanism both to introduce performance modeling aspects, such as communication resource sharing, and to interface portions of the model written at different levels of abstraction. However, Ptolemy, SystemC, and SpecC lack features that are necessary to orthogonalize functionality and architecture, such as the mapping between functional and architectural networks, or between different refinement levels. Ptolemy only represents functionality, and each module and channel in a SystemC or SpecC model can be represented only at one level of abstraction at a time. Metropolis, on the other hand, uses the refinement and mapping constructs to model several levels simultaneously within a single representation. Second, these other languages lack the ability to represent constraints explicitly. Finally, the use of the semantics of the underlying C/C++ language (in particular unrestricted method calls and pointers) hinders automated synthesis and optimization.

Tools and frameworks based on SystemC as their native representation, such as N2C from CoWare, SystemStudio from Synopsys, and ROSES [666], are thus limited to communication refinement in the form of replacement of channel models, and synthesis from subsets. CoWare and ROSES also offer some interesting communication optimization capabilities, which could be integrated as one of the Metropolis design steps.

The modeling environment for software and hardware (MESH) environment [667] is also oriented at modeling embedded systems and enabling hardware/ software trade-offs, and it exploits the separation between functionality and architecture. However, it does not represent communication explicitly using high-level primitives, but only using unsynchronized shared variables. Although arguably this is the final implementation in most technologies, we believe that it is inappropriate to introduce it early in the design space exploration domain, because it induces unnecessary dependencies on the scheduling of hardware and software computation tasks. Moreover, MESH is a two-layer model only (software above hardware) and has no mechanism to represent communication refinement, sophisticated communication architectures with multiple bus layers, parameterized architectures and mappings, and multiple mapping options.

Commercial hardware/software co-verification tools from other companies (e.g., Mentor, Vast, Virtio, and Axys) are essentially ISS, linked to various hardware simulators. They attack the functional and performance modeling problem for software-dominated embedded systems but do not even tackle the issues of high-level modeling of hardware, of separation of concerns, and of refinement. In particular, building a new platform model is an expensive and time-consuming task. Also, setting up a performance simulation and various design space exploration activities, such as assigning a function to a task or using a different peripheral for a given hardware acceleration, are time-consuming operations. Simulating functional models annotated with performance information, as in Metropolis, is much easier but less precise. In practice, co-verification of this kind can be targeted as back-ends by Metropolis.

16.2   THE METROPOLIS META-MODEL

The Metropolis meta-model is a language to specify networks of concurrent objects, each taking actions sequentially. The behavior of a network is formally defined by the execution semantics of the language [668]. A set of networks can be used to represent all the aspects described in the previous section for conducting designs, i.e., function, architecture, mapping, refinement, abstraction, and platforms.

In providing a specification, one needs a capability of describing the following three aspects: actions, constraints, and their refinement. This is true whether it is a specification of a system (behavior or architecture) or of the environment. The Metropolis meta-model has this capability, as presented in this and the next two sections.

A behavior can be defined as concurrent occurrences of sequences of actions.[1] Some action may follow another action, which may take place concurrently with other actions. The occurrences of these actions constitute the behavior of a system to which the actions belong. An architecture can be defined as the capacity to perform actions upon request, at a cost. Some actions may realize arithmetic operations, whereas others may transfer data. Using these actions, one can implement the behavior of the system.

A description of actions can be made in terms of computation, communication, and coordination. The computation defines the input and output sets, a mapping from the former to the latter, the condition under which inputs are acquired, and the condition under which outputs are made observable. The communication defines a state and methods. The state represents a snapshot of the communication. For example, the state of communication carried out by a stack may represent the number of elements in the stack and the contents of the elements. The communication methods are defined so that they can be used to transfer information. The methods may evaluate and possibly alter the communication state. For the example of the stack, methods called pop and push may be defined. In addition, one may define a method for evaluating the current number of elements available in the stack. Actions for computation and communication often need to be coordinated. For example, one may want to prohibit the use of the pop

---

[1] Here, behavior is used as a general term to refer to what a given system does. In the succeeding sections, we define the term more restrictively as a single sequence of observable actions.

method, whereas an element is added to the stack by the push method. Coordination may be specified either by formulas, declaring the desired coordination among actions, or by algorithms, whose execution guarantees the coordination.

In the meta-model, special types of objects called process and medium are used to describe computation and communication, respectively. For coordination, a special type called scheduler can be used to specify algorithms. Alternatively, one can write formulas in linear temporal logic [669], if one prefers to specify the coordination declaratively, without committing to particular algorithms.

When an action takes place, it incurs cost, modeled by the architectural item to which it is mapped, as described in Section 16.2.4. Cost for a set of actions is often subject to certain constraints. For example, the time interval between two actions may have to be smaller than some bound, or the total power required by a sequence of actions may need to be lower than some amount. The meta-model provides a mechanism by which a quantity, such as time or power, can be defined and associated with actions, and constraints on the quantity can be specified in a form of predicate logic. In the rest of this section, we present in more detail these key points of the meta-model and illustrate them with examples.

16.2.1   Function Modeling

The function of a system is described as a set of objects that concurrently take actions while communicating with each other. We call such an object process in the meta-model and associate with it a sequential program called thread. A process communicates through ports defined in it. A port is specified with an interface, declaring a set of methods that can be used by the process through the port. In general, one may have a set of implementations of the same interface, and we refer to objects that implement port interfaces as media. Any medium can be connected to a port if it implements the interface of the port. This mechanism allows the meta-model to separate computation performed by processes from communication among them. This separation is essential to facilitate the description of the objects to be reused for other designs. Figure 16-2 shows a network of two producer processes and one consumer process that communicate through a medium.

Once a network of processes is given, the behavior of the network is precisely defined by the meta-model semantics as a set of executions. First, we define an execution of a process as a sequence of events. Events are associated with the beginning and ending of actions, modeled as code fragments (e.g., basic blocks or function calls). For example, for process X in Figure 16-2, each beginning and ending of each call to R.read( ) is an event. An execution of a network is then defined as a sequence of vectors of events, with each vector having at most one

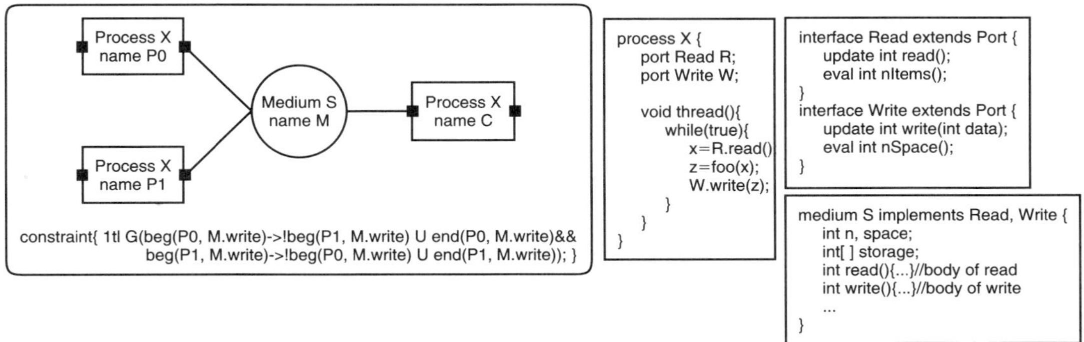

16-2   Functional model of two producers and one consumer.

FIGURE

event for each process, and defines the set of events that happen "together" (with together usually, but not necessarily, meaning at the same time). The meta-model can also model non-deterministic behavior, which is useful to abstract a part of the design. Non-determinism takes the form of a process not performing any visible event in some vector that is part of an execution, modeling "idling" or internal actions. It can also be modeled as a variable taking "any value" from its domain. Thus there may be more than one possible execution of the network in general. The set of (possibly non-deterministic) executions is further restricted by constraints, which are logic formulas defining the set of legal executions [668].

16.2.2   Constraint Modeling

There are three types of constraints in the meta-model, and each type is expressed in a different logic. Coordination constraints are represented by formulas of linear temporal logic (LTL), whereas quantity constraints are represented in the logic called logic of constraints (LOC). (Strictly speaking, quantity axioms are represented in an LOC extension called extended logic of constraints [ELOC].) The choice of different logics is driven by different requirements for three types of constraints and by our guiding principle to choose the most appropriate logic for the task at hand. We have surveyed existing, well-studied logics to find one suitable for each purpose. In the case of coordination constraints, LTL fulfilled the requirements. However, in the case of quantity constraints and axioms, we could not find a logic that fits our needs, so we have proposed LOC for this purpose [670].

We have designed LOC to meet the following goals, which we believe are essential for a constraint specification formalism to gain wide acceptance:

- it must be based on a solid mathematical foundation, to remove any ambiguity in its interpretation.

- it must feel natural to the designer, so that typical constraints are easy to specify.

- it must be compatible with existing functional specification formalisms, so that language extensions for constraint specification can be easily defined.

- it must be powerful enough to express a wide range of interesting constraints.

- it must be simple enough to be analyzable, at least by simulation, and ideally by automatic formal techniques.

In case of conflicting goals (e.g., expressiveness versus simplicity), we believe that LOC presents a reasonable compromise.

For example, the constraint in Figure 16-2 specifies the mutual exclusion of the two producers when one of them calls the write method of the medium. Constraints allow one to describe coordination of processes, or to relate behavior of networks through mapping or refinement as presented later.

In the following examples, we assume that the set of event names is $E = \{in, out\}$ and that a real-valued annotation is defined. Intuitively, we assume that $t(e[i])$ corresponds to the time of the i-th occurrence of an event e. The following common constraints are now easy to express:

- rate: "a new *out* will be produced every P time units":

$$t(out[i+1]) - t(out[i]) = P,$$

- latency: "*out* is generated no more than L time units after *in*":

$$t(out[i]) - t(in[i]) \le L,$$

- jitter: "every *out* is no more than J time units away from the corresponding tick of the real-time clock with period P":

$$|t(out[i]) - i * P| \le J,$$

- throughput: "at least X *out* events will be produced in any period of T time units":

$$t(out[i + X]) - t(out[i]) \leq T,$$

+ burstiness: "no more than X *in* events will arrive in any period of T time units":

$$t(in[i + X]) - t(in[i]) > T.$$

16.2.3   Architecture Modeling

Architectural models are distinguished by two aspects: the functionality that they are capable of implementing and the efficiency of the implementation. In the meta-model, the former is modeled as a set of services offered by an architecture to the functional model. Services are just methods, bundled into interfaces [671].

To represent efficiency of an implementation, we need to model the cost of each service. This is done first by decomposing each service into a sequence of events and then by annotating each event with a value representing the cost of the event.

To decompose the services into sequences of events, we use networks of media and processes, just as in the functional model. These networks often correspond to physical structures of implementation platforms.

For example, Figure 16-3 shows an architecture consisting of n processes, T1, . . . , Tn, and three media, CPU, BUS, and MEM. The architecture in Figure 16-3 also contains so-called quantity managers, represented by diamond-shaped symbols, which we will introduce shortly. The processes model software tasks executing on a CPU, whereas the media model the CPU, bus, and memory. The services offered by this architecture are the execute, read, and write methods implemented in the Task processes. The thread function of a Task process repeatedly and non-deterministically executes one of the three methods. In this way, we model the fact that the Tasks are capable of executing these methods in any order. The actual order will become fixed only after the system function is mapped to the architecture, when each Task implements a particular process of the functional model.

Although a Task process offers its methods to the functional part of the system, the process itself uses services offered by the CPU medium, which, in turn, uses services of the BUS medium. In this way, the top-level services offered by the Tasks are decomposed into sequences of events throughout the architecture.

The meta-model includes the notion of quantity, used to annotate individual events with values measuring cost. For example, in Figure 16-3 there is a quantity named energy used to annotate each event with the energy required to process it. To specify that a given event takes a given amount of energy, we associate with that event a request for that amount. These requests are made to the object called quantity manager, which collects all requests and fulfills them, if possible.

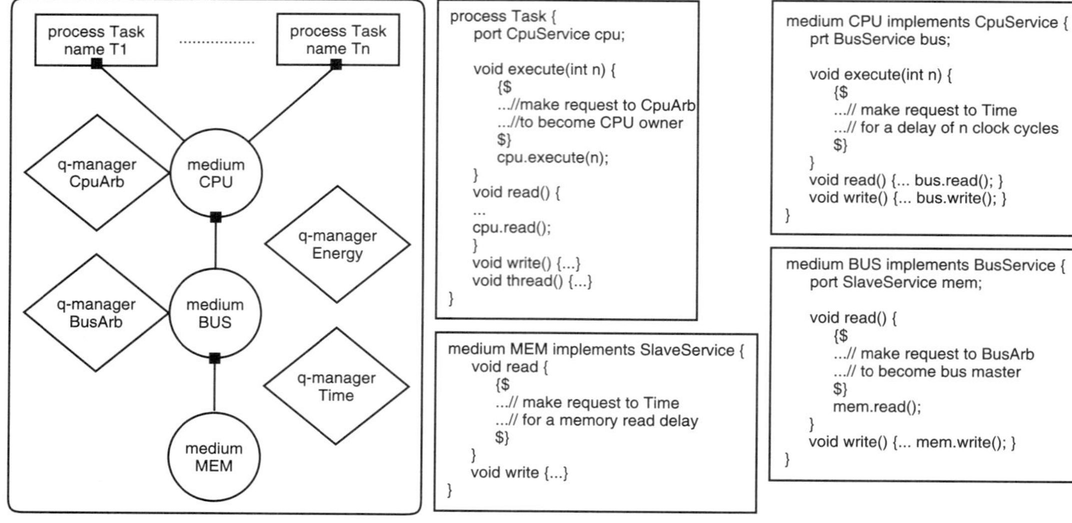

16-3

An architectural model.

FIGURE

Quantities can also be used to model shared resources. For example, in Figure 16-3 quantity CpuArb labels every event with the task identifier of the current CPU owner. Assuming that a process can progress only if it is the current CPU owner, the CpuArb manager effectively models the CPU scheduling algorithm.

The meta-model has no built-in notion of time, but the time can be modeled as yet another quantity that puts an annotation, in this case a time-stamp, to each event. Managers for common quantities, such as time, are provided as standard libraries with Metropolis and are understood directly by some tools (e.g., time-driven simulators) for the sake of efficiency. However, quantity managers can also be written by design flow developers, in order to support quantities that are relevant for a specific application domain.

16.2.4   Mapping

To evaluate the performance of a particular implementation, a functional model needs to be mapped to an architectural model. In the meta-model, it is possible to do this without modifying the functional and architectural networks. A new network can be defined to encapsulate the functional and architectural networks and relate the two by synchronizing events between them. This new network is called a mapping network, and one may consider it as a layer defined on top of the function and architecture, in which a mapping between the two is specified.

The synchronization mechanism roughly corresponds to an intersection of the sets of executions of the functional and architectural models. Functional model executions specify a sequence of events for each process but usually allow arbitrary interleaving of event sequences of the concurrent processes, as their relative speed is undetermined. On the other hand, architectural model executions typically specify each service as a timed sequence of events but exhibit nondeterminism with respect to the order in which services are performed, and on what data. The mapping eliminates from the two sets all executions except those in which the events that should be synchronized always appear simultaneously. Thus, the remaining executions represent timed sequences of events of the concurrent processes.

For example, Figure 16-4 shows a mapping network that combines the functional network from Figure 16-2 with the architectural network from Figure 16-3. Events of the two networks are synchronized using the constraint clause in the

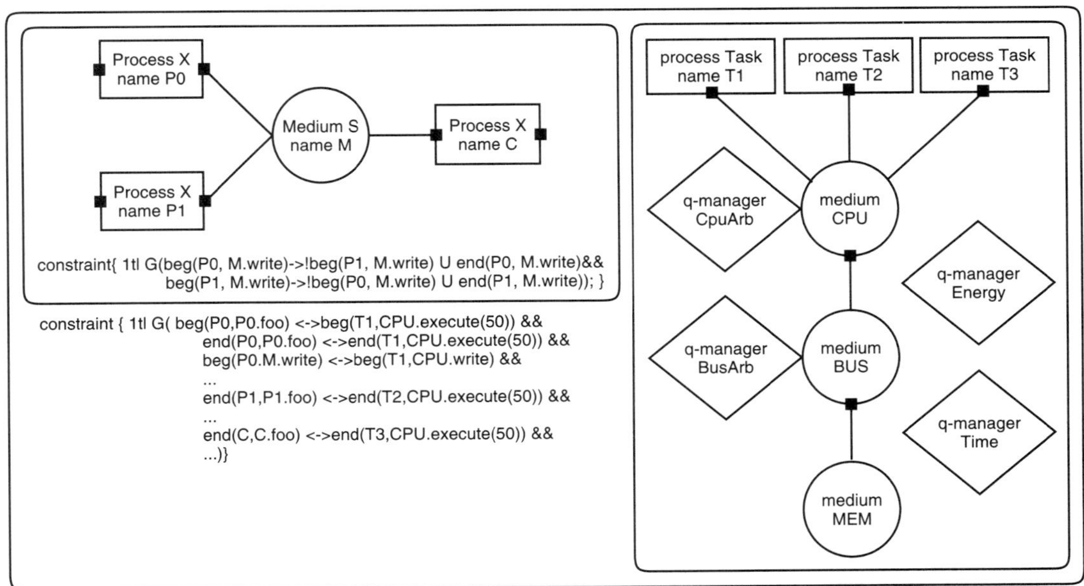

16-4   Mapping of function to architecture.

FIGURE

mapping network. For example, executions of foo, read, and write by P0 have been synchronized with executions of execute, read, and write by T1. Since P0 executes its actions in a fixed order whereas T1 chooses its actions non-deterministically, the effect of synchronization is that T1 is forced to follow the decisions of P0, whereas P0 "inherits" the quantity annotations of T1. In other words, by mapping P0 to T1, we make T1 become a performance model of P0. Similarly, T2 and T3 are made to be performance models of P1 and C, respectively.

16.2.5   Recursive Paradigm of Platforms

Suppose that a mapping network, such as the one shown in Figure 16-4, has been obtained as described above. One can consider such a network itself as an implementation of a certain service. The algorithm for implementing the service is given by its functional network, whereas its performance is defined by the architecture counterpart. In Figure 16-5, the interface on the top defines methods specifying a service, and the medium in the middle implements the interface at the

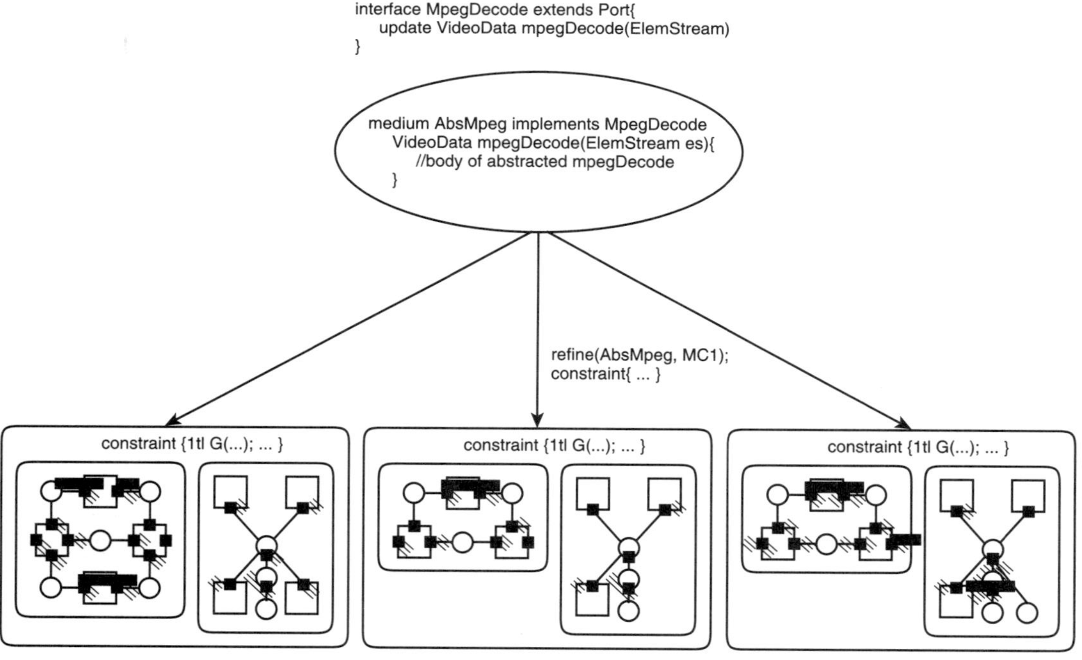

    Multiple mapping netlists refining the same service to form a platform.

desired abstraction level. Underneath the medium is a mapping network, providing a more detailed description of the implementation of the service. The meta-model can relate the medium and each network by using the meta-model construct refine and constraints. For example, the constraints may say that the begin event of mpegDecode of the medium is synchronized with the begin event of vldGet of the VLD process in the network, and the end of mpegDecode is synchronized with the end of yuvPut of the OUTPUT process, whereas the value of the variable YUV at the event agrees with the output value of mpegDecode.

In general, many mapping networks may exist for the same service with different algorithms or architectures. Such a set of networks, together with constraints on event relations for a given interface implementation, constitutes a platform. The elements of the platform provide the same service with different costs, and one is favored over another for given design requirements. This concept of platforms appears recursively in the design process. In general, an implementation of what one designer conceives as the entire system is a refinement of a more abstracted model of a service, which is in turn employed as a single component of the larger system. For example, a particular implementation of an MPEG decoder is given by a mapping network, but its service may be modeled by a single medium, with the events generated in the medium annotated with performance quantities to characterize the decoder. Such a medium may be used as a part of the architecture by a company that designs broadband set-top appliances. This medium serves as a model used by the appliance company to evaluate the set-top-box design, whereas the same medium is used as design requirements by the provider of the MPEG decoder. Similarly, the MPEG decoder may use in its architecture a model of a bus represented as a medium provided by another company or group. For the latter, the bus design itself is the entire system, which is given by another mapping network that refines the medium. This mapping network may be only one of the candidates in the platform of the communication service, and the designer of the MPEG decoder may explore various options based on his or her design criteria.

This recursive paradigm of platform-based design is described in a uniform way in the Metropolis meta-model. The key is the formal semantics to define precisely the behavior of networks and the mechanism of relating events of different networks with respect to quantities annotated with the events.

16.3   TOOLS

The Metropolis framework includes tools to perform a variety of tasks, some common, such as simulation, and some more specialized, such as property verification and synthesis, as described in this section.

16.3.1   Simulation

Traditionally, systems are verified by simulating their response to a given set of stimuli. However, non-determinism poses unique problems since there may be many valid responses. In Balarin et al. [672], we have proposed a generic simulation algorithm that selects, under the user's control, one of the acceptable behaviors for a given stimulus. The choice is driven by different objectives at various design stages. Initially, the algorithm optimizes simulation time to reveal the most trivial mistakes quickly. Later, the design is evaluated more thoroughly by simulating as many legal behaviors as possible, using, for example, a randomized algorithm.

In our approach we translate the meta-model and the simulation algorithm into an existing executable language. This feature is important to co-simulate designs captured in the meta-model together with existing designs that have been already specified in other languages. We have tested this approach in SystemC 2.0 [673], Java, and C++ with a thread library. Currently we are working toward an optimization of the simulation performance by using a co-routine-based approach, which reduces the overhead due to multithreading in the approaches above.

16.3.2   Property Verification

Formal property verification has long been a subject of investigation by academia and industry alike, but the state explosion problem restricts its usefulness to protocols or other higher levels of abstraction. At the implementation or other lower levels of abstraction, hardware and software engineers have utilized simulation monitors as basic tools to check simulation traces for debugging the designs.

Metropolis, with its formal semantics, allows full integration of both these approaches. Verification models are automatically generated for all levels of the design [674]. Verification languages, such as Promela, used by the Spin model checker [661], natively allow only simple concurrency modeling and thus become very clumsy to use when complex synchronization and architecture constraints are needed. Our translator automatically constructs the Spin verification model from the meta-model specification model, taking care of all the system level constructs. For example, we are able to generate a verification model automatically for the example in Figure 16-2 and to verify the non-overwriting properties of the medium. Furthermore, as the design is refined through structural transformation and architectural mapping, we are able to prove more properties, including throughput and latency. This kind of property verification typically requires several minutes on a 1.8-GHz Xeon machine with 1 GB of memory. When the state space complexity becomes too high, we use approximate verification and provide the user with a confidence factor on the passing result.

Simulation monitors are an attractive alternative. In Metropolis, monitors can be automatically generated from quantitative logic property specifications, thus relieving the designers of the tedious and error-prone task of writing monitors in the language of the simulators. The property to be verified is written using LOC [668], a logic particularly suitable for specifying performance constraints at the system level. From each LOC formula we automatically generate a monitor in C++, which analyzes the traces and reports any violations. Like any other simulation-based approach, this approach can only disprove an LOC formula (if a violation is found), but it can never prove it conclusively, as that would require analyzing traces exhaustively. The automatic trace analyzer can be used in concert with model checkers. It can perform property verification on a single trace when the other approaches failed because of excessive memory and space requirements.

We have applied the automatic LOC monitor technique to large designs with large and complex traces. In most cases with large traces (lines), the analysis completes in minutes and requires hundreds of bytes of data memory. We have found that the analysis time tends to grow linearly with the trace size, whereas the memory requirement remains constant regardless of the size of the traces [675].

16.3.3   Synthesis

We have developed an automatic synthesis technique called quasi-static scheduling (QSS) [676] to schedule a concurrent specification on computational resources that provide a limited amount of concurrency.

QSS considers a system to be specified as a set of concurrent processes communicating through FIFOs and generates a set of tasks that are fully statically scheduled except for data-dependent controls that can be resolved only at runtime. A task usually results from merging parts of several processes and shows less concurrency than the initial specification. Moreover, it allows inter-process optimizations that would have been difficult to achieve if processes were kept separated, such as replacing inter-process communication by assignments.

This technique proved particularly effective and allowed us to generate production quality code with improved performance. We applied it to a significant portion of an MPEG2 decoder [677], resulting in a 45% increase in performance when the entire decoder is considered.

The assumptions required by QSS on the input specification are a subset of what can be represented using the meta-model. We therefore naturally tried to reuse the same technique for meta-model specifications that satisfy those assumptions. When integrating such a tool within the Metropolis framework, we addressed two main problems:

1. How to verify whether a design satisfies a given set of rules.

2. How to convey all relevant design information to the external tool.

The main job of QSS is to schedule the communication, while doing little or no processing on the computation. Therefore, the first problem was solved by providing a library of interfaces and communication media that implement a FIFO model of communication. Those parts of the design that need to be scheduled and synthesized with QSS are required to use this kind of communication primitive only.

The second problem is solved by a customized back-end tool that reads in a design to be scheduled using QSS and generates a Petri net specification, the underlying model in QSS. QSS then uses the Petri net to produce a new set of processes. These new processes show no inter-process communication, which was removed by QSS, but only perform communication with the environment using the same primitives that we implemented in the library. The new code can thus be directly plugged into the meta-model specification of the entire system, as a refinement of the network that was selected for scheduling.

A very simple example of the meta-model is shown in Figure 16-6a. It is a system formed by one process and two media. The process reads integers from the input medium and writes them to the output medium after multiplying them by 2. The Petri net derived from this meta-model code is shown in Figure 16-6c. Such a Petri net can be presented as input to QSS.

16.4   THE PICTURE-IN-PICTURE DESIGN EXAMPLE

16.4.1   Functional Description

The application example that we consider here is called picture-in-picture (PiP), part of a digital set-top-box with PiP capabilities, which was originally developed by Philips in the context of the ESPRIT 25443 COSY project [677].

The set-top-box user has the possibility of watching a main channel and having at the same time a secondary small picture, inside the main picture, showing another channel. The user can tune parameters like position and size of the picture, frame color, and frame size for both the main and secondary channels.

```
process p1 {
   stack<int> in,out;
   void thread() {
      while (1)
         await [in] (in.n()>=1) {
            out.write (in.read()*2);
         }
   }
}

medium stack <type elem> {
   elem data [ ];
   int top;
elem read() {
   elem value;
   await [this] (this.top>=1) {
      value=data [top];
      top=top-1;
   }
   return value;
}
void write (elem e) {
   await [this] (true) {
      top=top+1;
      data[top]=e;
   }
}
}
```

a

```
process p1 {
   stack<int> in,out;
   void run () {
      int i, mul;
   START:
      await (in.p>=1);
      i=in.read(); mul=i*2; out.write (mul);
      endawait;
      goto START;
   }
}

medium stack <type elem> {
   unbounded place p;
   elem data [ ];
   int top;
   elem read () {
      elem value;
      await (this.p>=1);
      value=data [top]; p=p-1; top=top-1;
      endawait;
      return value;
   }
   elem write () {
      await (this.p>=0);
      p=p+1; top=top+1; data [top]=e;
      endawait;
   }
}
```

b

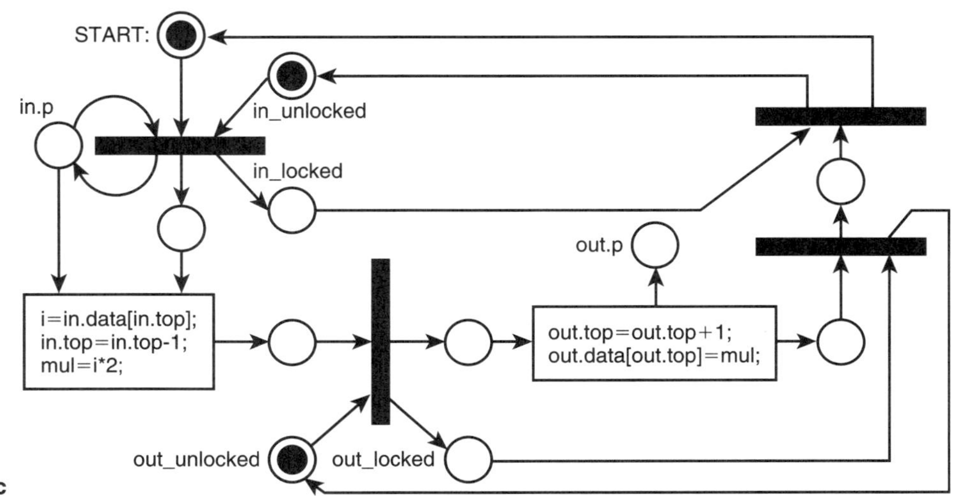

c

16-6 FIGURE   Example of derivation of a Petri net. (a) Meta-model. (b) Intermediate code model. (c) Petri net model.

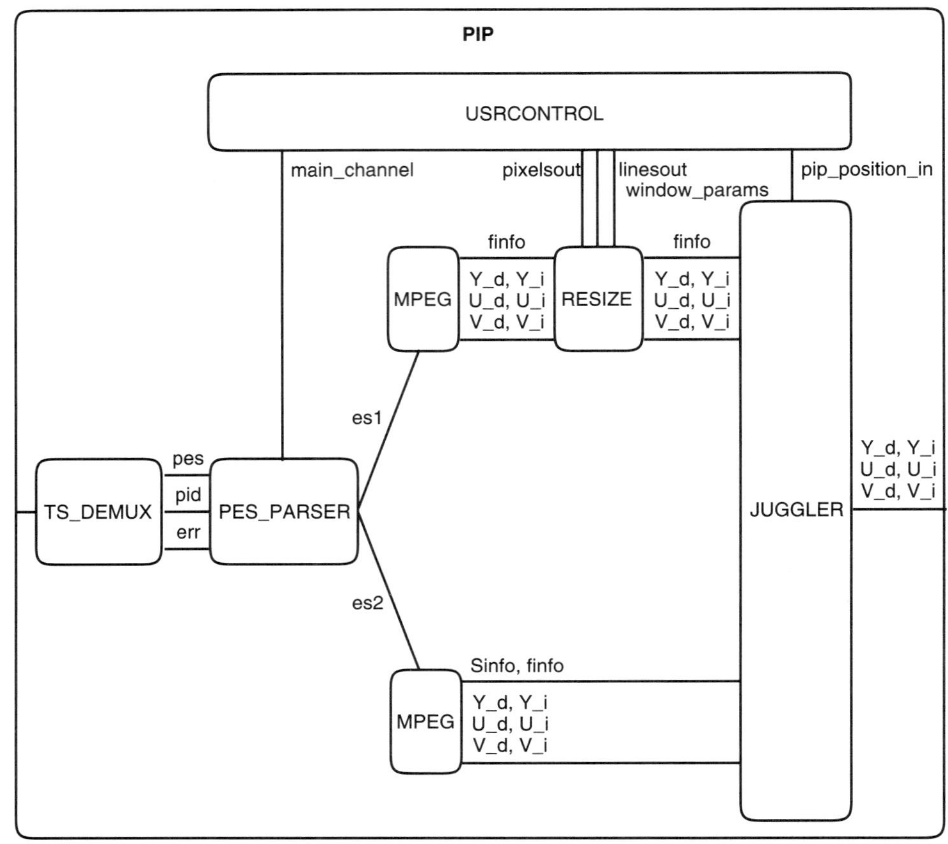

16-7   Block diagram representation of the picture-in-picture example.

FIGURE

The functional block diagram is shown in Figure 16-7. The transport stream is decoded by TS DEMUX. Its main functions are synchronization of the incoming stream and separation of payload packets (pes) from packet identification numbers (pid). Also error conditions can be generated by this block. The decoded stream is then passed to PES_PARSER, which selects the channel to visualize, depending on the user choice. Two streams can be chosen: one for the main channel and the other for the secondary picture. Each channel is then decoded by MPEG DECODER. The secondary channel is then resized. The RESIZE block takes user parameters to set the picture size, frame size, and color. Finally, the JUGGLER mixes the two decoded streams, while also setting the position of the secondary channel on the main one. Each block is a hierarchical netlist, which contains sub-netlists, processes, and media. The entire functional description contains 47 process instances and 294 channel instances.

The functional description is captured by the the Y-chart application programming interface (YAPI) model of computation developed by Philips [678], which is an extension of KPNs.

This model represents communication with a very simple FIFO-based API, including three functions: p.read(D,n), p.write(D,n), and select(p1,n1,p2,n2 . . .). The semantics assumes that each FIFO has infinite space and Write operations are nonblocking; they take as parameters the payload and the number of tokens to be written on the destination port p. Read operations are blocking (if there are not enough tokens in the FIFO to execute them) and take the same set of parameters. The select function takes as parameters pairs of port and number of tokens. It returns the index of any port (non-deterministically chosen, if more than one) where a read/write operation for the specified number of tokens can currently be completed. Select gives the user the power of checking the channel status, implementing non-determinism.

The YAPI platform is implemented in the Metropolis meta-model as a set of three elements: the medium type yapichannel, the process type yapiprocess, and the netlist type yapinetlist. By defining yapinetlist as connections of yapiprocesses through single-input/single-output yapichannels, the user can force the resulting system to have the YAPI semantics. All the netlist components are defined by extension of the YAPI library components. The yapiprocess declares a function called execute, which is used to define the behavior of the process. The designer first extends yapiprocess to define a sub-class and then overloads the execute function to specify the behavior of this particular process. The behavior can call the functions for the communication primitives described above through ports; the functions are implemented in the yapichannel class library, and its instances are connected to the ports.

The instances of the yapichannel are refined to implement the YAPI semantics using bounded resources. This is done as explained in Section 16.2, so that data transactions between reader and writer are subject to a particular protocol. In this example, we use the one called task-transaction-level (TTL) platform [679], defined by Philips. TTL is also provided as a library domain in Metropolis. Central to this platform is the definition of a finite length buffer and a set of APIs to access it. A ttlmedium implements four methods: claim_space, release_data, claim_data, and release_space. Their meanings are, respectively, to check whether there is enough space in the FIFO for writing, actually to write data, to check whether there are sufficient data for reading, and finally to release space previously occupied by tokens received by the reader.

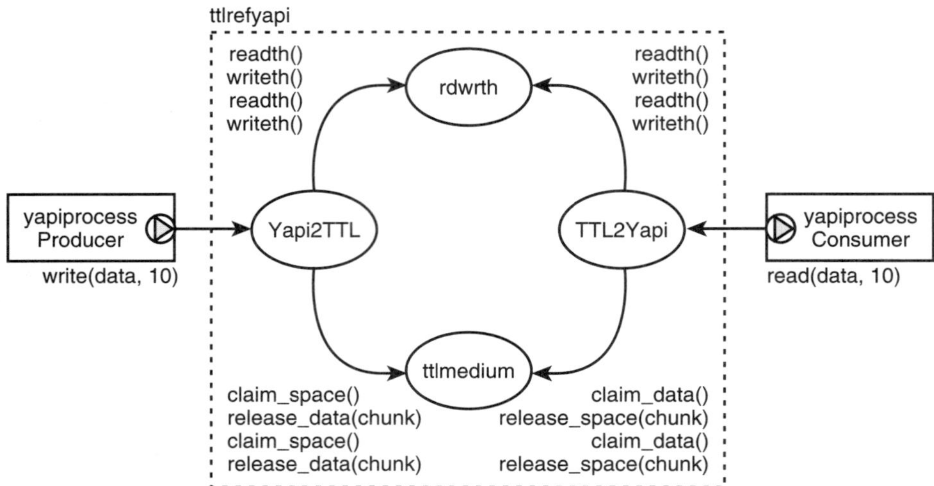

16-8 FIGURE   Structure of the netlist used to refine a yapichannel using a bounded FIFO ttlmedium.

In addition to the ttlmedium, the library also defines media, encapsulated in a netlist, that implement the YAPI communication interfaces using the services offered by a ttlmedium. By connecting ports of the original yapiprocesses to these media, the same execute functions of the processes can be used but this time with the transaction protocol specified by the TTL platform. The complete netlist that is used to refine a yapichannel is shown in Figure 16-8. Yapi2TTL and TTL2Yapi are two media that implement the YAPI API in terms of the lower level TTL API. A YAPI write is broken into a sequence of TTL claim_space and release_data calls. In order for a write to be completed on a finite-lenth buffer possibly smaller than the data size of a single read, tokens must be written and read in a ping-pong fashion. This handshake protocol is implemented using a shared synchronization medium called rdwrth. The designer can explore a set of refinement scenarios by changing a few parameters, such as the FIFO length (number of tokens), the token size, and also the handshake protocol.

The various available configurations of the netlist to simulate different scenarios can be selected by passing parameters to the constructor of the top netlist, as shown in Figure 16-9. The constructor takes as parameters the netlist name and a scenario that the user can choose. The scenario is propagated hierarchically to all the subnetlists. Three scenarios are defined in our example: YAPI, TTL.0, and TTL.1. If YAPI is selected, then all the interconnection channels are

```
public netlist PIPScenario {
   public PIPScenario (String n) {
      super (n);

      // Design only at the Yapi level
      // PIPSimNet net=new PIPSimNet ("PiPDesign", "YAPI");

      // Design upto the TTL level
      // TTL refinement scenario 0
      PIPSimNet net=new PIPSimNet ("PiPDesign", "TTL_0");

      // TTL refinement scenario 1
      // PIPSimNet net=new PIPSimNet ("PiPDesign", "TTL_1");

      addcomponent (net, this, "PiPDesign");
   }
}
```

16-9

PiP top level netlist constructor.

FIGURE

instantiated as yapichannel. In case one of the TTL scenarios is selected, after instantiation of yapichannels, a refinement function is called that refines each channel into a ttlrefyapi netlist with appropriate parameters.

16.4.2   Architectural Platform

Architectural platforms offer a set of services to implement the algorithms described in the functional part. Depending on the level of abstraction, services may have different granularities. At the lowest level, for instance, the services provided to implement a function are instructions and logic gates. In this example we decided to model architectures at a high level of abstraction where the quality of the implementations such as performance is annotated between the begin and end events of each function that constitutes the services of the architecture.

The architecture structure is depicted in Figure 16-10. The essential elements that compose the architecture are shown in the scheduled netlist. A number of software tasks (Ti in the figure) share a CPU under control of a real-time operating system (RTOS), which simulates concurrency between them. There is one main memory, which is accessible through the bus.

Tasks implement services that have to be considered as the building blocks to implement the functional specification. The implementation could be just a model

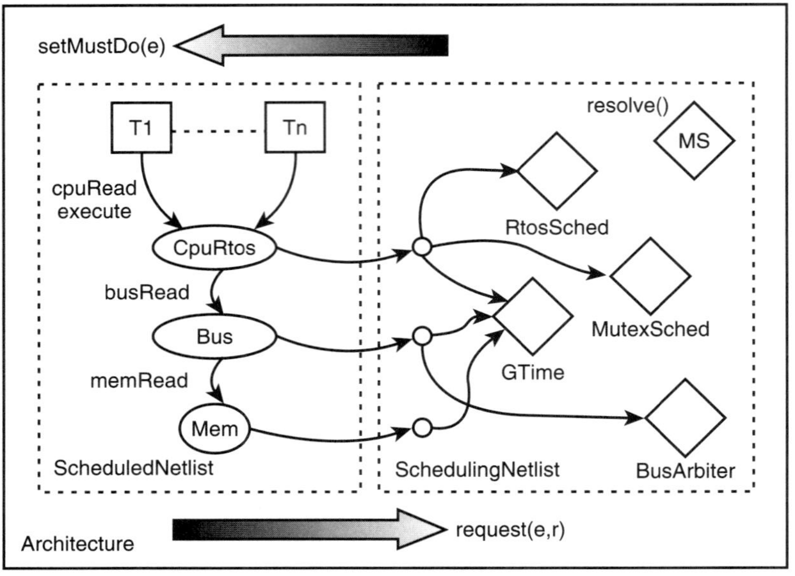

16-10

FIGURE

Block diagram of the architecture platform netlist in Metropolis.

for the performances of software library functions that are provided as intellectual property blocks (IPs) by the platform developer.

In our case, as developers of the models of the YAPI and TTL platforms, we decided to offer services that match the TTL services described in Section 16.4.1. Each task implements the TTL primitives claim_space, release_data, claim_data, and release_space. These methods are implemented using services provided by the CPU and the RTOS.

Tasks need two kinds of services: scheduling, in order to decide who can use the shared CPU, and actual computation. Figure 16-11 shows the interface definition and an example of how a software task can use the services to implement a specific function. The function in this case checks the number of tokens in the FIFO qn on the target slave T. (Each slave can have multiple queues for multiple types of input and output operations.) The task has a port to the real-time operating system of type RtosService, which is also shown in Figure 16-11. The first API service call resumes the task. Note that this call (unlike other calls to RTOS and YAPI or TTL services) will not be present in the final implementation code, since it is just a modeling artifact (used in simulation and formal verification).

```
interface RtosService extends Port {
   update void mutexLock (int mn) ;
   update void mutexUnLock (int mn) ;
   update void sleep (int TI, int p, int ncycles);
   update void resume (int TI, int p) ;
   update void suspend (int TI, int p) ;
   update void schedulerLock () ;
   update void schedulerUnlock () ;
}
```

```
public process SwTask {
   port RtosService RTOS ;
   port CpuService CPU;
   parameter int TASKID;
   parameter int TASKPRIORITY;
   SwTask (String instName, int TI, int p) {
      super (instName) ;
      TASKID=TI;
      TASKPRIORITY=p;
   }
   void query_space (int qn, int T) {
      RTOS.resume (TASKID, TASKPRIORITY) ;
      RTOS.mutexLock (mutexref (qn, T));
      CPU.cpuRead (T,qn,1);
      RTOS.mutexUnlock (mutexref(qn,T));
      RTOS.suspend(TASKID,TASKPRIORITY);
   }
}
```

```
interface CpuService extends Port {
   update void cpuRequest (int cycles) ;
   update void cpuRead (int target, int qn, int n) ;
   update void cpuWrite (int target, int qn, int n) ;
   update void cpuRead (int target, int addr, int n, int [] data) ;
   update void cpuWrite (int target, int addr, int n, int [] data) ;
}
```

16-11

FIGURE

RTOS interface and CPU interface that are used by a software task mapped on the CPU.

Then the task requests an exclusive access to the memory where the FIFO is mapped. After reading from the specified location, the mutex is released, and the task is suspended. (Again, this last call is just a modeling artifact, except in case of a co-routine-based cooperative scheduler.)

In this example, mutexreq is a special user-defined request class. The CPU offers services to read and write from and to peripherals. It also provides a service to perform computation (which keeps the CPU busy for a certain amount of clock cycles).

Interfaces define the set of services that are required by a task. Media implement interfaces. Inheritance is a way of defining multiple implementations of the same set of services. Together with polymorphism (as in the case of read and write functions in the CpuService interface), it provides a tool for design space exploration. Consider, for example, the case of a bus model. It has to implement the bus master interface, which declares two methods for reading and writing on the bus. Any bus that defines (meaning implements) the bus master interface can be

```
public process SwTask {
    port RtosService RTOS;
    port CpuService CPU;
    parameter int TASKID;
    parameter int TASKPRIORITY;
    SwTask (String instName, int TI, int p) {
        super(instName);
        TASKID=TI;
        TASKPRIORITY=p; }

    void filter(int nsamples) {
        RTOS.resume (TASKID, TASKPRIORITY);
        CPU.cpuRead (nsamples);
        CPU.execute (nsamples*2);
        CPU.cpuWrite(nsamples);
        RTOS.suspend(TASKID, TASKPRIORITY);
    }
    void anotherfnc () {}

    void thread() {
        while (true) {}
            await{
            (true; ; ) filter(nondeterminism(int));
            (true; ; ) anotherfnc ();
        }
    }
}
```

```
public netlist Architecture {
    public Architecture (String name, String scenario,...) {
        super(name);
        if (scenario="FCFSBus")
        ...
        else if (scenation="TimeSliceBus")
        ...
    }
}
```


16-12

FIGURE

Left: An example of a task model as a set of services offered by the architecture. Right: An example of how inheritance can be used to explore different platform instances.

used. The platform developer then defines architecture scenarios that instantiate one bus implementation or another, depending on the scenario parameters, such as in Figure 16-12. The architectural platforms are in general declared with a number of parameters, such as the clock speed of a CPU, the types of algorithms used for RTOS schedulers or bus arbiters, and the types and configurations of memories, which are set to particular values when an instance of the architectural netlist is instantiated.

16.4.3   Mapping Strategy

A mapping netlist relates a functional netlist and an architecture netlist. The mapping step in the design flow is usually done by system developers that use an architecture netlist developed by a platform provider. The architecture netlist has a set of parameters that can be set depending on the mapping scenario.

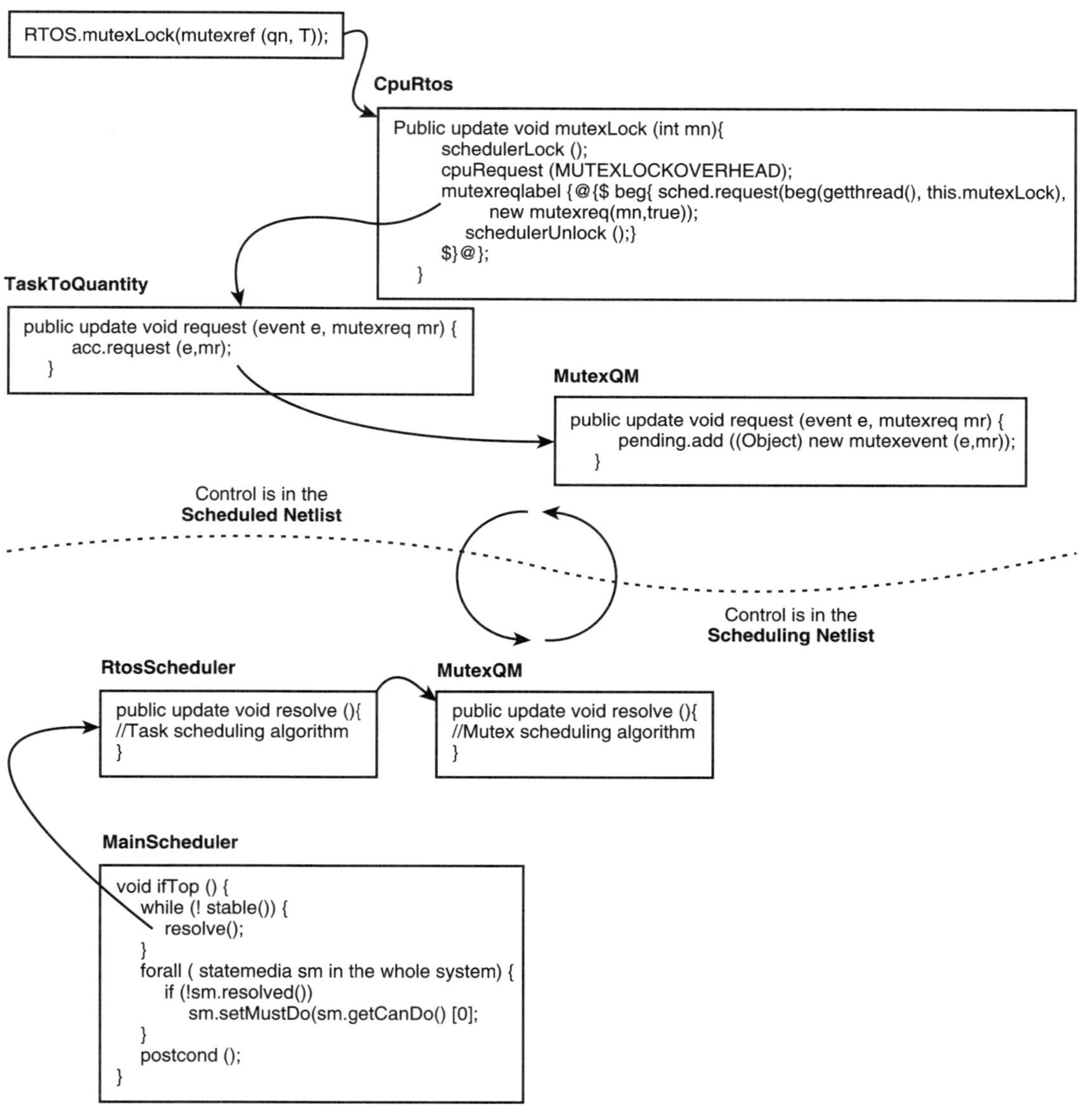

16-13   Execution model of the architecture platform.

FIGURE

Ttl map class	Synchronization Constraints
| ```
class TtlMap{
   TTLmediumnetlist TtlNet;
   int FifoLength;
   int TokenSize;
   int DataSize;
   int FifoTarget;
   int NtokenAddr;
   int WrTaskId;
   int RdTaskId;
}
``` | ```
constraint{
   ltl synch (
      beg (process_writer, bf.release_data),
      beg (task_writer, task_writer.release_data):
      beg (task_writer, task_writer.release_data).T==Ttl [i].FifoTarget,
      beg (task_writer, task_writer.release_data).qn==Ttl [i].NtokenAddr
   )
   ...
}
``` |

16-14

Part of the code that implements the mapping.

FIGURE

Partitioning of functional blocks and assignment to architectural resources are also part of the mapping scenario.

Figure 16-14 shows a sketch of the code that implements the mapping. As we mentioned in Section 16.4.1, communication plays a central role in this kind of application. Particular attention is paid to the ttlmedium netlist, and in particular to the Yapi2TTL and TTL2Yapi media. The Ttlmap class is defined to store all the important parameters of these two media, whose values are defined in the functional description: the token size and the number of tokens in the FIFO channel. It also stores mapping information, such as the buffer memory identifier (including the address range used by the TTL buffers) and the task identification numbers. By appropriately setting these parameters, designers can explore different mapping scenarios in terms of memory allocation and partitioning of writer and reader processes on architectural components.

The mapping netlist constructor takes four main sets of parameters:

+ the function netlist scenario (explained in Section 16.4.1).

+ the architectural parameters.

+ mapping information, such as tasks priorities, task identifiers, and memory mapping information.

+ the synchronization of events between the functional and architectural netlists.

The system designer specifies the mapping information by setting values to the fields of a class Ttlmap, which is defined as a part of the architectural platform in advance by the platform developer. Figure 16-14 shows an example of the class, together with the synchronization constraints we presented in Section 16.2.4.

Here, process_writer is the process in function netlist that calls the method release_data of the bounded FIFO in the ttlmedium netlist, whereas task_writer is a task in the architecture netlist to which we want to assign the functional process. The task_writer task offers the architectural implementation of release_data as a service. The formula states that the begin event of function release_data of process process_writer in the functional netlist has to be synchronized with the begin event of the release_data implementation of task task_writer in the architecture netlist. It also maps the memory space based on the information given in a Ttlmap instance. The mapping netlist includes function calls in the constructor that generate the synchronization constraints. The function takes as arguments a pair of events to be synchronized and an appropriate instance of the Ttlmap. In this way, the system designer does not need to write the constraint formulas manually.

16.5   CONCLUSIONS

We have presented the Metropolis environment for the design of complex electronic systems. It is based on a meta-model with formal semantics, which can be used to capture designs directly or import them from other specification languages. It also supports simulation, formal analysis, and synthesis tools. The framework is conceived to encompass different application domains. Each application domain needs different models, tools, and flows. Heterogeneity is supported at the utmost level in Metropolis to allow this kind of customization. The meta-model structure has been designed with orthogonalization of concerns in mind: function and architecture and communication and computation are clearly separated. The formal semantics of the meta-model permits embedding of models of computation in a rigorous framework, thus favoring design reuse and design chain support. The features of the system should facilitate the dialog among designers with different knowledge domains. Our view is not to impose a language or a flow on designers, but rather to make their preferred approach more robust and rigorous. We also offer a set of analysis and synthesis tools that are examples of how the framework can be used to integrate flows.

We plan to add tools as we address different application domains. At this time, we are exploring the automotive, wireless communication, and video application domains in collaboration with our industrial partners. As we understand better what are the critical parts of the design and what needs to be supported to facilitate design hand-offs, we plan to tune the meta-model and to increase the power of the infrastructure for the support of successive refinement as a major productivity enhancement.

We also plan to make Metropolis and its components open domain to expose the ideas to the academic and industrial community.

ACKNOWLEDGMENTS

We gratefully acknowledge the support of the MARCO GSRC program that partially supported the development of Metropolis. Cadence Berkeley Labs researchers have been instrumental in developing Metropolis and the ideas on which Metropolis is based. We wish to thank our partners PARADES, the Berkeley Wireless Research Center, Intel, Cypress Semiconductors, Nokia, Philips, ST Microelectronics, BMW, and Magneti Marelli for providing support and collaboration to the various phases of Metropolis development. We count on them to continue providing feedback on the implementation and the design flows supported in Metropolis.

A special thanks goes to the many people involved in the Metropolis and related projects. In particular, we would like to thank Jerry Burch, Luca Carloni, Rong Chen, Xi Chen, Robert Clariso Viladrosa, Jordi Cortadella, Erwin de Kock, Doug Densmore, Alberto Ferrari, Daniele Gasperini, Gregor Göossler, Timothy Kam, Arjan Kenter, Mike Kishinevsky, Alex Kondratyev, Wido Kruijtzer, Dunny Lam, Alberto La Rosa, Radu Marculescu, Grant Martin, Trevor Meyerowitz, John Moondanos, Andy Ong, Roberto Passerone, Claudio Pinello, Alessandro Pinto, Sanjay Reikhi, Ellen Sentovich, Marco Sgroi, Vishal Shah, Greg Spirakis, Laura Vanzago, Ted Vucurevich, Howard Wong-Toi, and Guang Yang.

Glossary

active power consumption
> power that is consumed in a circuit when there is a change in the inputs.

adaptive body-biasing
> a technique used to change the threshold voltage by applying a voltage bias to the body of a transistor.

application programmer interface (API)
> software interface consisting of functions that support hardware-independent application function implementation.

application programming interface (API)
> the interface to a software package.

application-specific processor
> the broadest and most generic term for processors tuned for various embedded application classes. Such devices may include both processors developed by traditional manual methods and automatically generated configurable processors. Examples of manually developed application-specific processors include the ARM9E (ARM with basic DSP extensions), and the MIPS application-specific extension (ASE) for graphics. Manually developed processor designs are typified by multiyear development cycles and require manual adaptation of software tools. The large development cost and time for manual adaptation typically means that these designs cannot be narrowly application-specific and may not be able to achieve high application efficiency.

arbiter
> a hardware component in a communication architecture that regulates access to shared communication resources by making use of resource sharing algorithms.

battery capacity
> the amount of charge that a battery delivers when discharged under an actual discharge workload and operating conditions.

branch history table (BHT)
table of 1-bit or 2-bit counters that records recent outcomes of branches, used for making future branch predictions.

branch prediction
predicting the target of a branch before the branch executes, typically in the instruction fetch stage.

branch target address cache (BTAC)
see *branch target buffer*.

branch target buffer (BTB)
a cache that records the taken target addresses of recently executed branches.

bridge
a hardware component of the communication architecture that enables the interconnection of two (or more) busses.

burst transfer
a communication protocol feature that permits a master and slave to execute multiple word transfers across a bus without being subject to arbitration decisions for each word.

bus-invert coding
a bus encoding technique that transmits either the original data or the inverted data based on which one reduces the total number of data transitions with respect to the data transmitted in the previous cycle.

bypass
a bus for communicating register values directly between data-dependent instructions, bypassing the register file.

cache
a small and fast memory close to the processor for holding recently and frequently referenced instructions and/or data.

cache coherence
method for ensuring that multiple replicas of a memory block, contained in the caches of multiple processors, are identical.

cache hit
the instruction/data requested by the processor is found in its instruction/data cache.

cache locking
preventing eviction of a memory block from the cache. Locking ensures a cache hit and is particularly useful in real-time systems that require determinism.

cache miss
the instruction/data requested by the processor are not found in its instruction/data cache.

cache way-prediction
a prediction technique used to access only one of the ways of an associative cache to reduce dynamic energy.

CDMA (code division multiple access)

spread spectrum communication technique achieved by multiplying data by a pseudorandom sequence before transmission on a channel and after reception.

CISC

complex instruction set computer.

closed embedded system

informally, a closed embedded system is characterized as being highly deterministic. It typically runs a fixed set of tasks, is not interactive, does not use desktop-derivative operating systems or virtual memory, and may use a real-time operating system (RTOS) for scheduling periodic real-time tasks. Also see *open embedded system*.

co-design finite state machine (CFSM)

a model of computation composed of finite state machines, extended with an arithmetic datapath, communicating via one-place lossy buffers.

co-exploration

the process of designing processor and memory architectures simultaneously.

communication architecture

a hardware fabric consisting of a network topology and associated communication protocols that allows the integration of a set of components that exchange data and control information.

communication mapping

the assignment of communication transactions to physical paths in the network topology.

communication pattern

a "standard recipe" used for implementing (often by refinement) a communication, e.g., using polling or interrupt.

communication protocol

the specification of the conventions and logical mechanisms that enable system components to communicate over a communication architecture.

communication refinement

a formally based design flow in which more abstract models are transformed into more concrete ones by means of successive steps, each of which preserves some of the properties of the abstract specification (e.g., losslessness), while introducing new aspects (e.g., energy or time).

compiler

a software tool that translates a program in a high-level language to an equivalent program in a target language, typically machine code.

component-based SoC design

in component-based SoC design methodologies, the goal is to allow the integration of heterogeneous components and communication protocols by using abstract interconnections.

composition

the principle that allows one to infer system properties from the properties of its components by enforcing certain interaction constraints on them. For example, it guarantees that a deadlock-free component will remain so after integration with another component. It is the basic tool for incremental system modeling.

computer artifact

model around which a class of computer systems or computer system elements are organized.

configurable, extensible processor

configurable, extensible processors have all of the attributes of a configurable processor and, most importantly, allow modifications and extensions to the processor's instruction set, register set, and execution pipelines in unique and proprietary ways by the application developer to include features never considered by the original processor designer. Automated software support coupled to automatic hardware integration is an essential element of a practical and useful configurable, extensible processor.

configurable processor

a design and tool environment that allows significant adaptation by changing major processor functions, to tune the processor to specific application requirements. Typical forms of configurability include additions, deletions, and modifications to memories, to external interface widths and protocols, and to commonly used processor peripherals. To be useful, configuration of the processor must meet two important criteria: (1) the configuration mechanism must accelerate and simplify the creation of useful configurations, and (2) the generated processor must include complete hardware descriptions (typically RTL descriptions written in Verilog or VHDL), software development tools (compilers, debuggers, assemblers, and real-time operating systems) and verification aids. Configurable processors may be implemented in many different hardware forms ranging from ASICs with hardware implementation cycles of many weeks to FPGAs with implementation cycles of just minutes.

context switch

the change from the execution of one process to the execution of another process or from one communication to another one. The context switch is associated with an overhead to store and reload the processor state (registers) or to change from one transmission to another.

control hazard

refers to possible instruction fetch disruptions caused by branch instructions in the pipeline.

cosimulation wrapper

implements the cosimulation interfaces required by multilanguage mixed-level global SoC executable models. There are two types of cosimulation wrappers: the simulator wrapper and the abstraction level/protocol wrapper. The first one allows communication among different simulators, and the second one handles protocol and abstraction level conversions.

custom operating system (OS)

the OS gives implementation to the application software of high-level communication primitives. The OS also gives sophisticated services such as task scheduling and interrupt management. Due to code size, power consumption, and performance reasons, it is not realistic to use a generic OS in multiprocessor SoC platforms. A custom OS implements an application-specific program interface and is streamlined and preconfigured for the software task(s) that runs on each processor of the target multiprocessor SoC platform.

data conflict

when two or more data accesses attempt to share the same memory location.

data hazard

possible pipeline stalls caused by data dependences between instructions in the pipeline.

data link

abstraction of communication channel control. It provides a reliable link over an unreliable physical channel.

data locality

the property of finding the reused data in the higher levels of memory hierarchy (e.g., SPM or L1 cache).

data prefetching

bringing data into higher levels of the memory hierarchy before data are actually needed. The goal here is to hide the memory latency.

dataflow equations

a kind of set constraints of the form $X = Y$ where X and Y are set expressions. Dataflow equations are used in static analyses such as liveness analysis.

dataflow (also known as Kahn process) network

a model of computation composed of processes communicating via infinite FIFO queues using nonblocking write and blocking read operations. A process in a dataflow network cannot test a FIFO for emptiness, thus ensuring uniqueness of computed results regardless of the scheduling of concurrency.

deadlock

a situation in which a set of active concurrent tasks cannot proceed because each holds non-preemptable resources that must be acquired by some other task.

design artifact or design tool

model around which a computer program used to assist the design of a computer system is organized.

design elements

models that can be integrated according to a common representation in one or more design tools or analytical models to form more complex computer system or MoCs.

design-time scheduling

unlike conventional off-line scheduling, the design-time scheduling only locally fixes the mapping and ordering of the thread nodes inside a thread frame and provides several different scheduling results at different performance and cost tradeoff, represented as a Pareto curve.

development environment

a set of software tools that supports software development for a platform.

DPM (dynamic power management)

method to reduce energy consumption by on-line control of frequency and/or supply voltage of components, including shutoff.

drowsy caches

caches that preserve the data contents but do not permit access when one is operating at a lower supply voltage to reduce leakage energy.

dynamic branch prediction
generating branch predictions at run time, using an adaptive hardware branch predictor (also see *branch prediction*).

dynamic scheduling
rescheduling/reordering instructions at run time using hardware mechanisms (also see *out-of-order execution*).

dynamic voltage scaling (DVS)
a technique that reduces (increases) the voltage of the circuit resulting in reduced (increased) energy consumption but slower (faster) clock speed.

EMI
electromagnetic interference.

event model
model of the sequence of events in time.

event-activated process (communication)
activation of a process (communication) by the occurrence of an event.

false sharing
the property of sharing a cache line but not a data item in that cache line.

flit
logical unit of information transfer in a packet-based network on a chip.

formal model of computation
a model of computation with a mathematical basis.

FPGA (field programmable gate array)
circuit whose function can be programmed on the field, by controlling transistor (or antifuse) switches with a bit map.

GALS
globally asynchronous locally synchronous circuits.

hard core
fully implemented hardware model (including physical layout), that is stable, unchanging, and ready for manufacturing (sometimes referred to as "hard macrocell"; also see *soft core*).

hardware abstraction layer (HAL)
software that isolates higher level software from hardware details. The HAL gives an abstraction of underlying hardware architecture to the OS. It contains all the software that is dependent on the hardware: boot code, context switch, I/O drivers, interrupt service routines (ISRs), and so on.

hardware multithreading
hardware support for simultaneously executing multiple software threads on a single microprocessor, especially support for multiple architectural register contexts.

hardware wrapper
> a processor/IP adapter, channel adapters, and an internal bus. The number of channel adapters depends on the number of channels that are connected to the corresponding hardware virtual component.

hardware-dependent software
> software, such as device drivers, that deals with details of hardware operation.

heterogeneous multiprocessor
> a multiprocessor with more than one type of processing element.

heterogeneous multiprocessor platform
> a multiprocessor computation platform is called a heterogeneous multiprocessor platform if it comprises different types of processors that can either be functionally different or of different types for many other reasons.

Hiperlan/2
> wireless local area network standard proposal.

hit-under-miss
> the ability for a cache to continue accepting new load/store requests in spite of a prior cache miss that is still being serviced.

ILP
> integer linear programming.

in-order execution
> instructions are executed sequentially, in program order; thus, a stalled instruction causes all later instructions to stall as well.

instruction issue
> the process of sending an instruction and its source operand values (once all of them are available) to a function unit to begin execution.

instruction window
> in a dynamically scheduled processor (i.e., a processor that is capable of out-of-order execution), the instruction window is the collection of instructions considered for execution.

instruction-level parallelism (ILP)
> independent instructions in the program that may execute in parallel.

instruction-set architecture (ISA)
> defines architectural state (registers and memory) and instructions for modifying architectural state; also the "contract" between hardware and software.

intellectual property (IP)
> reusable hardware component, represented either as synthesizable HDL, or as hardware netlist, or as layout, part of a system-on-chip.

layered model
> a model of computation in which the details of the lower layers of computation are hidden from the higher. The lower layers are not implementations of the higher layers, as in

traditional hierarchical relationships, rather they specify services, sometimes shared, available to the higher levels.

LFSR (linear feedback shift register)
shift register with feedback through EXOR gates that produces a pseudorandom sequence of vector; used for built-in self-test.

load-use stall
some number of cycles a load-dependent instruction must wait for the load value.

MAC (medium access control)
the layer in the ISO/OSI standard that determines how multiple transmitters and receivers share the communication medium; part of the data link layer.

master
a system component that is capable of initiating a communication transaction (read or write).

memory management unit (MMU)
the hardware unit in a microprocessor responsible for translating virtual memory addresses into physical memory addresses and detecting memory protection violations.

memory wall
the increasing difference in speed between processors and memory.

memory wrapper
manages a global memory shared by several processors of a multiprocessor SoC architecture. It is composed of a memory port adapter, channel adapters, and an internal communication bus. The memory port adapter includes several memory-specific functions such as control logic, address decoder, bank controller, and other functions that depend on the type of memory (e.g., refresh management in DRAM).

memory-memory ISA
an ISA that allows instructions besides loads and stores to use memory operands (versus register operands).

message passing
a parallel processing methodology in which processes pass messages to communicate.

message passing interface (MPI)
a message passing software standard.

middleware
software that sits between hardware-dependent software and application software.

model of computation (MoC)
abstract representations of computing systems; defines how a new state evolves from a previous state, what operations produce the new state, and how state is propagated to and from the operations.

modeling framework
provides the mechanism by which the various components comprising a model interact. It is a set of constraints on the components and their composition semantics and, therefore, it defines a model of computation that governs the interaction of its constituent components.

MTTF (mean time to failure)
expected failure time for a given reliability distribution.

multiple-instruction issue
the ability to issue multiple ready instructions to function units in the same cycle (also see *instruction issue* and *superscalar*).

multiply-accumulate (MAC)
a single operation that fuses a multiply operation with an add operation.

multiprocessor system
a tightly coupled system in which the global state and workload information on all processors are kept current by a centralized scheduler. Even when each processor has its own scheduler, the decisions and actions of the schedulers of all the processors are coherent.

multiprocessor system-on-chip (MPSoC)
system-on-chip that includes more than one programmable processor.

network routing
establishes the path followed by messages; delivery control.

network switching
abstraction of communication channel control; implements the end-to-end delivery control.

network topology
the set of communication channels (busses) and their interconnections that define the physical structure of a communication architecture.

nonblocking cache
a cache that accepts new load/store requests in spite of one or more prior outstanding misses (also see *hit-under-miss*).

nondeterminism
effects that will cause different behaviors (e.g., latency, execution time) even with the same system input. Interrupts, semaphores, and events can be the reasons for nondeterminism.

NP-complete
the set of problems that are the hardest problems in the complexity class NP in that they are the ones most unlikely to be solvable in polynomial time. Examples of NP-complete problems include the Hamiltonian cycle and traveling salesman problems. At present, all known algorithms for NP-complete problems require time exponential in the problem size. It is unknown whether there are any faster algorithms. Therefore, NP-complete problems are usually solved using approximation algorithms, probabilistic methods, or heuristics.

open embedded system
informally, an open embedded system is characterized as nondeterministic. The software it runs may change over time, it may be interactive, and it may use desktop-derivative operating systems and virtual memory (also see *closed embedded system*).

orthogonal frequency division multiplexing (OFDM)
a modulation scheme that uses a large number of different carrier frequencies, for the sake of robustness with respect to multipath propagation; based on direct and inverse Fourier transform for demodulation and modulation, respectively.

OSKit (Operating System Kit)
Tensilica's API and porting methodology for automatic configuration of real-time operating system environments tuned to individual application-specific versions of the company's Xtensa processor.

out-of-order completion
permitting instructions to write results into the register file not necessarily in the order that instructions appear in the program.

out-of-order execution
executing instructions as soon as their operands become available, instead of sequentially (also see *dynamic scheduling*).

Pareto curve
a set of Pareto-optimal points, each point representing a solution for a multiobjective function that is optimal in at least one tradeoff direction when all the other directions are fixed.

phase-ordering problem
an optimizing compiler performs several optimization phases. Different orderings of the phases can yield different performance of the generated code. This is because one phase can benefit or suffer from being performed after another phase. The phase-ordering problem is to determine the best ordering of the optimization phases.

phit
the unit of information that can be transferred across a physical channel in a single cycle.

physical model
representation of how design elements that relate to observable and measurable properties of computer systems may be integrated.

pipelining
a technique whereby new instructions are fetched, decoded, and so on before prior instructions have completed, resulting in overlapped execution of multiple instructions.

platform
an architecture designed to support a set of tasks; at least a hardware structure and may also include software; layer of abstraction that allows an application to be developed without one knowing the details of the underlying layers.

precedence constraints
the constraints on the order in which tasks may execute.

preemptive scheduling
resource sharing strategy whereby processes can interrupt the execution of lower priority processes.

processing element
any processor in a multiprocessor; may be programmable or dedicated-function.

program counter (PC)
the memory address of the next instruction to be fetched and executed.

programmatic model
> executable descriptions that abstract the key features of the way model elements behave both individually and as an integrated collection in the formation of a system but do not need to relate to mathematical models.

protocol
> set of formal rules describing the interaction between two or more entities.

protocol stack
> abstraction of communication control in seven layers.

QoS
> quality of service.

rated current
> the fixed rate of discharge at which standard capacity of the battery is calculated.

reactive systems
> the systems that maintain an ongoing interaction with their environment.

real-time computing
> computing to meet deadlines.

real-time operating system (RTOS)
> an operating system that stresses predictability, with specific support for efficiently and reliably scheduling real-time tasks.

real-time system
> system whose correctness depends not only on the results but also on the time at which these results are produced.

reliability
> probability that a system (or component) fails as a function of time.

return address stack (RAS)
> branch prediction hardware specifically designed for predicting targets of return instructions.

RISC
> reduced instruction set computer.

run-time scheduling
> the on-line step of combining the Pareto curves from the design-time stage to get the final global schedule, by selecting appropriate working points for each thread frame.

run-to-completion
> resource sharing strategy whereby process execution is not interrupted.

scalar
> a pipeline with a peak execution rate of one instruction per cycle.

scenario

typical execution path of a thread frame containing information, such as branch selection and (data-dependent) loop iteration number, and exposing it to the run-time scheduler.

schedulability analysis

analysis that determines whether a set of processes can be scheduled such that all deadlines are met.

SDRAM

widely used dynamic RAM memory that uses long word lines that are buffered in the memory; can use efficient burst (packet) transmission to transport whole word lines.

set constraints

constraints of the form $X \subseteq Y$ where X and Y are set expressions. Such constraints are used in set-based static analyses.

shared-memory multiprocessor

a multiprocessor in which the processing elements all read and write the same pool of memory.

SIMD (single instruction multiple data)

a form of instruction-level parallelism in which one instruction performs the same operation on multiple words (half-words, bytes) in parallel.

simultaneous multithreading (SMT)

the ability to execute multiple programs on a single microprocessor at the same time (also see *hardware multithreading*).

slave

a system component that is capable of responding to communication transaction initiated by a master.

soft core

a synthesizable hardware model in the form of a netlist or HDL code that is flexible and easy to optimize; sometimes referred to as "synthesizable core" (also see *hard core*).

software wrapper

consists of two parts: a custom operating system (OS) and a hardware abstraction layer (HAL).

spatial locality

a property exploited by caches to improve cache hit rates: if a word is referenced, then it is likely that neighboring words will also be referenced soon. This property is exploited by fetching a large contiguous block of words from the main memory when a reference to a particular word misses, anticipating references to other words in the block.

SpecC

a C-based language for embedded system modeling, adding to the base language facilities to model concurrency, synchronization, communication, abstraction of interfaces, refinement, and extended finite state machines.

speculation

fetching, decoding, and executing instructions following a predicted branch instruction, before the true outcome of the branch is known.

SPM (scratch pad memory)
a small on-chip memory managed by software (compiler and/or application).

sporadic events
event sequences that do not follow a periodic pattern.

standard capacity
the amount of charge that a battery delivers when discharged at a fixed rate, under standard operating conditions.

standby power consumption
power that is consumed in a circuit when there is no change in the inputs.

static analysis
a technique used by a compiler to determine facts about a program that are useful for optimizations.

static branch prediction
generating branch predictions at compile time, using program profiling or heuristics.

static priority
scheduling resource sharing strategy whereby processes are executed according to a fixed process priority or with communication using fixed priorities.

static scheduling
reordering instructions at compile time to improve run-time performance.

superscalar
refers to a pipeline with a peak execution rate exceeding one instruction per cycle.

supply gating
technique used for disconnecting the power supply to the circuit to reduce leakage energy.

synchronous
a model in which there is a global event at which all state advance occurs.

SystemC
a C++ class library used to model embedded systems composed of hardware and software, in a manner that is akin to the SpecC modeling style but does not create a new language or require a new compiler.

system-on-chip (SoC)
an integrated circuit with all (or most of) the components necessary to form a complete electronic system.

task allocation
the reservation of resources necessary for task execution.

task assignment
mapping of tasks to processors on which they will be executed; assignment is also called binding.

task graph
a directed graph representing a computation; nodes denote tasks, and edges denote precedence constraints between tasks.

task partitioning
grouping of tasks in a task graph such that the tasks that are tightly connected are put in the same group, whereas the number of connections from one group to the other is kept low.

task scheduling
the reservation of time slots for tasks for execution on processors.

task-level parallelism
parallelism that comes from several tasks or processes that can execute concurrently.

temporal locality
a property exploited by caches: if a word is referenced, then it is likely that it will be referenced again soon; this property is exploited by retaining recently used instructions/data in caches.

thread
consists of thread frames. It is an independent piece of code to complete a specific function.

thread frame
consists of thread nodes. By definition, the nondeterministic behaviors can only happen at the boundary of thread frames. The design-time scheduling is done inside each thread frame, whereas the run-time scheduler considers it as an atomic scheduling unit.

thread node
consists of CDFG nodes and arcs. It is the atomic scheduling unit of our design-time scheduler.

tick
a simulation cycle.

TIE (Tensilica Instruction Extension language)
Tensilica's comprehensive instruction-set-extension description language, which includes the description of new instruction formats, instruction semantics, additional registers and register files, instruction scheduling, mapping into C/C++ data types and intrinsic functions, and instruction-test suites.

tightly coupled memory (TCM)
the term used by ARM architects to describe on-chip memory that is explicitly managed by software, in contrast to hardware-managed caches.

time-activated process (communication)
activation of a process (communication) according to a fixed time pattern rather than by the occurrence of events.

timed model
a model in which references to time are all based on physical properties, resulting in a representation of time advancement that implies magnitude as well as order.

type constraints
constraints of the form $X = Y$ or $X \leq Y$ where X and Y are type expressions. Such constraints are used in type-based static analyses.

untimed model
a model in which references to time imply logical sequence only.

VCC

a tool from Cadence Design Systems that allows the designer to model separately at a high level the functionality of an application and a hardware/software architecture on which it can be executed, creating a simulation model for the full embedded system based on the function to architecture mapping information.

virtual architecture

an abstract model of the final SoC architecture composed of hardware, software, and functional virtual components and the global communication topology.

virtual component

a hardware, software, or functional component with two separated interfaces: internal and external interfaces. All the components hierarchically contained in the virtual component use the internal interface to communicate. Communication with the external world uses the external component interface. The internal and external interfaces can be different in terms of communication protocol, abstraction level, and specification language.

virtual component wrapper

adapts accesses between the internal components and the communication channels connected to the external interface of the virtual component.

virtual simple architecture (VISA)

a simple abstraction of the processor, and the basis for worst-case timing analysis.

von Neumann architecture

a style of machine in which control flows from one instruction to the next, instructions read and modify state, and data dependences among instructions are inferred from the order of instructions in the program.

worst-case execution time (WCET)

the execution time of a task that is guaranteed never to be exceeded, for all possible program inputs.

worst-case timing analysis

deriving the worst-case execution time for a task on a particular processor.

Xtensa

Tensilica's processor instruction set and methodology for automated generation of configurable, extensible processors.

XTMP (Xtensa Modeling Protocol)

Tensilica's software API and environment for rapid development of multiple-processor system simulation models, including heterogeneous cycle-accurate instruction-set simulations for each processor and simple control of debug, memory layout, and simulator actions.

YAPI

a model of computation based on concurrent processes, as in dataflow, extended with a primitive (called "select"), which allows a process to test which one among a set of input FIFOs has available data. YAPI models nondeterministic networks, while ensuring efficient implementation of nonstatically determinable data arrival rates.

References

[1] Kunimatsu, A., Ide, N., Sato, T., Endo, Y., Murakami, H., Kamei, T., Hirano, M., Ishihara, F., Tago, H., Oka, M., Ohba, A., Yutaka, T., Okada, T., and Suzuoki, M. Vector unit architecture for emotion synthesis, IEEE Micro, 20(2), 40–47, 2000.

[2] Culler, D. E., and Singh, J. P., with Gupta, A. Parallel Computer Architecture: A Hardware/Software Approach, Morgan Kaufman, San Francisco, 1999.

[3] Waingold, E., Taylor, M., Srikrishna, D., Sarkar, V., Lee, W., Lee, V., Kim, J., Frank, M., Finch, P., Barua, R., Babb, J., Amarasinghe, S., and Agarwal, A. Baring it all to software: raw machines, IEEE Computer, 30(9), 86–93, 1997.

[4] Haskell, B. G., Puri A., and Netravali, A. Digital Video: An Introduction to MPEG-2, Kluwer Academic Publishers, Boston, 1996.

[5] Keutzer, K., Newton, A. R., Rabaey, J. M., and Sangiovanni-Vincentelli, A. System-level design: orthogonalization of concerns and platform-based design, IEEE Transactions on Computer-Aided Design of Integrated Circuits and Systems, 19(12), 2000.

[6] Birnbaum, M. and Sachs, H. How VSIA answers the SOC dilemma, IEEE Computer, 32(6), 42–50, 1999.

[7] Dutta, S., et al. Viper: a multiprocessor SOC for advanced set-top box and digital TV systems, IEEE Design & Test of Computers, September/October 2001.

[8] Schlansker, M. S. and Rau, B. R. EPIC: explicitly parallel instruction computing, IEEE Computer, 33(2), 37–45, 2000.

[9] Kuroda, T. Optimization and control of vDD and vT for low power, high speed CMOS design, in International Conference on Computer Aided Design, pp. 28–34, November, 2002.

[10] Duarte, D., Tsai, Y. F., Vijaykrishnan, N., and Irwin, M. J. Evaluating run-time techniques for leakage power reduction techniques, in Asia-Pacific Design Automation Conference, January, 2001.

[11] Brodersen, R., Horowitz, M., Markovic, D., Nikolic, B., and Stojanovic, V. Methods for true power minimization, in International Conference on Computer Aided Design, pp. 35–42, November, 2002.

[12] Rabaey, J., Chandrakasan, A., and Nikolic, B. Digital Integrated Circuits: A Design Perspective. Prentice Hall, Englewood Cliffs, NJ, 2003.

[13] Usami, K., and Horowitz, M. Clustered voltage scaling techniques for low-power design, in International Symposium on Low Power Electronics and Design, pp. 3–8, April, 1995.

[14] Martin, S., Flautner, K., Mudge, T., Blaauw, V. Combined Dynamic Voltage Scaling and Adaptive Body Biasing for Lower Power Microprocessors under Dynamic Workloads. *ICCAD*, pp. 712–725, November, 2002.

[15] Itoh, K., Sasaki, K., and Nakagome, Y. Trends in low-power RAM circuit technologies, Proceedings of IEEE, 83, 524–543, 1995.

[16] Kim, S., Vijaykrishnan, N., Kandemir, M., Sivasubramaniam, A., Irwin, M. J., and Geethanjali, E. Power-aware partitioned cache architectures, in Proceedings of the 2001 International Symposium on Low Power Electronics and Design (ISLPED'01), pp. 64–67, 2001.

[17] Chandrakasan, A., Bowhill, W. J., and Fox, F. Design of High-Performance Microprocessor Circuits. New York: IEEE Press, 2001.

[18] Inoue, K., Ishihara, T., and Murakami, K. Way-predicting set-associative cache for high performance and low energy consumption, in Proceedings of the 1999 International Symposium on Low Power Electronics and Design (ISLPED'99), pp. 273–275, 1999.

[19] Powell, M. D., Agarwal, A., Vijaykumar, T. N., Falsafi, B., and Roy, K. Reducing set-associative cache energy via way-prediction and selective direct-mapping, in Proceedings of the 34th Annual ACM/IEEE International Symposium on Microarchitecture, pp. 54–65, 2001.

[20] Albonesi, D. H. Selective cache ways: on-demand cache resource allocation, in Proceedings of the 32nd Annual ACM/IEEE International Symposium on Microarchitecture (MICRO32), pp. 248–259, 1999.

[21] Kin, J., Gupta, M., and Mangione-Smith, W. H. The filter cache: an energy efficient memory structure, in Proceedings of the 30th Annual ACM/IEEE International Symposium on Microarchitecture (MICRO-30), pp. 184–193, 1997.

[22] Villa, L., Zhang, M., and Asanovic, K. Dynamic zero compression for cache energy reduction, in Proceedings of the 33rd Annual ACM/IEEE International Symposium on Microarchitecture (MICRO-33), pp. 214–220, 2000.

[23] Catthoor F., et al. Custom Memory Management Methodology, Kluwer Academic Publishers, Boston, 1998.

[24] Yang, S., Powell, M. D., Falsafi, B., Roy, K., and Vijaykumar, T. N. An integrated circuit/architecture approach to reducing leakage in deep-submicron high-

performance I-caches, in Proceedings of Seventh International Symposium on High-Performance Computer Architecture (HPCA-7), pp. 147–157, 2001.

[25] Kaxiras, S., Hu, Z, and Martonosi, M. Cache decay: exploiting generational behavior to reduce cache leakage power, in Proceedings of the 28th Annual International Symposium on Computer Architecture (ISCA-28), pp. 240–251, 2001.

[26] Chen, G., Shetty, R., Kandemir, M., Vijaykrishnan, N., Irwin, M. J., and Wolczko, M. Tuning garbage collection in an embedded java environment, in Proceedings of the Eighth International Symposium on High-Performance Computer Architecture (HPCA-8), pp. 80–91, 2002.

[27] der Meer, P. R. V. and Staveren, A. V. Standby-current reduction for deep sub-micron VLSI CMOS circuits: smart series switch, in The ProRISC/IEEE Workshop, pp. 401–404, December, 2000.

[28] Nikolic, B. State-preserving leakage control mechanisms, Gigascale Silicon Research Center Annual Report, September, 2001.

[29] Agarwal, A., Li, H., and Roy, K. Drg-cache: a data retention gated-ground cache for low power, in Proceedings of the 39th Conference on Design Automation (DAC-39), pp. 473–478, 2002.

[30] Zhou, H., Toburen, M. C., Rotenberg, E., and Conte, T. M. Adaptive mode control: a static-power-efficient cache design, in The 10th International Conference on Parallel Architectures and Compilation Techniques (PACT'01), pp. 61–70, September, 2001.

[31] Flautner, K., Kim, N. S., Martin, S., Blaauw, D., and Mudge, T. Drowsy caches: simple techniques for reducing leakage power, in Proceedings of the 29th Annual International Symposium on Computer Architecture (ISCA-29), pp. 148–157, 2002.

[32] Kim, N. S., Flautner, K., Blaauw, D., and Mudge, T. Drowsy instruction caches—leakage power reduction using dynamic voltage scaling, in Proceedings of the 33rd Annual ACM/IEEE International Symposium on Microarchitecture (MICRO-35), November, 2002.

[33] Qin, H. and J. Rabaey, J. Leakage suppression of embedded memories, Gigascale Silicon Research Center Annual Review, 2002.

[34] Heo, S., Barr, K., Hampton, M., and Asanovic, K. Dynamic fine-grain leakage reduction using leakage-biased bitlines, in Proceedings of the 29th Annual International Symposium on Computer Architecture (ISCA-29), pp. 137–147, 2002.

[35] Koyama, T., Inoue, K., Hanaki, H., Yasue, M., and Iwata, E. A 250-MHz single-chip multiprocessor for audio and video signal processing, IEEE Journal of Solid-State Circuits, 36(11), 1768–1774, 2001.

[36] Olukotun, K., Nayfeh, B. A., Hammond, L., Wilson, K., and Chang, K. The case for a single-chip multiprocessor, in The 8th Conference on Architectural Support for Programming Languages and Operating Systems (ASPLOS IIV), pp. 2–11, 1996.

[37] Takahashi, M., Takano, H., Kaneko, E., and Suzuki, S. A shared-bus control mechanism and a cache coherence protocol for a high-performance on-chip

multiprocessor, in 2nd IEEE Symposium on High-Performance Computer Architecture (HPCA'96), pp. 314–322, February, 1996.

[38] Woo, S. C., Ohara, M., Torrie, E., Singh, J., and Gupta, A. The splash-2 programs: characterization and methodological considerations, in Proceedings of the 22nd Annual International Symposium on Computer Architecture, pp. 24–36, 1995.

[39] Moshovos, A., et al. JETTY: filtering snoops for reduced energy consumption in SMP servers, in Proceedings of HPCA, pp. 85–96, January, 2001.

[40] Saldhana C. and Lipasti, M. Power efficient cache coherence, in Proceedings of the Workshop on Memory, Performance Issues, in conjunction with ISCA, 2001.

[41] Ekman, M., Dahlgren, F., and Stenstrm, P. Evaluation of snoop-energy reduction techniques for chip-multiprocessors, in Proceedings of the Workshop on Duplicating, Deconstructing, and Debunking, in conjunction with ISCA, 2002.

[42] Ho, R., Mai, K., and Horowitz, M. The future of wires, Proceedings of the IEEE, 89(4), 490–504, 2001.

[43] Osborne, S., Erdogan, A., Arslan, T., and Robinson, D. Bus encoding architecture for low-power implementation of an AMBA-based SoC Platform, IEEE Proceedings: Computers and Digital Techniques, 149(4), 152–156, 2002.

[44] Ramprasad, S., Shanbhag, N., and Hajj, I. Signal coding for low power: fundamental limits and practical realizations, in International Symposium on Circuits and Systems, vol. 2, pp. 1–4, May, 1998.

[45] Stan, M. and Burleson, W. Bus-invert coding for low-power I/O, IEEE Transactions on VLSI Systems, 3(1), 49–58, 1995.

[46] Stan, M. and Burleson, W. Low-power encodings for global communication in CMOS VLSI, IEEE Transactions on VLSI Systems, 5(4), 444–455, 1997.

[47] Su, C. L., Tsui, C. Y., and Despain, A. M. Saving power in the control path of embedded processors, IEEE Design and Test of Computers, 11(4), 24–30, 1994.

[48] Mehta, H., Owens, R. M., and Irwin, M. J. Some Isiues in Gray code addressing, in Great Lakes Symposium on VLSI, pp. 178–180, March, 1996.

[49] Musoll, E., Lang, T., and Cortadella, J. Working-zone encoding for reducing the energy in microprocessor address buses, IEEE Transactions on VLSI Systems, 6(4), 568–572, 1998.

[50] Aghaghiri, Y., Fallah, F., and Pedram, M. ALBORZ: address level bus power optimization, in International Symposium on Quality Electronic Design, pp. 470–475, 2002.

[51] Benini, L., De Micheli, G., Macii, E., Sciuto, D., and Silvano, C. Asymptotic zero-transition activity encoding for address busses in low-power microprocessor-based systems, in Great Lakes Symposium on VLSI, pp. 77–82, March, 1997.

[52] Aghaghiri, Y., Fallah, F., Pedram, M. Reducing transitions on memory buses using sector-based encoding technique, in International Symposium on Low Power Electronics and Design, pp. 190–195, 2002.

[53] Cheng, W-C. and Pedram, M. Power-optimal encoding for a DRAM address bus, IEEE Transactions on VLSI Systems, 10(2), 109–119, 2002.

[54] Benini, L., De Micheli, G., Macii, E., Poncino, M., and Quer, S. Reducing power consumption of core-based systems by address bus encoding, in IEEE Transactions on VLSI Systems, 6(4), 554–562, 1998.

[55] Benini, L., Macii, A., Macii, E., Poncino, M., and Scarsi, R. Synthesis of low-overhead interfaces for power-efficient communication over wide buses, in Design Automation Conference, pp. 128–133, June, 1999.

[56] Cheng, W-C. and Pedram, P. Chromatic encoding: a low power encoding technique for digital visual interface, in Design, Automation and Test in Europe Conference, pp. 694–699, 2002.

[57] Kim, K-W., Baek, K-H., Shanbhag, N., Liu, C-L., and Kang, S-M. Coupling-driven signal encoding scheme for low-power interface design, in International Conference on Computer Aided Design, pp. 318–321, 2000.

[58] Sotiriadis, P. and Chandrakasan, A. Low power bus coding techniques considering inter-wire capacitances, in Custom Integrated Circuits Conference, pp. 507–510, 2000.

[59] Hui, Z., George, V., and Rabaey, J. Low-swing on-chip signaling techniques: effectiveness and robustness, IEEE Transactions on VLSI Systems, 8(3), 264–272, 2000.

[60] Svensson, C. Optimum voltage swing on on-chip and off-chip interconnect, IEEE Journal of Solid-State Circuits, 36(7), 1108–1112, 2001.

[61] Bertozzi, D., Benini, L., and De Micheli, G. Low power error resilient encoding for on-chip data buses, in Design Automation and Test in Europe Conference, pp. 102–109, 2002.

[62] Worm, F., Thiran, P., Ienne, P., and De Micheli, G. An adaptive low-power transmission scheme for on-chip networks, in International Symposium on System Synthesis, pp. 92–100, 2002.

[63] Benini, L. and De Micheli, G. Networks on chips: a new SoC paradigm. IEEE Computer, 35, 70–78, 2002.

[64] Patel, C., Chai, S., Yalamanchili, S., and Shimmel, D. Power constrained design of multiprocessor interconnection networks, in IEEE International Conference on Computer Design, pp. 408–416, 1997.

[65] Zhang, H., Wan, M., George, V., and Rabaey, J. Interconnect architecture exploration for low-energy configurable single-chip DSPs, in IEEE Computer Society Workshop on VLSI, pp. 2–8, 1999.

[66] Dally, W. J. and Towles, B. Route packets, not wires: on-chip interconnection networks, in DAC 2001, pp. 684–689, June, 2001.

[67] Ye, T., Benini, L., and De Micheli, G. Packetized on-chip interconnect communication analysis for MPSoC, in Design Automation and Test in Europe Conference, pp. 344–349, 2003.

[68]   Kadayif, I., Kandemir, M., and Karakoy, M. An energy saving strategy based on adaptive loop parallelization, in Proceedings of the Design Automation Conference, New Orleans, LA, June, 2002.

[69]   Wolfe, M. High Performance Compilers for Parallel Computing. Boston: Addison Wesley, CA, 1996.

[70]   Carriero, N., Gelernter, D., Kaminsky, D., and Westbrook, J. Adaptive Parallelism with Piranha. Technical Report 954, Yale University, New Haven, CT, 1993.

[71]   Kandemir, M., Zhang, W., and Karakoy, M. Runtime code parallelization for on-chip multiprocessors, in Proceedings of the 6th Design Automation and Test in Europe Conference, Munich, Germany, March, 2003.

[72]   Kadayif, I., Kandemir, M., and Sezer, U. An integer linear programming based approach for parallelizing applications in on-chip multiprocessors, in Proceedings of the Design Automation Conference, New Orleans, LA, June, 2002.

[73]   Basu, K., Choudhary, A., Pisharath, J., and Kandemir, M. Power protocol: reducing power dissipation on off-chip data buses, in International Symposium on Microarchitecture, pp. 345–355, 2002.

[74]   Benini, L., De Micheli, G., Macii, A., Macii, E., and Poncino, M. Reducing power consumption of dedicated processors through instruction set encoding, in Great Lakes Symposium on VLSI, pp. 8–12, February, 1998.

[75]   Benini, L., De Micheli, G., Macii, E., Sciuto, D., and Silvano, C. Address bus encoding techniques for system-level power optimization, in Design Automation and Test in Europe, pp. 861–866, February, 1998.

[76]   Chen, J., Jone, W., Wang, J., Lu, H., and Chen, T. Segmented bus design for low-power systems, in IEEE Transactions on VLSI Systems, 7(1), 25–29, 1999.

[77]   Komatsh, S., Ikeda, M., and Asada, K. Low power chip interface based on bus data encoding with adaptive code-book method, in Great Lakes Symposium on VLSI, pp. 368–371, 1999.

[78]   Lv, T., Henkel, J., Lekatsas, H., and Wolf, W. An adaptive dictionary encoding scheme for SOC data buses, in Design Automation and Test in Europe Conference, pp. 1059–1064, 2002.

[79]   Mamidipaka, M., Hirschberg, D., and Dutt, N. Low power address encoding using self-organizing lists, in International Symposium on Low Power Electronics and Design, pp. 188–193, 2001.

[80]   Kadayif, I., Kandemir, M., Vijaykrishnan, N., Irwin, M. J., and Sivasubramaniam, A. EAC: a compiler framework for high-level energy estimation and optimization, in Proceedings of the 5th Design Automation and Test in Europe Conference (DATE'02), March, 2002.

[81]   Benini, L., Bogliolo, A., De Micheli, G. A Survey of Design Techniques for SystemLevel Dynamic Power Management, IEEE Transactions on Very Large Scale Integration Systems, vol. 8, no. 3, pp. 299–316, June 2000.

[82]   Ho, R., Mai, K., Horowitz, M. The Future of Wires, Proceedings of the IEEE, January 2001.

[83]   Dally, J. and Poulton, W. Digital Systems Engineering, Cambridge University Press, 1998.

[84]   Bertsekas, D., Gallager, R. Data Networks, Prentice Hall, 1991.

[85]   Walrand, J. and Varaiya, P. High Performance Communication Networks, Morgan Kaufman, 2000.

[86]   Sylvester, D. and Keutzer, K. A Global Wiring Paradigm for Deep Submicron Design, IEEE Transactions on CAD/ICAS, Vol. 19, No. 2, pp. 242–252, February 2000.

[87]   Theis, T. The future of Interconnection Technology, IBM Journal of Research and Development, Vol. 44, No. 3, May 2000, pp. 379–390.

[88]   Deutsch, A. Electrical Characteristics of Interconnections for High-Performance Systems, Proceedings of the IEEE, vol. 86, no. 2, pp. 315–355, February 1998.

[89]   Bakoglu, H. Circuits, Interconnections, and Packaging for VLSI, AddisonWesley, 1990.

[90]   Worm, F., Ienne, P., Thiran, P., and De Micheli, G. An Adaptive Lowpower Transmission Scheme for On-chip Networks, ISSS, Proceedings of IEEE Integrated System Synthesis Symposium, 2002.

[91]   Yoshimura, R., Koat, T., Hatanaka, S., Matsuoka, T., Taniguchi, K. DSCDMA Wired Bus with Simple Interconnection Topology for Parallel Processing System LSIs, IEEE Solid-State Circuits Conference, p. 371, Jan. 2000.

[92]   Benini, L., and De Micheli, G. System Level Power Optimization: Techniques and Tools, ACM Transactions on Design Automation of Electronic Systems, vol. 5, no. 2, pp. 115–192, April 2000.

[93]   Prince, B. Report on Cosmic Radiation Induced SER in SRAMs and DRAMs in Electronic Systems, May 2000.

[94]   Cohen, N., Sriram, T., Leland, N., Moyer, D., Butler, R., and Flatley, S. Soft Error Considerations for Deep Submicron CMOS Circuit Applications, IEDM, Proceedings of IEEE International Electron Device Meeting, pp. 315–318, 1999.

[95]   Nicolaidis, M. Time Redundancy Based Soft Error Tolerance to Rescue Nanometer Technologies, Proceedings VTS, 1999.

[96]   Benini, L., De Micheli, G., Macii, E., Poncino, M., Quer, S. Power Optimization of Corebased Systems by Address Bus Encoding, IEEE Transactions on Very Large Scale Integration Systems, vol. 6, no. 4, pp. 5785–5788, Dec. 1998.

[97]   Duato, J., Yalamanchili, S., Ni, L. Interconnection Networks: An Engineering Approach. Morgan Kaufmann, 2003.

[98]   Aldworth, P. System-on-a-Chip Bus Architecture for Embedded Applications, IEEE International Conference on Computer Design, pp. 297–298, 1999.

[99]   Cordan, B. An Efficient Bus Architecture for System-on-chip Design, IEEE Custom Integrated Circuits Conference, pp. 623–626, 1999.

[100]   Remaklus, W. On-chip Bus Structure for Custom Core Logic Design, IEEE Wescon, p. 714, 1998.

[101]   Weber, W. CPU Performance Comparison: Standard Computer Bus Versus Silicon Bacplane, www.sonics.com, 2000.

[102]   Winegarden, S. A. Bus Architecture Centric Configurable Processor System, IEEE Custom Integrated Circuits Conference, pp. 627–630, 1999.

[103]   Ackland, B., et al. A Single Chip, 1.6Billion, 16b MAC/s Multiprocessor DSP, IEEE Journal of Solid State Circuits, vol. 35, no. 3, March 2000.

[104]   Agrawal, A. Raw Computation, Scientific American, August 1999.

[105]   Patel, C., Chai, S., Yalamanchili, S., Shimmel, D. Power Constrained Design of Multiprocessor Interconnection Networks, IEEE International Conference on Computer Design, pp. 408–416, 1997.

[106]   Lee, S.Y., Song, S.J., Lee, K., Woo, J.H., Kim, S.E., Nam, B.G., Yoo, H.J. An 800 MHz Starconnected On-chip Network for Application to Systems On a Chip, IEEE Solid State Circuits Conference, pp. 468–469, 2003.

[107]   Bertozzi, D., Benini G., and De Micheli, L. LowPower ErrorResilient Encoding for On-chip Data Busses, DATE, International Conference on Design and Test Europe Paris, 2002, pp. 102–109.

[108]   Hegde, R. and Shanbhag, N. Toward Achieving Energy Efficiency in Presence of Deep Submicron Noise, IEEE Transactions on VLSI Systems, pp. 379–391, vol. 8, no. 4, August 2000.

[109]   Adriahantenana, A. and Greiner, A. Micronetwork for SoC: Implementation of a 32bit SPIN Network, Design Automation and Test in Europe Conference, p. 1129, 2003.

[110]   Guerrier, P., Grenier, A. A Generic Architecture for On-chip Packets-witched Interconnections, Design Automation and Test in Europe Conference, pp. 250–256, 2000.

[111]   Dall'Osso, M., Biccari, G., Giovannini, L., Bertozzi, D., Benini, L. Xpipes: A Latency Insensitive Parameterized Network-on-chip Architecture for Multiprocessor SoCs, International Conference on Computer Design, pp. 536–539, 2003.

[112]   Dally, W. and Towles, B. Route Packets, Not Wires: On-chip Interconnection Networks DAC, Proceedings of the 38th Design Automation Conference, pp. 684–689.

[113]   Goossens, K., van Meerbergen, J., Peeters A., and Wielage, P. Networks on Silicon: Combining Best Efforts and Guranteed Services, Design Automation and Test in Europe Conference, pp. 423–427, 2002.

[114]   Dumitra, T., Kerner, S., and Marculescu, R. Towards On-chip Fault Tolerant Communication, ASPDAC Proceedings of the Asian South Pacific Design Automation Conference, pp. 225–232, 2003.

[115]   Nilsson, M., Millberg, E., Oberg, J., Jantsch, A. Load Distribution with the Proximity Congestion Awareness in a Networks on Chip, Proceedings of Design Automation and Test in Europe, March 2003, pp. 1126–1127.

[116]   Kumar, S., Jantsch, A., Soininen, J., Forsell, M., Millberg, M., Oberg, J., Tiensyrj, K., and Hemani, A. A Network On Chip Architecture and Design Methodology, Proceedings of IEEE Computer Society Annual Symposium on VLSI, April 2002, pp. 105–112.

[117]   Ye, T., Benini, L., and De Micheli, G. Packetization and Routing Analysis of On-chip Multiprocessor Networks, Journal of System Architecture, 2004.

[118]   Ye, T., Benini, G., De Micheli, L. Packetized On-chip Interconnect Communication Analysis, DATE, International Conference on Design and Test Europe, 2003, pp. 344–349.

[119]   Wingard, D. Micro Network Based Integration for SOCs, Design Automation Conference, pp. 673–677, 2001.

[120]   Lahiri, K., Raghunathan, A., Lakshminarayana, G. LOTTERYBUS: A New High-Performance Communication Architecture for System-on-Chip Designs, Design Automation Conference, pp. 15–20, 2001.

[121]   Andrews, G. Foundations of Multithreaded, Parallel and Distributed Programming, AddisonWesley, 2000.

[122]   Culler, D., Pal Singh, J., Gupta, A. Parallel Computer Architecture: a Hardware/Software Approach. MorganKaufman, 1999.

[123]   Lee, E., Sangiovanni-Vincentelli, A. A Framework for Comparing Models of Computation, IEEE Transactions on Computer-Aided Design of Circuits and Systems, vol. 17, no. 12, pp. 1217–1229, Dec. 1998.

[124]   Leopold, C. Parallel and Distributed Computing: a Survey of Models, Paradigms and Approaches, WileyInterscience, 2001.

[125]   Janka, R., Willis, L., Baumstark, L. Virtual Benchmarking and Model Continuity in Prototyping Embedded Multiprocessor Signal Processing Systems, IEEE Transaction onf Software Engineering, vol. 28, no. 9, pp. 836–846, Sept. 2002.

[126]   Kanevsky, A., Skjellum, A., Rounbehler, A. MPI/RT an Emerging Standard for High-performance Realtime Systems, IEEE International Conference on System Sciences, vol. 3, pp. 157–166, 1998.

[127]   ElRewini, H., Ali, H., Lewis, T. Task Scheduling in Parallel and Distributed Systems PrenticeHall, 1994.

[128]   Panda, R., Dutt, N., Nicolau, A., Catthoor, F., Vandercappelle, A., Brockmeyer, E., Kulkarni, C., De Greef, E. Data Memory Organization and Optimization in Applicationspecific Systems, IEEE Design & Test of Computers, vol. 18, no. 3, pp. 5658, MayJune 2002.

[129]   RTEMS Embedded Real-Time Opensource Operating System. http://www.rtems.com.

[130] VSPWorks. Small Footprint Kernel Optimized for DSP. http://www.windriver.com.

[131] Gauthier, L., Yoo, S., Jerraya, A. Automatic Generation and Targeting of Application-specific Operating Systems and Embedded Systems Software, IEEE ComputerAided Design of Integrated Circuits and Systems, vol. 20, no. 11, pp. 1293–1301, Nov. 2001.

[132] Verhulst, E. The Rationale for Distributed Semantics as a Topology Independent System Design Methodology and Its Implementation In the Virtuoso RTOS, Design Automation for Embedded Systems, vol. 6, pp. 277–294, 2002.

[133] Lavagno, L., Dey, S., Gupta, R. Specification, Modeling and Design Tools for System-on-chip, AsiaPacific Design Automation Conference, pp. 21–23, Jan. 2002.

[134] Muchnick, S. Advanced Compiler Design and Implementation. Morgan & Kaufman, 1997.

[135] Catthoor, F., Wuytack, S., De Greef, E., Balasa, F., Nachtergaele, L., Vandecappelle, A. Custom Memory Management Methodology: Exploration of Memory Organization for Embedded Multimedia System Design, Kluwer, 1998.

[136] Janka, R., Judd, R., Lebak, J., Richards, M., Campbell, D. VSIPL: An Object-based Open Standard API for Vector, Signal and Image Processing, IEEE Conference on Acoustic, Speech and Signal Processing, pp. 949–952, vol. 2, 2001.

[137] Groetker, T., Liao, S., Martin, G., Swan, S. System Design with SystemC, Kluwer, 2002.

[138] Benini, L., Bertozzi, D., Bruni, D., Drago, N., Fummi, F., Poncino, M. SystemC Co-Simulation of Multi-Processor Systems-on-Chip, IEEE International Conference on Computer Design, pp. 494–499, 2002.

[139] Altman, E., Ebcioglu, K., Gachwind, M., Sathaye, S. Advances and Future Challenges in Binary Translation and Optimization, Proceedings of the IEEE, vol. 89, no. 11, pp. 1710–1722, Nov. 2001.

[140] Benini, L. and De Micheli, G. Networks on Chips: A New SoC Paradigm, IEEE Computers, January 2002, pp. 7078.

[141] Benini L. and De Micheli, G. Powering Networks on Chips: Energy Efficient and Reliable Interconnect Design for SoCs, ISSS, Proceedings of the International Symposium on System Synthesis, Montreal, October 2001, pp. 3338.

[142] Ditzel, D. Transmeta's Crusoe: Cool Chips for Mobile Computing, Hot Chips Symposium, Stanford, 2000.

[143] Leiserson, C. Fattrees: Universal Networks for Hardware Efficient Supercomputing, IEEE Transactions on Computers, vol. 34, no. 10, pp. 892–901, October 1985.

[144] Montanaro, J. A., et al. 160 MHz, 32b, 0.5 W CMOS RISC Microprocessor, IEEE Journal of Solid-State Circuits, vol. 31, no. 11, pp. 1703–1714, Nov. 1996.

[145] Zhang, H., Prabhu, V., George, V., Wan, M., Benes, M., Abnous, A., Rabaey, J. A. 1 V Heterogeneous Reconfigurable DSP IC for Wireless Baseband Digital Signal

Processing, IEEE Journal of Solid-State Circuits, vol. 35, no. 11, pp. 1697–1704, Nov. 2000.

[146]   Srivatsa, J. Embedded Microprocessors: Pushing Towards Higher Performance, Embedded Processors, iSuppli Corp., March 2002.

[147]   Hand, T. Real-time Systems Need Predictability, Computer Design (RISC Supplement), August 1989, pp. 57–59.

[148]   Anantaraman, A., Seth, K., Patil, K., Rotenberg, E., and Mueller, F. Virtual Simple Architecture: Exceeding the Complexity Limit in Safe Real-time Systems, In Proceedings of the 30th International Symposium on Computer Architecture, July 2003, pp. 350–361.

[149]   Rau, B. R. and Fisher, J. A. Instruction-level Parallel Processing: History, Overview, and Perspective, Journal of Supercomputing, Special Issue, 7, July 1993, pp. 9–50.

[150]   Burks, A. W., Goldstine, H. H., von Neumann, J. Preliminary Discussion of the Logical Design of an Electronic Computing Instrument, Report to the U.S. Army Ordinance Department, 1946.

[151]   Cisco Inc. Cisco's Toaster 2 Chip Receives the Microprocessor Report Analyst's Choice 2001 Award for Best Network Processor, http://www.cisco.com.

[152]   Shah, N. and Keutzer, K. Network Processors: Origin of Species, In Proceedings of the 17th International Symposium on Computer and Information Sciences, October 2002.

[153]   Suryanarayanan, D. A Methodology for Study of Network Processing Architectures, MS Thesis, North Carolina State University, July 2001.

[154]   Mitra, T. and Chiueh, T. Dynamic 3D Graphics Workload Characterization and the Architectural Implications, In Proceedings of the 32nd International Symposium on Microarchitecture, November 1999, pp. 62–71.

[155]   GeForce4 Ti Product Overview, http://www.nvidia.com/docs/lo/1467/SUPP/PO_GeForce4_Ti_92502.pdf

[156]   Watlington, J. A. Video & Graphics Processors: 1997, MIT Media Laboratory, May 1997.

[157]   Katevenis, M. G. H. Reduced Instruction Set Computer Architectures for VLSI, CS Division Report No. UCB/CSD 83/141, University of California, Berkeley, CA, October 1983.

[158]   Patterson, D. A. Reduced Instruction Set Computers, CACM, 28(1), January 1985, pp. 8–21.

[159]   Hennessy, J., Jouppi, N., Gill, J., Baskett, F., Strong, A., Gross, T., Rowen, C., and Leonard, J. The MIPS machine, In Proceedings of the IEEE Compcon, February 1982, pp. 2–7.

[160]   Hinton, G., Sager, D., Upton, M., Boggs, D., Carmean, D., Kyker, A., and Roussel, P. The Microarchitecture of the Pentium 4 Processor, Intel Technology Journal, 1st Quarter, 2001.

[161]   Tendler, J. M., Dodson, J. S., Fields, J. S., Jr., Le, H., and Sinharoy, B. POWER4 System Microarchitecture, IBM Journal of Research and Development, 46(1), January 2002, pp. 5–25.

[162]   Ball, T. and Larus, J. Branch prediction for free, In Proceedings of the SIGPLAN '93 Conference on Programming Language Design and Implementation, June 1993, pp. 300–313.

[163]   Smith, J. E. A Study of Branch Prediction Strategies, In Proceedings of the 8th Annual International Symposium on Computer Architecture, May 1981, pp. 135–148.

[164]   Hwu, W. W., Conte, T. M., and Chang, P. P. Comparing Software and Hardware Schemes for Reducing the Cost of Branches, In Proceedings of the 16th Annual International Symposium on Computer Architecture, June 1989, pp. 224–233.

[165]   McFarling, S. and Hennessy, J. L. Reducing the cost of branches, In Proceedings of the 13th Annual Symposium on Computer Architecture, June 1986, pp. 396–403.

[166]   McFarling, S. Combining branch predictors, WRL Technical Note TN-36, 1993.

[167]   Pan, S., So, K., and Rahmeh, J. T. Improving the accuracy of dynamic branch prediction using branch correlation, ACM Computer Architecture News, vol. 20, October 1992, pp. 76–84.

[168]   Yeh, T-Y. and Patt, Y. N. Two-level adaptive branch prediction, In Proceedings of the 24th ACM/IEEE International Symposium on Microarchitecture, November 1991, pp. 51–61.

[169]   Lee, J. K. F. and Smith, A. J. Branch prediction strategies and branch target buffer design, IEEE Computer, vol. 17(1), January 1984, pp. 6–22.

[170]   Wulf, W. A. and McKee, S. A. Hitting the Memory Wall: Implications of the Obvious, Computer Architecture News, 23(1), March 1995, pp. 20–24.

[171]   Tullsen, D. M., Eggers, S. J., Emer, J. S., Levy, H. M., Lo, J. L., and Stamm, R. L. Exploiting Choice: Instruction Fetch and Issue on an Implementable Simultaneous Multithreading Processor, In Proceedings of the 23rd Annual International Symposium on Computer Architecture, May 1996.

[172]   Smith, B. J. Architecture and applications of the HEP multiprocessor computer system, In SPIE Real-Time Signal Processing IV, August 1981, pp. 241–248.

[173]   Tullsen, D. M., Eggers, S. J., and Levy, H. M. Simultaneous Multithreading: Maximizing On-Chip Parallelism, In Proceedings of the 22nd Annual International Symposium on Computer Architecture, June 1995.

[174]   Yamamoto, W. and Nemirovsky, M. Increasing superscalar performance through multistreaming, In Proceedings of the Conference on Parallel Architectures and Compilation Techniques, June 1995, pp. 49–58.

[175]   Lo, J. L., Eggers, S. J., Emer, J. S., Levy, R., Stamm, R. M., and Tullsen, D. M. Converting thread-level parallelism to instruction-level parallelism via simultaneous multithreading, ACM Transactions on Computer Systems, 15(3), 1997, pp. 322–354.

[176]   http://www-306.ibm.com/chips/news/2000/000605/index.html,
http://news.designtechnica.com/article1321.html,
http://www.mips.com/content/PressRoom/PressReleases/2003–12–22,
http://www.arm.com/markets

[177]   Ubicom, The Ubicom IP3023 Wireless Network Processor, White Paper, April 2003.
(http://www.ubicom.com/pdfs/products/ip3000/processor/WP-IP3023WNP-01.pdf)

[178]   http://www.arm.com/arm/documentation

[179]   http://www.arm.com/armtech/ARM10_Thumb

[180]   http://www.arm.com/armtech/ARM11

[181]   http://www.arm.com/armtech/ARM7_Thumb

[182]   http://www.arm.com/armtech/ARM9_Thumb

[183]   http://www.arm.com/armtech/ARM9E_Thumb

[184]   Furber, S. Arm System-on-Chip Architecture, Addison-Wesley Pub. Co., 2nd edition,
August 2000.

[185]   Kaeli, D. and Emma, P. Branch history table prediction of moving target branches
due to subroutine returns, In Proceedings of the 18th International Symposium on
Computer Architecture, May 1991.

[186]   http://www-3.ibm.com/chips/techlib/techlib.nsf/products/PowerPC_440_Embedded_Core

[187]   http://www.mips.com/content/PressRoom/PressKits/files/mips32_24k.pdf

[188]   Pentium Processor Architecture Overview
(http://www.intel.com/design/intarch/techinfo/pentium/PDF/archfeat.pdf

[189]   Bondi, J. O., Nanda, A. K., Dutta, S. Integrating a misprediction recovery cache
(MRC) into a superscalar pipeline, In Proceedings of the 29th International
Symposium on Microarchitecture, December 1996, pp. 14–23.

[190]   Halfhill, T. R. Ubicom's New NPU Stays Small, Microprocessor Report, April
2003.

[191]   Alverson, R., Callahan, D., Cummings, D., Koblenz, B., Porterfield, A., and Smith, B.
The Tera Computer System, In Proceedings of the International Conference on
Supercomputing, June 1990, pp. 1–6.

[192]   Arnold, R., Mueller, F. B., Whalley, D., and Harmon, M. Bounding worst-case
instruction cache performance, IEEE Real-Time Systems Symposium, December
1994, pp. 172–181.

[193]   Healy, C. A., Whalley, D. B., and Harmon, M. G. Integrating the timing analysis of
pipelining and instruction caching, IEEE Real-Time Systems Symposium, December
1995, pp. 288–297.

[194]   Kim, S., Min, S., and Ha, R. Efficient worst case timing analysis of data caching,
Real-Time Technology and Applications Symposium, June 1996.

[195]   Lim, S-S., Bae, Y. H., Jang, T., Rhee, B-D., Min, L., Park, C. Y., Shin, H., and Kim, C.S. An accurate worst case timing analysis for RISC processors, IEEE Real-Time Systems Symposium, December 1994, pp. 97–108.

[196]   Li, Y-T. S., Malik, S., and Wolfe, A. Efficient microarchitecture modeling and path analysis for real-time software, IEEE Real-Time Systems Symposium, December 1995, pp. 298–397.

[197]   Li, Y-T. S., Malik, S., and Wolfe, A. Cache modeling for real-time software: Beyond direct mapped instruction caches, IEEE Real-Time Systems Symposium, December 1996, pp. 254–263.

[198]   Mueller, F. Timing analysis for instruction caches, Real-Time Systems, 18(2/3), May 2000, pp. 209–239.

[199]   Bate, I., Conmy, P., Kelly, T., and McDermid, J. Use of modern processors in safety critical applications, The Computer Journal, 44(6), 2001, pp. 531–543.

[200]   Engblom, J. On hardware and hardware models for embedded real-time systems, IEEE Workshop on Real-Time Embedded Systems, December 2001.

[201]   Pillai, P. and Shin, K. G. Real-time dynamic voltage scaling for low-power embedded operating systems, In Proceedings of ACM Symposium on Operating Systems Principles, December 2001, pp. 89–102.

[202]   ARM, Performance of the ARM9TDMI™ and ARM9E-S™ cores compared to the ARM7TDMI™ core, White Paper, September 2000. (http://www.arm.com/pdfs/comparison-arm7-arm9-v1.pdf)

[203]   ARM (D. Cormie), The ARM11™ microarchitecture, White Paper, April 2002. (http://www.arm.com/pdfs/ARM11%20Microarchitecture%20White%20Paper.pdf)

[204]   Snyder, C. D. ARM Family Expands at EPF, Microprocessor Report, June 2002.

[205]   Goering, R. Productivity may stumble at 100 nm, Electronic Engineering Times, September, 23, 2003.

[206]   Wilson, R. Chip industry tackles escalating mask costs, Electronic Engineering Times, June 17, 2002.

[207]   Wilson, L., ed. The National Technology Roadmap for Semiconductors: 1997 Edition, Semiconductor Industry Association, San Jose, California.

[208]   Bell, G., Newell A. The PMS and ISP descriptive systems for computer structures, in Proceedings of the Spring Joint Computer Conference (1970), AFIPS Press.

[209]   http://gcc.gnu.org

[210]   Van Praet, J., Goosens, G., Lanner, D., De Man, H. Instruction set definition and Instruction selection for ASIP, in High Level Synthesis Symp (1994), p. 1116.

[211]   Zimmermann, G. The Mimola Design System-A Computer Aided Digital Processor Design Method, in Proceedings of the 16th Design Automation Conference (1979).

[212]   Corporaal, H, Lamberts, R. TTA processor synthesis, in First Annual Conference of ASCI (May, 1995).

[213]   Pees, S., Hoffmann, A., Zivojnovic, V., and Meyr, H. LISA—Machine Description Language for Cycle-Accurate Models of Programmable DSP Architectures, in *Proceedings of the Design Automation Conference (1999)*.

[214]   http://www.arc.com

[215]   Aditya, S., Rau, B. R., Kathail, V. Automatic Architectural Synthesis of VLIW and EPIC Processors, in 12th International Symposium on System Synthesis (1999).

[216]   Faraboschi, P., Brown, G., Fisher, J., Desoli, G. Lx: A technology platform for customizable VLIW embedded processing, in *Proceedings of the 27th Annual International Symposium on Computer Architecture* (June, 2000).

[217]   Wang, A., Killian, E., Maydan, D., Rowen, C. Hardware/Software Instruction Set Configurability for System-on-Chip Processors, *Proceedings of the Design Automation Conference* (June, 2001), Las Vegas, Nevada.

[218]   Gonzalez, R. Configurable and Extensible Processors Change System Design. Hot Chips 11 (1999).

[219]   http://www.eembc.org

[220]   http://www.arm.com

[221]   http://www.necel.com

[222]   http://www.mips.com

[223]   http://www.semiconductors.philips.com

[224]   http://www.st.com

[225]   http://www.ti.com

[226]   http://www.mentor.com/seamless

[227]   Christensen, C. The Innovator's Dilemma, Harvard Business School Press, Boston, 1997.

[228]   Rowen, C., Maydan, D. Automatic Processor Generation for System-on-Chip, in 27th European Solid-State Circuits Conference (September 2001), Villach, Austria.

[229]   Chang, H., Cooke, L., Hunt, M., Martin, G., McNelly, A., Todd, L. Surviving the SOC Revolution—A Guide to Platform-Based Design, Kluwer Academic Publishers, 1999.

[230]   Philips Semiconductors. Nexperia, *http://www.semiconductors. philips. com/platforms/nexperia/*.

[231]   de Oliveira, J. A. and van Antwerpen, H. The Philips Nexperia Digital Video Platform, in G. Martin and H. Chang (eds). Winning the SoC Revolution, Kluwer Academic Publishers, 2003, pp. 67–96.

[232]   Dutta, S., Jensen, R., Rieckmann, A. Viper: A Multiprocessor SOC for Advanced Set-Top Box and Digital TV Systems, IEEE Design and Test of Computers, Sep.–Oct. 2001, pp. 21–31.

[233]   Buttazzo, G. Real-Time Computing Systems—Predictable Scheduling Algorithms and Applications, Kluwer Academic Publishers, 2002.

[234]   Tindell, K., Clark, J. Holistic Schedulability for Distributed Hard Real-Time Systems, Euromicro Journal, vol. 40, 1994, pp. 117–134.

[235]   Mentor Graphics, Seamless Co-Verification Environment, *http://www.mentor.com/seamless/*.

[236]   Axys Design Automation, MaxSim Developer Suite, http://www.axysdesign.com.

[237]   Krolikoski, S., Schirrmeister, F., Salefski, B., Rowson, J., Martin, G. Methodology and technology for virtual component driven hardware/software co-design on the system-level, IEEE International Symposium on Circuits and Systems ISCAS '99, June 1999, pp. 456–459.

[238]   AbsInt. Worst Case Execution Time Analyses, *http://www.absint.com/wcet.htm*.

[239]   Li, Y-T. S., Malik, S. Performance Analysis of Real-Time Embedded Software, Kluwer Academic Publishers, 1999.

[240]   Theiling, H. and Ferdinand, C. H. Combining Abstract Interpretation and ILP for Microarchitecture Modelling and Program Path Analysis, in: Proc. 19th IEEE Real-Time Systems Symposium, Madrid, Spain, Dec. 1998, pp. 144–153.

[241]   Wolf, F. Behavioral Intervals in Embedded Software, Kluwer Academic Publishers, 2002.

[242]   Wolf, F., Ernst, R., Ye, W. Path clustering in software timing analysis, IEEE Transactions on VLSI Systems, Dec. 2001, pp. 773–782.

[243]   Hergenhan, A., Rosenstiel, W. Static timing analysis of embedded software on advanced processor architectures in: Proc. of Design, Automation and Test in Europe (DATE00, Paris, March 2000), pp. 552–559.

[244]   Ferdinand, C. H. et al. Reliable and Precise WCET Determination for a Real-Life Processor, Embedded Software Workshop, Springer LNCS 2211, Lake Tahoe, USA, Oct. 2001, pp. 469ff.

[245]   Kim, S-K., Min, S., and Ha, R. Efficient worst case timing analysis of data caching, in: Proc. of IEEE Real-Time Technology and Applications Symposium, 1996, pp. 230–240.

[246]   Lim, S., Bae, Y., Jang, G., Rhee, B., Min, S., Park, C., Shin, H., Park, K., Moon, S., and Kim, C. An accurate worst case timing analysis for RISC processors, Transactions on Software Engineering, July 95, pp. 593–604.

[247]   Knudsen, P. V. and Madsen, J. Integrating Communication Protocol Selection with Hardware/Software Codesign, IEEE Transactions on CAD, vol. 18(8, Aug. 1999), pp. 1077–1095.

[248]   Balarin, F. STARS of MPEG decoder: a case study in worst-case analysis of discrete-event systems, in: Proc. 9th International Symposium on Hardware/Software Codesign (CODES01), Copenhagen, April 2001, pp. 104–108.

[249]   Bhattacharyya, S. S., Murthy, P. K., and Lee, E. A. Software Synthesis from Dataflow Graphs, Kluwer Academic Publishers, 1996.

[250]   Lee, E. A. and Messerschmitt, D. G. Static scheduling of synchronous data flow programs for digital signal processing, IEEE Transactions on Computers, vol. 36(1), Jan. 1987, pp. 24–35.

[251]   Kopetz, H. Real-Time Systems—Design Principles for Distributed Embedded Applications, Kluwer Academic Publishers, 1997.

[252]   Sonics SiliconBackplane MicroNetwork Overview, http://www.sonicsinc.com/sonics/products/siliconbackplane, Sonics Inc.

[253]   Wingard, D. MicroNetwork-based integration for SOCs, in: Proc. Design Automation Conference (DAC01), 2001, June 2001, pp. 673–677.

[254]   Liu, C. L. and Layland, J. W. Scheduling algorithms for multiprogramming in a hard-real time environment, Journal of the ACM, vol. 20(1), 1973, pp. 46–61.

[255]   Sha, L., Rajkumar, R., and Sathaye, S. S. Generalized Rate-Monotonic Scheduling Theory: A Framework for Developing Real-Time Systems, in: Proc. of the IEEE, vol. 82(1), 1994, pp. 86–82.

[256]   Yen, T., and Wolf, W. H. Performance estimation for real-time distributed embedded systems, IEEE Transactions on Parallel and Distributed Systems, vol. 9(11), Nov. 1998, pp. 1125–1136.

[257]   Lehoczky, J. Fixed priority scheduling of periodic task sets with arbitrary deadlines, in: Proc. Real-Time Systems Symposium, 1990, pp. 201–209.

[258]   Audsley, N. C., Burns, A., Richardson, M. F., Tindell, K., and Wellings, A. J. Applying new scheduling theory to static priority preemptive scheduling, Journal of Real-Time Systems, vol. 8(5), 1993, pp. 284–292.

[259]   Blazewicz, J. Modeling and Performance Evaluation of Computer Systems, Chapter Scheduling Dependent Tasks with Different Arrival Times to Meet Deadlines. North-Holland, Amsterdam, 1976.

[260]   Ziegenbein, D., Uerpmann, J., Ernst, R. Dynamic Response Time Optimization for SDF Graphs, in: Proc. International Conference on Computer-Aided Design (ICCAD2000), San Jose, Ca, USA, 2000, pp. 135–141.

[261]   Gresser, K. An Event Model for Deadline Verification of Hard Real–Time Systems, in: Proc. of 5th Euromicro Workshop on Real–Time Systems, Oulu, Finland, 1993, pp. 118–123.

[262]   Pop, P., Eles, P., and Peng, Z. Bus access optimization for distributed embedded systems based on schedulability analysis, in: Proc. Design, Automation and Test in Europe (DATE00), Paris, France, 2000, pp. 567–574.

[263]   Pop, T., Eles, P., and Peng, Z. Holistic Scheduling and Analysis of Mixed Time/Event-Triggered Distributed Embedded Systems, in: Proc. 10th International Symposium on Hardware/Software Codesign (CODES02), Estes Park, CO, USA, 2002, pp. 187–192.

[264]   Thiele, L., Chakraborty, S., and Naedele, M. Real-time Calculus For Scheduling Hard Real-Time Systems, International Symposium on Circuits and Systems (ISCAS 2000), Geneva, Switzerland, vol. 4, March 2000, pp. 101–104.

[265] Chakraborty, S., Erlebach, T., Künzli, S., and Thiele, L. Schedulability of Event-Driven Code Blocks in Real-Time Embedded Systems, in: Proc. 39th Design Automation Conference (DAC02), New Orleans, USA, 2002, pp. 161–621.

[266] Richter, K., Ernst, R. Event Model Interfaces for Heterogeneous System Analysis, in: Proc. 5th Design, Automation and Test Conference (DATE02), Paris, France, 2002, pp. 506–513.

[267] Richter, K., Ziegenbein, D., Jersak, M., and Ernst, R. Model Composition for Scheduling Analysis in Platform Design, in: Proc. 39th Design Automation Conference (DAC02), New Orleans, LA, USA, 2002, pp. 287–292.

[268] Richter, K., Jersak, M., Ernst, R. A Formal Approach to MpSoC Performance Verification, IEEE Computer, vol. 36(4), April 2003, pp. 60–67.

[269] Richter, K., Racu, R., Ernst, R. Scheduling Analysis Integration for Heterogeneous Multiprocessor SoC, In IEEE Real-Time Systems Symposium (RTSS), December 2003, pp. 236–245.

[270] www.spi-project.org

[271] Ernst, R., Ziegenbein, D., Richter, K., Thiele, L., and Teich, J. Hardware/software codesign of embedded systems—The SPI Workbench, in: Proc. IEEE Workshop on VLSI, Orlando, USA, June 1999, pp. 9–17.

[272] Tindell, K. An Extendible Approach for Analysing Fixed Priority Hard Real-Time Systems, Journal of Real-Time Systems, vol. 6(2), pp. 133–152, 1994.

[273] Ziegenbein, D., Richter, K., Ernst, R., Thiele, L., and Teich, J. SPI—a system model for heterogeneously specified embedded systems, IEEE Transactions Very Large Scale Integration (VLSI) Systems, 2002, pp. 379–389.

[274] Dally, W. J. and Towles, B. Route Packets Not Wires: On-Chip Inter-connection Networks, in *Proc Design Automation Conf.*, pp. 684–689, June 2001.

[275] Ho, R., Mai, K. W., and Horowitz, M. A. The Future of Wires, *Proc IEEE*, vol. 89, pp. 490–504, Apr. 2001.

[276] Sylvester, D. and Keutzer, K. A Global Wiring Paradigm for Deep Submicron Design, *IEEE Trans Computer-Aided Design*, vol. 19, pp. 242–252, Feb. 2000.

[277] OMI 324 PI Bus, Rev 0. 3d. OMI Standards Draft, 1994 Siemens AG.

[278] AMBA 2. 0 Specification. http://www.arm.com/armtech/AMBA.

[279] CoreConnect Bus Architecture. http://www.chips.ibm.com/products/coreconnect/.

[280] Turner, J. and Yamanaka, N. Architectural Choices in Large Scale ATM Switches, *IEICE Trans on Communications*, vol. E-81B, pp. 120–137, Feb. 1998.

[281] Adriahantenaina, A., Charlery, H., Greiner, A., Mortiez, L., and Zeferino, C. A. SPIN: A Scalable, Packet Switched, On-Chip Micro-Network, in *Proc Design Automation & Test Europe (DATE) Conf.*, pp. 70–73, 2003.

[282] Karim, F., Nguyen, A., and Dey, S. An Interconnect Architecture for Networking System-on-Chips, *IEEE Micro*, vol. 22, pp. 36–45, Oct. 2002.

[283]   Crossbow Technologies. http://www.crossbowip.com.

[284]   Tanenbaum, A. S. *Computer Networks*. Englewood Cliffs, N. J.: Prentice Hall, 1989.

[285]   Benini, L. and Micheli, G. D. Networks on Chips: A New Paradigm for SoC Design, *IEEE Computer*, vol. 35, no. 1, pp. 70–78, 2002.

[286]   Sonics Integration Architecture. Sonics. In: http://www. sonicsinc. com.

[287]   Wingard, D. and Kurosawa, A. Integration Architecture for System-on-a-Chip Design, in *Proc Custom Integrated Circuits Conf.*, pp. 85–88, 1998.

[288]   Lahiri, K., Lakshminarayana, G., and Raghunathan, A. LOTTERYBUS: A New Communication Architecture for High-Performance System-on-Chip Designs, in *Proc Design Automation Conf.*, pp. 15–20, June 2001.

[289]   Yoshimura, R., Boon, K. T., Ogawa, T., Hatanaka, S., Matsuoka, T., and Taniguchi, K. DS-CDMA Wired Bus With Simple Interconnection Topology for Parallel Processing System LSIs, in *Proc Int Solid-State Circuits Conf.*, pp. 370–371, 2000.

[290]   Viterbi, A. J. *CDMA: Principles of Spread Spectrum Communication*. Prentice Hall, 1995.

[291]   Zhu, X. and Malik, S. A Hierarchical Modeling Framework for On-Chip Communication Architectures, in *Proc Int Conf Computer-Aided Design*, Nov. 2002.

[292]   Virtual Component Interface Standard Version 2 http://www.vsia.org/resources/VSIAStandards.htm.

[293]   Open Core Protocol International Partnership (OCP-IP). http://www.ocpip.org.

[294]   Rowson, J. A., and Sangiovanni-Vincentelli, A. Interface Based Design, in *Proc Design Automation Conf.*, pp. 178–183, June 1997.

[295]   Carloni, L. P., McMillan, K. L., and Sangiovanni-Vincentelli, A. L. Theory of Latency-Insensitive Design, *IEEE Trans Computer-Aided Design*, vol. 20, Sept. 2001.

[296]   Passerone, R., Rowson, J. A., and Sangiovanni-Vincentelli, A. Automatic Synthesis of Interfaces Between Incompatible Protocols, in *Proc Design Automation Conf.*, pp. 8–13, June 1998.

[297]   Yoo, S., Nicolescu, G., Lyonnard, D., Baghadi, A., and Jerraya, A. A. A Generic Wrapper Architecture for Multi-Processor SoC Cosimulation and Design, in *Proc International Symposium on Hardware/Software Codesign*, pp. 195–200, Apr. 2001.

[298]   Lyonnard, D., Yoo, S., Baghdadi, A., and Jerraya, A. A. Automatic Generation of Application-Specific Architectures for Heterogeneous Multiprocessor System-on-Chip, in *Proc Design Automation Conf.*, pp. 518–523, June 2001.

[299]   Hines, K., and Borriello, G. Optimizing Communication in Embedded System Cosimulation, in *Proc International Symposium on Hardware/Software Codesign*, pp. 121–125, Mar. 1997.

[300]   Gerin, P., Yoo, S., Nicolescu, G., and Jerraya, A. A. Scalable and Flexible Cosimulation of SoC Designs with Heterogeneous Multi-Processor Target Architectures, in *Proc Asia and South Pacific Design Automation Conference (ASP-DAC)*, pp. 62–68, June 2001.

[301]   Li, Y., and Henkel, J. A Framework for Estimating and Minimizing the Energy Dissipation of HW/SW Embedded Systems, in *Proc Design Automation Conf.*, pp. 188–193, June 1998.

[302]   Simunic, T., Benini, L., and De Micheli, G. Cycle Accurate Simulation of Energy Consumption in Embedded Systems, in *Proc Design Automation Conf.*, pp. 867–872, June 1999.

[303]   Lajolo, M., Raghunathan, A., Dey, S., Lavagno, L., and Sangiovanni-Vincentelli, A. Co-Simulation Based Power Estimation for System-on-Chip Design, *IEEE Trans VLSI Systems*, vol. 10, pp. 253–266, June 2002.

[304]   Yen, T., and Wolf, W. Communication Synthesis for Distributed Embedded Systems, in *Proc Int Conf Computer-Aided Design*, pp. 288–294, Nov. 1995.

[305]   Daveau, J., Ismail, T. B., and Jerraya, A. A. Synthesis of System-Level Communication by an Allocation Based Approach, in *Proc Int Symp System Level Synthesis*, pp. 150–155, Sept. 1995.

[306]   Dey, S., and Bommu, S. Performance Analysis of a System of Communicating Processes, in *Proc Int Conf Computer-Aided Design*, pp. 590–597, Nov. 1997.

[307]   Knudsen, P., and Madsen, J. Integrating Communication Protocol Selection with Partitioning in Hard-ware/Software Codesign, in *Proc Int Symp System Level Synthesis*, pp. 111–116, Dec. 1998.

[308]   Gasteier, M., and Glesner, M. Bus-based Communication Synthesis on System Level, in *ACM Trans Design Automation Electronic Systems*, pp. 1–11, Jan. 1999.

[309]   Dave, B., Lakshminarayana, G., and Jha, N. K. COSYN: Hardware-Software Co-Synthesis of Embedded Systems, in *Proc Design Automation Conf.*, pp. 703–708, June 1997.

[310]   Kirovski, D., and Potkonjak, M. System-Level Synthesis of Low-Power Hard Real-Time Systems, in *Proc Design Automation Conf.*, pp. 697–702, June 1997.

[311]   Lieverse, P., Wolf, P. V. D., and Deprettere, E. A Trace Transformation Technique for Communication Refinement, in *Proc International Symposium on Hardware/Software Codesign*, Apr. 2001.

[312]   Lieverse, P., Wolf, P. V. D., Vissers, K., and Deprettere, E. A Methodology for Architecture Exploration of Heterogeneous Signal Processing Systems, in *J VLSI Signal Processing*, Nov. 2001.

[313]   Givargis, T. D., Vahid, F., and Henkel, J. Trace-Driven System-Level Power Evaluation of System-on-a-Chip Peripheral Cores, in *Proc Asia and South Pacific Design Automation Conference (ASP-DAC)*, pp. 306–311, Jan. 2001.

[314] Lahiri, K., Raghunathan, A., and Dey, S. System-Level Performance Analysis for Designing On-Chip Communication Architectures, *IEEE Trans Computer-Aided Design*, vol. 20, pp. 768–783, June 2001.

[315] Lahiri, K., Raghunathan, A., and Dey, S. Fast System-Level Power Profiling for Battery-Efficient System Design, in *Proc International Symposium on Hardware/Software Codesign*, pp. 157–162, May 2002.

[316] Lahiri, K., Raghunathan, A., and Dey, S. Efficient Exploration of the SoC Communication Architecture Design Space, in *Proc Int Conf Computer-Aided Design*, pp. 424–430, Nov. 2000.

[317] Drinic, M., Kirovski, D., Meguerdichian, S., and Potkonjak, M. Latency Guided On-Chip Bus Network Design, in *Proc Int Conf Computer-Aided Design*, pp. 420–423, 2000.

[318] Pinto, A., Carloni, L. P., and Sangiovanni-Vincentelli, A. Constraint-Driven Communication Synthesis, in *Proc Design Automation Conf.*, pp. 783–788, June 2002.

[319] Kernighan, B., and Lin, S. An Efficient Heuristic Procedure for Partitioning Graphs, *Bell Systems Technical Journal*, vol. 49, pp. 291–307, 1970.

[320] Lahiri, K., Raghunathan, A., Lakshminarayana, G., and Dey, S. Communication Architecture Tuners: A Methodology for the Design of High Performance Communication Architectures for System-on-Chips, in *Proc Design Automation Conf.*, pp. 513–518, June 2000.

[321] Mehra, R., Guerra, L. M., and Rabaey, J. A Partitioning Scheme for Optimizing Interconnect Power, *IEEE J Solid-State Circuits*, vol. 32, pp. 433–443, Mar. 1997.

[322] Stan, M., and Burleson, W. Low Power Encodings for Global Communication in CMOS VLSI, *IEEE Trans VLSI Systems*, vol. 5, pp. 49–58, Dec. 1997.

[323] Benini, L., Micheli, G. D., Macii, E., Poncino, M., and Quer, S. Power Optimization of Core-Based Systems by Address Bus Encoding, *IEEE Trans VLSI Systems*, vol. 6, pp. 554–562, Dec. 1998.

[324] Sotiriadis, P. P., and Chandrakasan, A. P. A Bus Energy Model for Deep Sub-Micron Technology, *IEEE Trans VLSI Systems*, vol. 10, pp. 341–350, June 2002.

[325] Uchino, T., and Cong, J. An Interconnect Energy Model Considering Coupling Effects, *IEEE Trans Computer-Aided Design*, vol. 21, pp. 763–776, July 2002.

[326] Taylor, C. N., Dey, S., and Zhao, Y. Modeling and Minimization of Interconnect Energy Dissipation in Nanometer Technologies, in *Proc Design Automation Conf.*, pp. 754–757, June 2001.

[327] Heydari, P., and Pedram, M. Interconnect Energy Dissipation in High-Speed ULSI Circuits, in *Proc Asia and South Pacific Design Automation Conference (ASP-DAC)*, pp. 132–137, Jan. 2002.

[328] Sotiriadis, P. P., and Chandrakasan, A. P. Bus Energy Minimization by Transition Pattern Coding (TPC) in Deep Sub-Micron Technologies, in *Proc Int Conf Computer-Aided Design*, pp. 322–327, Nov. 2000.

[329]   Kim, K. W., Baek, K. H., Shanbhag, N., Liu, C. L., and Kang, S. M. Coupling-Driven Signal Encoding Scheme for Low-Power Interface Design, in *Proc Int Conf Computer-Aided Design*, pp. 318–321, Nov. 2000.

[330]   Henkel, J., and Lekatsas, H. A^{2}BC: Adaptive Address Bus Coding for Low Power Deep Sub-Micron Designs, in *Proc Design Automation Conf.*, pp. 744–749, June 2001.

[331]   Henkel, J., and Lekatsas, H. ETAM++: Extended Transition Activity Measure for Low Power Address Bus Designs, in *Proc Int Conf VLSI Design*, pp. 113–120, Jan. 2002.

[332]   Shin, Y., and Sakurai, T. Coupling-Driven Bus Design for Low-Power Application-Specific Systems, in *Proc Design Automation Conf.*, pp. 750–753, June 2001.

[333]   Machiarullo, L., Macii, E., and Poncino, M. Wire Placement for Crosstalk Energy Minimization in Address Buses, in *Proc Design Automation & Test Europe (DATE) Conf.*, pp. 158–162, Mar. 2002.

[334]   Benini, L., and Micheli, G. D. Powering Networks on Chips, in *Proc Int Symp System Level Synthesis*, pp. 33–38, 2001.

[335]   Chen, J. Y., Jone, W. B., Wang, J. S., Lu, H. I., and Chen, T. F. Segmented Bus Design for Low Power, *IEEE Trans VLSI Systems*, vol. 7, pp. 25–29, Mar. 1999.

[336]   Hsieh, C.-T., and Pedram, M. Architectural Power Optimization by Bus Splitting, *IEEE Trans Computer-Aided Design*, vol. 21, pp. 408–414, Apr. 2002.

[337]   Lahiri, K., Raghunathan, A., Dey, S., and Panigrahi, D. Battery Driven System Design: A New Frontier in Low Power Design, in *Proc ASP-DAC/Int Conf VLSI Design*, pp. 261–267, Jan. 2002.

[338]   Lahiri, K., Raghunathan, A., and Dey, S. Battery-Efficient Architecture for an 802. 11 MAC Processor, in *Proc IEEE Int Conf on Communications*, vol. 2, pp. 669–677, Apr. 2002.

[339]   Panigrahi, D., Chiasserini, C. F., Dey, S., Rao, R. R., Raghunathan, A., and Lahiri, K. Battery Life Estimation for Mobile Embedded Systems, in *Proc Int Conf VLSI Design*, pp. 55–63, Jan. 2001.

[340]   Doyle, M., Fuller, T. F., and Newman, J. S. Modeling of Galvanostatic Charge and Discharge of Lithium/Polymer/Insertion cell, *J Electrochem Soc*, vol. 140, pp. 1526–1533, June 1993.

[341]   ETSI TS 101 475. Available at http://www.etsi.org

[342]   Balarin, F., et al. HW/SW Co-Design of Embedded Systems: the Polis Approach, Kluwer Academic Publisher, 1997.

[343]   Kienhuis, B., et al. An Approach for Quantitative Analysis of Application-specific Dataflow Architectures, *Proc IEEE Int Conf On Application-Specific Systems, Architectures and Processors, 1997*.

[344]   Lee, E. A., and Messerschmitt, D. G. Pipeline Interleaved Programmable DSP's: Synchronous Dataflow Programming, *IEEE Trans on Acoustics, Speech and Signal Processing, 1987.*

[345]   Krolikoski, S. J., et al. Methodology and technology for virtual component-driven hardware/software co-design at the system level, *Proc IEEE Int Symp on Circuits and Systems, ISCAS 99, June 1999.*

[346]   Gajsky, D., Zhu, J., Domer, R., Gerstlauer, A., and Zhao, S. SpecC: Specification language and Methodology, Kluwer Academic Publishers, 2000.

[347]   Solden, S. Architectural Services Modeling for Performance in HW-SW Co-Design, *Proc SASIMI, Japan, 2001.*

[348]   http://www.systemc.org.

[349]   Demmeler, T., et al. Enabling Rapid Design Exploration through Virtual Integration and Simulation of Fault Tolerant Automotive Application, *SAE, 2002.*

[350]   Da Silva, J. L., et al. Wireless Protocols Design: Challenges and Opportunities, *Proc Int Workshop on HW/SW Codesign, 2000.*

[351]   de Kock, E. A., et al. YAPI: Application Modeling for Signal Processing Systems, *Proc DAC 2000.*

[352]   Kahn, G. The semantics of a simple language for parallel programming, *Information Proc., L, J., Rosenfeld, Ed North Holland Publishing, Co, 1974.*

[353]   Buck, J. T., et al. Ptolemy: A Framework for Simulating and Prototyping Heterogeneous Systems, *International Journal of Computer Simulation, 1994.*

[354]   Buck, J. T., and Vaidyanathan, R. Heterogeneous Modeling and Simulation of Embedded Systems in El Greco, Proc. Int. Workshop on HW/SW Codesign, 2000.

[355]   Pahlavan, K., Zahedi, A., and Krishnamurthy, P. Wideband Local Access: Wireless LAN and Wireless ATM, IEEE Communications magazine, Nov. 1997.

[356]   Kunkel, J. COSSAP: A stream driven simulator. In IEEE Int. Workshop Microelectron. Commun., Interlaken, Switzerland, March 1991.

[357]   George, V., Zhang, H., and Rabaey, J. The design of a Low Energy FPGA, *Proc Int Symposium on Low Power Electronics and Design, 1999.*

[358]   Krishnan, V., and Torrellas, J. A chip multiprocessor architecture with speculative multi-threading. IEEE Transactions on Computers, Special Issue on Multi-threaded Architecture, September 1999.

[359]   Nayfeh, B. A., Hammond, L., and Olukotun, K. Evaluating alternatives for a multiprocessor microprocessor. In Proc. the 23rd Intl. Symposium on Computer Architecture, pp. 66–77, Philadelphia, PA, 1996.

[360]   Catthoor, F., Wuytack, S., De Greef, E., Balasa, F., Nachtergaele, L., and Vandecappelle, A. Custom Memory Management Methodology: Exploration of

Memory Organization for Embedded Multimedia System Design, Boston: Kluwer, 1998.

[361]   Vijaykrishnan, N., Kandemir, M., Irwin, M. J., Kim, H. Y., and Ye, W. Energy-driven integrated hardware-software optimizations using SimplePower. In Proc. the International Symposium on Computer Architecture (ISCA'00), June 2000.

[362]   Panda, P. R., Dutt, N., and Nicolau, A. Memory Issues in Embedded Systems-On-Chip: Optimizations and Exploration. Kluwer Academic Publishers, Norwell, MA, 1999.

[363]   Grun, P., Dutt, N., and Nicolau, A. Memory Architecture Exploration for Programmable Embedded Systems, Kluwer Academic Press, Norwell, MA, 2003.

[364]   Hennessy, J. L., and Patterson, A. D. Computer Architecture—A Quantitative Approach. Morgan Kaufman, San Francisco, CA, 1994.

[365]   Cuppu, V., Jacob, B. L., Davis, B., and Mudge, T. N. A performance comparison of contemporary dram architectures. In Proc. International Symposium on Computer Architecture, Atlanta, GA, 1999, pp. 222–233.

[366]   Lam, M., Rothberg, E., and Wolf, M. E. The cache performance and optimizations of blocked algorithms. In Proc. 4th International Conference on Architectural Support for Programming Languages and Operating Systems, 1991, pp. 63–74.

[367]   Panda, P. R., Nakamura, H., Dutt, N. D., and Nicolau, A. Augmenting loop tiling with data alignment for improved cache performance. IEEE Transactions on Computers 48, 1999, pp. 142–149.

[368]   Panda, P. R., Dutt, N. D., and Nicolau, A. On-chip vs. off-chip memory: The data partitioning problem in embedded processor-based systems. ACM Transactions on Design Automation of Electronic Systems 5, 2000, pp. 682–704.

[369]   Kandemir, M. T., Ramanujam, J., Irwin, M. J., Vijaykrishnan, N., Kadayif, I., and Parikh, A. Dynamic management of SPM space. In Proc. Design Automation Conference, Las Vegas, NV, 2001, pp. 690–695.

[370]   Jouppi, N. P. Improving direct-mapped cache performance by the addition of a small fully associative cache and prefetch buffers. In Proc. International Symposium on Computer Architecture, Seattle, WA, 1990, pp. 364–373.

[371]   Bajwa, R. S., Hiraki, M., Kojima, H., Gorny, D. J., Nitta, K., Shridhar, A., Seki, K., and Sasaki, K. Instruction buffering to reduce power in processors for signal processing. IEEE Transactions on VLSI Systems 5, 1997, pp. 417–424.

[372]   Bellas, N., Hajj, I. N., Polychronopoulos, D. C., and Stamoulis, G. Architectural and compiler techniques for energy reduction in high-performance microprocessors. IEEE Transactions on VLSI Systems 8, 2000, pp. 317–326

[373]   Steinke, S., Wehmeyer, L., Lee, B. S., and Marwedel, P. Assigning program and data objects to scratch-pad for energy reduction. In Proc. Design, Automation & Test in Europe, Paris, France, 2002.

[374] Chen, G., Chen, G., Kadayif, I., Zhang, W., Kandemir, M., Kolcu, I., and Sezer, U. Compiler-directed management of instruction accesses, In Proc. Euromicro Symposium on Digital System Design, Architectures, Methods and Tools (DSD'03), Antalya, Turkey, September, 2003.

[375] Panda, P. R., Dutt, N. D., and Nicolau, A. Incorporating DRAM access modes into high-level synthesis. IEEE Transactions on Computer Aided Design 17, 1998, pp. 96–109.

[376] Khare, A., Panda, P. R., Dutt, N. D., and Nicolau, A. High-level synthesis with SDRAMs and RAMBUS DRAMs. IEICE Transactions on fundamentals of electronics, communications and computer sciences E82-A, 1999, pp. 2347–2355.

[377] Sudarsanam, A., and Malik, S. Simultaneous reference allocation in code generation for dual data memory bank asips. ACM Transactions on Design Automation of Electronic Systems 5, 2000, pp. 242–264.

[378] Saghir, M. A. R., Chow, P., and Lee, C. G. Exploiting dual data-memory banks in digital signal processors. In Proc. International conference on Architectural Support for Programming Languages and Operating Systems, Cambridge, MA, 1996, pp. 234–243.

[379] Cruz, J. L., Gonzalez, A., Valero, M., and Topham, N. Multiple-banked register file architectures. In Proc. International Symposium on Computer Architecture, Vancouver, Canada, 2000, pp. 315–325.

[380] Chang, K. H., and Lin, Y. L. Array allocation taking into account SDRAM characteristics. In Proc. Asia and South Pacific Design Automation Conference, Yokohama, 2000, pp. 497–502.

[381] Grun, P., Dutt, N., and Nicolau, A. Memory aware compilation through accurate timing extraction. In Proc. Design Automation Conference, Los Angeles, CA, 2000, pp. 316–321.

[382] Grun, P., Dutt, N., and Nicolau, A. MIST: An algorithm for memory miss traffic management. In Proc. IEEE/ACM International Conference on Computer Aided Design, San Jose, CA, 2000, pp. 431–437.

[383] Panda, P. R. Memory bank customization and assignment in behavioral synthesis. In Proc. of the IEEE/ACM International Conference on Computer Aided Design, San Jose, CA, 1999, pp. 477–481.

[384] Ramachandran, L., Gajski, D., and Chaiyakul, V. An algorithm for array variable clustering. In Proc. European Design and Test Conference, Paris, 1994, pp. 262–266.

[385] Jha, P. K., and Dutt, N. Library mapping for memories. In Proc. European Design and Test Conference, Paris, France, 1997, pp. 288–292.

[386] Wuytack, S., Catthoor, F., Jong, G. D., and Man, H. D. Minimizing the required memory bandwidth in VLSI system realizations. IEEE Transactions on VLSI Systems 7, 1999, pp. 433–441.

[387]   Grun, P., Dutt, N., and Nicolau, A. Apex: Access pattern based memory architecture customization. In Proc. International Symposium on System Synthesis, Montreal, Canada, 2001, pp. 25–32.

[388]   Shackleford, B., Yasuda, M., Okushi, E., Koizumi, H., Tomiyama, H., and Yasuura, H. Memory-CPU size optimization for embedded system designs. In Proc. Design Automation Conference, 1997.

[389]   Tomiyama, H., Halambi, A., Grun, P., Dutt, N., and Nicolau, A. Architecture description languages for systems-on-chip design. In Proc. 6th Asia Pacific Conference on Chip Design Languages, Fukuoka, 1999, pp. 109–116.

[390]   Halambi, A., Grun, P., Tomiyama, H., Dutt, N., and Nicolau, A. Automatic software toolkit generation for embedded systems-on-chip. In Proc, ICVC'99, Korea, 1999.

[391]   Panda, P. R., Catthoor, F., Dutt, N. D., Danckaert, K., Brockmeyer, E., Kulkarni, C., Vandercappelle, A., and Kjeldsberg, P. G. Data and memory optimization techniques for embedded systems, ACM Transactions on Design Automation of Electronic Systems, Vol. 6, No. 2, April 2001.

[392]   Halambi, A., Grun, P., Ganesh, V., Khare, A., Dutt, N., and Nicolau, A. Expression: A language for architecture exploration through compiler/simulator retargetability. In Proc. DATE'99, Munich, Germany, 1999.

[393]   Gonzales, A., Aliagas, C., and Valero, M. A data cache with multiple caching strategies tuned to different types of locality. In Proc. International Conference on Supercomputing, Barcelona, Spain, 1995, pp. 338–347.

[394]   Kadayif, I., Kandemir, M., Vijaykrishnan, N., Irwin, M. J., and Ramanujam, J. Morphable cache architectures: potential benefits. In Proc. Workshop on Languages, Compilers and Tools for Embedded Systems, Snow Bird, Utah, 2001.

[395]   Su, C. L., and Despain, A. M. Cache designs for energy efficiency. In Proc. the 28th Hawaii International Conference on System Sciences (HICSS'95), January 04–07, 1995, Hawaii, USA.

[396]   Muchnick, S. Advanced Compiler Design and Implementation, Morgan-Kauffmann, 1997.

[397]   Tseng, C.-W., Anderson, J., Amarasinghe, S., and Lam, M. Unified compilation techniques for shared and distributed address space machines, In Proc. International Conference on Supercomputing, Barcelona, Spain, July 1995.

[398]   Wolfe, M. High Performance Compilers for Parallel Computing. Addison Wesley, CA, 1996.

[399]   Kadayif, I., Kandemir, M., and Sezer, U. An integer linear programming based approach for parallelizing applications in on-chip multiprocessors. In Proc. Design Automation Conference, New Orleans, LA, June 2002.

[400]   Kandemir, M., Zhang, W., and Karakoy, M. Runtime code parallelization for on-chip multiprocessors, In Proc. the 6th Design Automation and Test in Europe Conference, Munich, Germany, March, 2003.

[401]   Li, W., and Pingali, K. A singular loop transformation framework based on nonsingular matrices, International Journal of Parallel Programming, April 22(2):183–205.

[402]   Leung, S.-T., and Zahorjan, J. Optimizing data locality by array restructuring, Technical Report, Computer Science Department, University of Washington, Seattle, WA, 1995.

[403]   Kandemir, M., Ramanujam, J., and Choudhary, A. A compiler algorithm for optimizing locality in loop nests. In Proc. the ACM International Conference on Supercomputing, Vienna, Austria, July 1997.

[404]   O'Boyle, M., and Knijnenburg, P. Nonsingular data transformations: definition, validity and application in Proc. International Conference on Supercomputing, Vienna, Austria, 1997.

[405]   Chilimbi, T. M., Hill, M. D., and Larus, J. R. Cache conscious structure layout. In Proc. Programming Languages Design and Implementation, May 1999.

[406]   Kandemir, M., and Kadayif, I. Compiler-directed selection of dynamic memory layouts. In Proc. the 9th International Symposium on Hardware/Software Co-design, April 2001, Denmark, pp. 219–224.

[407]   Todd Mowry, C. Tolerating latency in multiprocessors through compiler-inserted prefetching. ACM Transactions on Computer Systems, 16(1): 55–92, February 1998.

[408]   Kandemir, M., Ramanujam, J., and Choudhary, A. Exploiting shared SPM space in embedded multiprocessor systems. In Proc. Design Automation Conference, New Orleans, LA, June 2002.

[409]   Kandemir, M., Choudhary, A., Ramanujam, J., and Banerjee, P. On reducing false sharing while improving locality on shared memory multiprocessors. In Proc. International Conference on Parallel Architectures and Compilation Techniques, Newport Beach, CA, October 1999, pp. 203–211.

[410]   Unnikrishnan, P., Chen, G., Kandemir, M., Karakoy, M., and Kolcu, I. Loop transformations for reducing data space requirements of resource-constrained applications. In Proc. 10th Annual International Static Analysis Symposium, June 11–13, 2003, San Diego, CA.

[411]   Lefebvre, V., and Feautrier, P. Automatic storage management for parallel programs. Research Report PRiSM 97/8, France, 1997.

[412]   Grun, P., Balasa, F., and Dutt, N. Memory size estimation for multimedia applications. In Proc. CODES/CACHE, 1998.

[413]   Strout, M., Carter, L., Ferrante, J., and Simon, B. Schedule-independent storage mapping in loops. In Proc. ACM Conference on Architectural Support for Programming Languages and Operating Systems, October 1998.

[414]   Wilde, D., and Rajopadhye, S. Memory reuse analysis in the polyhedral model. Parallel Processing Letters, 1997.

[415] Lim, A. W., Liao S.-W., and Lam, S. M. Blocking and array contraction across arbitrarily nested loops using affine partitioning. In Proc. ACM SIGPLAN Notices, 2001.

[416] Song, Y., Xu, R., Wang, C., and Li, Z. Data locality enhancement by memory reduction. In Proc. 15th ACM International Conference on Supercomputing, June, 2001.

[417] Catthoor, F., Danckaert, K., Kulkarni, C., Brockmeyer, E., Kjeldsberg, P. G., Achteren, T. V., and Omnes, T. Data Access and Storage Management for Embedded Programmable Processors, Kluwer Academic Publishers, 2002.

[418] Fraboulet, A., Kodary, K., and Mignotte, A. Loop fusion for memory space optimization. In Proc. 14th International Symposium on System Synthesis, Montreal, Canada, September 30–October 3, 2001.

[419] Thies, W., Vivien, F., Sheldon, J., and Amarasinghe, S. A unified framework for schedule and storage optimization. In Proc. the SIGPLAN Conference on Programming Language Design and Implementation, Snowbird, UT, June 2001.

[420] Tu, P., and Padua, D. Automatic array privatization. In Proc. 6th Workshop on Languages and Compilers for Parallel Computing, Lecture Notes in Computer Science, pp. 500–521, Portland, OR, August 1993.

[421] Barthou, D., Cohen, A., and J.-F., Collard. Maximal static expansion. In Proc. 25th Annual ACM Symposium on Principles of Programming Languages, San Diego, CA, 1998.

[422] Kadayif, I., Kandemir, M., and Karakoy, M. An energy saving strategy based on adaptive loop parallelization. In Proc. Design Automation Conference, New Orleans, LA, June 2002.

[423] Carriero, N., Gelernter, D., Kaminsky, D., and Westbrook, J. Adaptive parallelism with Piranha. Technical Report 954, Yale University, February 1993.

[424] Irigoin, F., and Triolet, R. Super-node partitioning. In Proc. the ACM Symposium on Principles and Practice of Programming Languages. San Diego, CA, January 1988, pp. 319–329.

[425] Jain, P., Devadas, S., Engels, D., and Rudolph, L. Software-assisted cache replacement mechanisms for embedded systems. In Proceedings, ICCAD'01, IEEE, 2001.

[426] Li, Y., and Wolf, W. Hardware/software co-synthesis with memory hierarchies, in IEEE Transactions on CAD/ICAS, 18(10), October 1999, pp. 1405–1417.

[427] LSI Logic Corporation Milpitas, CA, CW33000 MIPS Embedded Processor User's Manual, 1992.

[428] Mishra, P., Grun, P., Dutt, N., and Nicolau, A. Processor-memory co-exploration driven by a memory-aware architecture description language. In Proc. VLSI Design, Bangalore, 2001.

[429]   Panda, P. R., Dutt, N. D., and Nicolau, A. Local memory exploration and optimization in embedded systems. IEEE Transactions on Computer Aided Design 18, 1999, pp. 3–13.

[430]   Schmit, H., and Thomas, D. E. Synthesis of application-specific memory designs, IEEE Transactions on VLSI Systems, 5(1), March 1997, pp. 101–111.

[431]   Shiue, W.-T., and Chakrabarti, C. Memory exploration for low power embedded systems. In Proc. the Design Automation Conference, 1999.

[432]   Cassidy, A. S., Paul, J. M., and Thomas, E. D. Layered. Multi-Threaded, High-Level Performance Design. In *Design Automation and Test in Europe, DATE*, pp. 954–959, March 2003.

[433]   Cesario, W., Gauthier, L., Lyonnard, D., Nicolescu, G., and Jerraya, A. An XML-based Meta-Model for the Design of Multiprocessor Embedded Systems. In *VHDL International User's Forum*, pp. 75–82, October 2000.

[434]   Gerstlauer, A., Yu, H., and Gajski, D. RTOS Modelling for System-Level Design. In *Design Automation and Test in Europe, DATE*, pp. 132–137, March 2003.

[435]   Liu, C. W. and Layland, J. W. Scheduling Algorithms for Multiprogramming in a Hard Real-Time Environment. *Journal of the ACM*, 20(1):46–61, March 1973.

[436]   Sifakis, J. Modelling Real-Time Systems—Challenges and Work Directions. In *EMSOFT, Lecture Notes in Computer Science 2211*, pp. 373–389, October 2001.

[437]   Vestal, S. MetaH Support for Real-Time Multiprocessor Avionics. In *5th IEEE Workshop on Parallel and Distributed Real-Time Systems*, pp. 132–137, 1997.

[438]   Stankovic, J. A., Spuri, M., Ramamritham, K., and Buttazzo, C. G. *Deadline Scheduling for Real-Time Systems: EDF and Related Algorithms*. Kluwer Academic Publishers, 1998.

[439]   Open SystemC Initiative (OSCI). http://www.systemc.org.

[440]   Desmet, D., Verkest, D., and Man, D. H. Operating System-based Software Generation for Systems-on-Chip. In *Design Automation Conference, DAC*, pp. 396–401, June 2000.

[441]   Sifakis, J., Tripakis, S., and Yovine, S. Building Models of Real-Time Systems from Application Software. *IEEE Special Issue on Modelling and Design of Embedded Systems*, pp. 100–111, January 2003.

[442]   Tomiyama, H., Cao, Y., and Murakami, K. Modelling Fixed-Priority Preemptive Multi-Task Systems in SpecC. In *SASIMI*, pp. 93–100, October 2001.

[443]   Grotker, T., Martin, G., Liao, S., and Swan, S. *System Design with SystemC*. Kluwer Academic Publishers, New York, 2002.

[444]   Abadi, M., and Lamport, L. Composing specifications. *ACM Transactions on Programming Languages and Systems*, 15(3):73–132, January 1993.

[445]   Abadi, M., and Lamport, L. Conjoining specifications. *ACM Transactions on Programming Languages and Systems*, 17(3):507–534, May 1995.

[446]   Gonzalez, M., and Madsen, J. Abstract RTOS Modeling in SystemC. In *Proceedings of the 20th NORCHIP Conference*, pp. 43–49, November 2002.

[447]   SystemC Workgroup. Master-Slave Communication Library: A SystemC Standard Library. http://www.systemc.org, Version 2. 0 Beta-3 2001.

[448]   Burns, A., and Wellings, A. *Real-Time Systems and Programming Languages*. Addison-Wesley, 2001.

[449]   Sun, J., and Liu, J. Synchronization protocols in distributed real-time systems. In *the 16th International Conference on Distributed Computing Systems*, pp. 38–45, May 1996.

[450]   Sha, L., Rajkumar, R., and Lehoczky, P. J. Priority-Inheritance Protocols: An Approach to Real-Time Synchronization. *IEEE Transactions on Computers*, 39:1175–1185, 1990.

[451]   Baker, T. Stack-based Scheduling of Real-Time Processes. *Advances in Real-Time Systems*, 3:64–96, March 1991.

[452]   Madsen, J., Virk, K., and Gonzalez, M. Abstract RTOS Modelling for Multiprocessor System-on-Chip. In *Proceedings of the International Symposium on System-on-Chip*, November 2003.

[453]   Bokhari, H. S., and Nicol, M. D. Balancing Contention and Synchronization on the Intel Paragon. In *IEEE Concurrency*, pp. 74–83, April–June 1997.

[454]   Liu, J. W. S. *Real-Time Systems*. Prentice-Hall, 2000.

[455]   Schmitz, M., Al-Hashimi, B., and Eles, P. A Co-Design Methodology for Energy-Efficient, Multi-Mode Embedded Systems with the consideration of Mode Execution Probabilities. In *Design Automation and Test in Europe, DATE*, pp. 960–965, March 2003.

[456]   De Man, H. System Design Challenges in the Post-PC Era. In *Proceedings of the 37th Design Automation Conference*, Los Angels, 2000.

[457]   Claasen, T. High Speed: Not the Only Way to Exploit the Intrinsic Computational Power of Silicon. In *Proc Int Solid-State Circuits Conf.*, pp. 22–25, San Fransisco, CA, Feb. 1999.

[458]   Martin, G., and Chang, H., editors. *Winning the SoC Revolution: Experiences in Real Design*. Kluwer Academic Publishers, 2003.

[459]   Marchal, P., Wong, C., Prayati, A., Cossement, N., Catthoor, F., Lauwereins, R., Verkest, D., and De Man, H. Dynamic Memory Oriented Transformations in the MPEG4 IM1-player on a Low Power Platform. In *Proc Intnl Wsh on Power Aware Computing Systems(PACS)*, Cambridge MA, Nov. 2000.

[460]   Yang, P., Wong, C., Marchal, P., Catthoor, F., Desmet, D., Verkest, D., and Lauwereins, R. Energy-aware Runtime Scheduling for Embedded Multiprocessor SoCs. *IEEE Design and Test of Computers*, 19(3), Sept. 2001.

[461] Ramamritham, K., and Stankovic, A. J. Scheduling Algorithms and Operation Systems Support for Real-Time Systems. *Proceedings of the IEEE*, 82(1):55–67, Jan. 1994.

[462] Audsley, N., Burns, A., Davis, R., Tindell, K., and Wellings, A. Fixed Priority Preemptive Scheduling: An Historical Perspective. *Real-Time Systems*, 8(2):173–198, 1995.

[463] Sha, L., Rajkumar, R., and Lehoczky, P. J. Priority Inheritance Protocols: An Approach to Real-Time Synchronization. *IEEE Transactions on Computers*, 39(9):1175–1185, Sept. 1990.

[464] Thoen, F., and Catthoor, F. *Modeling, Verification and Exploration of Task-level Concurrency in Real-Time Embedded Systems*. Kluwer Academic Publishers, 1999.

[465] Balarin, F., et al. *Hardware-Software Co-Design of Embedded Systems: the POLIS Approach*. Kluwer Academic Publishers, 1997.

[466] Benner, T., and Ernst, R. An Approach to Mixed Systems Co-Synthesis. In *Proceedings of the International Workshop on Hardware/Software Code-sign(CODES)*, pp. 9–14, 1997.

[467] Prayati, A., Wong, C., Marchal, P., et al. Task Concurrency Management Experiment for Power-efficient Speed-up of Embedded MPEG4 IM1 Player. In *International Conference on Parallel Processing*, 2000.

[468] Liu, C. L., and Layland, W. J. Scheduling Algorithms for Multiprogramming in a Hard Real-Time Environment. *Journal of the Association for Computing Machinery*, 20(1):46–61, Jan. 1973.

[469] Leung, J.-T., and Whitehead, J. On the Complexity of Fixed-Priority Scheduling of Periodic Real-Time Tasks. *Performance Evaluation*, 2:237–250, 1982.

[470] Chetto, H., Silly, M., and Bouchentouf, T. Dynamic Scheduling of Real-Time Tasks Under Precedence Constraints. *Real-Time Systems*, 2(3):181–194, Sept. 1990.

[471] Lehoczky, J. P., Sha, L., and Strosnider, K. J. Enhanced Aperiodic Responsiveness in Hard Real-Time Environments. In *Proceedings of the IEEE Real-Time System Symposium*, pp. 261–270, 1987.

[472] Spuri, M., and Buttazzo, C. G. Scheduling Aperiodic Tasks in Dynamic Priority Systems. *Real-Time Systems*, 10(2):179–210, 1996.

[473] Baker, P. T. Stack-Based Scheduling of Realtime Processes. *Real-Time Systems*, 3(1):67–99, 1991.

[474] El-Rewini, H., Ali, H. H., and Lewis, T. Task Scheduling in Multiprocessing Systems. *IEEE Computer*, 28(12):27–37, Dec. 1995.

[475] Hoang, P. D., and Rabaey, M. J. Scheduling of DSP Programs onto Multiprocessors for Maximum Throughput. *IEEE Transactions on Signal Processing*, 41(6):2225–2235, June 1993.

[476]   Yen, T.-Y., and Wolf, W. Communication Synthesis for Distributed Embedded Systems. In *IEEE/ACM International Conference on Computer-Aided Design*, pp. 288–94, 1995.

[477]   Yen, T.-Y., and Wolf, W. Sensitivity-Driven Co-Synthesis of Distributed Embedded Systems. In *Proceedings of International Symposium on System Synthesis*, pp. 4–9, Sept. 1995.

[478]   Gruian, F., and Kuchcinski, K. Low-Energy Directed Architecture Selection and Task Scheduling. In *EUROMICRO'99*, pp. 296–302, 1999.

[479]   Jha, K. N. Low Power System Scheduling and Synthesis. In *IEEE/ACM International Conference on Computer-Aided Design*, pp. 259–63, 2001.

[480]   Hong, I., Potkonjak, M., and Srivastava, B. M. On-Line Scheduling of Hard Real-Time Tasks on Variable Voltage Processor. In *IEEE/ACM International Conference on Computer-Aided Design*, pp. 653–656, San Jose, CA, 1998.

[481]   Okuma, T., Ishihara, T., and Yasuura, H. Real-Time Task Scheduling for a Variable Voltage Processor. In *Proceedings of International Symposium on System Synthesis*, pp. 24–29, 1999.

[482]   Chandrakasan, A., Gutnik, V., and Xanthopoulos, T. Data Driven Signal Processing: An Approach for Energy Efficient Computing. In *Proceedings of International Symposium on Low Power Electronic Device*, pp. 347–352, 1996.

[483]   Transmeta. Crusoe processor. http://www.crusoe.com/crusoe/index.html.

[484]   Quan, G., and Hu, X. Energy Efficient Fixed-Priority Scheduling for Real-Time Systems on Variable Voltage Processors. In *Proceedings of the 38th Design Automation Conference*, 2001.

[485]   Burd, T. D., Pering, T. A., Stratakos, A. J., and Brodersen, W. R. A Dynamic Voltage Scaled Microprocessor System. *IEEE J Solid-State Circuits*, 35(11):1571–1580, Nov. 2000.

[486]   Hong, I., Kirovski, D., Qu, G., Potkonjak, M., and Srivastava, B. M. Power Optimization of Variable Voltage Core-Based Systems. In *Proceedings of the 35th Design Automation Conference*, pp. 176–181, San Francisco, CA, 1998.

[487]   Luo, J., and Jha, N. Power-consious Joint Scheduling of Periodic Task Graphs and Aperiodic Tasks in Distributed Real-time Embedded Systems. In *IEEE/ACM International Conference on Computer-Aided Design*, pp. 357–364, San Jose, USA, Nov. 2000.

[488]   Luo, J., and Jha, N. Static and Dynamic Variable Voltage Scheduling Algorithms for Real-Time Heterogeneous Distributed Embedded Systems. In *7th ASPDAC and 15th Int'l Conf on VLSI Design*, pp. 719–726, Jan. 2002.

[489]   Zhang, Y., Hu, X. S., and Chen, D. Z. Task Scheduling and Voltage Selection for Energy Minimization. In *Proceedings of the 39th Design Automation Conference*, 2002.

[490]   Azevedo, A., et al. Profile-based dynamic voltage scheduling using program checkpoints. In *Proceedings of the Design Automation and Test in Europe*, pp. 168–175, 2002.

[491] Adve, S., et al. The Interaction of Architecture and Compilation Technology for High-performance Processor Design. *IEEE Computer Magazine*, 30(12):51–58, Dec. 1997.

[492] Catthoor, F., Danckaert, K., Kulkarni, C., Brockmeyer, E., Kjeldsberg, P. G., Achteren, T., and Omnes, T. *Data Access and Storage Management for Embedded Programmable Processors*. Kluwer Academic Publishers, Boston, 2002.

[493] Ma, Z., Wong, C., Delfosse, E., Vounckx, J., Catthoor, F., Himpe, S., and Deconinck, G. Task Concurrency Analysis and Exploration of Visual Texture Decoder on a Heterogeneous Platform. In *IEEE Workshop on Signal Processing Systems (SIPS)*, pp. 245–250, Seoul, South Korea, 2003.

[494] Marchal, P., Jayapala, M., Souza, S. D., Yang, P., Catthoor, F., and Deconinck, G. Matador: an exploration environment for system-design. *Journal of Circuits, Systems and Computers*, 11(5):503–535, 2002.

[495] Simunic, T. *Energy Efficient System Design and Utilization*. PhD thesis, Dept. of Electrical Engineering, Stanford University, 2001.

[496] Himpe, S., Deconinck, G., Catthoor, F., and Meerbergen, J. MTG* and Grey-Box: modeling dynamic multimedia applications with concurrency and non-determinism. In *Proc Forum on Design Languages(FDL)*, Marseille, France, Sept. 2002.

[497] Pareto, V. *Manuale di Economia Politica*. Piccola Biblioteca Scientifica, Milan, 1906. Translated into English by Ann Schwier, S., (1971), Manual of Political Economy, MacMillan, London.

[498] Windver. Vspworks. www.windriver.com/products/vspworks/index.html.

[499] Yang, P., and Catthoor, F. Pareto-Optimization-Based Run-Time Task Scheduling for Embedded Systems. In *ISSS + CODES*, Newport Beach, CA, 2003.

[500] Shin, Y., Choi, K., and Sakurai, T. Power Optimization of Real-Time Embedded Systems on Variable Speed Processors. In *IEEE/ACM International Conference on Computer-Aided Design*, pp. 365–368, 2000.

[501] CPLEX mixed integer optimizer. http://www.ilog.com/products/cplex/product/mip.cfm.

[502] Nemhauser, G. L., and Wolsey, L. A. Integer and Combinatorial Optimization. John Wiley and Sons, New York, 1988.

[503] Motwani, R., Palem, K., Sarkar, V., and Reyen, S. Combining register allocation and instruction scheduling. Technical Report STAN-CS-TN-95–22, Department of Computer Science, Stanford University, 1995.

[504] Kremer, U. Optimal and near-optimal solutions for hard compilation problems. Parallel Processing Letters, 7(4), 1997.

[505] Ruttenberg, J., Gao, G. R., Stoutchinin, A., and Lichtenstein, W. Software pipelining showdown: Optimal vheuristic, s., methods in a production compiler. In Proceedings of the ACM SIGPLAN '96 Conference on Programming Language Design and Implementation, pp. 1–11, 1996.

[506]   Goodwin, D. W., and Wilken, K. D. Optimal and near-optimal global register allocation using 0–1 integer programming. Software–Practice & Experience, 26(8):929–965, August 1996.

[507]   Kong, T., and Wilken, K. D. Precise register allocation for irregular register architectures. In Proceedings of the 31st Annual ACM/IEEE International Symposium on Microarchitecture (MICRO-98), pp. 297–307. IEEE Computer Society, November 30–December 2 1998.

[508]   Appel, A. and George, L. Optimal spilling for CISC machines with few registers. In Cindy Norris, and James, R., Fenwick, B., editors, Proceedings of PLDI'01, ACM SIGPLAN Conference on Programming Language Design and Implementation, pp. 243–253, 2001.

[509]   Stoutchinin, A. An integer linear programming model of software pipelining for the MIPS R8000 processor. In PaCT'97, Parallel Computing Technologies, 4th International Conference, pp. 121–135. Springer-Verlag (LNCS 1277), 1997.

[510]   Liberatore, V., Farach-Colton, M., and Kremer, U. Evaluation of algorithms for local register allocation. Lecture Notes in Computer Science, 1575:137–152, 1999.

[511]   Sjodin, J. and von Platen, C. Storage allocation for embedded processors. In Proceedings of CASES 2001, pp. 15–23, 2001.

[512]   Avissar, O., Barua, R., and Stewart, D. Heterogeneous memory management for embedded systems. In Proceedings of the ACM 2nd Int'l Conf. on Compilers, Architectures and Synthesis for Embedded Systems CASES, November 2001.

[513]   Wilken, K. D., Liu, J., and Heffernan, M. Optimal instruction scheduling using integer programming. Proceedings of PLDI'00, ACM SIGPLAN Conference on Programming Language Design and Implementation, 35(5):121–133, May 2000.

[514]   Saputra, H., Kandemir, M., Vijaykrishnan, N., Irwin, M. J., Hu, J., Hsu, C.-H., and Kremer, U. Energy-conscious compilation based on voltage scaling. In Proceedings of the 2002 Joint Conference on Languages, Compilers and Tools for Embedded Systems & Software and Compilers for Embedded Systems (LCTES/SCOPES '02), pp. 2–11, June 2002.

[515]   Naik, M. and Palsberg, J. Compiling with code-size constraints. In LCTES'02, Languages, Compilers and Tools for Embedded Systems joint with SCOPES'02, Software and Compilers for Embedded Systems, pp. 120–129, Berlin, Germany, June 2002.

[516]   Williams, P. H. Model Building in Mathematical Programming. Wiley & Sons, 1999.

[517]   Milner, R. A theory of type polymorphism in programming. Journal of Computer and System Sciences, 17:348–375, 1978.

[518]   Sestoft, P. Replacing function parameters by global variables. Master's thesis, DIKU, University of Copenhagen, September 1989.

[519]   Sestoft, P. Analysis and Efficient Implementation of Functional Programs. PhD thesis, DIKU, University of Copenhagen, October 1991.

[520]   Palsberg, J. Closure analysis in constraint form. ACM Transactions on Programming Languages and Systems, 17(1):47–62, January 1995. Preliminary version in Proceedings of CAAP'94, Colloquium on Trees in Algebra and Programming, Springer-Verlag (LNCS 787), pp. 276–290, Edinburgh, Scotland, April 1994.

[521]   Wand, M. and Williamson, G. B. A modular, extensible proof method for small-step flow analyses. In Daniel Le M'etayer, editor, Proceedings of ESOP 2002, 11th European Symposium on Programming, held as Part of the Joint European Conference on Theory and Practice of Software, ETAPS 2002, Grenoble, France, April, 2002, pp. 213–227. Springer-Verlag (LNCS 2305), 2002.

[522]   Nielson, F. The typed lambda-calculus with first-class processes. In Proceedings of PARLE, pp. 357–373, April 1989.

[523]   Wright, A. and Felleisen, M. A syntactic approach to type soundness. Information and Computation, 115(1):38–94, 1994.

[524]   Nipkow, T. and von Oheimb, D. Javalight is type-safe–definitely. In Proceedings of POPL'98, 25th Annual SIGPLAN–SIGACT Symposium on Principles of Programming Languages, pp. 161–170, San Diego, California, January 1998.

[525]   Wand, M. and Steckler, P. Selective and lightweight closure conversion. In Proceedings of POPL'94, 21st Annual Symposium on Principles of Programming Languages, pp. 434–445, 1994.

[526]   Meyer, A. R., and Wand, M. Continuation semantics in typed lambda-calculi. In Proceedings of Logics of Programs, pp. 219–224. Springer-Verlag (LNCS 193), 1985.

[527]   Damian, D. and Danvy, O. Syntactic accidents in program analysis: On the impact of the CPS transformation. In Proceedings of ICFP'00, ACM SIGPLAN International Conference on Functional Programming, pp. 209–220, 2000.

[528]   Palsberg, J. and Wand, M. CPS transformation of flow information. Journal of Functional Programming. To appear.

[529]   Tarditi, D., Morrisett, G., Cheng, P., Stone, C., Harper, R., and Lee, P. TIL: A type-directed optimizing compiler for MIn, L., 1996 ACM SIGPLAN Conference on Programming Language Design and Implementation, pp. 181–192, Philadelphia, PA, USA, May 1996. ACM Press.

[530]   Morrisett, G., Walker, D., Crary, K., and Glew, N. From system F to typed assembly language. In Proceedings of POPL'98, 25th Annual SIGPLAN–SIGACT Symposium on Principles of Programming Languages, pp. 85–97, 1998.

[531]   Tang, Y. M. and Jouvelot, P. Separate abstract interpretation for control-flow analysis. In Proceedings of TACS'94, Theoretical Aspects of Computing Software, pp. 224–243. Springer-Verlag (LNCS 789), 1994.

[532]   Heintze, N. Control-flow analysis and type systems. In Proceedings of SAS'95, International Static Analysis Symposium, pp. 189–206. Springer-Verlag (LNCS 983), Glasgow, Scotland, September 1995.

[533]   Wells, J. B., Allyn Dimock, Robert Muller, and Franklyn Turbak. A typed intermediate language for flow-directed compilation. In Proceedings of TAPSOFT'97, Theory and Practice of Software Development. Springer-Verlag (LNCS 1214), 1997.

[534]   Nielson, F., Nielson, H. R., and Hankin, C. Principles of Program Analysis. Springer-Verlag, 1999.

[535]   Palsberg, J. Equality-based flow analysis versus recursive types. ACM Transactions on Programming Languages and Systems, 20(6):1251–1264, 1998.

[536]   Palsberg, J. and O'Keefe, P. M. A type system equivalent to flow analysis. ACM Transactions on Programming Languages and Systems, 17(4):576–599, July 1995. Preliminary version in Proceedings of POPL'95, 22nd Annual SIGPLAN–SIGACT Symposium on Principles of Programming Languages, pp. 367–378, San Francisco, California, January 1995.

[537]   Palsberg, J. and Pavlopoulou, C. From polyvariant flow information to intersection and union types. Journal of Functional Programming, 11(3):263–317, May 2001. Preliminary version in Proceedings of POPL'98, 25th Annual SIGPLAN–SIGACT Symposium on Principles of Programming Languages, pp. 197–208, San Diego, California, January 1998.

[538]   Lerner, S., Grove, D., and Chambers, C. Composing dataflow analyses and transformations. In Proceedings of the 29th ACM SIGPLAN-SIGACT Symposium on Principles of Programming Languages, pp. 270–282. ACM Press, 2002.

[539]   Palsberg, J. Type-based analysis and applications. In Proceedings of PASTE'01, ACM SIGPLAN/SIGSOFT Workshop on Program Analysis for Software Tools, pp. 20–27, Snowbird, Utah, June 2001. Invited paper.

[540]   ITRS, available at http://public.itrs.net/

[541]   Nagari, A., et al. A 2. 7V 11. 8 mW Baseband ADC with 72 dB Dynamic Range for GSM Applications, 21st Custom Integrated Circuits Conference, San Diego, 1999.

[542]   http://www.semiconductors.philips.com/platforms/nexperia/

[543]   Oka, M. and Suzuoki, M. Designing and Programming the Emotion Engine, IEEE Micro, vol. 19:6, pp. 20–28, Nov/Dec 1999.

[544]   Paulin, P., Karim, F., and Bromley, P. Network Processors: A Perspective on Market Requirements, Proc. of DATE, 2001.

[545]   Culler, D. E., and Singh, J. P. Parallel Computer Architecture, Morgan Kaufmann Publishers, 1999.

[546]   Patterson, D. A., and Hennessy, J. L. Computer Organization and Design—The Hardware/Software Interface, Morgan Kaufmann Pub., 1998.

[547]   Kuskin, J., et al. The Stanford FLASH Multiprocessor, Proc. of the 21st International Symposium on Computer Architecture, 1994.

[548]   Gauthier, L., Yoo, S., and Jerraya, A. A. Automatic Generation and Targeting of Application Specific Operating Systems and Embedded Systems Software, IEEE TCAD, Vol. 20 N11, r., November 2001.

[549]   Keutzer, K. A Disciplined Approach to the Development of Platform Architectures, Synthesis and System Integration of Mixed Technologies (SASIMI), Nara, Japan, October 18–19, 2001.

[550]   Ernst, R., et al. The COSYMA environment for hardware/software cosynthesis of small embedded systems, Microprocessors and Microsystems, 1996.

[551]   Balarin, F., et al. Hardware-Software Co-design of Embedded Systems: The POLIS approach, Kluwer Academic Press, 1997.

[552]   Gajski, D., et al. SpecC Specification Language and Methodology, Kluwer Academic Publishers, 2000.

[553]   Chang, H., et al. Surviving the SOC Revolution: a Guide to Platform-based Design, Kluwer Academic Publishers, 1999.

[554]   Keutzer, K., et al. System-level design: orthogonalization of concerns and platform-based design, IEEE TCAD, Dec. 2000.

[555]   Schirrmeister, F., Meindl, M., and Krolikoski, S. IP Authoring and Integration for HW/SW Co-Design and Reuse—Lessons Learned, The Ninth Electronic Design Processes (EDP) Workshop, Monterey, CA, 2002.

[556]   Cesário, W. O., et al. Multiprocessor SoC Platforms: A Component-Based Design Approach, IEEE Design & Test of Computers, Vol. 19, Issue: 6, Nov.–Dec., 2002.

[557]   Sgroi, M., et al. Addressing the System-on-Chip Interconnect Woes Through Communication-Based Design, Proc. of 38th Design Automation Conference, Las Vegas, June 2001.

[558]   IBM Inc., Blue Logic Technology, http://www.chips.ibm.com/bluelogic/

[559]   Virtual Socket Interface Alliance, http://www.vsi.org.

[560]   Leijten, J. A. J., et al. PROPHID: A Heterogeneous Multi-Processor Architecture for Multimedia, Proc. of ICCD, 1997.

[561]   CoWare Inc., N2C: http://www.coware.com/

[562]   Wingard, D. MicroNetwork-Based Integration for SOCs, Proc. of DAC, Las Vegas, June 2001.

[563]   Krolikoski, S. J., Schirrmeister, F., Salefski, B., Rowson, J., and Martin, G. Methodology and technology for virtual component driven hardware/software co-design on the system-level, Proceedings of the 1999 IEEE International Symposium on Circuits and Systems (ISCAS), pp. 456–459, vol. 6, May–June, 1999.

[564]   Synopsys web page. http://www.synopsys.com

[565]   Passerone, R., and Rowson, J. A. Automatic synthesis of interfaces between incompatible protocols, In proceedings of 35th Design Automation Conference (DAC), June 1998.

[566]   Smith, J., and De Micheli, G. Automated Composition of Hardware Components, In Proceedings of 35th DAC, June 1998.

[567]   Oberg, J. ProGram: A Grammar-Based Method for Specification and Hardware Synthesis of Communication Protocols, PhD thesis, Department of Electronics, Electronics System Design, Royall Institute of Technology, Kista, Sweden, 7 May 1999.

[568]   Narrayan, S., and Gajski, D. D. Synthesis of System-Level Bus Interfaces, In Proceedings of EDAC, 1994.

[569]   Bergamaschi, R. A., and Lee, W. R. Designing Systems-on-Chip Using Cores, In Proceedings of Design Automation Conference, June 2000.

[570]   Vercauteren, S., Lin, B., and Man, H. D. Constructing Application-Specific Heterogeneous Embedded Architectures from Custom HW/SW Applications, In Proceedings of DAC, June 1996.

[571]   Brunel, J. Y., et al. COSY Communication IP's, In Proceedings of 37th Design Automation Conference, June 2000.

[572]   OSCI: http://www.systemc.org/

[573]   Chatelain, A., Placido, G., LaRosa, A., Mathys, Y., Lavagno, L. High-level Architectural Co-Simulation Using Esterel and C, CODES01, Copenhagen, 2001.

[574]   Cesário, W. O., et al. Colif: A design representation for application-specific multiprocessor SOCs, IEEE Design & Test of Computers, Vol. 18, Issue: 5, pp. 8–20, Sept.–Oct. 2001.

[575]   Lyonnard, D., Yoo, S., Baghdadi, A., and Jerraya, A. A. Automatic Generation of Application-Specific Architectures for Heterogeneous Multiprocessor System-on-Chip, Proc. of DAC, Las Vegas, 2001.

[576]   Yoo, S., Nicolescu, G., Lyonnard, D., Baghdadi, A., and Jerraya, A. A. A Generic Wrapper Architecture for Multi-Processor SoC Cosimulation and Design, Int. Symposium on HW/SW Codesign (CODES) 2001.

[577]   Diaz-Nava, M., and Okvist, G. S. The Zipper prototype: A Complete and Flexible VDSL Multi-carrier Solution, ST Journal special issue xDSL, September 2001.

[578]   van Eijndhoven, J. T. J., Sijstermans, F. W., Vissers, K. A., Pol, E. J. D., Tromp, M. J. A., Struik, P., Bloks, R. H. J., van der Wolf, P., Pimentel, A. D., and Vranken, H. P. E. TriMedia CPU64 Architecture, Proceedings of the 1999 International Conference on Computer Design, Austin, TX, October 1999, pp. 586–592. http://www.semiconductors.philips.com/trimedia/.

[579]   Koga, T., Iinuma, I., Hirano, A., Ijima, Y., and Ishiguro, T. Motion compensated interframe coding for videoconferencing, in *Proceedings, National Communications Conference*, 1981, pp. 531–535.

[580]   Po, L. M., and Ma, W. C. A novel four-step search algorithm for fast block motion estimation, *IEEE Transactions on Circuits and Systems for Video Technology*, 6(3), June 1996, pp. 313–317.

[581]   Tham, J. Y., Ranganath, S., Ranganath, M., and Kassim, A. A. A novel unrestricted center-biased diamond search algorithm for block motion estimation, *IEEE*

Transactions on Circuits and Systems for Video Technology, 8(4), August 1998, pp. 369–377.

[582]   Chen, M. J., Chen, L. G., and Chiueh, T. D. One-dimensional full search motion estimation for video coding, *IEEE Transactions on Circuits and Systems for Video Technology*, 4(5), October 1994, pp. 504–509.

[583]   Kappagantula, S., and Rao, K. R. Motion compensated predictive interframe coding, *IEEE Transactions on Communication*, 33(9), September 1985, pp. 1011–1015.

[584]   Wolf, W., Ozer, B., and Lv, T. Smart cameras as embedded systems, *IEEE Computer*, 35(9) September 2002, pp. 48–53.

[585]   Yang, K. M., Sun, M. T., and Wu, L. A family of VLSI designs for the full-search block matching motion estimation, *IEEE Transactions on Circuits and Systems for Video Technology*, 2(2), June 1992, pp. 393–398.

[586]   Bhaskaran, V., et al. Multimedia architectures: from desktop systems to portable appliances, *Proceedings of Multimedia Hardware Architectures, SPIE*, vol. 3021, February 1997.

[587]   Knebel, P., et al. HP's PA7100LC: A low-cost superscaler PA-RISC processor, *COMPCON*, 1993.

[588]   Greenley, D., et al. UltraSPARC: the next generation superscaler 64-bit SPARC, *COMPCON*, 1995.

[589]   Peleg, A., et al. Intel MMX for Multimedia PCs, *Communications of the ACM*, 40(1), January 1997.

[590]   MMX(tm) Technology Architecture Overview. Available at http://www.intel.com/technology/itj/q31997/articles/art_2.htm.

[591]   Pentium Manuals. Available at http://x86.ddj.com/intel.doc/586manuals.htm.

[592]   Cyrix 233 MHz MediaGX(tm). Available at http://bwrc.eecs.berkeley.edu/CIC/announce/1998/MediaGX-233.html.

[593]   Mobile AMD-K6 Processor. Available at http://www.baznet.freeserve.co.uk/AMD-K6.htm.

[594]   Hardware/Software Interactions on the Mpact Media Processor. Available at http://www.hotchips.org/archive/hc8/hc8pres_pdf/6.2.pdf.

[595]   Rathnam, S., and Slavenburg, G. An Architectural Overview of the Programmable Multimedia Processor, TM-1, *COMPCON*, 1996.

[596]   Philips Media Processors. Available at http://www.semiconductors.philips.com/platforms/nexperia/media_processing/products/media_proc_ic/index.html.

[597]   Bannon, P., and Jain, A. MVI Instructions Boost Alpha Processor, *Electronic Engineering Times*, 74, February 3, 1997.

[598]   Real World Signal Processing. Available at http://dspvillage.tcom,i.,/docs/allproducttree.jhtml.

[599]   Pirsch, P., and Stolberg, H. J. VLSI Architectures for Multimedia, *IEEE International Conference on Circuits and Systems*, 1998.

[600]   Guttag, K., et al. A Single-Chip Multiprocessor for Multimedia: The MVP, *IEEE Computer Graphics & Applications*, November 1992.

[601]   Tremblay, M., et al. VIS Speeds New Media Processing, *IEEE Micro*, August 1996.

[602]   The Silicon Graphics Power Challenge. Available at http://www.top500.org/ORSC/1996/node23.html.

[603]   Dutta, S., et al. Viper: A Multiprocessor SOC for Advanced Set-Top Box and Digital TV Systems, *IEEE Design and Test of Computers*, September/October 2001.

[604]   The constantly shifting promise of reconfigurability. Available at http://www.qstech.com/pdf/the_constantly_shifting_1.pdf.

[605]   Platform-based design: A choice, not a panacea. Available at http://www.eetimes.com/reshaping/platformdesign/OEG20020911S0061.

[606]   Processor Architectures for Multimedia: A survey. Available at http://www.cdartmouth,s.edu/~nazareth/academic/CS107.pdf.

[607]   Tensilica. Available at http://www.tensilica.com/.

[608]   ARC. Available at http://www.arc.com/.

[609]   Embedded Systems: How to extend configurable CPU performance. Available at http://www.eetimes.com/in_focus/embedded_systems/OEG20020208S0043.

[610]   Research focuses on application-specific reconfigurable blocks. Available at http://www.eetimes.com/reshaping/reconfig/OEG20020911S0068.

[611]   Reconfigurable architectures for video processing: Gries, M., MS thesis, Technical University of Hamburg-Harburg, Germany, 12/1996. Available at http://www-cad.eecs.berkeley.edu/~gries/abstracts/diplom.html.

[612]   Defining platform-based design. Available at http://www.eedesign.com/features/exclusive/OEG20020204S0062.

[613]   Claasen, T. System on a Chip: Changing IC Design Today and in the Future, *IEEE Micro*, May/June 2003.

[614]   Chang, H. SOC Design Methodologies, in *Winning the SOC Revolution*, Norwell, MA: Kluwer Academic Publishers, 2003.

[615]   Roentgen, M., et al. The Changing Landscape of SoC Design, *Custom Integrated Circuits Conference*, May 1999.

[616]   CoreConnect Bus Architecture. Available at http://www.chips.ibm.com/products/core-connect.

[617]   Rye, K. K., et al. A Comparison of Five Different Multiprocessor SoC Bus Architectures, *Euromicro Symposium on Digital Systems Design*, September, 2001.

[618]   Zorian, Y. Emerging Trends in VLSI Test and Diagnosis, *IEEE Computer Society Annual Workshop on VLSI*, April 2000.

[619]   IEEE P1500 Standard for Embedded Core Test (SECT). Available at http://grouper.ieee.org/groups/1500/.

[620]   Marinissen, E., et al. Towards a Standard for Embedded Core Test: An Example, *IEEE International Test Conference*, September 1999.

[621]   Jason Fritts, *Architecture and Compiler Design Issues in Programmable Media Processors*, Ph. D. dissertation, Princeton University, January 2000.

[622]   Aldworth, P. J. System-on-a-Chip Bus Architecture for Embedded Applications, *IEEE International Conference on Computer Design*, 1999.

[623]   Cordan, B. An Efficient Bus Architecture for System-on-a-Chip Design, *IEEE Custom Integrated Circuits Conference*, May 1999.

[624]   Pirsch, P., et al. Architectural approaches for multimedia processors, *Proceedings of Multimedia Hardware Architectures, SPIE*, vol. 3021, February 1997.

[625]   ISQED Speakers back platform based design. Available at http://www.eedesign.com/news/OEG20030325S0045.

[626]   Ullman, J. D., *Computational Aspects of VLSI*, Computer Science Press, Rockville, MD, 1984.

[627]   Savage, J. E., *Models of Computation, Exploring the Power of Computing*, Addison Wesley, 1998.

[628]   Jantsch, A., *Modeling Embedded Systems and SoC's*, Morgan Kaufmann, San Francisco, CA, 2004.

[629]   Curriculum 68, A Report of the ACM Curriculum Committee on Computer Science, Communications of the ACM, Vol. 11, No. 3, 1968.

[630]   Petri, C. Fundamentals of a theory of asynchronous information flow, Proc. IFIP Congress 62, 1962. pp. 386–390.

[631]   Hoare, C. A. R. Communicating Sequential Processes, Communications of the ACM, Vol. 21, No. 8 Aug. 1978.

[632]   Caspi, P., Pilaud, D., Halbwachs, N., and Plaice, J. LUSTRE: A declarative language for programming synchronous systems, ASM Symposium on Principles of Programming Languages (POPL), 1987. pp. 178–188.

[633]   Kahn, G. The semantics of a simple language for parallel programming, in Proceedings of the IFIP Congress 74. Amsterdam, The Netherlands. North-Holland, 1974.

[634]   Harel, D. Statecharts: A visual formalism for complex systems. *Science of Computer Programming*, pp. 231–274. 1987.

[635]   Lynch, N., and Fischer, J. On describing behavior and implementation of distributed systems. *Theoretical Computer Science* 13(1), 1981.

[636]   Agha, G. *Actors: A model of concurrent computation in distributed systems*, MIT Press, London, England. 1986.

[637]   Lee, E., and Sangiovanni-Vincentelli, A. A Framework for Comparing Models of Computation, *IEEE Trans CAD,* Dec. '98.

[638]   Booch, G., Rumbaugh, J., and Jacobson, I. *The Unified Modeling Language User Guide.* Reading, MA: Addison-Wesley. 1999.

[639]   Boussinot, F., and de, R., Simone, The Esterel Language, *Proceedings of the IEEE,* Sept. 1991, pp. 1293–1304.

[640]   Benveniste, A., and Berry, G. The synchronous approach to reactive and real-time systems, *Proceedings of the IEEE,* Sept. 1991, pp. 1270–1282.

[641]   Caspi, P., and Pouzet, M. Synchronous Kahn Networks, International Conference on Functional Programming. 1996. pp. 226–238.

[642]   Dennis, J. B. First version data flow procedure language, MIT Laboratory for computer Science, Cambridge, MA, Tech Memo MAC TM61, May 1975.

[643]   Harel, D., Lachover, H., Naamad, A., Pnueli, A., Politi, M., Sherman, R., Shtul-Trauring, A., and Trakhtenbrot, M. STATEMATE: a working environment for the development of complex reactive systems. *IEEE Trans Software Engineering,* April 1990, pp. 403–414.

[644]   Cassandras, C. G., *Discrete Event Systems.* Aksen Associates, Boston MA 1993.

[645]   Milner, R., *Communication and Concurrency.* Englewood Cliffs, NJ, Prentice-Hall, 1989.

[646]   Seitz, L. C. System Timing. *Introduction to VLSI Systems.* C. Mead, Conway, L. Reading, MA: Addison-Wesley, 1980.

[647]   Zeigler, B., Praehofer, H., and Kim, T. *Theory of Modeling and Simulation 2nd Edition.* San Diego: Academic Press. 2000.

[648]   Eker, J., Janneck, J., Lee, E., Liu, J., Liu, X., Ludvig, J., Neuendorfer, S., Sachs, S., and Xiong, Y. Taming Heterogeneity—the Ptolemy Approach, *Proceedings of the IEEE.* January 2003, pp. 127–144.

[649]   Selic, B. Turning Clockwise: Using UML in the Real-Time Domain, *Comm of the ACM,* pp. 46–54. Oct. 1999.

[650]   Keutzer, K., Malik, A., Newton, A. R., Rabaey, J., and Sangiovanni-Vincentelli, A. System-level design: orthogonalisation of concerns and platform-based design, *IEEE Transactions on CAD,* December 2000, pp. 1523–1543.

[651]   Paul, J., Bobrek, A., Nelson, J., Pieper, J., and Thomas, D. Schedulers as Model-Based Design Elements in Programmable Heterogeneous Multiprocessors, *DAC,* June, 2003.

[652]   Paul, J., Peffers, S., and Thomas, D. Frequency Interleaving as a Co-Design Scheduling Paradigm, *International Workshop on Hardware/Software Co-Design,* May 2000.

[653]   Cassidy, A., Paul, J., and Thomas, D. Layered, Multi-Threaded, High-Level Performance Design. *DATE* 2003.

[654]   Paul, J., and Thomas, D. A layered, codesign virtual machine approach to modeling computer systems, *DATE 2002*.

[655]   www.ece.cmu.edu/~mesh

[656]   Paul, J. M., Peffers, S. N., and Thomas, D. E. A Codesign Virtual Machine for Hierarchical, Balanced Hardware/Software System Modeling, DAC, June 2000.

[657]   Are Single-Chip Multiprocessors in Reach?, *IEEE Design & Test*, Jan–Feb 2001.

[658]   Benini, L., and De Micheli, G. Networks on chips: A New SoC Paradigm, *Computer*, J02, a., pp. 70–78.

[659]   Keutzer, K., Malik, S., Newton, A. R., Rabaey, J., and Sangiovanni-Vincentelli, A. System level design: orthogonalization of concerns and platform-based design. IEEE Transactions on Computer-Aided Design, December 2000.

[660]   Lee, E., and Sangiovanni-Vincentelli, A. A framework for comparing models of computation. IEEE Transactions on Computer-Aided Design of Integrated Circuits and Systems, 17(12):1217–1229, Dec 1998.

[661]   Holzmann, J. G. The model checker spin. IEEE Trans. on Software Engineering, 23(5):279–258, May 1997.

[662]   Balarin, F., Chiodo, M., Giusto, P., Hsieh, H., Jurecska, A., Lavagno, L., Passerone, C., Sangovanni-Vincentelli, A., Sentovich, E., Suzuki, K., and Tabbara, B. Hardware-software co-design of embedded systems: the Polis approach. Kluwer Academic Publishers, Boston: Dordrecht, 1997.

[663]   Buck, J., Ha, S., Lee, E. A., and Messerschmitt, G. D., Ptolemy: a framework for simulating and prototyping heterogeneous systems. International Journal of Computer Simulation, special issue on Simulation Software Development, January 1990.

[664]   Grotker, T., Martin, G., Liao, S., and Swan, S. System design with SystemC. Kluwer Academic Publishers, Boston: Dordrecht, 2002.

[665]   Gajski, D., Zhu, J., and Domer, R. The SpecC language. Kluwer Academic Publishers, Boston: Dordrecht, 1997.

[666]   Cesario, W., Baghdadi, A., Gauthier, L., Lyonnard, D., Nicolescu, G., Paviot, Y., Yoo, S., Jerraya, A. A., and Diaz-Nava, M. Component-based design approach for multicore SoCs. In Proceedings of the 39th ACM/IEEE Design Automation Conference, June 2002.

[667]   Paul, J. A. M., and Thomas, E. D. A layered, codesign virtual machine approach to modeling computer systems. In Design Automation and Test in Europe, March 2002.

[668]   Balarin, F., Lavagno, L., Passerone, C., Sangiovanni-Vincentelli, A., Sgroi, M., and Watanabe, Y. Modeling and designing heterogenous systems. In Cortadella, J., Yakovlev, A., and Rozenberg, G. editors, Concurrency and Hardware Design, pp. 228–273. Springer, 2002. LNCS2549.

[669] Pnueli, A. The temporal logic of programs. In Proc. 18th Annual IEEE Symposium on Foundations of Computer Sciences, pp. 46–57, 1977.

[670] Balarin, F., Watanabe, Y., Burch, J., Lavagno, L., Passerone, R., and Sangiovanni-Vincentelli, A. Constraints specification at higher levels of abstraction. In Proceeding of the 6th Annual IEEE International Workshop on High Level Design Validation and Test-HLDVT'01. IEEE Computer Society, November 2001.

[671] Solden, S. Architectural services modeling for performance in HW-SW co-design. In Proceedings of the Workshop on Synthesis and System Integration of MIxed Technologies SASIMI2001, Nara, Japan, October 18–19, 2001, pp. 72–77, 2001.

[672] Balarin, F., Lavagno, L., Passerone, C., Sangiovanni-Vincentelli, A., et al. Concurrent execution semantics and sequential simulation algorithms for the metropolis meta-model. In Proceedings of the Tenth International Symposium on Hardware/Software Codesign. CODES 2002 Estes Park, CO, USA, 6–8 May 2002, pp. 13–18. IEEE Comput. Soc. Press, 2002.

[673] Open SystemC Initiative. Functional Specification for SystemC 2.0, September 2001. avaliable at www.systemc.org.

[674] Chen, X., Chen, F., Hsieh, H., Balarin, F., and Watanabe, Y. Formal verification of embedded system designs at multiple levels of abstraction. International Workshop on High Level Design Validation and Test-HLDVT02, September 2002.

[675] Chen, X., Hsieh, H., Balarin, F., and Watanabe, Y. Automatic generation of simulation monitors from quantitative constraint formula. In Design Automation and Test in Europe, March 2003.

[676] Cortadella, J., Kondratyev, A., Lavagno, L., Passerone, C., and Watanabe, Y. Quasi-static scheduling of independent tasks for reactive systems. In Proceedings of the 23rd Int. Conference on Application and Theory of Petri Nets, June 2002.

[677] van der Wolf, P., Lieverse, P., Goel, M., Hei, D. L., and Vissers, K. An MPEG-2 Decoder Case Study as a Driver for a System Level Design Methodology. In Proceedings of the International Workshop on Hardware/Software Codesign, pp. 33–37, May 1999.

[678] de Kock, E. A., Essink, G., Smits, W. J. M., van der Wolf, P., Brunel, J.-Y., Kruijtzer, W. M., Lieverse, P., and Vissers, A. K. YAPI: Application Modeling for Signal Processing Systems. In Proceedings of the Design Automation Conference, June 2000.

[679] Gangwal, O. P., Nieuwland, A., and Lippens, P. A Scalable and Flexible Data Synchronization Scheme for Embedded HW-SW Shared-Memory Systems. In International Symposium on System Synthesis, October 2001.

Contributor Biographies

Aravindh V. Anantaraman (*avananta@ncsu.edu*) is a Ph.D. Student in the Department of Electrical and Computer Engineering at North Carolina State University, Raleigh. He received his Bachelor's Degree in Electrical and Electronics Engineering (2001) from the Birla Institute of Technology and Science, Pilani, India, and the MS Degree in Computer Engineering (2003) from North Carolina State University. His research interests are in computer architecture, with a focus on processor architectures for embedded systems.

Felice Balarin (*felice@cadence.com*) is a Research Scientist at Cadence Berkeley Labs. His research interests include design and verification methodologies for embedded systems. On these topics, he published numerous papers and co-authored two books. He holds a Ph.D. degree in Electrical Engineering from the University of California at Berkeley.

Luca Benini (*lbenini@deis.unibo.it*) is an Associate Professor at the Department of Electrical Engineering and Computer Science (DEIS) of the University of Bologna. He also holds visiting researcher positions at Stanford University and the Hewlett-Packard Laboratories, Palo Alto, CA. His research interests are in all aspects of computer-aided design of digital circuits, with special emphasis on low-power applications, and in the design of portable systems. On these topics he has published more than 200 papers in international journals and conferences and three books.

Bishnupriya Bhattacharya (*bpriya@cadence.com*) is an R&D Engineer in the Systems and Functional Verification Group at Cadence Design Systems, Inc.,

San Jose, CA. Her research interests lie in the areas of hardware-software mixed-language co-simulation; modeling and scheduling of dataflow graphs; and new generation EDA tools for system design, simulation and verification based on the emerging SystemC language. She has a Masters Degree in Electrical and Computer Engineering from the University of Maryland at College Park.

Francky Catthoor (*catthoor@imec.be*) is a Research Fellow at IMEC, Leuven, Belgium, where he has headed several research domains in the area of architectural and system-level methodologies for embedded multimedia and telecom applications, all within the DESICS division. He is also Professor at the K. U. Leuven.

Wander O. Cesário (*Wander.Cesario@imag.fr*) is a Research Engineer at TIMA Laboratory. His research interests include methodologies and related CAD algorithms for system-level design of embedded systems. He has a Ph.D. in microelectronics from the Institut National Polytechnique of Grenoble, France. He is a member of the IEEE.

Patrick David (*davidp@imec.be*) designs software tools derived from research performed at DESICS/IMEC. Since graduating from the University of Liege with a degree in Mathematics and Computer Science, he has occupied several software engineering positions in Bruxelles and Paris, mainly in the space industry.

Giovanni De Micheli (*nanni@Stanford.Edu*) is Professor of Electrical Engineering and Computer Science at Stanford University. He is the recipient of the 2003 IEEE Emanuel Piore Award for contributions to computer-aided synthesis of digital systems. He is a Fellow of ACM and IEEE and served as President of the IEEE CAS Society in 2003. His research interests include several aspects of design technologies for integrated circuits and systems, with particular emphasis on synthesis, system-level design, hardware/software co-design and low-power design.

Sujit Dey (*dey@ece.ucsd.edu*) is a Professor at UC San Diego, where he heads the Mobile Embedded Systems Design and Test Laboratory, developing adaptive algorithms, protocols, and configurable hardware-software architectures for cost-effective, quality-aware, and reliable wireless data services. He is affiliated with the California Institute for Telecommunications and Information Technology, the UCSD Center for Wireless Communications, and the Gigascale System Research Center. He has co-authored more than 130 publications, including several that have won Best Paper awards, and holds 9 US patents.

Nikil Dutt (*dutt@uci.edu*) is a Professor of CS with the Center for Embedded Computer Systems at the University of California, Irvine. His research interests are in embedded systems design automation, computer architecture, optimizing compilers, system specification techniques, and distributed embedded systems. He has served on the steering, organizing, and program committees of several premier CAD and Embedded System Design conferences, and serves on the advisory boards of ACM SIGBED and ACM SIGDA.

Santanu Dutta (*santanu.dutta@philips.com*) is a Design Engineering Manager and Technical Lead in the Connected Multimedia Solutions group of Philips Semiconductors, at its Innovation Center USA, leading a video competence group. He is an Adjunct Faculty at the Santa Clara University and is a Senior Member of the IEEE. His main research interests include design of high-performance video signal processing architectures, circuit simulation & analysis, and design & synthesis of low-power & testable digital systems.

Rolf Ernst (*r.ernst@tu-bs.de*) is Professor of Electrical Engineering at the Technical University of Braunschweig, Germany, where he heads the Institute of Computer and Communication Network Engineering. He received a diploma in Computer Science and a Ph.D. in Electrical Engineering from the University of Erlangen-Nuremberg, Germany and then worked with Bell Laboratories in Allentown, PA. His current research interests include embedded architectures, hardware-/software co-design, real-time systems, and embedded systems engineering.

Mercury Jair Gonzalez (*jair.gonzalez@cts-design.com*) is a Design Engineer at the Semiconductor Technology Centre of CINVESTAV Mexico. He received his M.Sc. Degree in Computer Systems from the Technical University of Denmark in 2002. His current research interests are within VLSI design methods, MP SoC design methods, Timming Analisys of Real-Time Systems.

Stefaan Himpe (*Stefaan.Himpe@esat.kuleuven.ac.be*) is currently working towards a Ph.D. Degree in Electrotechnical Engineering in the KUL, working on high-level platform independent source transformations in the context of Task Concurrency Management (TCM) under supervision of Geert Deconinck and with people from IMEC under supervision of Francky Catthoor. He received a Degree in Industrial Engineering from the Katholieke Hogeschool Brugge-Oostende (KHBO) in 1998, and in 2000, he received the MSc. Degree in Electrotechnical Engineering from the Katholieke Universiteit Leuven (KUL).

Harry Hsieh (*harry@cs.ucr.edu*) is currently an Assistant Professor at the University of California at Riverside. He received his Ph.D. from the University of California at Berkeley in 2000. His research interests include embedded systems, verification, and computer aided design.

Mary Jane Irwin (*mji@psu.edu*) is the A. Robert Noll Chair in Engineering in the Department of Computer Science and Engineering at Pennsylvania State University. She is a member of the Technical Advisory Board of the Army Research Lab and is the Editor-in-Chief of ACM's Transaction on Design Automation of Electronic Systems. Her research and teaching interests include computer architecture, embedded and mobile computing systems design, low power design, and electronic design automation.

Mahmut Kandemir (*kandemir@cse.psu.edu*) is an Assistant Professor in the Computer Science and Engineering Department at Pennsylvania State University. His main research interests are optimizing compilers, I/O intensive applications, and power-aware computing. He is a member of the IEEE and the ACM.

Kanishka Lahiri (*klahiri@ece.ucsd.edu*) is a Research Staff Member at NEC Laboratories America, Princeton, NJ. His research interests include architectures and design methodologies for on-chip communication, energy-efficient systems, and configurable, platform-based systems. He obtained a Ph.D. in Electrical and Computer Engineering from the University of California, San Diego, in 2003. He received a Best Paper Award at the IEEE/ACM Design Automation Conference in 2000 and a Special Services Award at the IEEE VLSI Test Symposium in 2003.

Luciano Lavagno (*lavagno@polito.it*) is an Associate Professor with Politecnico di Torino and a research scientist with Cadence Berkeley Laboratories. He received his Ph.D. from U.C. Berkeley in 1992. He has co-authored three books—two on asynchronous design and one on hardware/software co-design—and over 160 scientific papers. His research interests include synthesis of asynchronous circuits, concurrent design of mixed hardware and software embedded systems, and compilation tools for dynamically reconfigurable processors.

Rudy Lauwereins (*lauwerei@imec.be*) is Vice-President of IMEC, Belgium's Interuniversity Micro-Electronic Centre. He leads the research focused on architectures, design methods, and tools for wireless and multimedia applications as required for creating an ambient intelligent world with smart networked mobile devices. He is also Professor at the Katholieke Universiteit Leuven, Belgium.

Tiehan Lv (*lv@ee.princeton.edu*) is currently working toward the Ph.D. Degree in the Department of Electrical Engineering at Princeton University. He received the B.S. Degree in Electrical Engineering in Peking University, Beijing, China, in 1998, and the M.Eng Degree in Electrical Engineering from Peking University in 2000. His research interests include embedded systems, video/image processing, computer architecture, and low-power design.

Jan Madsen (*jan@imm.dtu.dk*) is Professor of Computer Based Systems at the Department of Informatics and Mathematical Modeling at the Technical University of Denmark, where he is heading the System-on-Chip group. He received the MSc. Degree in Electrical Engineering and a Ph.D. Degree in Computer Science from the Technical University of Denmark in 1986 and 1992, respectively. His research interests include high-level synthesis, hardware/software codesign, System-on-Chip design methods, and system level modeling, integration and synthesis for embedded computer systems.

Paul Marchal (*marchal@imec.be*) is working at IMEC and is a Ph.D. student in the field of dynamic memory management on multiprocessors for multimedia applications. He graduated in 1999 as Electrical Engineer at the KUL.

Mayur Naik (*mhn@cs.stanford.edu*) is a Doctoral Student in Computer Science at Stanford University. He holds an M.S. in Computer Science from Purdue University and is a Fellow of Microsoft Research. His research interests include logic and programming languages, with an emphasis on the foundations and applications of program analysis.

Vijaykrishnan Narayanan (*vijay@cse.psu.edu*) is an Associate Professor with the Computer Science and Engineering Department at Pennsylvania State University. He received his B.E Degree in Computer Science and Engineering from SVCE, University of Madras, and his Ph.D. Degree in Computer Science and Engineering from the University of South Florida, Tampa. His research interests are in the areas of energy-aware reliable systems, embedded Java, nano/VLSI systems and computer architecture.

Jens Palsberg (*palsberg@ucla.edu*) is a Professor of Computer Science at UCLA. His research interests span the areas of compilers, embedded systems, programming languages, software engineering, and information security. He is an associate editor of ACM Transactions of Programming Languages and Systems, and a member of the editorial board of Information and Computation.

Claudio Passerone (*claudio.passerone@polito.it*) is a Researcher in the Department of Electronics at the Politecnico di Torino. His research interests include system-level design, electronic system simulation, and reconfigurable hardware. He received a Ph.D. in Electrical and Communication Engineering from the Politecnico di Torino in 1998.

JoAnn M. Paul (*jpaul@ece.cmu.edu*) is Research Faculty in the Department of Electrical and Computer Engineering at Carnegie Mellon University. Her engineering experience obtained prior to her career in academia includes the design and development of safety-critical systems (both software and hardware) and real-time kernels (both runtime and debug). Her research focus is the design and modeling of single chip, semi-custom heterogeneous multiprocessing systems.

Alessandro Pinto (*apinto@eecs.berkeley.edu*) is working towards his Ph.D. at the University of California at Berkeley, researching in the area of communication synthesis and system-level design methodologies. He previously worked as Consultant at Ericsson Lab Italy, working on systems on-chip and system-level design. He received the Laurea degree in Electronic Engineering Summa Cum Laude from the University of Rome "La Sapienza".

Anand Raghunathan (*anand@nec-labs.com*) is a Senior Researcher at NEC Labs, Princeton, NJ, where he leads research efforts on advanced system-on-chip architectures and design methodologies. He has received five best paper awards at leading conferences, NEC's inventor of the year award, and IEEE's meritorious service award. He has served as Program Chair, as a member of the Program and Organizing Committees of several IEEE and ACM conferences, and on the Editorial Board of IEEE Transactions on CAD, IEEE Transactions on VLSI, and IEEE Design & Test of Computers. He is a Golden Core member of IEEE Computer Society and a Senior member of the IEEE.

Jens Rennert (*rennerts@gmx.net*) is Senior Systems Engineer at Micronas GmbH in Freiburg, Germany. He has worked with Philips Semiconductors in Hamburg, Germany, and in the United States, where he served as VLSI Video and System Architect. He then joined MobileSmarts Inc., working in the field of reconfigurable computing until 2004. His research interests include video and communications algorithms, hardware-software co-design, emulation and verification of large scale SOCs.

Eric Rotenberg (*ericro@ece.ncsu.edu*) is Assistant Professor of Electrical and Computer Engineering at North Carolina State University. He received the BS Degree in Electrical Engineering and MS and Ph.D. Degrees in Computer Sciences from the University of Wisconsin-Madison. From 1992 to 1994, he participated in the design of IBM's AS/400 computer in Rochester, Minnesota. His main research interests are in high-performance computer architecture.

Chris Rowen (*rowen@tensilica.com*) is President and Chief Executive Officer of Tonsilica. He founded Tonsilica in July 1997 to develop automated generation of application-specific microprocessors for high-volume communication and consumer systems. He was a pioneer in the development of RISC architecture at Stanford in the early '80s and helped start MIPS Computer Systems Inc., where he served as Vice President for Microprocessor Development. Most recently he was vice president and general manager of the Design Reuse Group of Synopsys Incorporated. He received a B.A. in Physics from Harvard University and M.S. and Ph.D. in Electrical Engineering from Stanford University.

Alberto Sangiovanni-Vincentelli (*alberto@eecs.berkeley.edu*) is Professor of Electrical Engineering and Computer Sciences at the University of California at Berkley. He co-founded Cadence and Synopsys, the two leading companies in the area of electronic design automation. He is the author of over 600 papers and 15 books in the area of design methodologies and tools. He has been a Fellow of the Institute of Electrical and Electronics Engineers since 1982 and a member of the National Academy of Engineering since 1997.

Donald E. Thomas (*thomas@ece.cmu.edu*) is Professor of Electrical and Computer Engineering at Carnegie Mellon University, working in the area of modeling and simulation of single chip programmable heterogeneous multiprocessing systems. He is currently Associate Editor of the ACM Transactions on Embedded Computing Systems, and is co-author of the book, *The Verilog Hardware Description Language Fifth Edition.*

Laura Vanzago (*laura.vanzago@st.com*) is a Senior System Engineer in the Research and Innovation Department of the Advanced System Technology division of STMicroelectronics in Agrate Brianza, Italy. In 2001 and 2002 she was a Visiting Industrial Fellow at the University of California at Berkeley, working on system level design methodologies. She holds a degree in Physics from the University of Milano, Italy. Her research interests are in the fields of system level design methodologies, embedded systems and wireless sensor networks.

Kashif Virk (*virk@imm.dtu.dk*) is working with Professor Jan Madsen towards his Ph.D. in Computer Systems Engineering at the Technical University of Denmark,

Lyngby, Denmark. He obtained his B.Sc. in Electrical Engineering in 1995 from the University of Engineering & Technology, Lahore, Pakistan, and his M.Sc. in Computer Systems Engineering from the Technical University of Denmark in 2001.

Johan Vounckx (*vounckx@imec.be*) is Director of Design Technology within the DESICS division of IMEC vzw. The design technology activities focus on design methodologies and platforms for power-aware embedded systems. Research topics include SOC-design methodologies, low-power design methodologies and memory optimisations and reconfigurable systems research.

Yosinori Watanabe (*watanabe@cadence.com*) is a Research Scientist at Cadence Berkeley Labs. His research interests focus on methodologies and synthesis and verification of system designs. Watanabe received a Ph.D. in Electrical Engineering and Computer Sciences from the University of California, Berkeley.

Chun Wong (*chun.wong@philips.com*) is a Research Scientist with Philips Research Laboratories, Eindhoven. He received his PhD. Degree in 2003 from Katholieke Universiteit Leuven—IMEC, his Master of Engineering Degree from Tsinghua University, and his Bachelor of Science Degree from the University of Science & Technology of China. His current research activities span over the information processing architectures.

Guang Yang (*gyang@ic.eecs.berkeley.edu*) is pursuing his Ph.D. degree in Electrical Engineering and Computer Science Department at the University of California at Berkeley. He received his B.Eng. Degree and M.Eng. Degree in Microelectronics from Tsinghua University, Beijing, People's Republic of China. His interests are system level design validation, including simulation, formal and semi-formal verification.

Peng Yang (*yangp@imec.be*) is a Ph.D. student working at IMEC, Belgium. He received his BS Degree from Fudan University, China, in 1996 and his ME Degree from Tsinghua University, China, in 1999, both in Electrical Engineering. His current research focuses on the task concurrency management for real-time embedded system.

Shengqi Yang (*shengqiy@princeton.edu*) is a Ph.D. candidate of Electrical Engineering, Princeton University. His research interests include architecture level/circuit level power modeling and analysis, VLSI design for multimedia system, embedded system design and verification, transistor level construction for digital VLSI system and analog circuit module.

Jiang Xu (*jiangxu@ee.princeton.edu*) is currently a Ph.D. candidate at Princeton University. He received his B.S. and M.S. Degrees in Electrical Engineering from Harbin Institute of Technology of China in 1998 and 2000. His research interests include Networks-on-Chip architecture and protocol, on-chip interconnection structure, and Networks-on-Chip and System-on-Chip design methodology.

Subject Index

Page numbers followed by *f* indicate figures; page numbers followed by *t* indicate tables.

S